EXPLORATION AND MINING GEOLOGY

EXPLORATION AND MINING GEOLOGY

William C. Peters

Department of Mining and
Geological Engineering
The University of Arizona

TN
260
$.P47$

JOHN WILEY & SONS

New York · Chichester · Brisbane · Toronto · Singapore

Library of Congress Cataloging in Publication Data:

Peters, William C.
Exploration and mining geology.

Bibliography
Includes index.
1. Mining geology. 2. Prospecting. I. Title.

TN260.P47   622'.1   77-14006
ISBN 0-471-68261-6

Printed in the United States of America

10 9 8 7

PREFACE

The purpose of this book is to furnish an overview of the geologist's work in mineral discovery and mineral production. Most of the methods are new, some of the concepts are recent, but the application of geology to mining is long-standing. Our predecessors found minerals and mined them; they were aware that mineral deposits have geologic limits, and they knew that each mine would someday be depleted of its minerals.

Depletion was a matter of concern for miners, for communities, and sometimes for entire nations, although not for mankind,—until recently. Now, from vantage points in space and in history, we have begun to see the actual dimensions of the earth and its resources and we have begun to realize how badly we have misused these resources. Mining geologists are needed now more than ever, and their work has never been more important.

For those entering a vocation in exploration and mining or preparing for the responsibilities of managing a nation's minerals, this book will serve as an introduction to their professional work. If the transition from college graduate to professional geologist or geological engineer has already begun, this book can be part of a reference library.

The book deals with "hard" minerals, the nonfluid sources of materials and energy in the continental masses and ocean basins. The geographic distribution of mineral resources, the geology of major mining districts, and current theories of ore deposition are mentioned as illustrations of the geologist's function in exploration and mining, but they are not treated in detail. Brevity in these topics and the exclusion of comments on fluid resources (petroleum deposits, naturally occurring brines, and geothermal power) do not indicate a lack of relevance but rather that they are complex and specialized subjects requiring thorough study from other books.

A basic knowledge of structural geology, mineralogy, petrology, and economic geology is assumed. The theme of the book is geologic, but economics and engineering topics are necessarily included.

A preliminary chapter deals with the exploration and mining geologist in historical perspective; it allows the reader to visualize a personal and comparable role in the modern scene. Part One concerns the geology of mineral deposits. A chapter on mineral bodies in the zone of weathering is included to emphasize the appearance of mineralization at the scene of field investigations. Part One is called The Geologic Base Line because it discusses the assumptions and measurements from which all subsequent work is derived.

Part Two introduces the engineering factors in mining and exploration. These are the elements of rock mechanics, soil mechanics, hydrology, and mining practice that a geological engineer takes into account when dealing with earth materials. Part Three considers the economic factors a geologist must use to work with engineers and managers in choosing favorable sites for exploration and mining.

Part Four relates to the collection and analysis of geological, geochemical, and geophysical data during the preliminary phases of reconnaissance, during detailed investigations, and in the final written and graphic presentation of geologic information. Part Five concerns exploration and mining programs with emphasis on the planning and carrying out of ore discovery and production work on behalf of government and private enterprise organizations.

Acknowledgments

Major portions of the book were reviewed by W. B. Beatty, Ben F. Dickerson III, Paul I. Eimon, Paul C. Gilmour, Robert E. Lehner, Robert A. Metz, and Robert E. Radabaugh. I am extremely grateful for their comments drawn from wide experience in managing exploration and mining projects.

Chapters dealing with specific topics have benefited considerably from suggestions by Ben L. Seegmiller (geotechnics), Albert J. Perry and David D. Rabb (approaches to mining and the economic framework), H. David Mac-Lean and Sheldon Breiner (geophysics), and R. J. P. Lyon (reconnaissance). Comments by Wayne S. Cavender were of special value in preparing the chapter on exploration programs. Stephen G. Peters and Daryl Thorburn provided helpful comments in their review of the chapters on the geologist's role in exploration and mining. Comments by Ian Campbell and Anthony L. Payne on an early outline for the book were especially welcome.

I am indebted to all of the above-mentioned geologists, geophysicists, and engineers. Any technical or conceptual errors that remain in the text are due to my own departure from suggestions and to my own writing.

Special thanks go to Helen R. Hauck for her editing and to Linda Cassens, Marjorie Fowler, and Betty Radabaugh, who typed innumerable drafts of the manuscript. The assistance of Donald Denek and his staff at John Wiley & Sons, Inc., in guiding this book through publication is especially appreciated.

William C. Peters

INTRODUCTION

There is one at every conference, the participant who interjects ". . . but first we should go back and define the term 'porphyry copper' (or 'depth,' or 'hydrothermal')." Sometimes this is the heroic person who rescues a floundering discussion; sometimes it is the hairsplitting pedant who derails an entire train of developing ideas. In any event, the ubiquitous "definer" thrives on terms that should have been agreed on at the onset of a discussion—or a book. One way to avoid the need for rescue and the hazard of derailment is to set ground rules, a short list of key-word meanings that can be used until someone has time to resolve all the shades in semantics.

In this book, most terms will be defined as they are encountered. This will occur soon enough. For now, there are a few ground rules that cannot wait; they involve the following couplets.

Ore Deposit—Mineral Deposit.

Ore is a material from which minerals as well as metals of economic value can be extracted at the present time. Definers may object on the grounds that one should use the term "industrial mineral deposit" or "nonmetallic deposit" when dealing with a source of minerals that will be used as such without being reduced to a metal. Still, the people who actually discover and mine such things as barite, fluorspar, and asbestos speak of "ore" and measure "ore reserves." "Mineral deposit" is a broad term; in this book it encompasses ore deposits, coal and oil shale deposits, the "nonmetallics" family of industrial rock deposits, and natural concentrations of economic minerals that are likely to be extracted at a profit in the near future.

Prospecting—Exploration.

In North American parlance and in this book the terms are interchangeable, with "exploration" the preferred term for an entire sequence of work ranging from reconnaissance (looking for a prospect) to the evaluation of the prospect and finally to the search for additional ore in a mine. In foreign literature, we find some more specific definitions—and some need for comment. In several countries, the Soviet Union, for example, the terms translate as "prospecting *for* mineral deposits" and "exploration *of* mineral deposits," with the understanding that we explore the prospects after they have been found (Kreiter, 1968, p. 114). In France and some other countries the meanings are just the opposite; "exploration" refers to a wide-ranging search for indications of mineralization, "prospecting" refers to a more localized study of the indications, and "reconnaissance" is the delineation of an orebody (Routhier, 1963, p. 1002).

Geoscientist—Geological Engineer.

A scientist discovers; an engineer designs. This is an important difference. After a few years' experience in the mining industry the gap between the two terms begins to fill. Ultimately, in the atmosphere of professional responsibility, it may disappear. The process is called education (with a small "e"). By setting aside the academic need for distinguishing between the geological aspects of exploration and mining, we can reduce the couplet to the traditional term "geologist." Economic geologists should not object to being included in such company, nor should "explorationists" and exploration-oriented mining engineers.

Of course, it is understood that the term "he" where it is used in this book refers to a member of the human race.

Most of the English-language definitions needed throughout a geologist's career can be found in the American Geological Institute's *Glossary of Geology* (Gary, McAfee, and Wolf, 1972) and in the U.S. Bureau of Mines' *Dictionary of Mining, Mineral, and Related Terms* (Thrush and Staff of the Bureau of Mines, 1968). For international work there is the excellent publication of the Royal Geological and Mining Society of The Netherlands, *Geological Nomenclature* (Schieferdecker, 1959), in English, Dutch, French, and German, and the more subjective little book, *Glossary of Mining Geology* (Amstutz, 1971), in English, Spanish, French, and German.

The units in this text are metric. A conversion table for English to SI (International System) and related metric units is given in Appendix F. "Thinking metric" may be a temporary problem for readers oriented to the

English system, but it avoids having to think of such things as ''100 meters of ⅞-inch pipe.''

W. C. P.

This book is dedicated to the miners, who create the materials of civilization from the substance of the earth. May their labors be well rewarded, their traditions respected, and the hazards of their working places diminish.

Without them all that follows in this book would be futile.

CONTENTS

CHAPTER 1
PERSPECTIVE

Mines are inevitably depleted of their minerals. The miner has worked in the shadow of this fact throughout history. In the remains of the gold-silver workings at Cassandra in Greece there is evidence that miners dug in search of faulted vein segments at some time prior to 300 B.C. Their Athenian contemporaries, faced with the depletion of silver and lead ore at Laurium, recognized the favorability of marble near a schist contact and sank more than 1,000 shafts through barren rock, some to depths of 100 m, in search of hidden orebodies. Geology, however primitive, was involved, but the science of mining geology was yet to be born—in the Saxon-Bohemian Erzgebirge, the mining profession's classic ground in Central Europe (Fig. 1-1).

1.1 MINING GEOLOGY'S PATRIARCH

During the sixteenth century, Georgius Agricola, a physician of Chemnitz, Saxony, published several essays on prospecting, mining, and metallurgy that dominated geologic thought for two centuries. He was mining's own Renaissance Man with the particular Renaissance gift for description and analysis. In *De Re Metallica,* Agricola (1556) presented the first comprehensive theory of epigenetic ore deposits, and he wrote about exploration in a way that favored field observation far above what we now refer to as "signals from black boxes":

> *Therefore a miner, since we think he ought to be a good and serious man, should not make use of an enchanted twig, because if he is prudent*

1

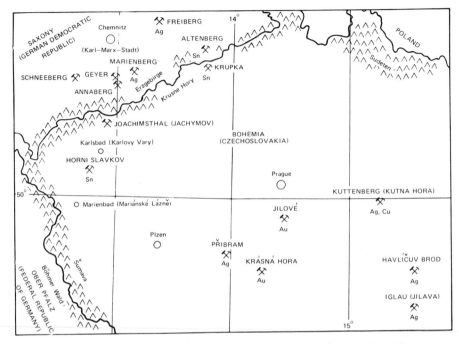

Figure 1-1. Saxon-Bohemian mines, sixteenth to eighteenth centuries. Newer geographic names are in parentheses.

and skilled in the natural signs, he understands that a forked stick is of no use to him, for as I have said before, there are the natural indications of veins which he can see for himself without the help of twigs. (From translation by Hoover and Hoover, 1912, p. 41.)

Agricola's rejection of enchanted twigs did not close his mind to the possibility of conditions existing in nature by which one might actually sense the nearness of ore. One of his comments borders on geophysics and geobotany; modern writers would refer to the thermal and toxic effects of oxidizing orebodies in words not too different from those of Agricola:

Further, we search for the veins by observing the hoarfrosts, which whiten all herbage except that growing over the veins, because the veins emit a warm and dry exhalation which hinders the freezing of the moisture, for which reason such plants appear rather wet than whitened by the frost. This may be observed in all cold places before the grass has grown to its full size, as in the months of April and May; or when the late crop of hay, which is called the cordum, *is cut with scythes in the month of September. Therefore in places where the grass has a dampness that is not congealed into frost, there is a vein beneath; also if the exhalations be excessively hot, the soil will produce only small and pale-colored plants. . . (From translation by Hoover and Hoover, 1912, p. 37–38.)*

There is reason to emphasize Agricola's work. His "miner," the little sage who points obligingly to rocks and pits in *De Re Metallica's* precious illustrations (Fig. 1-2), is the ancestral mining geologist. The things "he can see for himself" are the elements of mining geology. If the place-names in most manuals of mineralogy and textbooks on ore deposits were put on a map, the most closely "mapped" region would be Agricola's Saxon-Bohemian homeland.

Mining geology was born in the Erzgebirge, but it did not yet have the distinctive name and it did not grow for awhile. Even though minerals had a role in the European penetration of America, Africa, and Asia, there was so much emphasis on barter, plunder, and the skimming of ore from exposed orebodies that geology had little to contribute. A wealth of geologic information was being collected in the journals of explorers, but it lay dormant. Seventeenth-century study and conjecture on ore deposits were confined to the mining districts of Europe and in the work of scientist-philosophers who kept Agricola's views intact until they could be refined in the eighteenth-century mining academies.

A—Twig. B—Trench.

Figure 1-2. Prospecting methods in sixteenth-century Europe. (From Georgius Agricola, *De Re Metallica,* 1556.)

1.2. EIGHTEENTH-CENTURY EXPEDITIONS AND ACADEMIES

There came a time for geologic observation to be tested in regions as well as in mining districts. It was a time for *all* natural science to be tested; therefore, mines and minerals came to be described within the broader scope of such books as *Natural History of Cornwall* (Borlase, 1758).

Colonial nations and those with large undeveloped areas had particular reason to be interested in the geology of their lands; thus, when the first map of geologic features in eastern Canada was published in Paris in 1756, it included symbols for mineral deposits. In Russia, interest in the geology of mineral deposits was great enough to support a few governmental mining expeditions into Siberia. The work of the Russian expeditions was clearly spelled out to include reporting on occurrences, searching for ore in the known mining centers, and collecting data for the "confirmation of certain scientific and theoretical views" (Goldenberg, 1968, p. 346). Mineral maps were made with symbols to designate types of ore and geological associations. They were rather modern. So were some of the complications, such as the less-than-scientific interest of certain self-appointed observers. Instructions to Russian field parties in 1752 contain wording that would be just as appropriate in the twentieth century: "You are forbidden to enter into any relations and conversations with ore seekers—swindlers, such as Zubarev and others" (Goldenberg, 1968, p. 351). One enterprising Ivan Zubarev, eager to obtain land on the basis of its minerals, had become infamous for his stealthy visit to a St. Petersburg laboratory in which he "enhanced" some of his rock samples with melted gold and silver coins. Unfortunately for Zubarev, it was the laboratory of M. V. Lomonsov, an outstanding naturalist, and to him the first analysis seemed odd; a second analysis uncovered the fraud.

As mining expeditions and miners gained in stature, a need arose for specific education. Mining academies were founded, and mining geology became a subject in its own right. The Saxon-Bohemian miners led the way with the installation of a mining school at Jachymov (Joachimsthal), Czechoslovakia, in 1716 and at Freiberg, Saxony, in 1765. Other mining schools at Banska Stiavnica, Slovakia (1735), Kongsberg, Norway (1757), St. Petersburg (1773) and Paris (1783) enriched the scope of learning and added the influence of regions with different mineralogy and mining traditions. Mining science, mining law, geology, and mineralogy were added to the curriculum at Charles University in Prague in 1762 to begin the first department of mining at a major university.

Plutonists and Neptunists

At the close of the eighteenth century, the early sparks of the industrial revolution were struck from minerals, and they attracted scientific attention.

It was an atmosphere for geologic debate, and one debate, the plutonist-neptunist controversy, was so intense that it left its mark on all subsequent philosophies of mineral exploration. *Plutonists* followed the theory of James Hutton of Edinburgh that ore deposits are of igneous derivation and are brought into place as ore magmas from greater depth. *Neptunists* agreed with Abraham Gottlob Werner of Freiberg that ore deposits are derived from sediments in a primeval ocean and that veins represent cracks in the bottom of that ocean. Even though Werner's ideas were influenced by a few mine workings in the bedded Kupferschiefer deposits of Germany, his emphasis, and the emphasis of the time, was on veins. Veins were orebodies, and the economic conditions for mining stratabound deposits had not arrived. As with many geologic arguments (including modern-day variations on the same argument, as we shall see in Chapter 4), practical benefit came from field work by both plutonist and neptunist partisans seeking support for their views in areas of controversial evidence.

1.3. THE AGE OF INDUSTRIALIZATION

Geology entered the nineteenth century as an adolescent and argumentative science, but despite the practical atmosphere in the mining academies, it was still more philosophical than "applied." Mineralogy, on the other hand, was a quantitative science and more closely allied to mining; it had benefited from advances in chemistry and had, in fact, been a concern of chemistry for many years. Spectroscopy, the measurement of interfacial angles in crystals, and the invention of the polarizing microscope made the nineteenth century an important era in the understanding of mineral assemblages. Agricola's mining geology was still valid, but mining had spread so far beyond the Saxon-Bohemian Erzgebirge that some new dimensions were needed.

A new alloy of mineralogy and geology was cast into a solid theory of ore deposits by such chemist-geologists as Leonce Élie de Beaumont at the Paris School of Mines. Élie de Beaumont (1847) gave quantitative dimension to Agricola's epigenetic concept and to Hutton's idea of magmatic association; he described an affiliation between hydrothermal fluids and magmatic water that became—and remains—a basic element in theories of ore deposition. Werner's neptunists retreated—to regroup.

Nineteenth-century Geological Surveys

Geologists had proved their point: A connection exists between stratigraphic sequence, geologic structure, and mineral deposits. The natural history expeditions had proved something too: It was time for organized geologic mapping. Minerals were important, and coal was the most important of all.

In the first geologic map of England and Wales, 1815, at a scale of 1 inch to

5 miles, stratigraphic relationships were emphasized for the specific purpose of showing coal fields. Systematic geologic mapping of France, Germany, and Russia began in the 1830s, and in each area there was emphasis on mineral deposits.

By the middle years of the nineteenth century, geologic survey agencies had been established in nearly all industrialized countries; attention was given to the major mining fields—a matter of priority—and to speculation on concealed reserves. This was more than "natural history." The first memoir of the Geological Survey of Great Britain, in 1839, contained an exhaustive description rather than a passing mention of veins, dikes, and granitic masses in the Cornwall-Devon region. The U.S. government sponsored a geological survey of the copper and iron lands of Lake Superior in 1850. State geological surveys in Iowa, Wisconsin, and Illinois, established in the 1860s, made lead mining areas their first concern, and the California State Geological Survey gave its earliest attention to gold deposits.

The Geological Survey of Canada, established in 1842, operated in the Appalachians and in the more accessible parts of the Canadian Shield until 1867, the year of confederation. Then, with a new area of responsibility extending from the Great Lakes to the Pacific and northward into the Arctic, there began a series of epic reconnaissance traverses. The resulting geologic maps, with comments on mineral deposits, became immediate guidelines to prospecting. A successful Canadian formula for cooperative effort between government scientists and private prospectors began to take shape, and the shape became so attractive that it was copied by geological surveys throughout the world.

The U.S. Geological Survey did not become a formal organization until 1879, but it had antecedents in a series of expeditions. The earliest of these, to the Ozark Mountains, Great Plains, Rocky Mountains, Great Basin, and along the Mexican boundary, were more military and topographic than scientific; still, even as a camp follower, a study of geology and mineral deposits was included. In 1867, the Congress authorized an exploration of the 40th parallel by Clarence King, one of the most visible mining geologists of the nineteenth century, and subsequently provided for geological surveys of the Rocky Mountains by F. V. Hayden and Major J. W. Powell. The success of these "task force" surveys led to their consolidation into a permanent agency with King as director. The director was an intensely practical man, and he lost no time in assigning two graduates of the Freiberg Mining School to investigate geological conditions in mining districts so that the principles of ore deposition could be derived and used in finding new orebodies. Thus, when G. F. Becker's monograph on the Comstock lode and S. F. Emmons' monograph on the Leadville district, Colorado, appeared in 1882, there began a productive relationship between the U.S. Geological Survey and the mineral industry.

The U.S. Geological Survey did not act as a consultant to prospectors in the Canadian way, but it provided services—direct and indirect—to a large and growing number of geology-oriented mining engineers and to a smaller but equally dynamic group of mining geologists. Some of the early professional papers, folios, and bulletins gave such detailed geological information that they are still "bibles" for their particular mining districts.

One function of the nineteenth-century government geological surveys, more by circumstance than by design, was to provide a great "graduate school" for the mining profession. The list of engineers and geologists who gained their first field experience with national and colonial geological surveys is impressive in its numbers and even more impressive in the ultimate status of its "graduates."

New Depths and New Scenes

The scope of mining geology—access to deep information and new areas— literally exploded after 1850. Mining technology, with tools such as the power hoist and the Cornish pump, opened deeper ground so that theories of mineral zoning and structural control could be tested below 1,000 m in the Michigan copper mines, the gold-quartz reefs of Bendigo, Australia, and the lead-zinc-silver veins at Příbram, Bohemia. There was a new geography of mining. The names of a dozen Australian mines and of Leadville, Butte, and Bisbee in the American West were added to the predominantly European list of "textbook" localities. There was gold mining as never before, and it had more than its share of experimentation and innovation. Kalgoorlie, Cripple Creek, and the Witwatersrand became testing ground for geology and geologists.

In making the new geography of nineteenth-century mining, science generally came after the fact and as a guide to additional ore rather than to original discoveries. In spite of academicians and their concepts, most discoveries were made by "one-blanket, one-jackass prospectors" rather than by geologists. Prospectors were so much a part of the new regions that stories of their mineral discoveries have been told in countless books—part history and part folklore. The discovery of copper at Wallaroo, South Australia, actually may have involved tree roots that burned with a green flame, a copper carbonate pebble thrown up from a wombat hole, and a shaft site chosen by throwing a pick and digging where it landed (Blainey, 1969). The discovery of the Bunker Hill and Sullivan orebody in the Coeur d'Alene district of Idaho actually may have resulted from a prospector picking up a chunk of galena while chasing his runaway burro (Wolle, 1953).

Some of the tales are overdone, but there *were* chance discoveries, there *were* stampedes of opportunists, and the romanticized prospector of America and Australia *was* a player in the drama. The prospector was generally unlettered, but he knew rocks and he knew what he wanted. He was an exploration geologist without a diploma.

Geologists with Diplomas

Privately sponsored efforts in geology involved occasional mine appraisals by academicians, but the continuous work of finding and outlining orebodies was done by mining engineers. The mining engineer was geologist, surveyor, lawyer, chemist, and metallurgist—a general practitioner. He was *the* geologist in most mining camps and was the prime mover for exploration work until mining geologists and geological engineers arrived as accredited professionals in the twentieth century.

Consulting geologists were engaged from time to time to solve specific problems. In many instances, the consultants were respected scientists, such as Baron Ferdinand von Richthofen, who advised the management of the Comstock mines in 1864 in regard to deep extensions of the orebodies. In some other situations a geologist's report was sought only to provide a show of scientific legitimacy for an overbold speculation. If a geologist-accomplice could not be found, a prospect might be "dressed" or "salted" with minerals from elsewhere to set the stage for inspection by a reputable but gullible scientist. Happily, one such tactic backfired on its designers when geologists Clarence King and S. F. Emmons visited an exciting Wyoming diamond "discovery" in 1872 and exposed the entire affair as a swindle. Part of the evidence was geologic; part was even more direct—the swindlers had the gall to scatter about a few semicut stones.

There was a credibility barrier. Geologists often were less practical than mining engineers; and they sometimes drowned their clients in unrelated conjecture or, in appreciation of geology's blind spots, were purposely vague in their recommendations. An 1880 burlesque report by "Professor Noncommital" (he *had* to be a professor) can, unfortunately, still be regarded as timely humor in some mining camps.

There is undoubtedly a mine here if the ore bodies hold out. The gangue rock is favorable to the existence of ore and the overlapping seams of schistose show an undoubted tendency to productiveness in rock, which may be ore bearing. While I refrain from pronouncing with certainty on the Goosetherumfoodle mine, still I argue that as great expectations regarding the yield of this vein may be maintained as of any ground in the vicinity. The trend of the rock is S.S.E. and the direction of all the dips and angles show this to be a true lead, and as such liable to rich ore. Above all things the ground should be thoroughly prospected. I would advise the sinking of one hundred shafts ten feet apart through the hardest rock which can be found. If water be encountered it should be pumped out. If the rock prove rich the mine will prove valuable. If it prove very rich the mine will prove very valuable. It should be borne in mind that if it is necessary to sink deep on the vein the lode must be penetrated farther than if not. (Spence, 1970, p. 102, quoted from the Nevada Monthly, *Virginia City, v. 1, July 1880, p. 247.)*

The Appearance of Exploration Technology

While the technology of mining gained momentum and mining districts were being created, a technology of exploration began to appear. The diamond drill made its debut during the nineteenth century and passed its early trials—along with magnetic prospecting—in the Lake Superior iron ranges. The Comstock lode in Nevada was mapped by self-potential geophysical methods in 1880. T. Sterry Hunt (1873), using his field geologic experience in Canada, gave expression to geochemistry in a way very near to the modern sense of geochemical prospecting.

New technology and the mineral-rich frontiers created a need for engineer-geologists, and the need was met first of all by graduates of European mining schools. The academies at Freiberg, Berlin, Paris, and St. Petersburg sent their graduates to the developing regions of the world, and there was a return flow of students. Freiberg was the mecca; in the 1860s, about half of the enrollment came from America.

American mining schools began to have an impact on geology in the mid-nineteenth century, but the professors and the textbooks were still European until the end of the century when teachers with broad field experience began to build an American science of economic geology.

The nineteenth-century alloy of mineralogy and geology was being improved. The work of František Pošepný, professor at the Mining Academy in Příbram, Bohemia, had a profound influence on geologists throughout the world. In his most widely read treatise, *The Genesis of Ore Deposits,* published in English by the American Institute of Mining and Metallurgical Engineers in 1894, Pošepný drew upon field experience in Nevada and California as well as in Europe. In this book, he gave some convincing support to Élie de Beaumont's earlier igneous-hydrothermal ideas and opened the way to modern thoughts of hypogene and supergene mineralization. Louis de Launay (1913), one of the most perceptive of economic geologists, recognized the importance of regional ore characteristics and regional zoning patterns; on this basis, he introduced the terms "metallogenic province" and "metallogenic epoch." De Launay initiated the technique of metallogenic mapping; one of the most basic modern-day elements in regional mineral exploration, it is still strongly influenced by French geologists.

Waldemar Lindgren, associated with the U.S. Geological Survey during its formative years, with mining enterprises in Australia and the Americas, and a professor at the Massachusetts Institute of Technology, was the giant of American mining geology. In his (1913) textbook, *Mineral Deposits,* in the hundreds of papers written by him, and in the great number of discussions written about his ideas, there is a theme that ore deposits must be explained through *both* field observations and laboratory investigations. If some mineralogical association or some anomalous structural detail did not fit a particular model for ore deposition, Lindgren had to know why; he went back to the field or to the laboratory until he could resolve the contradictions. Our most widely used classification scheme for mineral deposits has this field and laboratory basis. The terms "hypothermal," "mesothermal," and "epithermal" are Lindgren's.

1.4. PROFESSIONAL IDENTITY

By the close of the nineteenth century, a science of economic geology had been shaped by government, private enterprise, and academia. The new science gained its own literature in the international journals *Zeitschrift für Praktische Geologie,* first published in Germany in 1893, and *Economic Geology,* beginning in the United States in 1905.

Early-day memoirs and professional papers of the governmental geological surveys provided detailed information in such testing grounds for mining geology as the Marquette Range, Michigan (1893), Aspen, Colorado (1893), Bisbee, Arizona (1904), and Sudbury, Ontario (1905). These works quickly became guides for detailed investigations. Regional summaries of mining activity and geological associations, also published by the national geological surveys, were guides to wider prospecting.

Mining company management began to credit geology with the finding of new orebodies at Tombstone, Arizona, at Bendigo, Victoria, and at many other great nineteenth-century mining camps. Geologists had gained credibility since Professor Noncommittal's time.

Exploration Organizations and Mine Geologists

Exploration organizations belong to the middle years of the twentieth century, but as a prelude there was the work of the Exploration Company, Ltd., founded in London in 1886 for prospecting in the American West and staffed by mining engineers. In the Guggenheim Exploration Company, another pioneering organization established in 1903, the record of searching-selecting-testing-evaluating could just as well apply to a modern-day exploration department; during 1910 and 1911, company engineers considered 1,605 mine propositions, made preliminary examinations on 268, continued with examinations on 74, and recommended 3 for purchase (Spence, 1970, p. 138).

Organized "exploration" or more accurately, wide-ranging prospect evaluation, had a particular appeal to newly formed companies and syndicates. The older mining companies were more often disillusioned by experiments with such fast-paced action; it seemed to them that expertise and capital resources did not count as much as audacity. Such was the experience of the Tharsis Company, a well-established European mining firm that appointed four prospecting engineers to comb the world's frontiers for new opportunities. Between 1899 and 1913, the opportunities came in a flood from the Americas and from Australia. All were rejected. The time had passed for nineteenth-century attitudes, so the Tharsis Company chose firmer ground and turned inward to the improvement of its existing mines in Spain (Checkland, 1967).

The first geological department at a large mining camp was established by the Anaconda Copper Mining Company in 1900. In part, the basis for organizing a group of geologists and assigning them to the Butte mines came from court litigation over boundary conflicts. When litigation ended in 1906 the department had proved so successful in finding orebodies to substantiate legal positions that it was expanded and given the task of exploration throughout the district. Reno H. Sales, Anaconda's chief geologist from 1906 until his retirement in 1948, developed a system of precise geologic mapping and a concept of structural control in orebodies that were widely copied by other geologists and resident mine geology departments. So many outstanding mining personnel received their early experience in Butte that Reno Sales' department, like the U.S. Geological Survey, became known as a great graduate school of mining geology.

The professional mining geologist and the exploration company had arrived, almost simultaneously, four and a half centuries after the science of mining geology arrived in the mountains of Central Europe.

Conditions of depletion and discovery had changed in the time between Agricola and Lindgren, but the most profound changes in the dimensions of depletion were only beginning. Mining geologists would soon have to supply minerals at a staggering rate and in an atmosphere of confrontation between technology and society. The stage was set for geoscience specialists within mining geology—and for the entrance of the geological engineer.

Geologists and Mines in the Current Context

Modern-day mining geologists may have been educated as geological engineers or as geoscientists. If the degree is in geological engineering, a curriculum in engineering science, geoscience, geotechnics, mining, and economics has provided him or her with a variety of capabilities that extend from exploration to the development and utilization of minerals. If the degree is in geoscience, a background in the physical and natural sciences, ore petrology, stratigraphy, and geomorphology affords a basis for a deeper understanding of the geological characteristics of mineral deposits. When geologists of either background enter the mining profession and make the best use of their experiences, their capabilities broaden and deepen according to their natural talents to a point where the wording on the diploma is immaterial. Deepening—becoming a specialist—may require postgraduate study and additional diplomas. In any event, college residence is a preparatory step; professional work and professional growth define the career.

Mining geologists have a powerful set of tools and concepts at their disposal, and many of these are associated with the unprecedented—even frightening—growth of all technology and science in this century. These are the tools and concepts that are described in the remainder of this book; they reflect an almost overwhelming range in genetic models for ore deposits, an integration of geology and geophysics, the appearance of exploration geochemistry, a revolution in logistics, and the appearance of an entire new group of computer-assisted sciences.

Exploration geophysics, outside of long-established magnetic methods, did not gain acceptance in mining until several decades after it had attained common use in the petroleum industry. A few geophysical discoveries of ore during the 1920s in Quebec, Newfoundland, and Sweden created a premature enthusiasm, which cooled after a surge of poorly conceived and unsuccessful trials. By the end of World War II, geophysical theory, interpretation, and tools had matured and geophysical exploration, now joined by geochemical exploration, entered a new series of trials. Success followed. The resurgence of mining geophysics involved surface, drill-hole, and airborne applications, but it was the last application that provided the most impressive steps toward mineral discoveries in new areas.

Aerial photography and the airborne transportation of personnel and materials began to be a part of mineral exploration in the 1920s and, like geophysics, came into prominence after (and partly resulting from) World War II. More recently, the great flexibility of helicopters and the spacecraft camera's ubiquitous "eye" have permitted exploration programs to operate in primitive areas of nationwide and even subcontinent size.

The route from Lindgren's *Mineral Deposits,* first published in 1913 and continued through many editions until 1933, to the current ideas of ore deposits has been described in two American works, *Ore Deposits of the Western States* (American Institute of Mining and Metallurgical Engineers, 1933) and *Ore Deposits of the United States, 1933–1967* (Ridge, 1968). Another work, one that deals specifically with concepts and their originators, is *History of the Theory of Ore Deposits* by Crook (1933). Landmark books along the way would be too numerous to include here; however, two have received so many nominations as twentieth-century classics that they must be mentioned: L. C. Graton's (1940) *Nature of the Ore-forming Fluid* and W. H. Newhouse's (1942) *Ore Deposits as Related to Structural Features.*

A special decade of important textbooks in economic geology and mining geology produced the still-valuable volumes by Emmons (1940), Schneiderhöhn (1941), Forrester (1946), McKinstry (1948), and Bateman (1950).

Among textbooks that reflect more recent advances in economic geology, there is the comprehensive two-volume *Les Gisements Métallifères* by Routhier (1963), the concise *Ore Deposits* by Park and MacDiarmid (1975), and the theoretical *Geochemistry of Hydrothermal Ore Deposits* by Barnes (1967). Two additional books dealing with the basis for modern theories of ore deposition bear special mention; these are Stanton's (1972) *Ore Petrology,* a comprehensive presentation on "ores as rocks," and Niccolini's (1970) *Gîtologie des Concentrations Minérales Stratiformes,* in which stratabound ore deposits are considered in the broadest regional sense as well as in the most minute detail.

"Exploration" was still primarily "prospect examination" until mid-century. A prospector reported the discovery of ore mineralization. A geologist or mining engineer assessed what had been found and, if he liked what he saw, sought the mineral rights for his company or client. He might further project geologic conditions into deeper ore, into an extended area, or into a nearby zone with comparable characteristics. Exploration by prospect examination and evaluation was quite successful in its time, and it is still a fundamental part of the exploration process. The emphasis has changed, however, to district-wide and regional efforts, and with the change in emphasis there has been a change in the role of the independent prospector.

The number of professional prospectors and small-scale miners has decreased during the last half century: their opportunities have shrunk with the depletion of near-surface high-grade ore. Exploration concessions of dozens or hundreds of square kilometers and mining leases of several square kilometers are beyond their financial grasp; so are deep drilling programs and the combination of capital and technology needed for mine development. The traditional prospector, an observant outdoorsman with a good level of intelligence but limited finances and limited formal education, had the hope of becoming an independent miner or at least a partner in a mine. He or she still has hope, but in a more restricted way; there will be prospectors as long as speculation in mining property carries a chance of profit by sale to a larger organization and as long as there are some high-grade orebodies. Shallow-seated high-grade uranium deposits were of interest to the individual prospector during the 1950s, and a resurgence of prospecting for gold took place in more recent years. These periods of excitement may have provided the last major arenas for prospecting in the "one-person, one-objective" sense.

However, there are other modern-day prospectors, the trained geologists who prefer to work as individuals or as spearheads for small groups of speculators. These individuals are often experienced consultants or former "company men" who have decided to follow their own geologic ideas until they gain an acceptable reward or decide to return to the shelter of a company's salary. Fully aware of the odds against finding and developing a major orebody on their own, they obtain ownership or partnership in promising areas and bring them to the attention of companies able to support the massive cost of further work. Such a man was Franc Joubin, a geologist with experience in uranium exploration and a strong belief in the favorability of the Blind River area, Ontario. He followed his own prospecting pattern and perfected his own theories of near-surface mobility in radioactive minerals until he was able to obtain the venture capital to open Canada's principal uranium district in 1955.

But for the most part, mineral exploration programs, whether within a private or a governmental framework, are organization affairs rather than individual efforts. They commonly involve the successive identification of favorable regions, favorable areas, and "targets." Prospectors and prospector-geologists still have a role, but it is a *group* of geologists with their concepts and field data that provides a central theme. Geophysics, geochemistry, and drilling technology provide the team with tools of discovery; laboratory and the computer facilities provide the tools of measurement.

The search for new ore districts is only a part of the geologist's concern. The challenge of depletion is every bit as important as it was in the mines of Cassandra and Laurium in ancient Greece, but it is now coupled with the realization that ore mineralization amounts to an orebody only within certain limiting conditions of conservation, economy, and technology; these are some of the other aspects of a geologist's work to be discussed in the chapters that follow.

PART ONE
THE GEOLOGIC BASE LINE

The term "ore finding" is sometimes used as a plain-language substitute for "exploration." It *is* plain, but inaccurate. A geologist is much more likely to find weathered evidence of ore than to find ore itself. With enough evidence—a geologic base line—a geologist can picture the ore that should lie at a depth of 1 m or 1,000 m, just as a surveyor can use a measured base line to visualize the location of a remote triangulation station. The picture is a conceptual model of the orebody, a model to be tested, revised, and tested again until it can be verified or abandoned. If the geologic base line is weak, the model will be weak, and the entire exploration effort will be shaky—regardless of the precision in subsequent measurements.

Chapter 2 pertains to the characteristics of available models—generalizations derived from orebodies and other mineral deposits that have been thoroughly investigated in deep mine workings and in drill holes. Chapter 3 returns to the surface and to the disguises taken by mineral deposits that have been weathered, eroded, and covered by postmineral processes. In Chapter 4, some arm waving at the global aspects of mineral provinces and a closer look at laboratory indications of the mineral-forming processes follow Lindgren's procedure of integrating laboratory and field evidence to understand *why* mineral associations appear as they do.

Taken together, an existing model, a surface environment, and a concept provide a geologic base line for exploration.

CHAPTER 2
MINERAL DEPOSITS IN THE SUBSURFACE ENVIRONMENT

The picture of an unseen mineral deposit—the model for testing—emphasizes minerals, not chemical elements. Otherwise, mineral deposits would be nothing more than intense geochemical anomalies and they could be pictured in the simple terms of *concentration clarkes,* coefficients by which the average percentage (the clarke) of individual elements in a part of the earth's crust would reach an economic level. Tables showing the abundance of useful elements in specific rock types and in petrologic provinces have a place in formulating theories of ore genesis, but they fall short of picturing the relationship between geology and technology that results in economic mineral deposits.

Models of ore deposits, and of coal and oil-shale deposits as well, are built from the mineral associations expected at a certain depth. The guides to ore deposits, weathered remnants discovered at the surface, belong to a system in a state of higher *entropy* than that of the model. It does not matter if we use the term "entropy" in its thermodynamic sense (degree of disorder) or in its stratigraphic sense (degree of mixing), the relationship is the same: the subsurface model, concentrated by some form of energy, does not appear at the surface until it receives the impact of the present-day surface environment.

2.1. ORE AND GANGUE MINERALOGY

The terms "gangue" and "wall rock," like the term "ore," have economic overtones. Even the strictest geologic definitions include the word "valueless." Enter technology once more: valueless at a certain time and place. Fluorite and barite, the gangue minerals of historic lead-zinc mining in the English Pennine ore field, are now major ore minerals, with galena and sphalerite as byproducts. Porphyry copper mines in western North America produce ore that would have been called wall rock or "waste" (below 0.4 percent copper) a decade ago. Gangue occurs within an orebody; wall rock forms the boundaries. "Country rock," another term in common use, is extensive and unmineralized, and presumably it has nothing to do with the mineral deposit. None of these terms can be defined in relation to a specific orebody without adding "under current conditions."

The inadequacy of such distinctions as "ore and gangue" and "ore and wall rock" has prompted many geologists to say that ore deposits are petrologic associations that should be studied in the same way as rocks; hence, the term "ore petrology" (Stanton, 1972). This view can save us from being shortsighted in exploration. Just as some rock types are found in several different environments, so are ore deposits of a certain mineralogy. Granite occurs in plutons, in stratiform complexes, and in migmatites. A pyrite–galena–sphalerite–quartz association may occur in foliated metamorphic rocks, in tactite, within an intrusion, or in relatively unaltered limestone. The range in possibilities cannot be ignored.

Rocks and ore deposits share another characteristic. Their components have entered new surroundings from time to time during geologic history where they have retained some aspects and changed other aspects in their mineralogical nature under the new geochemical conditions. In a sense, rocks and ore deposits are born with a *heritage*.

Heritage

In some deposits, placers, for example, the presence of a heritage and the identity of an immediate ancestor are quite apparent. Supergene sulfide bodies have identifiable ancestors; so do certain roll-front uranium orebodies that show evidence of having migrated downdip in sandstone beds through leaching, transportation, and redeposition. More generally, however, the heritage is mixed and it has been so thoroughly disguised by geologic processes that an association of minerals can be explained in several ways. The Mount Isa and Broken Hill orebodies in Australia and the Zambian copper belt ores were thought to be prime examples of hydrothermal metasomatic replacement in consolidated rock until a wave of observations in the 1950s exposed their sedimentary-metamorphic heritage. Other orebodies of similar shape and in comparable terrain throughout the world had been accepted as single-stage hydrothermal replacements; now the orebodies had to be reexamined. The terrain was reexamined as well, and this led the way to new exploration targets selected on a stratigraphic basis. Discoveries were made, including the Hilton orebody to the north of Mount Isa, on a reacquired property that had once been relinquished on the basis of an earlier metasomatic-hydrothermal model.

The side-by-side presence of orebodies that do not "belong together" is a part of an argument for complex heritage. If more than one mineralogical or geochemical association can be recognized, this may indicate several distinct stages of mineralization or it could just as well reflect the thermal history of a single stage. Textural, stratigraphic, and structural evidence must then be taken into account. This was done at Mibladen (Fig. 2-1), a complex lead deposit extending for 15 km in a 2-km-wide belt in central Morocco, where galena is found in disseminated zones, in stratiform bodies, and in veins (Emberger, 1969). In one group of open-pit workings, bedded zones of dense barite and disseminated zones of galena occur in Lias (Lower Jurassic) marl and shale. In another pit, galena occurs with crystalline barite in stratiform bodies in Lias marl and also with crystalline barite in a fault. In some underground workings where barren faults cut a stratiform zone of galena and crystalline barite, an overlying Cretaceous conglomerate contains pebbles of dense barite but there is no evidence of postconglomerate mineralization. In other underground workings, veins of galena and crystalline barite occur in the Cretaceous conglomerate above stratiform zones of galena and crystalline barite. The explanation: There were several stages of mineralization, beginning with syngenetic deposition of galena and massive barite and extending to *localized* stages of pre-Cretaceous and post-Cretaceous remobilization that produced new galena and crystalline barite.

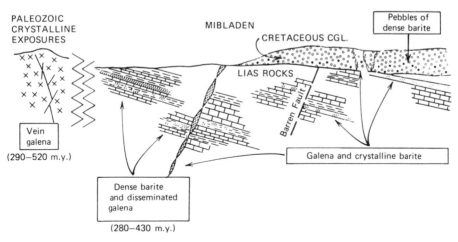

Figure 2-1.   Galena-barite associations at Mibladen, Morocco. (Data from Emberger, 1969, p. 37–57.)

Isotopic age determination and other forms of isotope geochemistry provide additional ways to identify successive stages in mineralization. On the basis of lead isotope studies, still another and earlier stage can be added to the Mibladen sequence. Samples taken from the bedded and vein deposits in the Lias (about 180 m.y.) show anomalous radiometric ages of 430 ± 60 to 280 ± 60 m.y. Other samples taken from veins in nearby exposures of pre-Lias crystalline "shield" show radiometric ages in about the same range, 520 ± 70 to 290 ± 50 m.y. Thus, the syngenetic lead mineralization in Lias rocks is likely to have been derived from older mineralization in the shield.

As with mineralogical lines of evidence, isotopic evidence for remobilization must be supported by geologic work. Current thoughts on ore genesis at the Homestake gold mine, South Dakota, provide an example. The orebodies, groups of pipelike ore shoots lying concordantly within isoclinically folded schist of the Precambrian Homestake formation, had long been thought to be of Tertiary age. Studies involving stable isotopes of sulfur, oxygen, and hydrogen upset the picture by indicating that the gold and associated minerals were indigenous to the Homestake formation. What about the closely associated gold mineralization in veins that cut Tertiary intrusions and Paleozoic rocks as well as the Precambrian rocks? Lead isotope studies of galenas from the host rocks and the veins and of feldspars from the Tertiary intrusions give evidence of two mineralization periods, a major period at 1,600 m.y. (Precambrian) and a minor one associated with Tertiary intrusions (Rye, Doe, and Delevaux, 1974). And we have the current concept for the Homestake ore:

1. Major orebodies were formed in dilation zones during Precambrian folding and metamorphism by the mobilization of syngenetic ore minerals. How did the minerals accumulate in the first place? Possibly in the same way as modern-day metalliferous brines accumulate on the floor of the Red Sea.

2. Minor orebodies in Precambrian, Paleozoic, and Tertiary rocks were derived in Tertiary time from the host rocks and—still a persistent possibility—the underlying basement.

The unraveling of the Homestake problem, like the geologic reassessment at Mount Isa and like nearly all conceptual breakthroughs in major mining districts, resulted in a new definition for "Homestake-type gold deposits," and this has prompted a new look at similar conditions in Morro Velho, Brazil, and in the Kolar gold fields of India (Sawkins and Rye, 1974). Perhaps these deposits are also of syngenetic origin.

Further examples of heritage in ore deposits and some associated views on *consanguinity* (relationship of ore material to the present host rock) are given in a concise form by Routhier and associates (1973). The idea behind the term "heritage" in ore deposits is however not so new. Schneiderhöhn (1941) long ago applied the term "regeneration" to processes in which deep-seated ore deposits could be dissolved and redeposited in veins. Lindgren (1933) recognized that deposits formed by circulating waters might appear as vein-type bodies as well as stratabound red-bed copper and sandstone-type uranium ore; his principal argument for most ore deposits, however, was tied to the concept of an ultimate magmatic source for hydrothermal solutions.

The concepts of ore-deposit heritage and consanguinity have begun to erode all our comfortable systems of genetic classification. Do secondarily enriched porphyry copper deposits belong in the same category as secondarily enriched red-bed copper deposits? Not if we consider the rock associations. Where does the subsurface environment become the zone of weathering? The interface should be at the base of the zone of oxidation; but the interface may have occupied other positions during geologic time. In this book, for convenience, the view is taken that mineral deposits belonging to the subsurface environment did not require present-day weathering conditions to reach economic grade. Mineral deposits in the zone of weathering, the subject of Chapter 3, have been changed—for better or worse—by leaching and secondary enrichment. Lateritic nickel and bauxite deposits, members of this group, are discussed in that chapter.

Economic Mineral Associations

Sphalerite and galena are common associates; they are likely to be accompanied by pyrite and there is a good chance of seeing some chalcopyrite as well. The association is so prevalent that it is abbreviated as BPGC (blende–pyrite–galène–chalcopyrite) on French geologic maps. Fluorite and barite are common accompanying minerals and, in fact, are so often found in BPGC veins that we can use the occurrence of one of these minerals as a clue in searching for the others. The resulting conceptual model can be exciting—it might be based upon the Walton mine, Nova Scotia, where a rich lead-zinc-copper-silver body was found at a depth of 75 m while mining a barite orebody.

We can use wolframite or perhaps topaz or tourmaline as a guide to cassiterite. An occurrence of molybdenite noted in a mine dump or on a map can serve as a guide to porphyry copper ore. We work with mineral associations (Table 2-1) in the same way the exploration geochemist works with *indicator elements* (one of the elements sought) and *pathfinder elements* (an accompanying element).

But there are some limits to this sort of thing. Mineral deposits have individual associations with lithology, local geological conditions, and (again) heritage. Thus, when an "association" is mentioned without the word "general" in the same sentence and the word "except" in the same paragraph, it can lead us astray. Even so, the modifying terms are sometimes ignored and geologists have wasted countless days in the field looking for

Table 2-1.   Economic Mineral Associations in the Subsurface Environment

The associations are broad, but they identify some of the common mineral assemblages to be expected when working with a particular commodity. Examples are given so that detail can be obtained from the literature on specific deposits.

Supergene orebodies, residually enriched deposits, recent evaporite deposits, and placers are excluded. These are listed in Table 3-4.

Abbreviations in common use are shown for each of the major ore minerals; the additional abbreviations used in describing ore and matrix assemblages are from Appendix A.

In the "ore assemblage" list, minerals containing the subject commodity are italicized. Minerals that are missing from some of the major deposits are shown in parentheses; in a few deposits these same minerals are important. Mining grades are

Commodity—Ore Minerals	Associations—Example	Ore Assemblages
BERYLLIUM		
brl beryl—14% BeO $Be_3Al_2Si_6O_{18}$	Pegmatites Black Hills, S.D. Ceara, Brazil	*brl* (in pegmatite assemblage) with Li, Nb-Ta, Sn minerals
bte bertrandite—42% BeO $Be_4Si_2O_7(OH)_2$	Volcanics Aguachile Mtn., Coahuila, Mexico Spor Mountain, Utah	bte
hlv helvite—13% BeO $(Mn,Fe,Zn)_8Be_6Si_6O_{24}S_2$ cbl chrysoberyl—20% BeO $BeAl_2O_4$		
bit barylite—16% BeO $BaBe_2Si_2O_7$	Pyrometasomatic Victorio Mtns., N.M. Iron Mountain, N.M.	*hlv*
	Veins Lake George, Colorado Lost River, Alaska	*cbl*, (*brl, hlv,* bte, cas)
CHROMIUM		
crt chromite—33%–58% Cr_2O_3 $(Fe,Mg)Cr_2O_4$	Layered Ultramafics Bushveld, South Africa Great Dyke, Rhodesia	*chromite*
	"Alpine" Ultramafics Khrom-Tau, USSR Guleman, Turkey	*chromite*
COPPER		
cp chalcopyrite—35% Cu $CuFeS_2$ bn bornite—69% Cu Cu_5FeS_4	Pluton Association Porphyry Copper Bingham Canyon, Utah Chuquicamata, Chile	py, *cp, bn,* mb, (sl, gn, en)

given for general context; many commodities are recovered at a much lower grade as coproducts and byproducts. And, of course, there are higher grade "bonanzas."

Reference works have been chosen on the basis of their descriptive treatment of the mineral associations. A commodity-by-commodity description of the geologic environment for each type of deposit is included in the U.S. Geological Survey Prof. Paper 820, *United States Mineral Resources,* edited by Brobst and Pratt (1973). Another encyclopedic reference book with detail on the mineralogy and morphology of deposits, commodity by commodity, is a German-language compilation edited by Bentz and Martini (1968). Additional information on specific elements and their economic mineral associations can be obtained from Volume IV A (geochemistry) of the *Encyclopedia of Earth Sciences,* edited by Fairbridge (1972), and the *Handbook of Geochemistry,* edited by Wedepohl (1958, with subsequent revisions).

Matrix Assemblage	Mining Grade	References
qtz, mica, feld, tm, tz	0.5%–3% beryl	General, Be: Beus (1966), Burt (1975)
sil, cal, fl, mont, MnOx	0.3%–0.8% BeO	Spor Mountain: Shawe (1968)
vesuvianite, gr, fl	0.2%–3.5% BeO	Pyrometasomatic: Warner and others (1959)
tm, mica, fl, diaspore	0.5%+ BeO	Lost River: Sainsbury (1968)
mafic norite to anorthosite rock	10%–50% Cr_2O_3	General, Cr: Wilson (1969) Bushveld: Visser and von Gruenewaldt (1970) Great Dyke: Worst (1960)
olivine, serp, and dunite rock		
qtz, ser, (k-spar, cal, sid, mag, fl, bar, anh, tm, and skarn assemblage)	0.3%–2%	General, Cu: Pelissonnier (1972) Porphyry copper: Lowell and Guilbert (1970), Sutherland Brown (1976)

Table 2-1 (con'd).   Economic Mineral Associations in the Subsurface Environment

Commodity—Ore Minerals	Associations—Example	Ore Assemblages
en enargite—49% Cu Cu_3AsS_4		
cc chalcocite—80% Cu Cu_2S	Vein and Replacement Butte, Montana Bor, Yugoslavia	py, *cp, cc, bn, en, cv,* *(tt, tn,* mb, sl, gn)
tt tetrahedrite ⎱ tennantite ⎰ —30–50% Cu $Cu_3(Sb,As)S_3$	Stratiform-Sedimentary Association Zambian copper belt Dzhezkazgan, USSR	cc, bn, cp, (Cu, gn, sl, py)
Cu native copper		
cv covellite—66% Cu CuS	Stratabound Massive Sulfides Rio Tinto, Spain Noranda, Quebec	py, *cp,* sl, gn, *(tt, bn)*

GOLD		

Au native gold	Volcanic-Subvolcanic
	  Telluride Veins
	Cripple Creek,
	  Colorado
	Kalgoorlie, Western
	  Australia
		cvt, slv, pet, hes, py,
		base-metal sulfides
		and sulfosalts

cvt calaverite—39% Au $AuTe_2$		
slv sylvanite—24% Au $(Au,Ag)Te_2$		

pet petzite—25% Au
Ag_3AuTe_2
	Precambrian
	  Conglomerates
	Witwatersrand, South
	  Africa
	Tarkwa, Ghana
		Au, py, (base-metal
		sulfides, U-Th
		minerals, Ag)

	Pluton-Vein and Dis-
	  seminated Associ-
	  ation
	Mother Lode,
	  California
	Carlin, Nevada
		Au, py, asp, po, (hm,
		mag, sch, cbr,
		stibnite, tt, base-
		metal sulfides)

	Stratabound-Metamorphic Homestake, South Dakota Morro Velho, Brazil	*Au*, py, po, asp, mag, hem, base-metal sulfides, wf, sch

Matrix Assemblage	Mining Grade	References
qtz, (bar, fl, cal, anh, rc, sid)	0.6%–4%	Butte: Meyer (1968)
qtz, chl, ser, (bar)		Stratiform: Bartholomé (1974), Wolf (1976)
qtz, carb, gyps, anh, (bar)		Stratabound: Hutchinson (1973), Wolf (1976) Noranda: Gilmour (1965)
qtz, (cal, dol, fl)	6–30 + g/t Au	General, Au: Koschman and Bergendahl (1968)
qtz, ser, chl	4–12 g/t Au	Witwatersrand: Pretorius (1975), Wolf (1976, v. 7)
qtz, (car, feld, chl, tm)		Nevada: Roberts, Radtke, and Coats (1971)
qtz, (tm, car, chl)		Homestake: Noble (1950)

Table 2-1 (con'd).   Economic Mineral Associations in the Subsurface Environment

Commodity—Ore Minerals	Associations—Example	Ore Assemblages
IRON		
mag magnetite—72% F Fe$_3$O$_4$ hm hematite—70% Fe Fe$_2$O$_3$ goe goethite—63% Fe Fe$_2$O$_3$·H$_2$O	Precambrian Iron Formation Lake Superior Cerro Bolivar, Venezuela	*hm, mag,* (*sid,* py)
sid siderite—48% Fe FeCO$_3$ chm chamosite—30% Fe 2SiO$_2$·Al$_2$O$_3$·FeO aq	Oolitic "Ironstone" Ores "Clinton," New York to Alabama "Minette," Lorraine, France	*hm, goe,* (*sid, mag, chm,* py)
	Pyrometasomatic Magnitnaya, USSR Iron Springs, Utah	*mag, hm,* (py, po, base- metal sulfides)
	Siderite Orebodies Siegerland, German Erzberg, Austria	*sid,* (py, base-metal sulfides)
	Magmatic Orebodies Pea Ridge, Missouri Kiruna, Sweden	*mag, hm,* (py, cp)
LEAD–ZINC		
gn galena—86% Pb PbS sl sphalerite—60%–67% Zn ZnS	Stratabound Massive Sulfides Kidd Creek, Ontario Mattagami, Quebec	py, *sl,* gn, cp, (po, mag)
	Pluton Association Pyrometasomatic Hanover, New Mexico Magdalena, New Mexico	*sl, gn,* (cp, mag, hm, py, po)
	Veins and Replacement Tintic, Utah Coeur d'Alene, Idaho	*gn, sl,* tt, py, cp, (po, asp, mag)
	Stratiform-Metamorphic Broken Hill, New South Wales Mt. Isa, Queensland	py, *sl, gn,* (cp, po, mag)

Matrix Assemblage	Mining Grade	References
jasp, cht, Fe silicates	25%–50% Fe	General, Fe: United Nations (1970), Lepp (1975) Precambrian: James and Sims (1973)
cal	30%–60% Fe	Oolitic: James (1966)
mica, apa, qtz, sil		Pyrometasomatic: Park (1972)
ank, bar, fl	30%–40% Fe	Erzberg: Hajek (1966)
apa, qtz, cal, FeMg silicates	30%–70% Fe	Missouri: Snyder (1969)
qtz, (cal, dol, ser)	5%–25% Pb + Zn	Stratabound massive sulfides: Wolf (1976, v. 6) Kidd Creek: Walker and others (1975)
gr, pyx, cal, dol, qtz		Hanover: Hernon and Jones (1968)
qtz, sid, bar, cal, dol, jasp, (rc, ank)		Tintic: Cook (1957)
dol, cal, bar, fl, (mica, chl)		Australia: Stanton (1972)

Table 2-1 (con'd).   Economic Mineral Associations in the Subsurface Environment

Commodity—Ore Minerals	Associations—Example	Ore Assemblages
	Stratabound-Limestone Southeast Missouri (Mississippi Valley) Bleiberg, Austria (Alpine)	py, *gn*, *sl*, (mc, cp)
LITHIUM spd spodumene—4%-8% Li_2O LiAlSi$_2$O$_6$ lep lepidolite—3%-6% Li_2O K$_2$Li$_3$Al$_4$Si$_7$O$_{21}$ amb amblygonite—8%-10% Li_2O LiAlFPO$_4$ pet petalite—4% Li_2O LiAlSi$_4$O$_{10}$	Pegmatites King's Mtn., North Carolina Bikita, Rhodesia Lake Beds and Brines Searles Lake, California Silver Peak, Nevada	*spd* or *lep*, (*amb*, *pet*, brl, cas)
MANGANESE plu pyrolusite—55%-63% Mn MnO$_2$ mnt manganite—50%-62% Mn MnO(OH) psl psilomelane—35%-60% n·MnO·MnO$_2$·mH$_2$O brt braunite—60%-69% Mn 3Mn$_2$O$_3$·MnSiO$_3$ rc rhodochrosite—40%-45% Mn MnCO$_3$ hau hausmannite—65%-72% Mn Mn$_3$O$_4$ rd rhodonite—32%-36% Mn MnSiO$_3$	Sedimentary-Marine Chiaturi, USSR Nikopol, USSR Volcanogenic-Sedimentary Coast Range, California Aroostook County, Maine Pluton-Associated Veins Butte, Montana Ilfeld, Germany Metamorphic Postmasburg, South Africa Minas Gerais, Brazil	*plu*, *mnt*, *rc*, (*psl*, brt, mc, py) *hau*, rc, (*brt*, base-metal sulfides) rc, *plu*, *mnt*, *hau*, *brt*, *rd*, (base-metal sulfides) *brt*, *hau*, (*mnt*, rc, po, base-metal sulfides)
MERCURY Hg native mercury cbr cinnabar—86% Hg HgS	Volcanic-Sedimentary Almadén, Spain Monte Amiata, Italy	*cbr*, py, stibnite, (mc, Hg, gn, sl)

Matrix Assemblage	Mining Grade	References
dol, cal, (fl, bar, qtz, jasp)		Stratabound-limestone: Brown (1967, 1970)
ab, qtz, musc	1%–4% Li_2O	General, Li: Norton (1973)
	0.015%–0.06% Li_2O	Beds and brines: Vine (1975) Silver Peak: Kunasz (1970)
bar, sil, gyp, glauconite	15%–40% Mn	General, Mn: Gonzalez-Reyna (1956), Hewett (1972) USSR: Sapozhnikov (1970)
bar, sil	10%–50% Mn	
qtz, bar, (fl)	7%–15% Mn	
gr, cht, qtz, (dol, apa)	20%–50% Mn	
qtz, sil, dol, cal, (fl, bit, bar)	0.2%–8% Hg	Almadén: Almela-Samper and others (1964)

Table 2-1 (con'd).   Economic Mineral Associations in the Subsurface Environment

Commodity—Ore Minerals	Associations—Example	Ore Assemblages
	Volcanic-Subvolcanic- Metamorphic New Almaden, California Terlingua, Texas	py, mc, *cbr,* (orpiment, realgar, tellurides, base-metal sulfides)

MOLYBDENUM
mb   molybdenite—60% Mo
   MoS_2
pow powellite—48% Mo
   $CaMoO_4$
	Porphyry Molybdenum
	Climax, Colorado
	Endako, British
	Columbia
		py, *mb,* (cp, mag)

	Porphyry Cu-Mo Porphyry copper association	(see porphyry copper)
	Pyrometasomatic Association Bishop district, California Azegour, Morocco	sch, *mb, pow* (cpy, bn, mag)

NICKEL–COBALT
pn   pentlandite—10%–40% Ni
   $(Fe,Ni)_9S_8$
nic niccolite—44% Ni
   NiAs
sgt siegenite—11%–53% Co
   $(Co,Ni)_3S_4$
crl carrolite—36% Co
   $CuCo_2S_4$
	Veins with Ag, Co, Ni, As
	Cobalt-Erzgebirge type
		(see Silver-Cobalt,
		Ontario, Erzgebirge)

	Norite and Ultramafic Fe-Cu-Ni Sudbury, Ontario Thompson, Manitoba	po, *pn,* py, (cp, mc, mag, Ni arsenides)
	Stratabound Cu-Co-Ni-U Shaba, Zaire Zambian copper belt	py, cp, bn, cc, *crl, sgt,* (pch)
	Manganese Nodules Sea floor (potential)	Cu:Ni:Co = 3:4:1

NIOBIUM–TANTALUM
tnt tantalite—45%–84% Ta_2O_5
   $(Fe,Mn)Ta_2O_6$
clb columbite (niobite)—45%–
   78% Nb_2O_5
   $(Fe,Mn)Nb_2O_6$
	Pegmatite
	Bernic Lake, Manitoba
		clb, tnt, mic, brl, spd,
		amb

	Carbonatite Oka, Quebec Araxá, Brazil	*pcl, per,* mag, ilm, rt, py, monazite, rare- earth minerals, (base- metal sulfides)

Matrix Assemblage	Mining Grade	References
opal, sil, bit, sulfur, qtz, carb, (gyp)		New Almaden: Bailey and Everhart (1964)
qtz, k-spar, bio, ser, (fl)	0.1%–0.6% Mo	General, Mo: Kirkemo, Anderson, and Creasey (1965)
(see porphyry copper)	0.003%–0.05% Mo	
gr, pyx, qtz, amph	0.2%–1.0% Mo	Bishop: Gray and others (1968)
(see Silver-Cobalt, Ontario, Erzgebirge)	0.3%–5.0% Ni + Co	General, Ni: Cornwall (1966)
sil rock minerals, dol, ank, (ser, chlo, talc)		Norite-Ultramafic: Naldrett (1973), Wilson (1969) Sudbury: Guy-Bray (1972)
qtz, dol, chl, (rt, tm)		Stratabound: Bartholomé (1974)
	0.35% av. Co	Manganese nodules: Horn (1972)
feld, mica, tm	0.2%–2.5% Nb_2O_5	Canada: Dawson (1974)
cal, dol, bar, (fl, apa, zircon)		Carbonatite: Heinrich (1966)

Table 2-1 (con'd).   Economic Mineral Associations in the Subsurface Environment

Commodity—Ore Minerals	Associations—Example	Ore Assemblages
pcl pyrochlore—23%–73% Nb_2O_5 $(Na,Nb)Nb_2O_6F$ per perovskite $CaTiO_3$ with Nb mic microlite—33%–77% Ta_2O_5 $(Na,Ca)_2Ta_2(O,OH,F)_7$		

PLATINUM METALS

spe sperrylite—56% Pt $PtAs_2$ fep ferroplatinum—75%–84% Pt brg braggite—59% Pt $(Pt,Pd,Ni)S$	Ultramafic Cr-Ni-Pt Association Merensky Reef, Bush- veld, South Africa	crt, Fe-Cu-Ni sulfides (Au, *spe, fep, brg*)
	Norite Fe-Ni-Cu-Pt Association Sudbury, Ontario Noril'sk, USSR	po, pn, cp, (mag, ilm, py, *spe*, base-metal sulfides, arsenides)

SILVER

Ag native silver agt argentite—87% Ag Ag_2S pu proustite—65% Ag Ag_3AsS_3 prg pyrargyrite—60% Ag Ag_3SbS_3 ste stephanite—69% Ag Ag_5SbS_4 plb polybasite—64%–72% Ag $(Ag,Cu)_{16}Sb_2S_{11}$	Silver-rich Base-metal Deposits Massive sulfide association Tintic–Coeur d'Alene Association	(see Lead–Zinc)
	Silver-Cobalt-Nickel Arsenide Veins Cobalt, Ontario Erzgebirge, GDR-CSSR	py, *agt*, cp, *gn*, Ag, Bi, (Fe-Ni-Co-Ag arsenides and sulfosalts, also pch)
	Volcanic-Subvolcanic Veins Pachuca, Hidalgo, Mexico Comstock Lode, Nevada	ag, *agt, plb*, (gn, cp, py, Au-*Ag* tellurides)

TIN–TUNGSTEN

cas cassiterite—79% Sn SnO_2 stn stannite—28% Sn Cu_2FeSnS_4	Pegmatite Kamativi, Rhodesia Manono, Zaire	*cas,* nio-tnt, spd, brl, (*wf, sch*)

Matrix Assemblage	Mining Grade	References
pyx, plag, (bio, olivine, qtz)	3–15 g/t Pt	General, Pt: Stumpfl (1974) Merensky Reef: Vermaak (1976)
silicate rock minerals	Byproduct (variable)	Sudbury: Cabri and Laflamme (1976)
(see Lead-Zinc)	100–1000+ g/t Ag	General, Ag: Boyle (1968)
qtz, cal, (dol, bar, fl)		Cobalt, Ontario: Berry (1971)
qtz, rc, rd, zeo		Volcanic, subvolcanic: Wisser (1966)
feld, mica, qtz, and complex pegmatite assemblage	0.2%–5% Sn + W	General, Sn: Fox (1967, 1969) General, W: Li and Wang (1955)

Table 2-1 (con'd).   Economic Mineral Associations in the Subsurface Environment

Commodity—Ore Minerals	Associations—Example	Ore Assemblages
wf wolframite—60%–75% WO_3 (Fe,Mn)WO_4	Vein Oruro, Bolivia Cornwall, England	cp, cas, py, *wf*, asp, (bismuthinite, base-metal sulfides)
hn huebnerite (Mn variety wf) frb ferberite (Fe variety wf) sch scheelite—80% WO_3 $CaWO_4$	Skarn Mill City, Nevada (W) Azegour, Morocco (W,Mo,Cu)	cp, py, po, mb, *sch*, pow, *cas*
TITANIUM ilm ilmenite—53% TiO_2 $FeTiO_3$	Carbonatites Nb-Ta association	(see Niobium-Tantalum carbonatites)
rt rutile—92%–98% TiO_2 TiO_2 tit titanite—41% TiO_2 $CaTiSiO_5$	Anorthosite Allard Lake, Quebec Sanford Lake, New York	mag, *ilm*, hem, (*rt*, sulfides, apatite)
URANIUM urn uraninite—47%–88% U UO_2 pch pitchblende (variety urn) cof coffinite—60% U $USiO_4$	Quartz Pebble Con-glomerates Witwatersrand, South Africa Blind River, Ontario	py, *urn*, (*thu*, *urt*, bra, Au, po, sl, mb, gn)
bra brannerite—28%–44% U (U,Th)Ti_2O_6 urt uranothorite—5%–15% U (Th,U,Fe)SiO_2H_2	Sandstones Colorado Plateau Wyoming basins	*urn*, *cof*, (V minerals, gn, sl, mb, cc)
thu thucolite—1%–5% U U-hydrocarbon	Veins Great Bear Lake, N.W.T. Erzgebirge, Czecho-slovakia	py, *pch*, (Sn-Ni-Co-V-As-Ag-Bi minerals, Cu-Pb-Zn sulfides)
	Pegmatites Bancroft, Ontario Rössing, S.W. Africa	*urn*, *urt*, betafite
	Magmatic Ross-Adams, Alaska Illimaussaq, Greenland	U-rich variety of monazite (Greenland)
	Pyrometasomatic Mary Kathleen, Queensland	*urn*, allanite, rare-earth silicates, py, po, cp

Matrix Assemblage	Mining Grade	References
qtz, topaz, tm, (sid, rc, apa)		Oruro: Sillitoe, Halls, and Grant (1975) Cornwall: Dines (1956)
qtz, pyz, diopside, (cal, fl, bar)		Nevada: Hotz and Willden (1964)
(see Nb-Ta, carbonatites)	10%–50% TiO$_2$	General, Ti: Lynd and Lefond (1975)
Anorthosite rock		Canada: Rose (1965)
qtz, ser, chl, (feld, carbon)	0.02%–0.15% U$_3$O$_8$	General, U: Bowie, Davis, and Ostle (1972) Witwatersrand: Pretorius (1975) Blind River: Roscoe (1968)
carbonaceous matter, gyp, cal, dol, (fl, bar, kao)	0.1%–1% U$_3$O$_8$	Sandstones: Fischer (1974)
qtz, dol, sid, (bar, fl, cal)	0.1%–2% U$_3$O$_8$	Veins: Smith (1974)
pegmatite and alaskite rock	0.1%–1%	Rössing: Berning and others (1976)
syenite and granite rock minerals	0.03% U (Greenland)	Illimaussaq: Sørensen (1973)
qtz, fl, apa, cal	complex	Mary Kathleen: Hughes and Munro (1965)

Table 2-1 (con'd).   Economic Mineral Associations in the Subsurface Environment

Commodity—Ore Minerals	Associations—Example	Ore Assemblages
	Syngenetic-Sedimentary Alum and Chatanooga shales Phosphate	organouranium complex, (py)
VANADIUM pat patronite—28%–39% V$_2$O$_5$ VS$_4$ ros roscoelite vanadium mica car carnotite—12% V$_2$O$_5$ K-U vanadate	Marine Sediments Idaho-Montana phosphate Minas Ragras, Peru	V complex in fluorapatite and shale (*pat*, py)
	Sandstone Colorado Plateau	*car, ros,* with U ore
	Anorthosite Bushveld, South Africa Allard Lake, Quebec	V in mag, ilm, (po, cp)

mineral associations that would only fit a "back-home" model or a mineral equilibrium diagram. The outcome may amount to something worse than wasted time: a missed orebody that (on hindsight) should have been perfectly obvious.

Table 2-1 is meant to be taken with "general" firmly in mind and with ample room for "except." Also, there is the proviso that minerals commonly occur in a zonal arrangement rather than all together, so the association depends on what part of the mine one is in. Other minerals and other zones are to be expected in depth, the important third dimension that, for economic reasons, may not have been fully investigated in the type localities. In most European geologic literature, Table 2-1 would be considered a table of mineral paragenesis (characteristic associations or occurrences). In English-language geology, the term "paragenesis" refers to a sequence of mineral deposition.

Paragenesis

The paragenesis (sequence of deposition) has certain overall characteristics in deposits of hydrothermal character and simple heritage: oxides are early; sulfides and arsenides of iron, nickel, cobalt, and molybdenum are somewhat later; sulfides of zinc, lead, silver, and iron-copper follow; cobalt, lead, and silver sulfosalts are still later; and the latest minerals are tellurides, stibnite, and cinnabar. Obviously, the overall sequence involves considerable overlap-

Matrix Assemblage	Mining Grade	References
qtz, feld, ill, kao	0.03% U (Alum Sh)	Alum shale (Ranstad, Sweden): Peterson (1967) Phosphate: Gautier (1970)
gyp, asphalt, shale wall rock (Minas Ragras)	Byproduct (variable)	General, V: Fischer (1973)
	1%–5% V_2O_5	Sandstone: Fischer and Stewart (1961)
pyx, amph, bio, olivine, plag	0.3%–2% V_2O_5	Bushveld: Willemse (1969)

ping and each deposit has its own unique paragenesis—its fingerprint. Nevertheless, there are enough similarities between deposits of similar environment to provide an indication of whether we can apply some "typical" orebody model to the small part of the mineralization we may have discovered.

The paragenetic sequence shown in Figure 2-2 is for the upper Mississippi Valley zinc-lead district, Wisconsin-Illinois-Iowa; in this sequence, the deposition of some minerals is repeated, some minerals have periods of weak and intense deposition, and fracturing during mineralization is evident. Complexities are to be expected, especially where an entire district is involved; in fact, even the relatively simple upper Mississippi Valley sequence took 25 years of detailed geologic mapping and thousands of samples to interpret (Heyl, 1968).

Studies of paragenesis are a part of all mine geologic investigations; with these studies, which rely heavily on the microscopic study of thin sections and polished sections of ore, field relationships can be tested, reexamined, and amplified with more field work until a reasonable origin and heritage of an orebody can be pieced together. For example, in a thin section of ore from a fluorspar mine in the Zuni Mountains, New Mexico (Fig. 2-3), the outlines of earlier quartz crystals are shown preserved in a fluorspar-hematite association. A thin section of ore from Divide, Montana (Fig. 2-4) shows that fluorite has replaced chalcedonic quartz and has inherited some of its colloform texture. Figure 2-5, a thin section of ore from Iron Mountain, New Mexico,

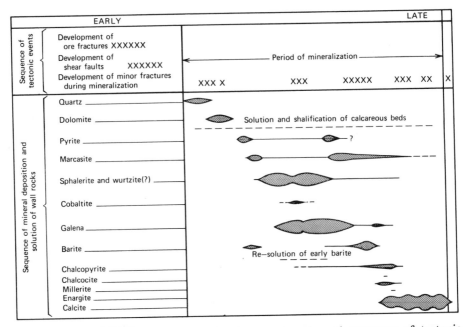

Figure 2-2. Paragenetic sequence of primary minerals and sequence of tectonic events and leaching of wall rocks, Upper Mississippi Valley base-metals district. (From A. V. Heyl, "The Upper Mississippi Valley base metal district," in *Ore Deposits of the United States, 1933–1967*, J. D. Ridge, ed., Fig. 14, p. 449, 1968, by permission American Institute of Mining, Metallurgical and Petroleum Engineers, Inc.)

shows galena replacing fluorite along cleavages. In all of these illustrations, some part of a paragenetic sequence is demonstrated from which some idea can be incorporated into the field work of an exploration program.

Optical mineralogy and thin-section petrography become part of mining geology when they involve the nonopaque minerals in gangue, wall rock, and country rock. Ore microscopy (mineragraphy) is a more specific technique in mining geology. The standard reference book in ore microscopy is *Ore Minerals and Their Intergrowths* by Ramdohr (1956). For an introduction to the subject, *Microscopic Study of Opaque Minerals* by Galopin and Henry (1972) is recommended. Most of the past work with polished sections of ore was done in connection with microchemical tests; these tests are still important, but current emphasis is on the electron microprobe, an instrument for nondestructive analysis of very small grains and inclusions of 0.001–0.2 mm in diameter. Instrumental mineralogy and petrography, basic tools in mining geology, are well presented in a handbook by C. S. Hutchinson (1974) and a treatise edited by Nichol (1975).

We do not want to lose sight of mineral associations in bodies of industrial

Figure 2-3.   Fluorite, Zuni Mountains, New Mexico, ordinary light, ×40. Outlines of quartz crystals preserved in fluorite (light gray) and fluorite-hematite mixture (black).

minerals and rocks. The same principles of ore and gangue mineralogy apply to these economically important, shirttail relatives of metallic ore deposits. An introduction to the mineralogy of "nonmetallics" can be obtained in the textbook by Bates (1960) and in *Industrial Minerals and Rocks* (Lefond, 1975).

2.2. ZONING PATTERNS

Suppose we are examining a group of lead-zinc mines and prospects in an old mining district. In the deepest of the shafts, there is a small amount of

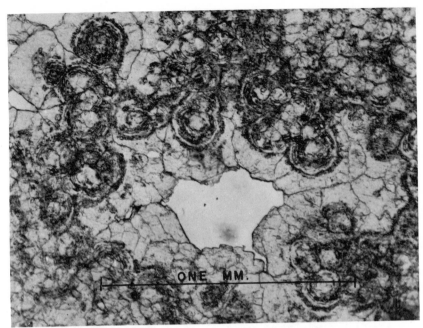

Figure 2-4. Fluorite, Divide, Montana, ordinary light, ×70. Colloidal texture in fluorite apparently inherited from replaced chalcedonic quartz.

chalcopyrite associated with the more common sphalerite and galena. Our curiosity is aroused and we take another look at the shallower workings. No additional chalcopyrite, but we note that the sphalerite is generally lighter in color—perhaps lower in iron—than in the deep shaft and there appears to be less pyrite in the shallow workings. Now we may be onto something; this could be part of a zonal pattern around a deep copper orebody.

Zonal guides to ore need not be such coarse affairs as this. They can be as delicate as parts-per-million changes in the amount of strontium (geochemical zoning) or as tenuous as a change from argillized to sericitized rock (wall-rock alteration zoning). All of these are a mining geologist's ''ringed targets,'' described and so-named by McKinstry (1948).

Once a part of an orebody has been found the consideration of zoning patterns becomes even more important. Zoning is three dimensional, and some kind of zonal arrangement may go right through the orebody—to its center and into its roots. If we can recognize and interpret the pattern, we have a guide to the bull's-eye in the target and we can decide whether a zone of weakened mineralization on the lowest level of a mine is really the bottom of the orebody. This is the kind of knowledge that may take years of careful investigation to obtain. And the investigation may have an intensely inter-

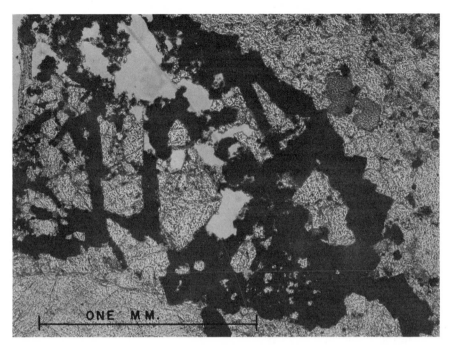

Figure 2-5. Galena-fluorite, Iron Mountain, New Mexico, ×60. Galena (black) replacing fluorite (gray) along cleavage.

ested audience. "Do we go deeper?" "Prepare to quit?" "How do *you* know the silver values will increase?"

A single crystal of galena shows zoning (mineralogy); so does the Appalachian mountain system (metallogeny). This chapter is more concerned with zoning at the immediate scale of mining geology—in orebodies (dimensions of a few meters to a few thousand meters) and in districts (dimensions of kilometers to tens of kilometers). Metallogeny and such mega-thoughts will be left to Chapter 4.

The "zonal theory" of Lindgren and W. H. Emmons was related to the paragenesis of minerals that formed in traversing hydrothermal fluids as pressure and temperature decreased away from a parent intrusive body. The theory is still valid (and actually not much improved upon) as long as it is applied to pluton-derived hydrothermal zoning. Emmons (1936) stated that silver, lead, zinc, and copper ores should be found in succession as we go inward and downward; and, in general, so they are. What has been added is the recognition that similar zonal patterns are also associated with sedimentary and metamorphic ore-forming processes—patterns that may have no relationship to a real or imagined intrusion.

Whatever the circumstance, no geologist really expects a zonal pattern to be symmetrical, except in idealized illustrations. We make allowance for complex preore geologic conditions, for multiple stages and sources of mineralization, and for postore adjustments. Even the distorting effects of former supergene patterns on a hypogene assemblage cannot be discounted when we consider that so many ore deposits are associated with unconformities that were once sites of weathering and erosion.

Mineral and Metal Zoning

For most of the mineral associations in Table 2-1, there are local characteristics in zoning. If we go to broad categories, we can even cite some generalized zoning patterns. In hydrothermal vein associations there is a "normal" paragenesis and a related zonal arrangement of sulfides: Co–Fe–Ni–Sn–Cu–Zn–Pb–Hg, with the earliest sulfide most abundant in the innermost and deepest zones. In those stratiform sedimentary deposits where the mineralization is complete and has not been thoroughly metamorphosed and redistributed, we would expect to find a sequence of chalcocite–bornite–chalcopyrite–galena–sphalerite–pyrite, with the first zone beginning closest to an ancient shoreline. The bornite–chalcopyrite–pyrite sequence is well represented in the Zambian copper belt.

Zoning patterns have been *almost* explained by the relative stability of complex ions, by differences in density, by electrode potentials, and by a few other approaches. For now, it is enough to say that there is a fascinating lack of agreement. The explanations are well worth reading, however, because they provide alternative models for unseen orebodies, something we need to know when deciding where to place another drill hole or how far to continue a hole in progress. For more detail on theories of zoning in hydrothermal deposits there are transactions of symposia on this and related topics edited by Kutina (1963) and Stemprok (1965). The zoning aspects of sedimentary and magmatic stratabound ore deposits are summarized by Stanton (1972) in his ore petrology textbook. Examples of zoning in sedimentary ore deposits have been collected by Amstutz and Bernard (1973).

Figures 2-6 and 2-7 provide examples of hydrothermal zoning and show some of its complications. Figure 2-7 is a diagrammatic section of one of the major rootlike chimney orebodies at Gilman, Colorado, from which strata-controlled manto orebodies extend updip (Radabaugh, Merchant, and Brown, 1968). A typical chimney orebody at Gilman has a massive pyrite core in which bodies of chalcopyrite, silver-bearing tetrahedrite and other silver-bearing minerals, and galena are distributed. The chimney orebodies are partly enclosed by thin zones of marmatite (high-iron sphalerite) and siderite. The manto orebodies are sinuous, but the mineralization in them is continuous. Manto ore is associated with a specific dolomite bed, where it comprises

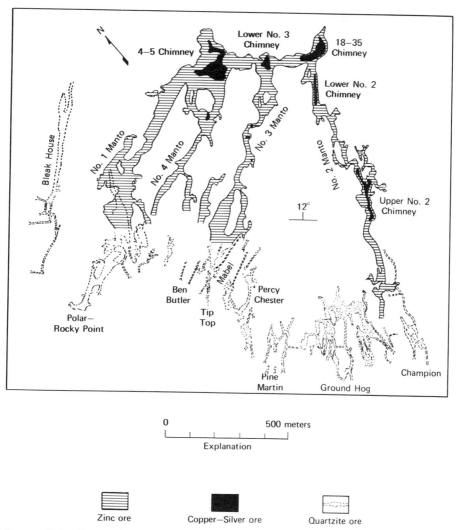

Zinc ore   Copper–Silver ore   Quartzite ore

Figure 2-6. Composite plan map of the principal orebodies in the Eagle Mine, Gilman, Colorado. [From R. E. Radabaugh, J. S. Merchant, and J. M. Brown, "Geology and ore deposits of the Gilman (Red Cliff, Battle Mountain) district, Eagle County, Colorado," in *Ore Deposits of the United States, 1933–1967*, J. D. Ridge, ed., Fig. 3, p. 654, 1968, by permission American Institute of Mining, Metallurgical and Petroleum Engineers, Inc.]

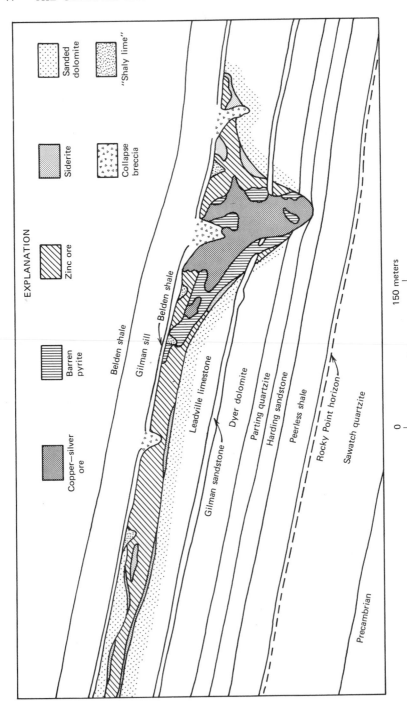

Figure 2-7. Diagrammatic section of typical chimney and limestone manto orebodies, looking northwest, Gilman, Colorado. [From R. E. Radabaugh, J. S. Merchant, and J. M. Brown, "Geology and ore deposits of the Gilman (Red Cliff, Battle Mountain) district, Eagle County, Colorado," in *Ore Deposits of the United States, 1933–1967*, J. D. Ridge, ed., Fig. 5, p. 660, 1968, by permission American Institute of Mining, Metallurgical and Petroleum Engineers, Inc.]

massive pyrite, marmatite, and siderite; there is a smaller amount of galena that increases in quantity updip. The iron content in marmatite decreases updip.

The well-studied paragenesis at Gilman points out a problem that must be faced in all attempts to explain zoning patterns: what we see in a zonal arrangement is the *final* result of successive, simultaneous, and alternating stages in mineral deposition. In the Gilman deposit the earliest mineral was siderite. It was partly replaced by pyrite. Marmatite, galena, and chalcopyrite replaced the earlier minerals and they replaced each other. Sulfosalts of copper and silver were in part contemporaneous with chalcopyrite and in part younger. A late stage of silver-rich galena filled spaces and replaced some of the earlier minerals. In a final stage all sulfide minerals and a new generation of siderite were deposited as spectacular linings in druses. "Sanding"— dissolution of the dolomite—took place all along. The zoning pattern at Gilman cannot be related in a simple way to the overall paragenesis or to a "normal" zoning succession for hydrothermal orebodies. This is one of the many orebodies that fits only part of the rules.

Geochemical Zoning

Figure 2-8 shows another kind of zoning—a geochemical pattern of major and minor elements around the tin deposit at Altenberg, German Democratic Republic (Tischendorf, 1973). The ore in this thoroughly investigated mine, (one of the Erzgebirge classics) has been worked so intensively since the fifteenth century that a funnel-shaped depression (unplanned block caving!) occurs at the surface. The depression can be seen in the cross section. The orebody is a stockwork of cassiterite–wolframite–molybdenite veinlets in "greisen," an alteration assemblage that will be defined shortly. Tin and bismuth have been found in anomalous amounts at a considerable distance from the greisen ore zone; they would make good "target rings" in exploring for another similar body. Molybdenum would be a good indicator of nearer ore. The gallium zone is relatively deep and narrow—an indicator of the orebody roots. The lithium pattern differs from the others in a way which suggests that it may have been enriched during the broad magmatic stage as well as in the later hydrothermal stage. A paragenetic sequence for elements has been established for the Erzebirge region: F–Ga–Li–Sn–As–Bi–Mo. This generally agrees with the local Altenberg pattern in that gallium and lithium would be expected centrally and at depth, with the later tin–bismuth– molybdenite mineralization spreading out in a higher level.

Mineral zoning as well as geochemical zoning is used as a guide to tin deposits in the Erzgebirge. Based on the idea that Mn–Fe ratios in wolframite should increase toward the higher temperatures of formation in the tungsten- and tin-bearing greisen zones, the huebnerite–ferberite ratio ($MnWO_4/FeWO_4$)

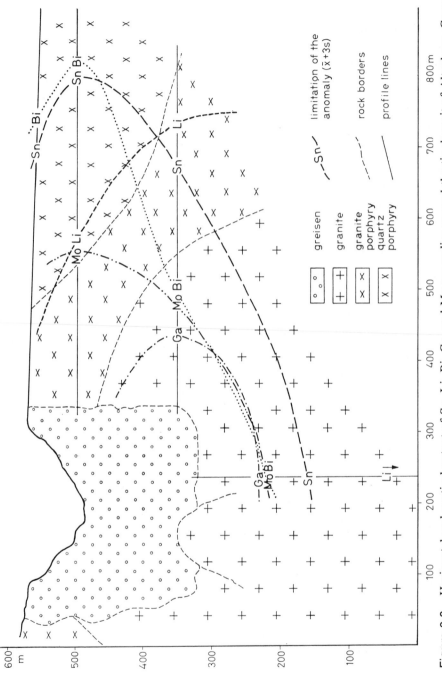

Figure 2-8.   Horizontal and vertical extent of Sn, Li. Bi. Ga, and Mo anomalies around the tin deposit of Altenberg, German Democratic Republic. [From G. Tischendorf, "The metallogenetic basis of tin exploration in the Erzgebirge," Fig. 3, p. B19, Inst. Mining Metall. (London) *Trans.*, v. 82, by permission.]

was plotted along the course of a small tungsten vein to find the highest value; then, exploration led to the discovery of an intrusive cupola with tin ore at a depth of 300 m beneath the chosen point.

Geochemical zoning in orebodies and in mining districts is really no different from metal and mineral zoning, except that it is often expressed in terms of trace elements (oligoelements). Laboratory analyses are used to find subtle changes that would be missed in mapping mines or logging drill core. This is part of geochemical prospecting, a topic discussed in Chapter 14. One of the most useful characteristics of geochemical zoning is that it is so well suited to quantitative characterization in graphs and isoline maps showing element content and showing ratios between elements or stable isotopes.

Sometimes a mineralogical pattern such as the Erzgebirge huebnerite-ferberite distribution is not recognized in a mine until *after* it has been pointed out in samples sent to the laboratory. It helps, for example, to know that the iron content of sphalerite increases in a certain direction; this may mean that the color of sphalerite will darken in the same direction. If uranium or thorium in trace amounts has an abundance pattern in fluorite, so may the color of the fluorite—becoming deeper purple where the radiation is higher. We might not attach any significance to the various species of carbonate gangue minerals in a deposit until a geochemical survey indicates a significant change in Mg-Ca or Mg-Fe ratios; a pattern can then be sought in ratios between dolomite, calcite, and ankerite. Something similar is often done in mapping wall-rock alteration zones; observations are made in the mine or in the field. Laboratory work provides a better label than "light-gray phase," and we go back to mapping "potassic alteration."

Wall-rock Alteration

Wall-rock alteration is another expression of geochemical change. In much of the literature, it is called "hydrothermal alteration," with the supposition that there is adequate evidence of hydrothermal solutions and epigenetic processes. The most striking hydrothermal alteration in its specific and original sense is associated with Lindgren's and Graton's mesothermal and epithermal deposits, where fluids and wall rock are likely to have been in greater geochemical contrast than in very deep seated, hypothermal and very shallow seated telethermal deposits. The hypothermal environment should not have been greatly different from the environment at the source of the fluids. Telethermal fluids should have already begun to take on surface characteristics.

Wall-rock alteration is much more common than the "hydrothermal" label would indicate. Even in syngenetic deposits, there is disequilibrium between earlier or later wall-rock minerals and the ore minerals, and this promotes reactions during diagenesis and metamorphism. In these deposits, the wall-rock alteration pattern is a postore development.

As with other zonal patterns around an orebody, wall-rock alteration is most useful where it displays several zones, as seen in Figures 2-9 and 2-10. This is an instance in miniature, involving rich silver-bearing veins at Cobalt, Ontario, that are generally less than a few centimeters in width. The veins are filled with native silver, nickel-cobalt arsenides, carbonate minerals, and minor quartz. In diabase, gray unaltered rock grades into a light-gray, "spotted chlorite" zone and next to the veins a chlorite–carbonate–albite zone. A decrease in SiO_2, Al_2O_3, and K_2O and an increase in CO_2 and Fe_2O_3 toward the vein reflect the composition of the two alteration facies. Ni, Co, As, and Sb increase toward the vein, and, as would be expected, silver increases—but only in the immediate vicinity of the vein. This compressed alteration pattern takes on some stature when we realize that it involves widths of 5 to 20 times that of the veins themselves. More stature: The Cobalt district has produced 400 million ounces (12.5 million kg) of silver.

Table 2-2 provides a few illustrations of zonal sequences in wall-rock alteration. There is no average width to these zones; they may measure in centimeters, as at Cobalt, or in a kilometer or more, as in some porphyry copper deposits. Also, there is no indication of intensity; like most things in geology, this ranges upward from "barely perceptible."

A more serious problem with Table 2-2 is that it might give the impression of wall-rock alteration being too clear cut. As with mineral zoning at Gilman, we see only the final result of a very complex series of events. Rock types, surges in mineralization, subsequent metamorphism, and, of course, weathering have generally done something to the picture.

Sometimes, alteration is cited in a chemical context: example, magnesium metasomatism. But most terms for types or facies of alteration are made from the roots of mineral names with "—ization" suffixed to indicate a process or "—ic" suffixed to indicate the diagnostic or dominant species. A few assemblages are so unique that special and longer established terms are used.

Argillic (argillization). Kaolin and montmorillonite. The two clay minerals may be separated into zones as well, with an inner zone of kaolinization. The term "advanced argillic alteration" is sometimes used for a dickite-kaolinite–pyrophyllite assemblage.

Phyllic (sericitization). Sericite, quartz, and pyrite. This is one of the most common and most easily recognized alteration assemblages.

Potassic. Potassium feldspar (as adularia or microcline) with biotite and commonly with hematite or magnetite, sericite, and anhydrite.

Propylitic (propylitization). Epidote, albite, chlorite, and carbonate, commonly with sericite and pyrite. There are several subtypes in the complex process of propylitization: chloritization, albitization, carbonatization, and zeolitization.

Silication. A process in which skarn or tactite silicates, such as garnet, diopside, wollastonite, and amphiboles are formed, often with magnetite or pyrite and sulfide ore minerals. The name is not to be misread as *silicification,* which refers to the formation of quartz or jasper.

Greisen. An aggregate of muscovite, feldspars, quartz, and topaz; often with tourmaline. It is most common in granites and pegmatites associated with tin and molybdenum ore.

A pair of terms, "marmorization" and "marbleization" are in common use; they are synonyms referring to a process of recrystallization in carbonate rocks. Like many of the above terms, they could also refer to a metamorphic process that has no particular connection with ore mineralization.

Here is a point to emphasize even though we have not yet considered how to map alteration in the field: Two adjacent rock types may be mistaken for two zones of alteration, and worse yet, there are other geologic processes that mimic zonal wall-rock alteration. Supergene alteration and weathering processes (ancestral and current) are notorious—they bleach biotite to "sericite" or change it to chlorite, make montmorillonite and kaolinite from feldspar, carry hematite and silica into new sites, and generally wreak havoc upon patterns inherited from the subsurface environment. As alteration processes go from one rock type to another, they change; thus, a boundary between propylitic and sericitic alteration may reflect a contact between mafic and felsic volcanic rocks rather than a zonal change related to a particular orebody. Metamorphism is to be reckoned with; the low-temperature muscovite–hornfels facies of contact metamorphism, with abundant chloride, sericite, and biotite may have nothing to do with an orebody. And there is *deuteric alteration,* the loosely defined, late magmatic process in which an intrusion "stews in its own juice"—a process which may or may not have been associated with the actual ore mineralization.

A good review of wall-rock alteration types and patterns is given by Boyle (1970) as part of an entire symposium on the subject. Schwartz (1959) also provides a useful list of alteration guides to ore.

Minerals and Zoning in the Depth Dimension

Unless our conceptual model is totally inappropriate (like looking for a massive sulfide assemblage in a pegmatite), it can be modified and used in new surroundings. This may mean extending a zoned model to a depth greater than has been investigated in any single mine. We might consider stacking several models on top of each other to see if they make a reasonable picture.

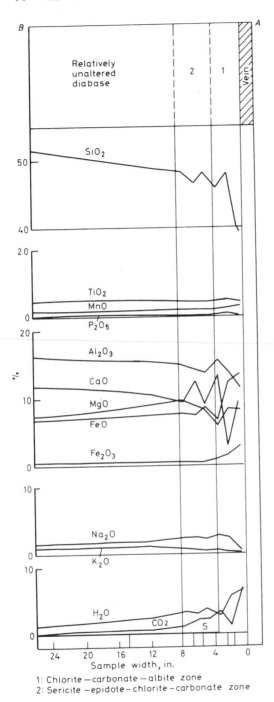

Figure 2-9. Chemical variations in alteration halos of diabase, Glen Lake adit, Cobalt, Ontario. (From A. S. Dass, R. W. Boyle, and W. M. Tupper, "Endogenic haloes of the native silver deposits, Cobalt, Ontario, Canada," in *Geochemical Exploration in 1972,* M. J. Jones, ed., Fig. 3, p. 29, 1973, by permission Institution of Mining and Metallurgy.)

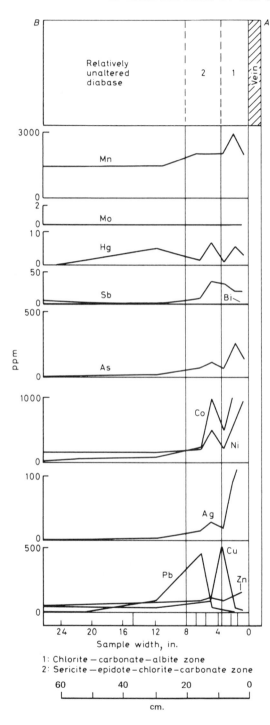

Figure 2-10. Distribution of minor elements in alteration halos of diabase, Glen Lake adit, Cobalt, Ontario. (From A. S. Dass, R. W. Boyle, and W. M. Tupper, "Endogenic haloes of the native silver deposits, Cobalt, Ontario, Canada," in *Geochemical Exploration in 1972*, M. J. Jones, ed., Fig. 4, p. 29, 1973, by permission Institution of Mining and Metallurgy.)

Table 2-2. Examples, Zonation of Wall-rock Alteration. Abbreviations from Table 2-1 and Appendix A

Locality	Host Rock	Zoning, toward Ore	Ore Assemblage
Western North America, porphyry coppers (Guilbert and Lowell, 1974)	sed and met Rx and Qm	Prop-Arg-Phyl-Pot	py, cp, bn, mb
	Carb and Qm	Mrbl-Skarn-(Pot)	
	Mafic and intermed. Rx	Prop-Bio-(Phyl)	
Butte, Montana (Meyer, 1968)	Qm	Prop-Arg-Ser-(Sil)-(advanced Arg)	py, cc, en, bn, cp
Tintic, Utah (Lovering, 1949)	Mp	Prop-Arg-qtz, ser, py	py, gn, sl
	Carb	Dol-Arg-py, Jasp-Ser	
Yellowknife, N.W.T. (Henderson, 1966)	Grnst	Chl-Carb-Carb, Ser (decr. Sil)	py, po, asp, cp, Au
Hollinger (Porcupine district), Ontario (Ferguson, 1966)	Qp and Sch	qtz, cal, ab-qtz, ank, ab-(chl)-(ser)	py, Au
Bolivian porphyry tins (Sillitoe and others (1975)	Qtz latite, Dac, Volcs	Prop-Ser-(qtz,Tm)	py, cas, sl, asp
Mattabi, Ontario (Franklin, Kasarda, and Poulsen, 1975)	felsic Volcs	complex: Dol-sid-Sil-Ser-Chlor	py, sl, cp, gn, tt, tr, aspy
Chibougamau, Quebec (Miller, 1961)	Anorth	Prop-Prop, Seric, cal-Prop, ank-sid	py, cp, po, sl

Greenside (Lake district), England (Schnellman and Scott, 1970)	intermed. Volcs	Chl-(Ser)-Sil	gn, sl, fl, ba
Southern Black Forest, Germany (Murad, 1974)	Gr	Arg-Phyl-Pot	gn, sl, fl, ba
Killingdal, Norway (Rui, 1973)	Metased and Metavolc	Chlor-qtz, musc	py, cp, sl
Cobalt, Ontario (Dass, Boyle, and Tupper, 1973)	Diabase	ser, ep, chl, car-chl, carb, ab	Ag, Ni-Co arsenides
	Gwke	qtz, ser, carb-Chl, carb, ser	
	Grnst	chl, ep, carb, ser-ser, carb, chl	
Kuroko deposits, Japan (Abe and Aoki, 1975)	felsic Volcs	Mont-Chlor, Ser, (py)-chlor, Ser, (qtz)	sl, gn, cp, py, po
Morococha, Peru	Ls and Qm	Mb-anhydrous silicates-hydrous sil plus anh	gn, sl, tt, tn, en, py, cp
Compacca, Peru (Hollister, 1975a)	Qmp and Sh	Prop-Arg-Phyl	imb, py, en
El Salvador, Chile (Gustafson and Hunt, 1975)	Qdp and Volcs	(Early) Prop-Pot (Late) Ser-andalusite (Very late) Arg	cp, bn, py, mb

Graton (1933) did this with the hydrothermal vein concept a long time ago. We will use a newer, but not necessarily better, illustration.

The current porphyry copper model—disseminated pyrite, chalcopyrite, and molybdenite, sometimes with peripheral sphalerite and galena, associated with a quartz monzonite stock and with a potassic alteration core zone—is generally right for the southwestern United States. It also applies to Cretaceous-Tertiary porphyry coppers in British Columbia. Add a stronger volcanic affiliation, and it applies to Cordilleran South America and to the Triassic-Jurassic deposits in British Columbia. Take the model elsewhere, into older and more deeply eroded terrain in the Appalachians, for example; it can apply to "porphyry-type" deposits of Maine, Quebec, and the Canadian Maritime provinces where there is a higher molybdenum-copper ratio, less pyrite, and little or no lead-zinc mineralization. The relationship: Perhaps the Appalachian deposits represent the roots of the Cordilleran model, as shown in Figure 2-11. Place the model in Precambrian terrain, and the term "porphyry deposit" can be applied to copper orebodies in South Africa and Australia (Jacobsen, 1975).

The Cordilleran porphyry copper model has been related to disseminated and stockwork copper deposits in the Philippines, where the association is with small (or assumed) diorite porphyry intrusions rather than with quartz monzonite stocks. In the Philippine deposits, pyrite and gold are more abundant than in Arizona and there is less molybdenite; perhaps the Philippine deposits are exposed at a higher level in a tall model that, according to some geologists, cuts through geologic time as well and extends from Appalachian "porphyry-type" Paleozoic roots to an ultimate top in the Cenozoic "Kuroko" copper-lead-zinc deposits of Japan (Wolfe, 1973). Many geologists believe that if Kuroko-type deposits and their volcanic wall rocks were metamorphosed, they would be typical stratabound massive sulfide deposits.

Gilmour (1971) and Hutchinson and Hodder (1972) have studied the implications of a possible Kuroko–stratabound massive sulfide–porphyry copper relationship. They noted the similarities that might link these deposits, but they found important differences as well. For example, massive sulfides are more likely to be associated with an early magmatic, eugeosynclinal environment, whereas porphyry coppers are more likely to have been formed in a late magmatic, subvolcanic environment. Hollister (1975b) has suggested that two porphyry copper models should be used, one (the Arizona model) for the cratonic environment and another (the diorite model, with a gold association) for areas of thin continental crust. Evidently, we are not dealing with a simple matter of stacking models on each other. But the pattern is stimulating at any rate. In a geographic sense, we have come full circle; geologists in the American Southwest, steeped in the original porphyry copper model, are now using a "Kuroko" model to explore Precambrian

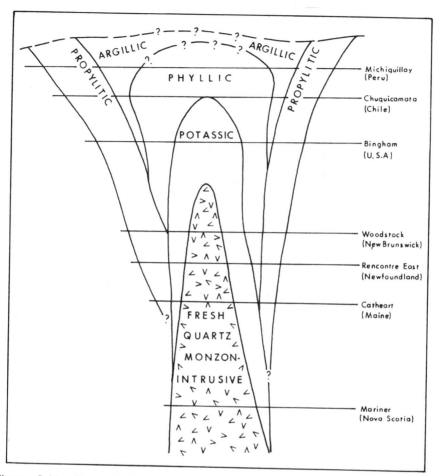

Figure 2-11. Theoretical section showing alteration halos around a mineralized quartz-monzonite intrusion, based on a composite of examples from the Appalachian, Cordilleran, and Andean orogens. (From V. F. Hollister, R. P. Potter, and A. L. Barker, "Porphyry type deposits in the Appalachian orogen," *Econ. Geology*, v. 69, Fig. 5, p. 627, 1974. Published by Economic Geology Publishing Company. Used by permission of the publisher.)

terrain for stratabound massive sulfides of the type that occurs at Jerome, Arizona.

2.3. COAL AND OIL SHALE PETROLOGY AND ZONING

Many of the comments on ore mineral associations can be extended to *macerals,* the "minerals" of coal deposits, and to the incombustible materials that correspond to gangue. There are characteristic maceral associations; these are coal rock types. Coal has a range in heritage: most is autochthonous (formed in situ), but there are allochthonous coals that have accumulated from transported vegetation. Metamorphism has an important role. Finally, coal deposits occur in typical stratigraphic sequences (cyclothems)—and they are zoned.

Why do we treat coal separately? It is a bow to tradition. Coal deposits are affairs for stratigraphy and sedimentary petrology, so a separate nomenclature grew up in the days when most ore deposits were categorized as "hard-rock" veins. The same kind of separate treatment can be cited for the study of saline deposits and phosphate rock deposits. But where can we draw a boundary between all of these and sedimentary ore deposits? Illinois Basin coal refuse

Table 2-3.   Coal Petrology Terms: Components, Macerals, and Lithotypes[a]

Microscopic Study			Megascopic Study
Transmitted Light (North American Practice)	Reflected Light (European Practice)		Hand Specimen
Components	Maceral Groups	Macerals	Lithotypes
Anthraxylon—mainly Vitrinite. A major component of "bright coal."	Vitrinite	Collinite Telinite	Vitrain—mainly Vitrinite
			Clarain—Vitrinite and exinite
Attritus—various macerals. A major component of "splint coal" and "dull coal"	Exinite	Alginite Cutinite Sporinite Resinite	
			Durain—Exinite and inertinite
Fusain—mainly inertinite	Inertinite	Sclerotinite Micrinite Semifusinite Fusinite	Fusain—mainly inertinite
			Cannel and Torbanite— sapropelic "nonbanded" coals

[a] Data from Williamson (1967) and from International Committee for Coal Petrology (1963).

piles contain enough sphalerite to be considered a potential source of zinc and cadmium. Uranium-bearing coal beds have been mined solely for their uranium content. Brines are recovered for their lithium content. Phosphate rock has been mined as vanadium ore. Vanadium and nickel are potentially recoverable from the fly ash from Athabasca tar sands. The traditional separations in geologic study will not be with us for long.

Petrologic characteristics begin to count in the very broadest category, in the separation of coal by its intended use into *energy* (power coal, steam coal, gasification coal) and metallurgy (coking coal). Coking quality depends largely on the proportions and physical properties of *vitrinite,* one of the groups of maceral. Other maceral groups have bearing on the mining, crushing, washing, and blending of coal.

Terminology (Table 2-3) is based on the examination of thin sections of coal in transmitted light and polished sections in reflected light. The three broadest terms, coal components in North American usage, are "anthraxylon" (translucent; in brilliant bands), "attritus" (opaque or translucent; dull and granular), and "fusain" (opaque and cellular). Figure 2-12 shows the general appearance of the three components in thin section.

The macerals, organic units or single fragments, are involved in all detailed coal petrography. *Collinite* is a structureless gel, *telinite* is a gel that retains

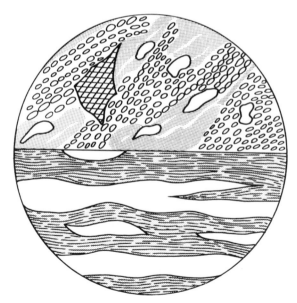

Figure 2-12. Composite diagram of coal from thin sections. *Above:* fusain, opaque cellular charcoallike material. Pores are filled with calcite. *Below:* anthraxylon bands and lenses separated by attritus. (×40 to ×100)

some cellular structures; these macerals, derived principally from wood and bark, fuse and swell on heating, hence their importance as members of the *vitrinite* group, especially in coking coal. The macerals in the *exinite* group, derived from algae, cuticles, spores, and resins, are generally transformed into gas and tar on heating. The macerals of the *inertinite* group are, as the name indicates, almost chemically inert. *Sclerotinite* is composed or rounded bodies or interlaced filaments, probably the remains of fungae; *micrinite* is granular; *fusinite* and *semifusinite* are cellular "mineral charcoal."

There are microlithotypes (associations of maceral groups) in coal, and there are lithotypes. Lithotypes, like other rocks, are recognizable in hand specimen.

Vitrain, rich in vitrinite macerals, occurs in brilliant bands up to 20 mm thick. It breaks with a conchoidal fracture.

Clarain, composed of various macerals, has a bright, silky luster caused by fine laminations. Clarain occurs in bands up to 3 mm thick.

Fusain, principally fusinite, is friable and sooty, like common charcoal. It occurs in variable bands and lenses or sometimes in irregular wedges lying at various angles on bedding planes. Since it is porous, it is sometimes impregnated with troublesome calcite or pyrite, a reason for its importance in the component system of petrologic classification.

Durain, composed of exinite and inertinite macerals, has the descriptive synonym "dull coal." It occurs in fine-textured bands several centimeters or more thick.

Cannel coal and *torbanite* are hydrogen-rich allochthonous coals with a waxy luster and conchoidal fracture. They are similar in hand specimen except that torbanite is brown rather than black.

The incombustible materials in coal deposits amount to layers of clastic sedimentary rocks (bone), lenses and concretions of carbonate or pyrite, and microscopic matter. The microscopic group is of particular importance because it is almost impossible to separate during preparation for market.

Zoning patterns in coal fields are generally shown in relation to coal *rank* (peat, lignite, bituminous, anthracite) rather than petrology, with the gradient shown in *isorank* lines (Fig. 2-13) or in *isovols, isocarbs,* or *isocals* for percentages of volatile matter, carbon, or calorific values, respectively. The basis for classification is a *proximate analysis,* a rapid method of determining the general characteristics of a coal. *Ultimate analysis* is a more painstaking procedure dealing with the actual chemical elements in a coal.

Coal rank, deriving from both chemical and physical characteristics, is in large part related to depth of burial—to load metamorphism. Coal rank, in addition, is a result of biochemical alteration, time, and local or regional metamorphism. The occurrence of higher rank coal in the southern part of the Illinois basin, shown in Figure 2-13, may be the result of a formerly greater depth of burial or the result of heat flow from the large igneous body that has been suspected to underlie small basic intrusions (Damberger, 1971).

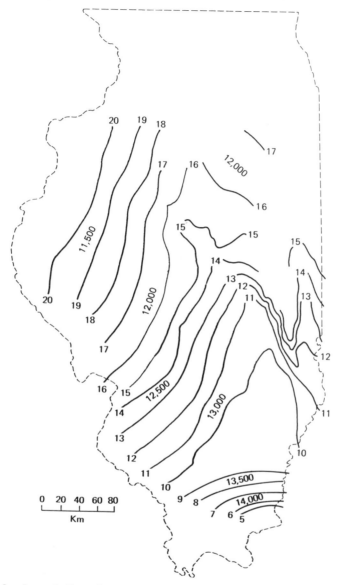

Figure 2-13. Isorank lines for Herrin (No. 6) coal, Illinois, expressed by the seam moisture in percent and the calorific value in Btu/lb. High volatile A bituminous coal has a calorific value above 14,000 Btu/lb or moisture below 6.2 percent. High volatile B bituminous coal has a calorific value between 13,000 and 14,000 Btu/lb or moisture between 10.5 and 6.2 percent. High volatile C bituminous coal has a calorific value below 13,000 Btu/lb or moisture between 22.5 and 10.5 percent. (After H. H. Damberger, "Coalification pattern of the Illinois Basin," *Econ. Geology*, v. 66, Fig. 2, p. 490, 1971. Published by Economic Geology Publishing Company. Used by permission of the publisher.)

A good introductory treatment of coal geology and petrology is given by Williamson (1967). Advanced aspects of coal geology are covered in *Coal and Coal-bearing Strata,* edited by Murchison and Westoll (1968). More detail on coal petrology and petrography can be obtained in a textbook by Stach (1975). A specific example of applied petrology can be found in a paper on Australian coal by Shibaoka and Smyth (1975).

An interesting petrologic association is shown in Figure 2-14. It extends in one direction from coal to native bitumens, such as gilsonite, and on to the Athabasca "tar" sands. Beyond that point, the material is crude oil, and we leave mining geology. Again, we bow to tradition, but not for long. When we trace the association in another direction, through oil shale to "black" shale, we return to stratabound ore deposits in carbonaceous sedimentary rocks.

For a review of oil shale geology, see Duncan and Swanson (1965). For native bitumens, there is a summary by Phizackerley and Scott (1967). As will be seen in Chapter 4, oil field brines have much to do with Mississippi Valley-type deposits and a few others that we are beginning to understand.

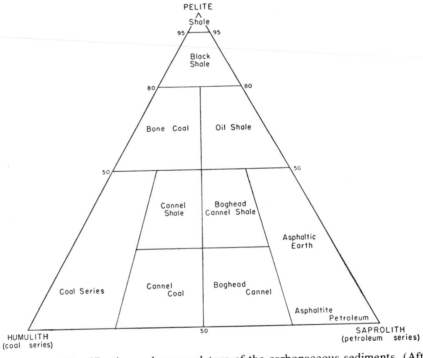

Figure 2-14.   Classification and nomenclature of the carbonaceous sediments. (After F. J. Pettijohn, *Sedimentary Rocks,* ed. 2, p. 489, 1957, by permission of Harper & Row, Publishers.)

2.4. THE MORPHOLOGY OF MINERAL DEPOSITS

Even in such pure science as this, we cannot escape economics. A chimney or "pipe" may be the wider than average part of a vein where a certain economic stoping width can be attained. A system of veins in a particular limestone bed may be taken collectively as a single stratabound deposit if that is the way they will be mined.

The boundaries of bulk low-grade copper orebodies, such as the porphyry copper body at Bingham Canyon, Utah, are so flexible that engineers design an "ultimate pit" with tongue in cheek; it is ultimate only until the next cutoff ore grade is calculated. The "porphyry copper" deposit at Majdanpek, Yugoslavia, is actually composed of irregular pyrite-chalcopyrite bodies in quartz, lenticular pyrite bodies at a limestone-andesite contact, disseminated pyrite-chalcopyrite zones in andesite, and lenticular-veinlet bodies of pyrite-chalcopyrite-molybdenite in schist and gneiss. A morphologic description of the Majdanpek orebody would either have to be very detailed or very flexible.

When we relate the morphology of a particular orebody to causes and controls (which will be done repeatedly in the next few pages) we have another reason for tongue-in-cheek statements. The "ultimate" explanation, commonly designed by an academician who has spent several days or weeks at the mine, may be valid only until a better explanation is made by a geologist who has been studying the deposit in detail for months or years. This is one of the things that makes resident geologic work at a mine so satisfying.

Control by Premineralization Conditions

Sometime in history, it must have been prior to the search for a marble-schist contact in the ancient workings at Laurium, miners learned that orebodies are related to specific geologic features. Then, perhaps just before Agricola's treatise was written, miners and philosophers began to think that geologic features must exercise some kind of *control* over the shapes and sizes of orebodies. The idea that favorable structure and favorable lithology form conduits and traps for ore has dominated mining geology ever since.

Structural and stratigraphic analysis is *the* classic technique in exploration, and it has been so successful that no serious exploration effort leaves it undone. Even if a mineral deposit was formed syngenetically in sediments, where the premineralization conditions are not reflected in a classic "source, plumbing, and trap" way, structural and stratigraphic conditions will certainly have had a bearing on the paleogeography that allowed ore minerals to accumulate.

Structural Control. If there can be a classic within a classic, this is it. More hidden orebodies have been found by structural analysis than in any other way, and the mining literature is rich in structure-oriented descriptions of mineral deposits. Three references on ore deposits and their structural environment are worth particular mention, and the information in them is as applicable today as when the books were published. *Ore Deposits Related to Structural Features,* edited by W. H. Newhouse (1942), is a collector's item as well as a classic. Two volumes, *Structural Geology of Canadian Ore Deposits* (Wilson, 1948) and *Geology of Canadian Ore Deposits* (Gilbert, 1957) contain hundreds of well-illustrated and detailed examples of structurally controlled orebodies.

Mining geologists must without question be structural geologists. They will most certainly have completed at least one college course in structural geology, and their personal libraries will include well-thumbed textbooks in structural geology. A comprehensive textbook on the structural geology of folded rocks by Whitten (1966) deserves special recommendation; in this book, field mapping methods, ways of presenting structural data, and a glossary of fold terminology are included. A very practical review of structural analysis methods for exploration geologists, with exercises, has been provided by Ragan (1973).

In the textbook literature on mining geology itself, the sections on "Fracture Patterns as Guides" and "Dislocated Orebodies" in McKinstry's (1948) *Mining Geology* have never been surpassed for down-to-earth value.

The initial function of structural control in hydrothermal mineralization is called "ground preparation," and this term describes what has commonly occurred; a structural vortex, a local deflection in trend, or a change in fracture pattern between brittle ("competent") and plastic ("incompetent") rocks prepares an ore-forming geochemical environment for fluids moving through a permeable system.

Many of the most time-honored guides to orebodies are related to a miner's intuition that something should happen to mineralization in a vein where it branches, changes course, or intersects another. The guide is based on experience, and experience is just as valid when used by geologists. Intersections in brittle rock are sites of brecciation, faulting will have opened or closed the "steps" in an irregular fracture surface, a series of "feeder" veinlets (gash joints) should lead to a larger vein, and "bleeder" veinlets may indicate that a thinning vein has lost its strength to another one that occurs en echelon. Brittle rocks will have "cracked" in the axial zones of folds, and fold systems will have been accompanied by axial fracture cleavage. There are many empirical and theoretical variations on the theme.

For traditional guides to ore, we can take examples from a traditional district: Cornwall (Fig. 2-15), where more than a thousand mines have been worked since Roman times and where six major mines are now in operation. The mining geology of southwestern England, a veritable textbook in itself, is covered in detail by Dines (1956) in a two-volume memoir. Papers by Hosking

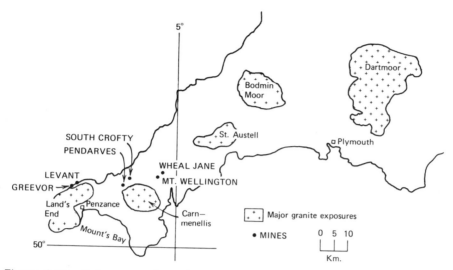

Figure 2-15. Index map of southwest England, with location of major operating mines.

(1965) and Garnett (1967) are specifically directed to structural analysis in the search for orebodies. Moore (1975) provides a structural interpretation for the regional pattern of orebodies. It is no wonder that Cornish miners, Cousin Jacks, were considered the best underground ore finders in American frontier mining camps; most orebodies were in veins, and the Cousin Jacks had seen all kinds in the old country.

The principal Cornish tin-copper veins are in granite, slate (killas), hornfels, and greenstone. The most productive system, cassiterite-bearing "main lodes," are cut and displaced by sphalerite and chalcopyrite-bearing "caunter lodes" and again by pyrite and galena-bearing "cross courses." A large number of porphyry dikes, or elvans, generally parallel to the main lodes, are important controls for ore mineralization and are in some instances mineralized in themselves. Orebodies of local importance occur in stockworks, in steeply plunging pipes, and in flat-lying "floors," as well as in the veins. Well-studied zonal patterns are expressed within orebodies, within groups of orebodies around exposed granite masses, and through the region. The overall zonal sequence, Sn–Cu–Zn–Pb–Ag–Sb, is locally complicated—as zonal sequences so often are in other places—by overlapping stages of mineralization. Lead-zinc orebodies, for example, are normally found away from the granites and their metamorphic aureoles; yet, these ores are also found as late-arriving "strangers" well within the tin zone.

Vein intersections and branches are major controls for ore shoots (richer zones) in Cornish orebodies, as in the Wheal Jane mine (Fig. 2-16) where the B lode is especially rich and thick at its intersection with the Moor Shaft lode. Figure 2-17 shows enrichment along a premineralization fault that cuts the Number 2 Branch lode in the Geevor mine. In Figure 2-18, again from the Geevor mine, ore on the Wethered lode increases significantly in both width and grade where it is joined by the thin, high-grade Coronation lode, and the result is greater than the sum of the two parts.

In some places, a sufficient number of veinlets occur in gash joints, in slaty cleavage, and in joined elvan dikes to make wall-rock zones minable for widths as great as 15 m. The cross courses are weakly mineralized for the most part, but some contain high-grade ore and others act as limits to vein

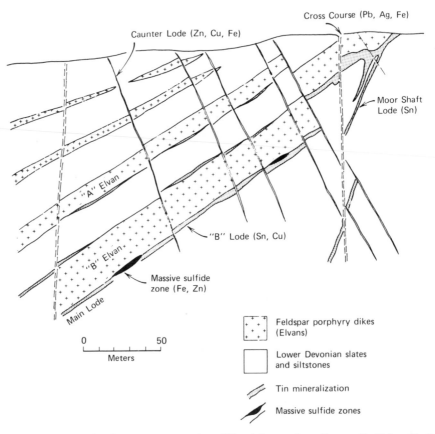

Figure 2-16. Idealized transverse section, Wheal Jane mine, Cornwall. (After B. D. Rayment, G. R. Davis, and J. D. Willson, "Controls to mineralization at Wheal Jane, Cornwall," *Transactions,* v. 80, Fig. 6, p. B230, 1971, by permission of Institution of Mining and Metallurgy.)

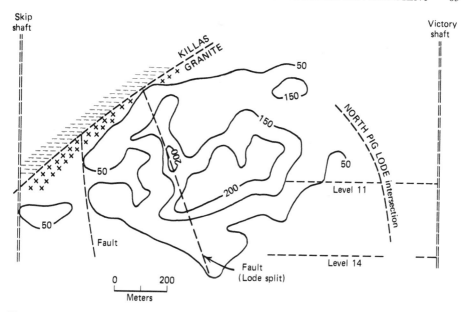

Figure 2-17.   Structural control of mineralization, Number 2 Branch lode, Geevor mine, Cornwall. Isolines are ft-lb SnO_2 per long ton. (After R. H. T. Garnett, "The underground pursuit and development of tin lodes," in *A Technical Conference on Tin*, W. Fox, ed., Vol. 1, Fig. 28. 1967. Copyright © 1967 by the International Tin Council and R. H. T. Garnett. Used by permission of the publisher and author.

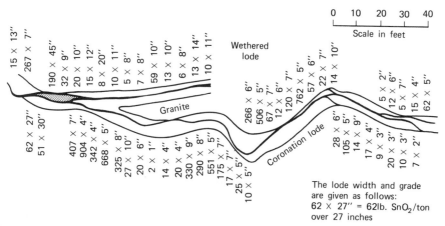

Figure 2-18.   Junction between the Wethered and Coronation lodes, Geevor mine, Cornwall. (After R. H. T. Garnett, "The underground pursuit and development of tin lodes," in *A Technical Conference on Tin*, W. Fox, ed., Vol. 1, Fig. 11. 1967. Copyright © by the International Tin Council and R. H. T, Garnett. Used by permission of the publisher and author.

mineralization, depending on how often the individual faults were mineralized and reopened.

Vein steepening and flattening, together one of the most widely recognized controls for ore shoots, are shown in Figure 2-19, where the Borehole lode in the Geevor mine is wider in the steeper stretches. This is explained by wall separation during normal faulting; if the vein had occupied a reverse fault, the flatter portions would be the wider.

An effect of wall-rock lithology on vein structure is apparent in Figure 2-16, which shows the principal B lode following B elvan at the Wheal Jane mine. Smaller lodes at Wheal Jane also follow the tops and bottoms of elvans. Undulations in the footwall of B elvan near Tippett's shaft (Fig. 2-20) have been interpreted by Rayment and others (1971) to be additional localizing features for richer and thicker mineralization. Not all of the mineralization at Wheal Jane can be related so directly to the elvan dikes; there is also

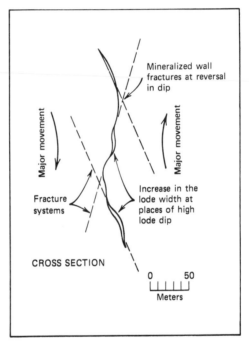

Figure 2-19. Widening of the Borehole lode in places of steep dip, Geevor mine, Cornwall. (After R. H. T. Garnett, "The underground pursuit and development of tin lodes," in *A Technical Conference on Tin,* W. Fox, ed., Vol. 1, Fig. 27. 1967. Copyright © 1967 by the International Tin Council and R. H. T. Garnett. Used by permission of the publisher and author.

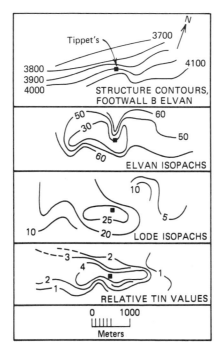

Figure 2-20.   Contour diagrams of a section of B lode, Wheal Jane mine, Cornwall. (After B. D. Rayment, G. R. Davis, and J. D. Willson, "Controls to mineralization at Wheal Jane, Cornwall," *Transactions,* v. 80, Fig. 10, p. B235, 1971, by permission of Institute of Mining and Metallurgy.)

evidence of an early phase of cassiterite that predates the dikes (Edwards, 1976).

Throughout Cornwall, it is common for a single vein to become a stockwork upon passing into a brittle elvan and for a wide vein in brittle granite to become narrower upon passing into relatively plastic hornfels. But there are uncommon situations; in a few Cornish tin mines (Wheal Vor, for example) veins that are productive in metasediments become barren when they pass into granite. Here is a point worth stressing: What is favorable in one mine or one district may be just the opposite in another.

Structural analysis has been a useful ore guide in Cornwall despite the complexities of local tectonic history. Six or more stages of folding have been involved in the regional slaty cleavage, axial fracture cleavage, conjugate shears, and "relaxation" faults that control the pattern of lodes and elvan dikes. Repeated opening, rebrecciation, and remineralization appear to have been necessary for the highest grade mineralization. In some mines as many as ten generations of mineral deposition have been identified. In others several distinct swarms of veins are concentrated in a single locality, as at the South Crofty mine where late cassiterite–chlorite–fluorite–quartz veins intersect an earlier swarm of feldspathic wolframite–arsenopyrite veins. The localization of multistage mineralization is thought to have come from repeated movement above a late granite "cusp" or cupola emplaced in an earlier granite body or, as suggested by Sibson and others (1975) from sporadic "seismic pumping" of fluids through deep faults.

The importance of reopening and of repeated mineralization within a single structure is emphasized at Wheal Jane, where the massive bodies of pyrite, sphalerite, and chalcopyrite associated with the tin-bearing B lode (Figure 2-16) were evidently derived from subsequent mineralization in the caunter lodes. This is also a good illustration of complex heritage in orebodies.

Lithologic Control. Localization of ore by premineralization lithology may be indirect, as in the Cornish examples, or it may be direct and specific in a mining district's "favorable beds" that can be named by everyone from the local equipment storekeeper to the general manager. Direct and specific localization has sedimentary connotations, but it does not exclude ore in volcanics (as in the Keeweenaw, Michigan, native copper deposits) or in stratiform intrusive bodies (copper ore in diabase sills at Ray, Arizona).

Naturally, we can enjoy the marathon epigenetic-syngenetic argument as soon as mention is made of strata being replaced by ore-bearing fluids. Even so, replacement of wall rock by ore is in evidence at so many mines that most geologists admit the existence of large replacement bodies. A replacement theory of origin is favored where ore mineralization occupies only part of a particular facies or stratigraphic unit, and the idea is further strengthened where crosscutting veins or pipes with similar ore mineralogy extend to other ore beds. Due to hydrothermal fluids emanating from an *intrusive* rock source? Not necessarily. We know that ore minerals are mobile and that they can go from a bed to a vein and vice versa. We also have good evidence that circulating ground water and released connate water can be mineralizing fluids.

One example of a replacement deposit will serve to illustrate the typical morphology. The Magma copper mine, Arizona, has an interesting variety of orebodies and ore controls. There is a stockwork contained within a plug of diorite porphyry, a major vein with local branches, and limestone replacement deposits, one of which is shown in Figure 2-21. The principal limestone replacement deposit is a group of tabular bodies of chalcopyrite, bornite, hematite, and pyrite within crystalline limestone lying parallel to the bedding. Some of the orebodies reach 28 m in thickness, although most measure 8 m or less. Individual bodies have lengths of as much as 300 m. The replacement orebodies are clustered near major veins, but the relationship is not clear;

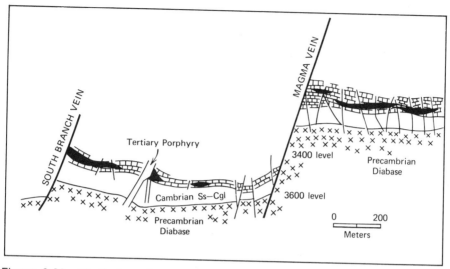

Figure 2-21. Vertical section through the limestone replacement deposit, Magma mine, Arizona. (Adapted and much simplified from D. F. Hammer and D. W. Peterson, "Geology of the Magma mine area, Arizona," in *Ore Deposits of the United States, 1933–1967*, J. D. Ridge, ed., Fig. 7, p. 1307, 1968, by permission American Institute of Mining, Metallurgical and Petroleum Engineers, Inc.)

bedded ore does not always come in contact with the vein, and some rich zones of bedded ore are adjacent to barren parts of the vein. (Nature can be infuriating.) Within the bedded ore, premineral faults and sharp monoclinal flexures are generally associated with ore shoots, an indication of structural control. Why are the orebodies only in the lowermost beds? Since the limestone above the ore as well as within the ore zone has been recrystallized, it is not entirely a matter of permeability; a major factor in the location of the orebodies may have been the accessibility of the lower beds to the mineralizing fluids. Some geologists who are familiar with the district suggest another possibility: The stratabound ore zones are of syngenetic origin in the lower dolomite and the mineralization was rearranged by igneous activity and deformation.

The apparent reasons for hydrothermal (or at least epigenetic) replacement deposits seeking out certain strata are almost as diverse as the examples—and there are thousands of these. Geochemical reactivity certainly has a bearing; so does permeability and so do textural and structural features. Certain rock types are generally favorable where they occur in a series with other rock types, but when we use such generalizations we are back in the middle of the syngenetic-epigenetic imbroglio. For many years, it was hard to explain how the "ore shale" in the Zambian copper belt could have been replaced in preference to the more permeable and potentially more receptive dolomite. But with a syngentic view—ore accumulating in clay around a carbonate reef—this condition is not perplexing at all.

When we deal with epigenetic stratiform and stratabound deposits (*stratabound* deposits do not have to be tabular in detail), we often deal with unconformities that can be called "contemporaneous surfaces" (Derry, 1973). Mills and Eyrich (1966) recognized that epigenetic ore deposits are commonly associated with unconformities and inherent changes in the porosity and permeability of the preore rock. In their discussion of orebody localization from a hydrothermal point of view the following categories are mentioned:

1. Primary rock openings in coarse clastic rocks or volcanic rocks above an unconformity.
   a. In old stream channels (Colorado Plateau uranium).
   b. In reactive rocks overlying coarse clastic rocks (southeast Missouri lead).
2. Induced openings created by folding or faulting that has in turn been localized by an unconformity.
   a. Near irregularities in an old erosion surface (southeast Missouri lead).
   b. In brecciated, fractured, and sheared zones (Walton, Nova Scotia, barite-sulfides).
3. Igneous rocks intruded along an unconformity (Sudbury, Ontario).
4. Induced openings—caves, sinks, channels—below an unconformity (upper Mississippi Valley zinc-lead).
5. Impermeable barriers above an unconformity with ore in carbonate rocks below the unconformity (Gilman, Colorado, zinc-copper)

Combined Control. Why do bedded orebodies appear next to barren points of the Magma vein? Why do the manto orebodies at Gilman leave a perfectly comfortable home in the Leadville Limestone and go into a chimney? Or look at it the other way: Why does a well-controlled chimney orebody decide to go updip in a particular bed for more than a thousand meters? No explanation can stand unchallenged because orebodies—like people—respond to their own unique environment and we are dealing with epigenetic deposits that depend on a delicate or conjectural balance of geochemical surroundings.

But first of all, "epigenesis" requires that fluids be able to enter an existing geological environment—and there is no such thing as a simple geological environment. Mineralizing fluids must have had a variety of options to exercise en route, and we learn more about the route as mines and drill holes penetrate new ground. In working with any mineral deposit, it is well for us to know what the options were.

An example of combined structural and stratigraphic control—optional conditions for ore deposition—can be seen at Salsigne, Europe's most important gold mine, near Carcassone, France. Three kinds of orebodies in the Salsigne mine have been identified by Tollon (1972): veins, replacement bodies in dolomitic limestone, and stratiform impregnations in sandstone. Veins occur in two groups: Fontaine de Santé and Ramèles. The former, principally in limestone wall rock, contains pyrite and pyrrhotite with ore-cemented breccia; the latter veins, in limestone, sandstone, and slate wall rock, are more regular fissure fillings, with additional arsenopyrite and chalcopyrite. Limestone replacement bodies of pyrite and pyrrhotite and gold-rich sandstone ore zones of arsenopyrite, chlorite, and bismuthinite are closely associated with the veins. The vein orebodies, from 1–10 m thick, are most important. The stratiform orebodies in limestone and the high-grade zones in sandstone are generally less than one meter thick and are limited in extent. The correlation between mineral associations and wall rock is explained by Tollon. Open joints in the sandstone were mineralized and completely filled in an early high-gold stage and they remained plugged against later mineralization; the limestone wall rock and limestone breccia of the Fontaine de Santé veins were corroded during the early stage of mineralization and then were completely filled during a second pyrite-pyrrhotite stage; open fissures of the Ramèles group, in limestone-standstone-slate walls, were incompletely filled during the two preceding stages of mineralization, leaving them open for a third chalcopyrite-bearing stage.

Features Developed during Mineralization

Zoning patterns are believed to have been formed during—or because of—the mineralization process. Now the morphology of orebodies can be considered in the same light: Certain physical changes in the ore–wall rock system must have taken place during mineralization.

Which ore mineralization? As far as the geologist is concerned, the orebody is a result of the final stages—epigenetic or syngenetic—that produced the orebody *now* being mined (or studied). Early epigene fluids may have worked upon a favorable bed and made it receptive to ore mineralization; we will have to accept this action as "during mineralization" as long as it appears to have affected the same route as that taken by the ore-forming fluids. The "sanding" and channeling of dolomite at Gilman began very early in the mineralization; collapse breccias, minor faults, and shallow synclinal folds appear to have had a connection with the resulting change in volume—so they all occurred "during mineralization." Of course, this is not the entire story at Gilman; a fracture pattern and a pre-Pennsylvanian karst surface supplied preore control for mineralization.

In epigenetic orebodies, brecciation and collapse appear to have been fairly common occurrences during mineralization. Breccia pipes, brecciated ore shoots, and "pebble dikes" are commonly filled with ore-cemented fragments composed of smaller ore-cemented fragments. Some of these features are of special interest in exploration, since they extend beyond the main part of an orebody—rather like fingers indicating that a hand must be somewhere near. Widespread interest in a particular geologic feature is generally accompanied by extensive theorizing, and breccia pipes have had more than their share of this. A brief comment on all prior theories (and proposing a new theory) was recently made by Mitcham (1974). A review of igneous associations with breccia bodies has been provided by Mayo (1976). Phillips (1973, 1974) has offered an interesting explanation for the breccia pipes and for the extensive brecciation commonly associated with porphyry copper deposits. He calls upon ''retrograde boiling''—boiling of the volatile components in a crystallizing residual magmatic liquid due to a relative increase in their vapor pressures—to provide a first stage of brecciation. Hydraulic fracturing would then carry the brecciation process into the wall rock and result in any array of breccia pipes.

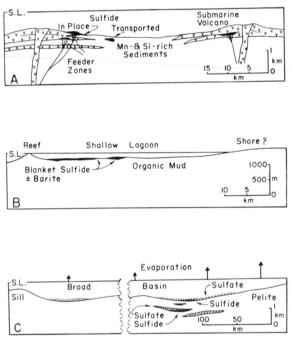

Figure 2-22. Diagrammatic sections to represent three possible environments of formation of massive sulfide deposits. (After W. F. Jenks, "Origins of some massive pyritic ore deposits in western Europe," *Econ. Geology*, v. 70, Fig. 5, p. 495, 1975. Published by Economic Geology Publishing Company. Used by permission of the publisher.

The widespread association of breccia pipes with epigenetic ore deposits is so impressive that nearly all intensely brecciated areas have been investigated time after time. Yet, many of them are barren. Every rule has exceptions, and ore deposits are geologic exceptions rather than geologic rules.

Orebodies are exceptional occurrences in ultramafic intrusions, carbonatites, and pegmatites; yet where they occur, there is little doubt that the morphology of the orebody reflects the shape of the enclosing body. Lenses and layers are parallel to the walls, zonal disseminations follow petrologic zones in the host rock—this, of course, *before* metamorphism and rejuvenation. Syngenetic and almost-syngenetic ore deposits in sedimentary and volcanic rocks follow many of the same rules; they begin as concordant bodies, but they are commonly distorted by later geologic processes. In addition, syngenetic stratiform and stratabound deposits may have epigenetic roots (veins or pipes) in older rocks.

Figure 2-22 shows three possible environments for the formation of stratiform massive sulfide deposits as envisioned by Jenks (1975). In Figure 2-22A, the sulfides have been fed by submarine volcanic-hydrothermal activity into a subsiding volcanic trough; "feeder" veinlets would be expected, as at Rio Tinto, Spain, where massive sulfide lenses are associated with a stratigraphically lower stockwork orebody. The Rio Tinto stockwork, 37 million tons of 0.8 percent copper ore is mined "porphyry copper style." The Japanese Kuroko deposits and a large number of stratabound massive sulfide deposits in the Canadian Shield also have stockwork ore zones. Figure 2-23, a cross section of the Kidd Creek orebody, Ontario, provides an example of a volcano-sedimentary orebody that has been metamorphosed, deformed, and overturned; note the chalcopyrite stringer ore (the roots) at the former base of the deposit.

Sulfides and barite that have accumulated in a semistagnant lagoon behind a reef or some similar barrier are illustrated in Figure 2-22B. Volcanic activity could have been involved, but not to the extent seen at Rio Tinto. The orebodies at Rammelsberg, Germany, Europe's oldest continuously operating lead-copper-silver producer, are of this type. Figure 2-24 is a cross section through the "new orebody" at Rammelsberg ("new" because is was discovered in 1859—the "old" orebody was discovered in 968). The stratiform sequence, pyrite–chalcopyrite–sphalerite–sphalerite with galena–galena–barite, has been buckled and pressed into a tight syncline in schist. At the base of the "old orebody," a network of veins may represent mobilized sulfides leading *from* the orebody or the original roots leading *to* it. The literature on Rammelsberg is impressive, and it is one of the "textbook" districts. A synopsis in English, with a selected bibliography, has been provided by Schot (1971).

In the example shown in Figure 2-22C, sulfate (anhydrite) has been precipitated as an evaporite in a moderately deep, protected environment and

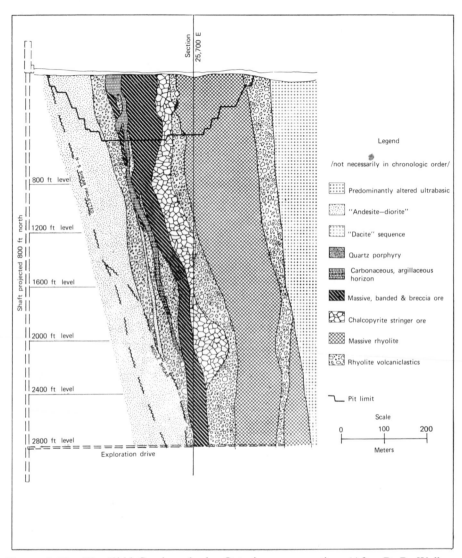

Figure 2-23.   The Kidd Creek orebody, Ontario, cross section. (After R. R. Walker and others, "The geology of the Kidd Creek mine," *Econ. Geology*, v. 70, Fig. 4, p. 84, 1975. Published by Economic Geology Publishing Company. Used by permission of the publisher.)

iron sulfide has accumulated in a more restricted, stagnant area. Pyrite orebodies mined in the Colline Metallifere area in Tuscany are thought to be of this origin, with morphology changed by remobilization of the ore into nearby faulted and brecciated zones.

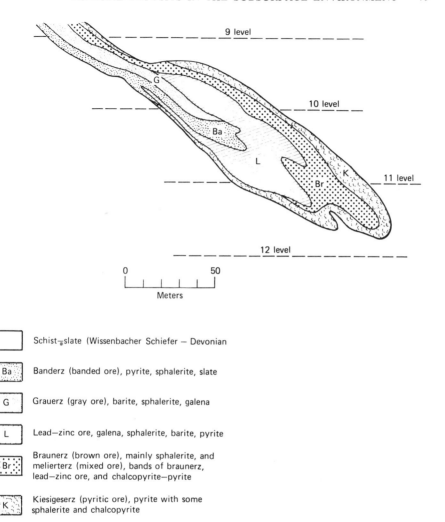

Figure 2-24. Cross section through the "Neues Lager," Rammelsberg, Federal Republic of Germany. (Data from mine visit and from E. H. Schot, 1971, p. 264–272.)

2.5. WARPING AND DISLOCATION OF MINERAL DEPOSITS

Assume that we are outlining an orebody from drill-hole data or in mine workings. Now, the orebody suddenly loses strength or, worse yet, disappears. Or we could take the equally intriguing assumption that someone else's orebody has died and we have been asked to resurrect it or at least find its lost twin.

Postmineralization faulting and folding—it is remarkable that so few international symposia have been devoted to this subject. Perhaps it is considered a topic for mine managers rather than for scientists. It certainly *is* a matter of importance to mine managers, but it is even more important to the geologists who are expected to untangle the mess before it puts an end to the mining operation. The point has already been made that a mining geologist must be a structural geologist. Now we can underscore the point. Problems in structural geology arise nearly every day in the life of a mining geologist; too often they cannot be met by everyday solutions.

The first step: place the problem in a geologic sequence. Did the orebody pinch down or disappear because of something inherent or because of a postmineral condition? Without this preliminary step, we might make the mistake of looking for a displaced orebody segment that never existed. The offending fault zone may have been an original boundary to the orebody.

Faulting

The premineral–postmineral choice needs some hedging because structural movement may have preceded, accompanied, *and* followed ore mineralization. If fault movement occurred repeatedly in the principal ore-bearing structure, we could have the situation shown in Figure 2-25 and described by Spurr (1923), a pioneer mining geologist who became well known for his grasp of complex structural problems. The solution to this particular fault problem led to the discovery of deeper ore in one of the major mines at Aspen, Colorado, that had apparently been exhausted. In many instances, movement in a "cross fault" that offsets an orebody should produce slickensided ore and drag ore (broken fragments of ore in fault breccia or gouge) to indicate that the movement took place after the particular ore association was formed. But look again. Is it the *same* mineral association as in the truncated orebody? The fragments could be from an earlier stage of mineralization. A truncated zoning pattern generally indicates postmineral fault movement, but the local agreement of zoning patterns with a fault would not actually preclude postore faulting. The zoning might have come from later metamorphism or local remobilization of the ore.

Suppose the major movement on a truncating fault appears to be premineral. While this naturally discourages the thought of a lost segment of the orebody, it does allow for the possibility of similar ore controls having existed in the displaced zone. Analysis of the fault and its nearby geologic patterns would still be worthwhile. This is what was done at Guanajuato, Mexico, and it resulted in the discovery of an entire series of important orebodies beneath mine workings that were nearly exhausted after more than 400 years of mining (Gross, 1975). At Guanajuato, the faulting, wall-rock alteration, and mineralization had each occurred in several stages; the keys to exploration were found in a geologic study that revealed two separate families of known orebodies and the premineral age of an assumed postmineral fault.

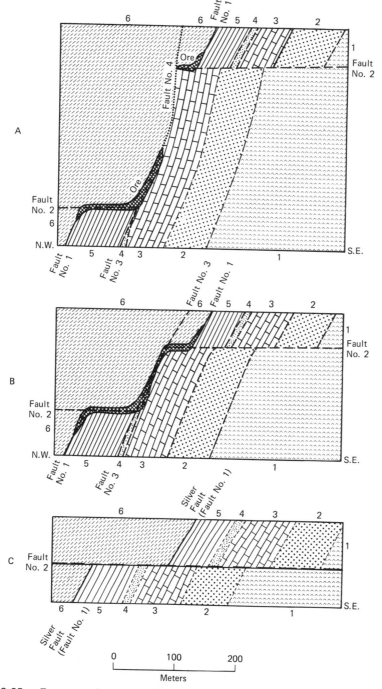

Figure 2-25. Cross sections through the Smuggler shaft, Aspen, Colorado. A = present structure; B = structure before latest movement on fault number 4 (nearly the same fault plane as fault number 3); C = structure before movement on fault number 3 and subsequent deposition of the ore. (From *The Ore Magmas* by J. E. Spurr, 1923, Fig. 69, p. 358. Copyright © 1923, McGraw-Hill, Inc. Used with permission of McGraw-Hill Book Company.)

Part of the mineral deposit has been displaced by faulting. Which direction? In the direction indicated by gash joints, shear joints, drag folds in bedding, the trail of drag ore, and slickensided surfaces. At first glance, there is a simple beauty to the analysis of slickensided surfaces; stroke the fault surface or inspect it: the opposite wall has moved parallel to the striations and in the smoothest direction. True enough, the striations and furrows should parallel the direction of the latest movement in *that particular* fault slice. Unfortunately, there are too many cases of multiple fault slices—each with striations in a different direction. The smoothness criterion also has some serious ambiguities. From a detailed study of actual fault surfaces, Tjia (1968) found that "chatter marks" and similar features associated with slickensides often gave a sensation of smoothness in the wrong direction. At any rate, directional criteria apply to an episode in faulting movement and a painstaking collection of episodes may re-create the fault's history.

How extensive was the movement on the fault? Stratigraphic or lithologic mapping is one of the best approaches, and it can be especially helpful if there are some discordant features (dikes, for example) in the sequence. Geochemical zones, wall-rock alteration patterns, and geophysical "signatures" can also help in placing reference points on both sides of a fault. Wall-rock alteration zones were used by Lowell (1968) in his discovery of the Kalamazoo orebody—the faulted "other half" of the San Manuel orebody in Arizona.

Geochemical studies in the Coeur d'Alene district, Idaho, indicate unique patterns of lead-zinc ratios on each side of the large Dobson Pass fault (Gott and Botbol, 1973, 1975). Movement on the fault was reconstructed by fitting the exposure of one intrusive body into a position above another exposed intrusive body and matching the geochemical patterns. With the faulted upper part restored to an ancestral stock, additional verification was found in further geochemical work. Thus, the extensive mineralization near the lower part of the ancestral stock can be expected to extend into the upper plate near the displaced portion of the stock.

In attempting to reconstruct the position of a faulted orebody, it must be remembered that hinge faults and "scissors" faults will have variable displacement along their strike. Also, intervals of sedimentation or volcanism may postdate some, but not all, of the fault movement. This kind of situation is shown in Figure 2-26. The total movement on the Esperanza fault, El Oro, Mexico, was A-A rather than the B-B as was indicated at first by the displacement of the andesite. Two-thirds of the fault movement had taken place before the andesitic volcanics were deposited.

Folding and Metamorphism

The effects of postmineral folding are not often so abrupt or so thoroughly disappointing as those of postmineral faulting, but they are much more

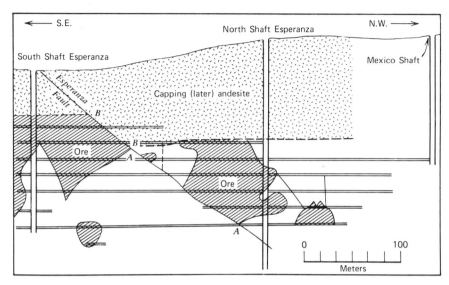

Figure 2-26. Esperanza mine, El Oro, Mexico. Longitudinal section of the San Rafael vein. Total movement, A-A, was determined by matching ore shoots in the vein rather than through measuring the displacement of the andesite. (From *The Ore Magmas* by J. E. Spurr, 1923, Fig. 65, p. 339. Copyright © 1923, McGraw-Hill, Inc. Used with permission of McGraw-Hill Book Company.)

common and, considering the widespread relation of ore to metamorphic terrain, more important. A great many fold-type "structural controls" have been reclassified as "structural deformation" in orebodies that have themselves been reclassified from epigenetic to syngenetic.

It has been noted that postore faulting carries the connotation of a missing segment, whereas preore faulting indicates something less direct—a *possible* repetition of the ore control and (with luck) ore. With postore and preore folding the unseen continuation of a folded mineral deposit is a more direct exploration target than the continuation of a folded host bed. Something else can be said about postore folds as well as faults and just about everything else that has been touched upon in this chapter: Tectonism may have taken place more than once. Folding may have begun before the rocks were lithified, and—this will sound familiar—it may be difficult to find evidence of anything other than the last event. Even the isotopic age-dating clocks may have been reset by the last major thermal effects of tectonism.

There is one profound difference between the effects of postore folding and postore faulting: Folding affects the entire orebody rather than just the zone of truncation. Therefore, evidence that an orebody has been folded should be reflected in the texture of the ore as well as in the shape of the body. Stanton (1972) summarizes the effects of "deformation and annealing" by postmineral folding in his textbook on ore petrology. Laboratory simulation of tectonism has also provided some valuable guidelines to the understanding of sulfide rheology (deformation and flow); for a description of compression experiments on ore mineral aggregates, an article by Atkinson (1974) is well worth reading. Atkinson describes the temperature and pressure conditions of cataclasis (fracturing and rotation) and of intracrystalline deformation.

Because sulfide ores are relatively plastic in comparison with most rocks, a folded orebody will have a much more complex geometry than its host rock. This is shown in Figures 2-27 and 2-28 in which the Rosebery, Tasmania, massive sulfide deposit has been reconstructed after a detailed analysis. Most of this study was done in the mine, but there was considerable attention given to a laboratory investigation of metamorphic ore textures.

We have noted the intense folding in the "new orebody" at Rammelsberg (Fig. 2-24). There is more: Three entire orebodies have been squeezed into a single overturned syncline. Postore faulting as well as folding was involved, and a small displaced segment of the "old orebody" has been located and mined beyond one of the faults.

There are some exciting challenges to be met in the reconstruction of folded orebodies; this is aptly called "unrolling" in the Zambian copper belt (van Eden, 1974), where it is used to study paleocurrent directions and paleogeography, prime factors in ore deposition along an ancient shoreline.

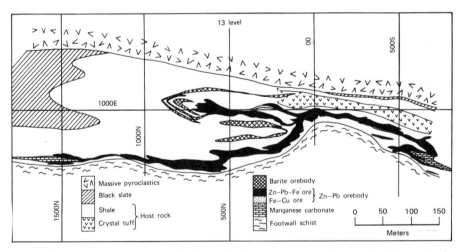

Figure 2-27. Geological plan of 13 level, Rosebery mine, Tasmania. (From R. L. Braithwaite, "The geology and origin of the Rosebery ore deposit, Tasmania," *Econ. Geology*, v. 69, Fig. 4, p. 1091, 1974. Published by Economic Geology Publishing Company. Used by permission of the publisher.)

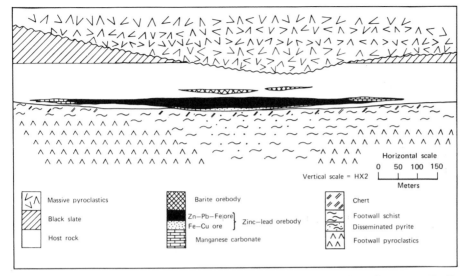

Figure 2-28. Prefolding reconstruction of the Rosebery ore deposit. (From R. L. Braithwaite, "The geology and origin of the Rosebery ore deposit, Tasmania," *Econ. Geology*, v. 69, Fig. 5, p. 1091, 1974. Published by Economic Geology Publishing Company. Used by permission of the publisher.)

Postmineral Intrusion

Another aspect of postmineral dislocation is encountered where an orebody stops at an intrusive contact. In the Homestake gold mine, postore (Tertiary) dikes cut the orebodies and wedge them apart. The orebodies are displaced perpendicular to the dike walls. We can extend this principle to other mines as long as we recognize the possibility of additional faulting dislocation (predike or postdike) and the possibility that a "dike" may not be an intrusion at all but may have been formed metasomatically without forcing the walls apart.

We have the premineralization-versus-postmineralization problem when dealing with orebodies that terminate against an intrusive rock. A dike can be part of the preore control, as at the Wheal Jane in Cornwall, or it can be distinctly postore, as in the Homestake mine.

A postore dike that has not undergone subsequent metamorphism should have some of the following features:

1. The orebody is affected by thermal metamorphism near the dike. This is expressed in mineralogy (re-equilibration), in isotopic-thermal characteristics (see Chapter 4), and in the destruction of fluid inclusions.

2. The chilled border of the dike is of different composition, grain size, or

thickness against the orebody than against the physically contrasting, silicate wall rock.

3. Minerals in the chilled border zone of the dike may contain some of the ore and gangue elements, but if ore minerals occur as such, they have a different crystal habit or some other varietal difference in comparison with those in the orebody. Zoning of ore minerals in the dike agrees with the dike walls rather than with zonal patterns in the orebody.

4. Xenoliths of ore in the dike have coronas of dike minerals or show peripheral replacement by the dike minerals.

Where a dike is preore in age and there has been no metamorphism subsequent to ore deposition, some of the following features should be apparent:

1. Veins, mineral zoning, and ore shoots swing parallel to the dike or extend into the dike. They are not truncated.

2. Wall-rock alteration associated with the orebody continues into the dike, in pattern if not in composition.

3. Dike minerals are selectively replaced by ore minerals of the same crystal habit as in the orebody.

4. Naturally, the postore dike features—thermal metamorphism, differential chilling, and so forth—should be lacking.

Metamorphism complicates most of these features. If the dike is postore, there is likely to have been a remobilization-replacement of dike minerals by ore minerals, but the remobilized ore association should differ (both in the dike and the orebody) from the original association. A metamorphosed postore dike may also contain veinlets of ore that have been squeezed into fractures, but these veinlets should contain identifiable fragments of broken ore and there should be textural evidence of plastic flowage. If the metamorphosed dike is preore, the diagnostic alteration pattern will have been distorted, but there still should be a "ghost" pattern in the distribution of newer minerals and there may be evidence of metamorphic sulfide-silicate reactions along the dike wall—still with some indication of the original pattern. All of this requires painstaking study. A more complete listing of dike-orebody-metamorphism age criteria has been provided by Mookherjee (1970).

The District Aspect—and Beyond

Postmineral faulting and folding has been considered at the scale most likely to be used by mine geologists: fault displacements and fold amplitudes of a few meters to a few kilometers. What about the displacement of entire districts? This is every bit as important, especially to a wide-ranging exploration geologist. "Unrolling" may be almost too much to handle, but all geologists will attempt to do this on a district scale many times during their careers.

The Coeur d'Alene district, Idaho, is a well-studied type locality for postore fault situations. Faults occur in most sizes and types, ranging from small normal, reverse, and strike-slip faults to the larger scale Dobson Pass fault and on to the postore Osburn fault with a lateral displacement of approximately 25 km. A thorough description of the Coeur d'Alene district can be found in U.S. Geological Survey professional papers by Hobbs and others (1965) and Fryklund (1964). Every geologist who works with district-scale faulting should be able to draw upon the Coeur d'Alene district for analogies.

A well-described example of regional postore faulting and appropriate geological work is given by Henderson and Brown (1966) and Campbell (1948). This pertains to the West Bay fault in the Yellowknife gold mining district, Northwest Territories, Canada. Displacement on the fault was calculated and successful exploration was done on the basis of dike patterns that were shifted 5 km horizontally and 0.5 km vertically.

The Tintina Trench (fault), one of the major structural units of Canada and Alaska, has a postore displacement calculated at 400 km and a preore "favorable controls" displacement of 1,600 km (Kuo and Follinsbee, 1974).

We have arrived at the plate tectonics scale of thinking. Perhaps the Boleo copper deposits in Baja California have been shifted by postore and preore faulting (or both) for 500 km from a position on the mainland coast of Mexico (Guilbert, 1971). Perhaps the gold-bearing conglomerates of northern Brazil and similar orebodies in the Tarkwa gold field of West Africa have been pulled apart from a single district in the ancestral Southern Hemisphere continent of Gondwanaland or the still larger ancestral continent of Pangea. Predrift reconstructions of mineral provinces in Gondwanaland have been furnished by Petrascheck (1968). Predrift positions for the pieces of Laurasia, the Northern Hemisphere part of Pangea, are less apparent; various interpretations—interesting to "play with" while creating your own matching mineral provinces—are given in books by Tarling and Tarling (1973) and Brock (1972). Continental and oceanic plate margins are thought to have formed new mineral belts as well as to have displaced old ones. More on this in Chapter 4.

Recent and Near-recent Effects

Postore warping can also be postoxidation warping, as at Ajo, Arizona, where a porphyry copper deposit was exposed in Tertiary time, enriched, covered by conglomerate, tilted, reexposed, and enriched again. Postore faulting can be postweathering faulting, as at the St. Anthony mine, Arizona, where the original sulfide-bearing vein was oxidized to a certain depth and then cut into two segments by the Mammoth fault (Fig. 2-29). A curious pattern of

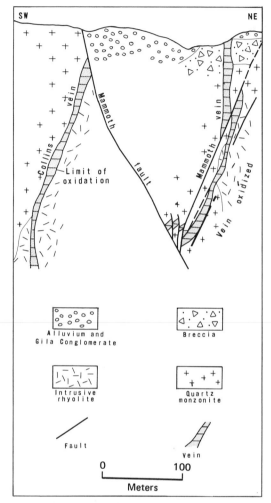

Figure 2-29. Generalized cross section, St. Anthony mine, Pinal County, Arizona. [From S. C. Creasey, "Geology of the St. Anthony (Mammoth) area, Pinal County, Arizona," in *Arizona Zinc and Lead Deposits,* Bulletin 156, Fig. 20, 1950, courtesy Arizona Bureau of Mines.]

mineralogy was developed at St. Anthony when a still later introduction of molybdenum and vanadium reacted with the deep-lying oxidized base-metal carbonates and sulfates in the Mammoth vein to produce wulfenite and vanadates.

There are still some postmineralization stages to be accounted for in subsurface orebody models—changes brought about weathering and supergene processes. These processes and their results are the subject of Chapter 3.

CHAPTER 3
MINERAL DEPOSITS IN THE ZONE OF WEATHERING

Exploration begins at the surface, in an arena of soil, weathered, rock, and landforms where there is access to mineralization but where the orebodies have adjusted, at least in part, to the new environment. These adjustments are the subject of the present chapter and the theme in the following example of ore discovery.

Harrison Schmitt, a consulting geologist familiar with ore minerals in the zone of weathering, examined a small molybdenum mine—the New Year's Eve mine—in southern Arizona in 1955. The molybdenum orebody was nothing remarkable, but what impressed Schmitt was the extensive brecciation with abundant goethite and some jarosite and hematite in the area around the mine. He mapped the area of intense iron staining and noted that it was accompanied by argillic-sericitic alteration. Some of the weathered exposures of intrusive and volcanic rock were laced with quartz veinlets. There were traces of green and blue copper stain—but only traces that in themselves could have illustrated an Arizona saying that "ten cents worth of copper will stain an entire mountain." Schmitt looked beyond the copper stain, and he found that the capping (disseminated gossan) characteristics were good enough for a conceptual model of an underlying porphyry copper-molybdenum deposit with supergene chalcocite enrichment. The rest of the story can be abbreviated. The Esperanza copper mine now occupies the oval area outlined by Schmitt; two newer pits, West Esperanza and Sierrita, are nearby. Reserves at the three mines: more than 500 million tons of ore.

The relationship of oxidized and impoverished outcrops to secondarily enriched orebodies and ultimately to primary ore is not always so clearly indicated. The Blue Dike prospect in White Canyon, Utah, did not seem to be much of a bargain when purchased in 1946 for $500 by Joe Cooper, a part-time prospector. Worse yet, Cooper's first shipment of oxidized copper ore was rejected because of impurities. Since a small uranium content was one of the impurities, he made the next shipment to a uranium processing plant; rejected—too much copper. Later, the Atomic Energy Commission agreed to buy some of the ore if it would encourage a search for a zone of less contaminated ore. Cooper accepted the challenge and extended the workings 60 m further into a sandstone cliff—deeper into a paleostream channel and through the zone of secondary copper enrichment—right into a body of high-grade uranium ore. Again, the rest can be abbreviated; the Blue Dike prospect, now the Happy Jack mine, became the largest producer in the important White Canyon uranium district.

Experanza and Happy Jack have relatively straightforward explanations. In both cases, primary mineralization had been leached in the zone of weathering so that three mineral zones remained: an oxidized zone with hematitic minerals, a zone of supergene copper ore, and a deeper zone of primary ore. There are, however, a great many orebodies in which something more complicated has occurred during weathering, orebodies that keep geologists from getting too smug about their ability to understand the near-surface environment. The Tynagh, Ireland, lead-zinc mine contains one of these.

Sphalerite, pyrite, and galena in a hypogene assemblage are ordinarily oxidized and either dispersed or changed to nonsulfide minerals during weathering. Not at Tynagh, where the ore has been water-soaked throughout the process of weathering. A residual orebody resembling a dense black mud overlies the lower grade hypogene mineralization, and the shallower orebody contains supergene galena, sphalerite, and pyrite. The new pyrite is ubiquitous—still being formed and replacing almost everything in its way, including the body of a newt that died in an exploration trench (Derry, 1971).

3.1. THE SURFACE ENVIRONMENT—WEATHERING AND EROSION

In becoming accessible to discovery, orebodies from deeper surroundings are exposed to the zone of circulating meteoric water, atmospheric pressure, and atmospheric temperature. Physical and chemical weathering operates in the direction of equilibrium, as stated in Le Chatelier's principle: A change in the external conditions of a system in equilibrium will shift the equilibrium in such a way as to counteract the effect of the change (the stress). Thus, the effect of lower pressure is met by an increase in volume; this is accomplished in orebodies by disintegration and by formation of lower density minerals. Lower temperature is accommodated by exothermic reactions, enough—as in

Agricola's day—to melt the early-morning frost. For example, bornite (specific gravity 5.06) is replaced in part by malachite (specific gravity 4.0), with a release of energy as heat.

If a mineral deposit has been formed at or near the surface and has remained in the same environment, no change in character would be expected. In fact, the deposit might continue to form. Some saline playa deposits, such as the sodium carbonate body at Lake Magadi, Kenya, are renewed while they are being mined. But change rather than stability is the normal geologic condition; a Pleistocene saline deposit certainly will have begun to adjust to changed climatic and erosional conditions. Some lateritic nickel ores, supergene copper sulfides, and "roll-front" uranium deposits are relatively well adjusted near the surface at the present time; but they, like the pyritic newt at Tynagh, have been formed at the expense of parent mineral concentrations that are not now in equilibrium.

The path from mineralization at an ancient land surface to reexposure at another land surface in a later time is likely to have been complex. Stratabound mineral deposits that have formed on paleosurfaces have had to adjust to the higher pressures and temperatures of deeper environments before being exposed again. In addition, the paleoenvironment would have undoubtedly differed in many respects from the present environment. If a gold placer deposit that accumulated in a Precambrian hydrosphere and atmosphere is exposed to present-day fluvial or marine processes, it is under stress.

Axiomatically, exposed mineral deposits have undergone changes in order to accommodate to the surface environment. None escapes.

Whatever the definition of "surface environment," it deals with erosion and weathering. Any definition of "weathering" deals with physical and chemical processes that work together. Erosion removes weathered material and exposes new material at a certain rate, which, if rapid enough, will reach into unweathered material and allow us to see fresh ore minerals in landslide scars, downcutting stream beds, and recently glaciated areas. If the processes of physical weathering are dominant, they provide the erosional medium with loose ore material that has not yet been chemically weathered; prospectors call it "float." If physical weathering is well balanced by chemical weathering, fresh mineral surfaces are provided for continued and thorough chemical attack; this is the balance needed for the accumulation of supergene sulfide ore at the base of the zone of oxidation. If physical weathering is weak, chemical reactions may go to near equilibrium in very restricted surroundings, as they do in a thin and well-developed residual soil.

Chemical Weathering

The processes of chemical weathering are of special importance; they are the changes that take place in an orebody's original *mineral* composition, by the removal of nonresistant minerals and the accumulation of others, and in its *chemical* composition, by the formation of new minerals with elements added from the zone of oxidation. Climatic and topographic conditions are determining factors in chemical weathering and in all of the associated processes just mentioned. When consideration was given to processes involved in forming a particular hypogene mineral assemblage in Chapter 2, the question was "at *which* stage?" Heritage was a factor; now heritage is again a factor when studying the erosion and weathering of mineral deposits. "In *which* environment?" Near-surface conditions and climatic controls have changed in Pleistocene and Holocene times just as surely as they have changed since the Precambrian.

The tools of chemical weathering at the surface are the atmosphere (oxygen, nitrogen, carbon dioxide), the hydrosphere (water, water vapor, ice), and the biosphere. In relating these to leached and secondarily enriched mineral deposits, we must take them into the *near*-surface environment as far as their effects can be traced—into ground-water and even ocean-water surroundings. Figure 3-1 shows the relative position of these and other natural environments in terms of Eh and pH.

Eh-pH Diagrams

Diagrams of this kind are widely used in attempting to explain mineral equilibria. Eh, like pH, is based on electrical measurement, but it refers to the oxidizing and reducing tendency (redox potential) of a system rather than to the hydrogen-ion concentration. As with "acidic" and "basic" terms for pH, the terms "oxidizing" and "reducing" refer to relative rather than to absolute positions on the scale.

Since minerals are important in outcrops and in the near-surface zones of orebodies, an appropriate demonstration of Eh-pH diagrams can be made with iron oxide mobility and stability. Figure 3-2 shows the conditions under which goethite or magnetite should exist in equilibrium in the presence of water and at a total dissolved iron concentration of 10^{-6} mole/liter. Hematite, Fe_2O_3, may form instead of goethite under slightly different conditions of vapor pressure and temperature, in a hot desert terrain, for example, but it still is a member of the ferric iron limonite group and it occupies approximately the same field.

First of all, and rather obviously, goethite and its reddish-brown rock color belong to oxidizing conditions, whereas magnetite and its gray rock color belong in a more reducing environment. Both minerals will dissolve in the acidic environment shown on the left of the diagram, and the iron will be deposited in some nearby alkaline soil or carbonate rock. In stream water with a pH of about 8 and an Eh of about 0.4, goethite is stable and will be formed when ferrous iron is released by weathering from silicate rocks or

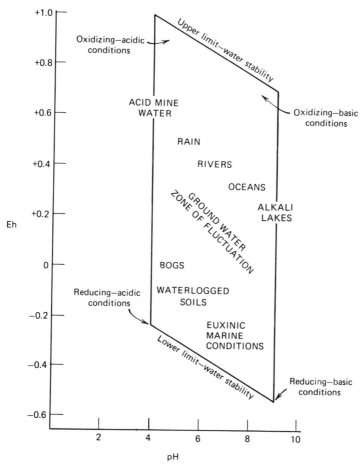

Figure 3-1. Generalized Eh-pH diagram showing the position of selected natural environments. The parallelogram shows the most usual limits of Eh and pH in the zone of weathering and in near-surface conditions.

from a magnetite body. In less oxygenated environments, say that of deep ground water with the same or slightly higher pH, magnetite need not be dissolved. But take the environment of stagnant bog water, conditions are both reducing and acid, and goethite and magnetite will both be leached—with goethite to be formed again at a spring or seep where oxygen is again available.

Just beneath an organic layer (A horizon) in soils, low pH and reducing Eh conditions exist, and iron is leached and carried to a zone of accumulation in lower levels of soil. However, under moist tropical conditions where organic

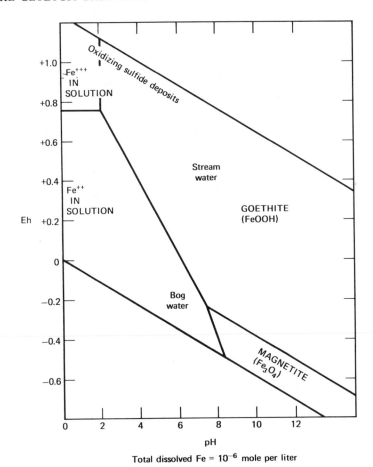

Figure 3-2.   Eh-pH diagram showing stability fields for goethite and magnetite. (After P. L. Cloke, "The geochemical application of Eh-pH diagrams," *Jour. Geol. Education,* v. 14, no. 4, p. 146, 1966. Copyright © 1976, National Association of Geology Teachers, Inc. Used by permission of the publisher.)

matter decays quickly, soil water does not become acid or reducing and an accumulation of goethite concretions develops. Later and more active erosion may expose some of the concretions as a lateritic crust, or cuirasse.

In an oxidizing pyritic sulfide deposit where low pH and high Eh conditions exist, goethite and sulfuric acid may form in combination with the high concentration of ferric iron in solution shown in the upper left-hand side of Figure 3-2. Some of the goethite may also go back into solution. When this solution percolates into the orebody, it can cause the oxidation and leaching of other sulfides, thus:

$$Cu_2S + 10Fe^{3+} + 4H_2O \rightarrow 2Cu^{2+} + SO_4^{2-} + 8H^+ + 10Fe^{2+}.$$

The outcrop remains as an iron-rich gossan, and the leached copper may be concentrated as supergene sulfide at a deeper level in various reactions with other sulfides, such as:

$$14Cu^{2+} + 5FeS_2 + 12H_2O \rightarrow 7Cu_2S + 5Fe^{2+} + 3SO_4^{2-} + 24H^+.$$

Another diagram, a composite (Fig. 3-3), shows the stability of lead, zinc, and copper sulfides and oxidation products at a certain temperature, pressure, and activity. The lower boundary of the zone of oxidation runs between the smithsonite and sphalerite fields; a minor chalcocite field lies above this boundary—in theory and also in the assemblages seen in nature. Note that galena, stable in the entire sulfide zone, forms anglesite in acid-oxidizing environments and forms cerussite in alkaline-oxidizing environments. Under alkaline and highly oxidizing conditions, tenorite is the stable copper mineral associated with cerussite; under alkaline but less oxidizing conditions, cuprite or native copper accompanies cerussite.

The stability fields shown on Eh-pH diagrams have most of the same limitations that plague all attempts to explain natural conditions in terms of quantitative chemistry. First, there are a great many more variables in nature than can be accommodated in a two-dimensional plot; three-dimensional block diagrams are sometimes used so that concentration of one particular element can be shown on the third axis, but even these diagrams are much too limited in scope. Vapor pressure of water, temperature, accompanying elements in solution, and isomorphous substitution of other metals in the subject minerals are a few of the things that would shift the field boundaries in various ways. Finally, these are *equilibrium* diagrams that represent something nature approaches but never attains.

Still, Eh-pH diagrams, based on theory and experiment, agree rather closely with what is observed in nature. They are good for illustrating natural assemblages that *should* be expected.

Then why go to all this work to tell us something we should have understood from observations in the field? Mining geologists had suspicions about duplicating field conditions in the laboratory long before J. E. Spurr (1927) wrote: "Geologists may write extravaganzas./But only God can make bonanzas."

But we really don't understand the details of what we observe in the field. For this reason, equilibrium diagrams, Eh-pH and all others, are helpful ways of explaining the relationships we see in part of an orebody and of envisioning the entire orebody. If something in a mine seems to be missing from a predicted assemblage, try to find it. Look again. If an association seen in the

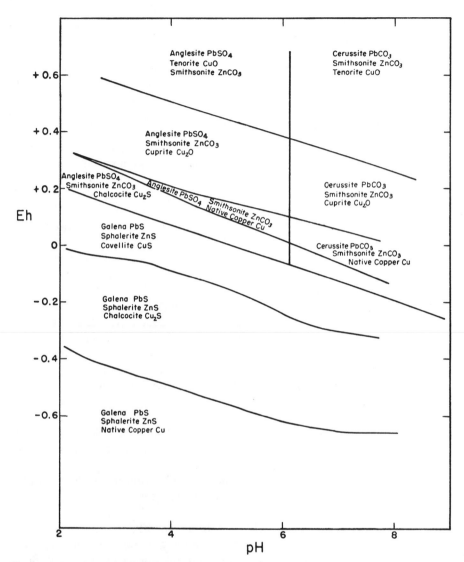

Figure 3-3. Composite diagram of stability of metal sulfides and oxidation products at 25°C and 1 atm total pressure in the presence of total dissolved carbonate = $10^{-1.5}$, total dissolved sulfur = 10^{-1}. (After R. M. Garrels, *Geochim. et Cosmochim. Acta,* v. 5, Fig. 5, p. 166, 1953. Copyright © 1953, Pergamon Press Ltd. Used by permission of the publisher.)

field does not fit with theory, perhaps the theory is wrong—but it is also possible that we have overlooked something. Look once more.

A comprehensive review of chemistry in the weathering environment is provided by Perel'man (1967) in his book *Geochemistry of Epigenesis.* Perel'man and other Russian geologists refer to all geologic processes originating at or near the surface of the earth as epigenesis, a different usage of the term than we are familiar with. Most textbooks on geochemistry have good coverage of weathering phenomena; Krauskopf (1967) has a special chapter on the oxidation of ore deposits. Equilibrium diagrams are related to weathering and to ore mineral associations in a textbook by Garrels and Christ (1965). For general reference (and good browsing) there is the *Encyclopedia of Geochemistry and Environmental Sciences,* edited by Fairbridge (1975).

3.2. LEACHING AND ASSOCIATED MINERALOGICAL CHANGES

Tables 3-1, 3-2, and 3-3 cover some general aspects of mineral mobility, mineral associations, and leaching products found in the zone of weathering.

Table 3.1.   Relative Chemical Stability of Primary Minerals in the Zone of Weathering[a]

	Very Stable	Stable	Fairly Stable	Unstable
Rock-forming Minerals and Accessories	quartz corundum spinels topaz tourmaline zircon	alkali feldspar sodic plagioclase muscovite andalusite garnet kyanite sillimanite	actinolite apatite chloritoid diopside epidote staurolite	amphiboles (most) biotite calcic plagioclase calcite chlorite dolomite feldspathoids glauconite gypsum olivine pyroxene
Ore and Economic Minerals	chromite diamond gold platinum rutile	barite cassiterite galena ilmenite magnetite monazite niobite-tantalite thorianite	hematite scheelite wolframite titanite	arsenopyrite chalcopyrite fluorite molybdenite pentlandite pyrite pyrrhotite sphalerite

[a] Data from Andrews-Jones (1968) and Perel'man (1967).

Table 3-2. Ore Mineral Associations in the Zone of Oxidation, Excluding Supergene and Relict Sulfides (Minerals in italics are common; minerals in parentheses are resistates)

Element	Oxides	Carbonates	Silicates	Sulfates	Other
Aluminum	"bauxite" gibbsite diaspore boehmite				
Antimony	senarmontite cervantite valentinite = "antimony bloom"				
Arsenic	arsenolite = "arsenic bloom"				Arsenic scorodite (hydrous Fe-arsenate)
Beryllium			(beryl)		
Chromium	(chromite)				
Cobalt	cobaltian "wad"				erythrite (hydrous arsenate) = "cobalt bloom"
Copper	cuprite tenorite "copper pitch"	malachite azurite	chrysocolla	chalcanthite antlerite bronchantite linarite	Copper atacamite (halide) turquoise (phosphate)
Gold					(Gold)
Iron	"limonite" goethite	siderite	nontronite	jarosite	vivianite (phosphate)

	hematite *lepidocrocite* (*magnetite*)				
Lead	minium = "red lead" massicot = "lead ocher" plattnerite	*cerussite*		*anglesite* plumbojarosite	*wulfenite* (molybdate) vanadinite (halide with **V**) pyromorphite (phosphate) mimetite (chlor-arsenate)
Manganese	"*wad*" *psilomelane* *pyrolusite* (hausmannite) (manganite)		(braunite)		
Mercury	montroydite				*Mercury* calomel (halide)
Molybdenum	*ferrimolybdite* ilsemannite				*powellite* (Ca-molybdate) *wulfenite* (Pb-molybdate)
Nickel			*garnierite*		*annabergite* (hydrous arsenate) = "nickel bloom"
Niobium-Tantalum	(niobite-tantalite)				
Platinum					(*Platinum*)
Rare Earths					
Silver				argentojarosite	(*Silver*) *cerargyrite*

95

Table 3-2. (con'd) Ore Mineral Associations in the Zone of Oxidation, Excluding Supergene and Relict Sulfides (Minerals in italics are common; minerals in parentheses are resistates)

Element	Oxides	Carbonates	Silicates	Sulfates	Other
Thorium	(thorianite)		(thorite)		(monazite) (Th-rare earth phosphate)
Tin	(cassiterite) = "wood tin" hydrocassiterite				
Titanium	(rutile) (ilmenite)				
Tungsten	tungstite				(scheelite) (Ca-tungstate)
Vanadium					vanadinite (halide with Pb) descloizite (complex vanadate) carnotite (K-vanadate)
Uranium	gummite uraninite		uranophane sklodovskite	uranopilite zippeite	autunite (Ca-phosphate) torbernite (Cu-phosphate) tyuyamunite (Ca-vanadate) carnotite (K-vanadate)
Zinc		smithsonite hydrozincite	hemimorphite = "calamine"	goslarite	
Zirconium			(zircon)		

Table 3-3. Leaching Products and Textures in Gossan—Based on Work by Blanchard (1968)

I. Mainly indigenous
   A. Cellular pseudomorphs, reflecting cleavage, fractures, and grain outlines
      1. Cellular boxworks, with angular walls (example: after galena, Fig. 3-9)
      2. Cellular sponge, with rounded walls (example: after sphalerite, Fig. 3-12)
   B. Cell filling or coating
      1. Flaky or shriveled limonite crusts (Fig. 3-11, after pyrrhotite)
      2. Granular limonite, sometimes in "sintered" crusts
      3. Fluffy limonite
      4. Hard pseudomorphs—limonite "dice" (after pyrite, Fig. 3-10)
II. Partly indigenous (cell filling) and partly exotic fringing limonite (Figs. 3-4 and 3-5)
   A. "Relief" limonite, fibrous to arborescent, with fibers visible under hand lens (example: after chalcocite)
   B. Partly sintered crusts, a weblike coating (example: after galena)
   C. Thin platy ribs (from galena cleavage) in pyramidal boxwork
   D. Surface coalescences—semiglazed to "sintered" coating, less than 1 cm in thickness. Original texture sometimes apparent in "ghost" outlines.
III. Exotic (transported); fine-grained impregnations and lenses to large sprawling masses extending for 100 m or more
   A. Granular and coagulated limonites, pulverulent to cinderlike
   B. Flat crusts, composed of thin coats of "paint" on surfaces
   C. Smeary-crusted limonites, with nodules composed of finer nodules; generally dark and sometimes iridescent
   D. Thick-walled limonite, in crude and erratic "pseudo-cellular" form
   E. Columnar limonite, with columns ranging from nodular blobs to stalactitic rods
   F. Caked crusts, resembling a miniature dessicated mud flat
   G. Surface coalescences; like II-D except without "ghost" texture

---

Stability of Primary Minerals

Table 3-1 deals with the geochemical stability of primary minerals in outcrops, not with their physical durability during transportation. Barite, for example, is a stable mineral that occurs in leached outcrops from which the associated fluorite, calcite, and most of the sulfide minerals have been dissolved, but it abrades very easily and pebbles of barite travel only a short distance in streams. Galena is also apparent in many leached outcrops of lead-zinc orebodies, not because it is as chemically inert as barite but because it tends to develop a shell of anglesite that protects it from further decomposition; like barite, it is a poor traveler in streams. Neither barite nor galena can be expected to provide evidence of mineralization at a great distance from an

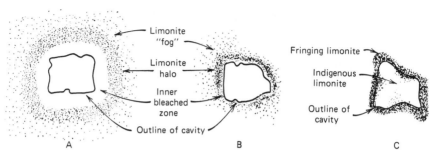

Figure 3-4. Sketches showing limonite "fog," limonite halo, fringing limonite, and indigenous limonite from the oxidation of pyrite. *Sketch A*. Rock of low neutralizing power (kaolinized alaskite porphyry) has permitted a bleached zone to form next to the cavity. (After R. Blanchard and P. F. Boswell, "Status of leached outcrops investigation," *Eng. Mining Jour.*, v. 125, Fig. 9, p. 374, 1928. Copyright © 1928, American Institute of Mining, Metallurgical and Petroleum Engineers, Inc. Used by permission of the publisher.)

Sketch B. Rock of moderate neutralizing power (slightly sericitized quartz monzonite) with a narrow bleached zone around the cavity. (After R. Blanchard, "Interpretation of leached outcrops," *Chem. Metall. Mining Soc. South Africa Jour.*, v. 39, no. 11, Fig. 16, p. 361, 1939, by permission Chemical, Metallurgical, and Mining Society of South Africa.)

Sketch C. Rock of moderately strong neutralizing power (limy shale) with indigenous limonite remaining in the cavity and fringing limonite outside the cavity. (After R. Blanchard, "Interpretation of leached outcrops," *Chem. Metall. Mining Soc. South Africa Jour.*, v. 39, no. 11, Fig. 16, p. 361, 1959, by permission Chemical, Metallurgical, and Mining Society of South Africa.)

exposed orebody. Some of the "rather stable" minerals, such as scheelite, are just as far traveling as the "very stable" minerals, in especially pure cold rivers; pebbles of scheelite sometimes appear as far as 15 km downstream and recognizable grains can travel as far as 70 km (Zeschke, 1964). Naturally, there are departures from the generalizations given in Table 3-1 wherever special Eh-pH conditions, extreme temperatures, and certain mineralogical and textural conditions exist. Molybdenite, for example, is an unstable mineral in alkaline environments (if fully exposed), but it is stable in the acid conditions found in oxidizing sulfide deposits. Vein quartz may, of course, enclose almost any mineral—pyrite, for example—and protect it from weathering. It pays to break open oxidized rock fragments and look inside.

Minerals in the Zone of Oxidation

Table 3-2 shows resistant minerals, indicated by parentheses, in a list of new minerals that have formed in the zone of oxidation. Inasmuch as supergene

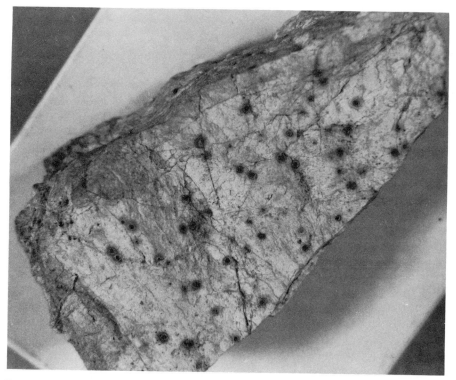

Figure 3-5. Hand specimen showing variations in limonite expression from the oxidation of pyrite in kaolinized rhyolitic volcanics, Stein's Pass, New Mexico. Most voids are surrounded by a limonitic halo, some are associated with fringing limonite, and a few contain indigenous limonite.

sulfide minerals and relict sulfide minerals are associated with special reducing and protected conditions, they are not listed, even though some of these, such as chalcocite, may be found in stable association with oxidized lead and zinc minerals, as suggested in Figure 3-3.

Gossan and Capping. Table 3-3 is intended as a "dictionary" of gossan and capping terms that have come into general use from the writings of Roland Blanchard (1968), Augustus Locke (1926), and other experts in the interpretation of leached outcrops. *Gossan* is a cellular accumulation of limonitic material derived from the leaching of sulfide-bearing veins and massive sulfide deposits. *Capping* is the leached and limonite-stained upper portion of disseminated sulfide orebodies, especially porphyry copper deposits. Blanchard used a 20 percent original sulfide content (by volume) to differentiate between the two terms. *Limonite* is a "basket" term understood by most geologists to comprise hematite, goethite, lepidocrocite (uncommon, and chemically the same as goethite), and jarosite, the hydrous iron-potassium sulfate. Much of the material in gossan is "limonitic jasper," a mixture of limonite and silica. Mixtures of limonite with clay minerals, manganese minerals, and sulfates or carbonates are also common.

Limonite Color. The color of limonitic gossan and capping is considered by many geologists to be an indicator of mineralogy in the parent orebody, and all students of leached capping pay a great deal of attention to differences in the reds, yellows, and browns of outcrops. But limonite will appear to have a certain color for various reasons: mineralogy, grain size, aggregation, moisture, and even the amount of sunlight and shadow on an outcrop. A fluffy powder, a compact mass, and a glazed surface may have the same limonitic composition but be quite different in color. The reddish or yellowish component in a crusted brown limonite, for example, will not be apparent until it is scratched. Blanchard, who specialized in leached outcrop studies for most of his life, was not a color-interpretation enthusiast. He agreed that certain generalizations can be made: reddish limonites point to hematite, yellowish limonites to jarosite, and dark-brown limonites to goethite. Yet, he recognized that most outcrops of orebodies are too complex in mineralogy and in history for such direct color interpretation. He preferred to use the textural characteristics of limonites as guides.

Still, the color of limonitic capping and gossan continues to be an important entry in field notes—with good reason. A change from one color to another means that there is a change in *something* that has to do with a possible orebody. Call it a colorimetric expression of geochemical zoning. We sometimes use colorimetric methods of chemical analysis that we understand even less.

One difficulty with limonite color lies in the terminology. "Friendly brown" is widely used in the American Southwest for the color of limonite (principally goethite) that has formed from a pyrite–chalcopyrite mixture, but no two geologists will agree on exactly what it should look like, other than "sort-of reddish brown." "Maroon to seal brown," indicating that a capping is hematitic and derived from chalcocite (in part, anyway), is a better term in that a comparison can be made with color patches on the Geological Society of America rock color chart (Goddard, 1963). Because color patches and powdered limonite are hard to compare, some mining companies have furnished their field geologists with samples of powdered limonite for which color and mineralogy have been determined in the laboratory.

Laboratory Investigation. Mineralogical and geochemical analyses have been used in evaluating limonitic capping and gossan but without much success because of the scavenging nature of iron oxides and manganese oxides in contact with most ions in solution. These studies have, however, had enough significance to make them a part of the broader field identification procedure. X-ray diffraction is used for identifying limonite minerals and associated minerals. Electron microprobe analysis helps identify trace metals. Zimmerman (1967) determined the chemical composition of gossans near Mount Isa by emission spectrography, X-ray fluorescence, and "wet" chemical analysis, and with this information he was able to map "visible geochemical anomalies" in the field. Polished-section mineralogical studies have been of special value in relating well-developed boxworks to the incipient boxworks seen in partly oxidized sulfide assemblages.

Transported Limonite and False Gossan. Before discussing the various oxidized ores and their derivatives, more should be said about varieties of cellular boxworks and limonitic outcrops. Hypogene boxworks are not common, but where they do occur, from hydrothermal leaching, collapse, and remineralization, for example, they may be identified as such by the hypogene minerals in the void linings; secondary minerals and textures will have been superimposed, so identification is not easy. Supergene limonite may be indigenous (formed in place), transported or exotic (displaced by millimeters to hundreds of meters), or it may be false gossan, a limonitic "noise" unrelated to real capping or gossan. Some kinds of false gossan are quite misleading; these include Australian "billy," which forms as iron-stained quartzitic zones at the base of lava flows, "pseudo-jaspers," or iron-stained jasperoid, residual networks of hypogene quartz-hematite veinlets, and limonite-soaked soil. Some additional and more easily recognized "noise" includes bog iron ore, desert varnish (desert patina), and, of course, all the limonitic products formed in the weathering of ferruginous sedimentary rocks and rock-forming ferromagnesian minerals. The red and ocherous indurated crust (cuirasse) of laterite is not likely to be mistaken for gossan, but it has the perverse capability of masking and obliterating all evidence of gossan and most geochemical-geophysical indications of ore. For a geologist working in the tropics, a study of laterite geochemistry and geomorphology is essential; *Tropical Geomorphology* by Thomas (1974) provides a good overview.

Leached Outcrop Characteristics. Some of the major ore and gangue minerals contribute to leached outcrops in the following ways:

Arsenopyrite produces pale olive-green, bluish-green, or violet scorodite and a characteristic granular-arborescent or clinkerlike limonitic boxwork of dark-brown color.

Calcite seldom yields a limonitic boxwork, but when it does, it is a fine cellular, rhombohedral network with delicate walls.

Chromite is relatively stable in most oxidizing environments, but it may form a coarse honeycomb boxwork (after octahedral crystals) with a finer cellular sponge in the voids or it may form a cellular sponge in which some voids are lined with arborescent limonitic knobs.

Cobalt sulfides and arsenides. Pink erythrite or "cobalt bloom" is the index product.

Copper minerals are most thoroughly studied in relation to gossan and capping. Green and blue oxidation products and copper pitch are characteristic, but they are removed by intense leaching, leaving several characteristic colors and forms of limonitic boxwork. For example,

Chalcopyrite: brown to reddish-brown cellular boxwork with a fine network or powder in voids (Fig. 3-6).

Bornite: ocherous-orange cellular boxwork with curved triangular voids (Fig. 3-7).

Tetrahedrite: chocolate-brown "contour" boxwork (Fig. 3-8).

Chalcocite: maroon to seal-brown "relief limonite" crusts; it does not form boxworks by itself but forms coatings and filling in voids left by other minerals.

Minerals of the porphyry copper association form an important capping type of leached outcrops from various ratios of pyrite, chalcopyrite, chalcocite, and molybdenite in several environments of reactive and nonreactive wall rock. Harrison Schmitt once stated in a lecture that there are at least 11,160 possible combinations of critical hypogene and supergene conditions

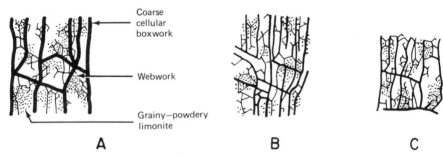

Coarse
cellular
boxwork

Webwork

Grainy—powdery
limonite

A   B   C

Figure 3-6. Boxwork derived from chalcopyrite, showing coarse cellular boxwork with a finer webwork and with grainy or powdery limonite in the voids, enlargement ×5. (*Sketch A* after R. Blanchard, *Chem. Metall. Mining Soc. South Africa Jour.*, v. 39, no. 11, Fig. 16, p. 361, 1939. Used by permission of the publisher.

Sketches B and *C* after R. Blanchard and P. F. Boswell, "Limonite products derived from bornite and tetrahedrite," *Econ. Geology*, v. 25, Fig. 2, p. 560, 1930. Published by Economic Geology Publishing Company, 1930. Used by permission of the publisher.)

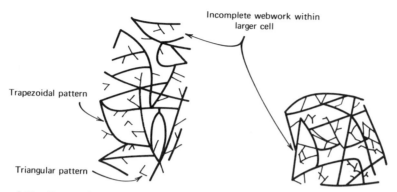

Figure 3-7. Boxwork derived from bornite, showing curved triangular cells and network, enlargement ×5. (After R. Blanchard and P. F. Boswell, "Limonite products derived from bornite and tetrahedrite," *Econ. Geology*, v. 25, Fig. 5, 1930. Published by Economic Geology Publishing Company. Used by permission of the publisher.)

involved in the formation of capping—and he showed that they could be enumerated.

Most of the techniques of capping evaluation were developed to estimate the original chalcopyrite-pyrite ratio and to estimate the thoroughness of the leaching that would drive early-formed chalcocite downward to an enriched supergene "blanket." Nowadays, with less need for chalcocite to make a mineable orebody, the interest has broadened to include all degrees of hypogene and supergene mineralization and nearly all ratios of chalcopyrite to pyrite. A brief summary of capping evaluation has been written by Anderson (1955), and well-described case histories of capping evaluation have been written for the porphyry copper deposits at Battle Mountain, Nevada (Sayers, Tippett, and Fields, 1968) and Ithaca Peak, Arizona (Eidel, Frost, and Clippinger, 1968). A compilation of capping and supergene enrichment

Figure 3-8. "Contour" boxwork derived from tetrahedrite, enlargement ×5. (After R. Blanchard and P. F. Boswell, "Limonite products derived from bornite and tetrahedrite," *Econ. Geology*, v. 25, Fig. 6, p. 571, 1930. Published by Economic Geology Publishing Company. Used by permission of the publisher.)

characteristics at 29 porphyry copper-type deposits is included in Sillitoe's (1976) summary report on the Mexican copper belt.

Dolomite yields a fluffy limonitic powder from contained iron impurities; this powder sometimes coalesces into an irregular sponge.

Fluorite is subject to both hypogene and supergene leaching, with silica boxworks showing the cubic crystal form or octahedral cleavage. Hypogene boxworks generally have later generations of fluorite, quartz, or galena in the voids. Supergene boxworks are composed of thin-walled siliceous jasper with open voids.

Galena is commonly found in leached outcrops as unoxidized cores in the larger fragments with shells of anglesite and cerussite. Where galena has disappeared, it leaves a thin-walled ocherous-orange (plumbo-jarosite) box-work with cubic or steplike voids representing the galena cleavage or with diamond-shaped voids after cerussite (Fig. 3-9). Early limonite, sintered crusts, and cellular sponge are sometimes found; they grade into each other and generally retain some suggestion of the cubic or diamond-shaped box-work.

Hematite and *magnetite* do not weather easily except in tropical regions and in aggregation with pyrite. The weathered product, ocherous goethite without any particular boxwork or sponge texture, is likely to mask all other forms and colors of gossan from associated minerals. The leaching of taconite to form hematite and goethite in "soft" iron ores is a slow and delicately balanced process accompanied by the removal of chert and other gangue minerals.

Manganese minerals form characteristic black and powdery products. They may also leave a light-brown fine cellular sponge.

Molybdenite forms earthy canary-yellow ferrimolybdite (molybdic ochre) or, in the presence of lead, yellow-orange wulfenite crystals. Ilsemannite, another product, has a striking blue color, but it is seldom seen except in thin

Figure 3-9. "Cleavage" boxwork and diamond-mesh boxwork derived from galena. The diamond-mesh pattern, developed from fine-grained "steel" galena, is suspected to have been derived from cerussite rather than directly from galena, enlargement ×3. (After R. Blanchard and P. F. Boswell, "Additional limonite types of galena and sphalerite derivation," *Econ. Geology*, v. 29. Fig. 6, p. 675, 1934. Published by Economic Geology Publishing Company. Used by permission of the publisher.)

Figure 3-10. Limonite "dice," North Star prospect, Peloncillo Range, New Mexico. The hard pseudomorphs remain in clusters after the matrix has disappeared.

and indistinct coatings. Powellite (fluorescent in ultraviolet light) has been reported in gossan. In the few instances where a boxwork is formed after molybdenite, it is tan, orange, or maroon and has rounded cells containing tiny curved flakes of limonite that have formed along the original cleavage.

Nickel sulfides form bluish-green annabergite in the presence of arsenic. Any boxwork is generally that of the commonly associated pyrrhotite.

Pyrite dissociates easily and yields sulfuric acid. It produces many forms of exotic limonite, especially columnar and clinkery varieties in various colors and sometimes with an iridescent sheen. It also forms indigenous cellular sponge and limonite "dice" (Fig. 3-10) where it is associated with reactive rock, such as limestone.

Pyrrhotite forms an irregular cellular sponge, sometimes crudely hexagonal, in which the cell walls are coated with shriveled limonite crusts (Fig. 3-11).

Quartz forms supergene silica as a component of limonitic jasper, and under desert conditions, it forms transported silcrete.

Siderite forms fluffy limonite or rhombohedral carbonate-walled boxworks.

Sphalerite and its oxidation products are very mobile. It contains enough

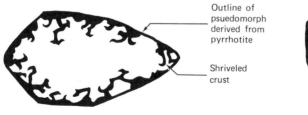

Outline of
psuedomorph
derived from
pyrrhotite

Shriveled
crust

Figure 3-11. Shriveled limonite crusts in boxwork voids, from the oxidation of pyrrhotite, enlargement ×15. (From R. Blanchard, *Interpretation of Leached Outcrops*, Fig. 26, p. 123, 1968, by permission Nevada Bureau of Mines and Geology.)

iron, however, to leave a coarse to fine cellular sponge or a fine angular "hieroglyphic" boxwork with small grains or rosettes in the voids (Fig. 3-12).

Uranium minerals are generally so mobile and form so many secondary minerals that only a few can be mentioned here. Luckily, many of them are fluorescent in ultraviolet light and, of course, are radioactive. Secondary uranium minerals are commonly yellow, orange, or red, and they occur as powdery blebs or as aggregates of needlelike platy crystals. Black secondary pitchblende is structureless, but it may be recognized by its association with yellow and green crusts of uranium sulfates and carbonates.

In attempting to relate oxidized minerals to their parent minerals, there is one very important fact to keep in mind: Leached outcrops are formed by mixtures of ore and gangue minerals which have reacted in more ways than we can conceive in the laboratory or show in diagrams. There are more bewildering mixtures of indigenous and transported limonite and more ambiguous boxworks than there are clear-cut occurrences. Leached outcrop evaluation is one of the most important jobs in exploration, and it calls for attention to detail. The time and effort are well spent because the next step, exploration at depth, will be expensive.

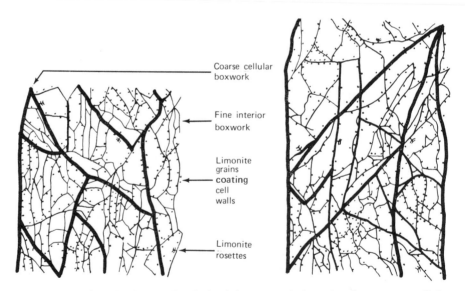

Coarse cellular boxwork

Fine interior boxwork

Limonite grains coating cell walls

Limonite rosettes

Figure 3-12. Limonite boxworks derived from sphalerite, showing coarse cellular boxwork, a finer interior boxwork, and grains and rosettes in the voids. (From R. Blanchard, *Interpretation of Leached Outcrops*, Fig. 66, p. 154, 1968, by permission Nevada Bureau of Mines and Geology.)

Depth of Weathering

Oxidation takes place in the near-surface zone, down to the locale of secondary sulfide enrichment, that is, down to the water table. This zone, the zone of vadose water, may extend to a depth of 20–30 m in temperate humid climates and to 200 m in deserts. Shvartsev (1972) reported oxidation depths as great as 200 m in permafrost terrain. There are exceptions, of course—localized weathering to a depth of 1,000 m in the Zambian copper belt (Mendelsohn, 1961), oxidation of sulfide minerals to a depth of 700 m at Tintic, Utah, and occurrences of oxidized copper minerals along faults at depths of nearly 1,000 m at Tsumeb, Namibia (southwestern Africa) (Park and MacDiarmid, 1975). As with so many other measurements and definitions cited in this chapter and the preceding one: *Which* water table? January or July? Holocene or Pleistocene? According to Blanchard (1968, p. 42), the most active oxidation of sulfide minerals takes place just above the water table (or just above a perched water table) where *both* water and oxygen are abundant. This critical zone is likely to sweep quite a range in depth in the course of time.

3.3. MINERAL DEPOSITS CONCENTRATED IN THE ZONE OF WEATHERING

We have considered the dispersive effects of weathering—dispersive and impoverishing as far as orebodies are concerned, at any rate. But sometimes the weathering process has given us an enrichment or "sweetening" of low-grade ore, and it may also have created orebodies from weakly dispersed minerals that would not in themselves amount to ore. Table 3-4 gives a few examples and references.

The most important among these deposits are the residually enriched iron ores and the lateritic aluminum and nickel ores. Prior to the development of methods for processing taconite ore, most of the world's iron ore came from residually enriched bodies of "direct shipping" or "soft" hematite and limonite containing 50–60 percent iron. Such ores continue to be of major importance, especially in regions where processing costs are high and transportation is expensive. Lateritic iron ores are of less importance now, but they include major resources for the future.

Bauxite is nearly the sole industrial aluminum ore; it is of widespread occurrence but is associated with areas that have had a tropical climate at some time in the Cenozoic. Kaolinite, rather than bauxite, will form unless weathering takes place in a moist, warm, essentially alkaline environment. Other conditions need for bauxite formation include well-drained topography that has remained above the permanent water table for a long period of time.

Table 3-4.   Examples of Economic Mineral Deposits Concentrated from Rock and Ore Exposed to the Surface Environment

A. Resistates, concentrated by the destruction of host rock and gangue, generally without the formation of newly economic minerals
    1.   "Lag" deposits, essentially in place
        Barite; residual fragments and boulders in clay. Washington County, Missouri (Brobst and Wagner, 1967)
        Gold; grains, crystals, and veinlets in the oxidized zone of vein deposits. Worldwide, especially in the "free-milling" gold operations of the historic United States frontier (Koschman and Bergendahl, 1968)
        Ilmenite, from weathered granite and mafic plutons. "Residual iron sands" of Japan (Tsusue and Ishihara, 1975)
    2.   Elluvial deposits, in debris slopes (Wolf, 1976, v. 3)
        Beryl; crystals from weathered pegmatites, Ankola district, Uganda
        Cassiterite, Malaya
        Chromite, New Caledonia
        Diamonds, Botswana
        Gold, worldwide
        Platinum, Ural Mountains, USSR
        Scheelite; boulders in residual soil, Atolia-Randsburg district, California
        Wolframite, Kiangsi Province, China
    3.   Alluvial placer deposits (Wolf, 1976, v. 3)
        Cassiterite; Cornwall, England
        Chromite; Rhodesia (Stanley, 1961)
        Gold; worldwide (Koschman and Bergendahl, 1968)
        Magnetite; ilmenite; Malaysia
        Columbite; monazite; ilmenite; Bear Valley, Idaho (Mackin and Schmidt, 1956)
        Garnet; Emerald Creek, Idaho
        Wolframite; Kiangsi Province, China
        Phosphate rock; "river pebble" deposits of Florida
        Platinum; Ural Mountains, USSR
        Rutile; West Africa
    4.   Marine and beach placer deposits
        Gold; Nome, Alaska
        Platinum; Goodnews Bay, Alaska (Stumpel, 1974)
        Chromite; Oregon (Griggs, 1945)
        Rare earth minerals; monazite; rutile; zircon; New South Wales, Australia
        Ilmenite; rutile; zircon; Florida (Pirkle and Yoho, 1970)
        Cassiterite; Thailand (Fox, 1967, 1969)
        Diamonds; Namibia
        Magnetite; Ariake Bay, Japan

Table 3-4. (con'd)   Examples of Economic Mineral Deposits Concentrated from
Rock and Ore Exposed to the Surface Environment

B. Residually Enriched Deposits, concentrated from rock by leaching of impurities
and the formation of stable ore minerals
1. Laterites, widely occurring in tropical areas
Bauxite; mixture of hydrous aluminum oxides associated with clay minerals
and limonitic iron. Source rocks: syenites, argillaceous carbonate rocks,
basalt, andesitic volcanic ash (Shaffer, 1975)
Lateritic iron; limonite pellets associated with clay. Source rocks: mafic
igneous and metamorphic rocks (United Nations, 1970)
Lateritic nickel, with *cobalt, chromium,* and *manganese* content, in "garnierite" and mixtures of ferrous hydrosilicates associated with lateritic
aluminum and iron oxides. Source rocks: serpentinized periodotite
(Webber, 1972)
Lateritic manganese; psilomelane, manganite, and polianite with lateritic
iron and aluminum oxides. Source rocks: manganese-bearing schist
("gondite") and mafic rocks (Baud, 1956)
2. Enriched iron and manganese ores, occurring in temperate as well as tropical
areas
"Soft" iron ore; goethite and hematite from leaching of taconite and other
cherty iron beds (Berge, 1971)
Oxidized siderite; goethite and hematite from leaching of siderite bodies.
Bilbao, Spain (United Nations, 1970)
Residual manganese ore; formed from nearly all primary manganese ores
(Prinz, 1967)
Residual cobalt deposits; associated with certain low-grade residual manganese ores. Southern Appalachians (Pierce, 1944)
3. Enriched clay and phosphate deposits
Kaolin; widespread (Vachtl, 1968)
Phosphate; "brown rock" and "white rock" of Tennessee (Burwell, 1950),
enriched western United States phosphorite (Hale, 1967), and guano
deposits of western Pacific Islands (British Sulphur Corp., 1964)
C. Supergene Orebodies and Enriched Zones, with new minerals formed from
dissolved material
1. Supergene sulfide deposits formed in reducing environment by leaching of
primary ore
Copper; chalcocite and covellite, especially in the "chalcocite blanket" of
porphyry copper deposits (Titley and Hicks, 1966)
Nickel; violarite (FeNi$_2$S$_4$) with pyrite and primary nickel minerals. Mt.
Windarra, Western Australia (Watmuff, 1974)
Silver; acanthite (low-temperature argentite) and possibly pyrargyrite and
proustite. Tonopah, Nevada (Nolan, 1935)
Zinc; wurtzite and light-colored sphalerite. Widely occurring but not in
major concentrations
Lead; secondary galena. Accompanies residual galena in the Tynagh
"mud" ore, Ireland (Morrissey and Whitehead, 1969)

Table 3-4. (con'd)   Examples of Economic Mineral Deposits Concentrated from
Rock and Ore Exposed to the Surface Environment

Mercury; supergene cinnabar. Uncommon in major concentrations. New
Idria, California

Cadmium; greenockite and hawleyite, rare except in local concentrations in
the Tynagh "mud" ore

2.   Supergene deposits formed in the zone of oxidation

Copper; carbonates, oxides, halides, silicates, and native copper, especially
in desert regions. Ray, Arizona (Phillips, Cornwall, and Rubin, 1971)

Silver; cerargyrite (horn silver) and native silver. Tonopah, Nevada (Nolan,
1935)

Lead; anglesite and cerussite. Tintic, Utah (Cook, 1957)

Zinc; smithsonite and calamine. Leadville, Colorado (Heyl, 1964)

Uranium; carnotite and tyuyamunite where associated with vanadium, as at
Uravan, Colorado (Motica, 1968). Uraninite and coffinite in "roll-front"
orebodies at interface with reducing environment, as in the Texas coastal
region (Fischer, 1974). Autunite in leached uraninite veins (Barbier, 1974)

D.   Recent Evaporites and Associated Brines

1.   Lake beds

Sodium carbonate; Lake Magadi, Kenya (Eugster, 1970)

Sodium sulfate, sodium chloride, magnesium, and *lithium* compounds;
Great Salt Lake, Utah

Borates; playa lakes in California

Lithium brines; Silver Peak, Nevada (Kunasz, 1970; Vine, 1975)

Lateritic nickel ore represents about 75 percent of known nickel reserves, and lateritic nickel orebodies are being mined. Nickel resources in manganese nodules occur on the sea floor, and a large additional tonnage may be developed and mined in the near future, but the eventual mining of low-grade lateritic nickel deposits is more certain.

Metallurgical recovery from lateritic nickel ore is difficult because the absence of definite nickel-bearing minerals in most deposits makes it virtually impossible to upgrade them by physical means. Direct hydrometallurgical processing is used on raw ore in the lateritic nickel operations of Cuba and New Caledonia, and some of the highest grade material (garnierite) in New Caledonia and Oregon is treated by pyrometallurgy. Cobalt and chromium occur with lateritic nickel, and there is an overlying iron crust; cobalt is recovered, but the metallurgical recovery of chromium has not yet been worked out. Figure 3-13, an idealized section through a lateritic nickel orebody, shows the typical zoning with depth and—to amplify the point about metallurgy—a proposal for two types of treatment. The thickness of the lateritic zone above unaltered periodotite: about 3–50 m.

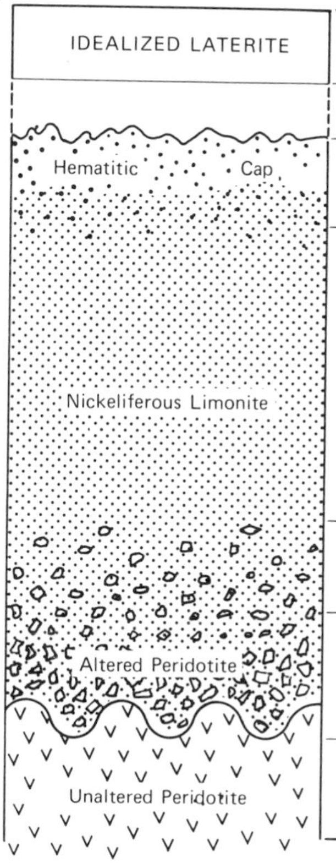

IDEALIZED LATERITE	APPROXIMATE ANALYSIS—%					EXTRACTIVE PROCEDURE
	Ni	Co	Fe	Cr_2O_3	MgO	
Hematitic · Cap	<0,8	<0,1	>50	>1	<0,5	Overburden to Stockpile
Nickeliferous Limonite	0,8 to 1,5	0,1 to 0,2	40 to 50	2 to 5	0,5 to 5	Hydrometallurgy
Altered Peridotite	1,5 to 1,8	0,02	25 to 40	1	5 to 15	Hydrometallurgy or Pyrometallurgy
	1,8 to 3	to 0,1	10 to 25	to 2	15 to 35	Pytometallurgy
Unaltered Peridotite	0,25	0,01 to 0,02	5	0,2 to 1	35 to 45	Left *in situ*

Figure 3-13. Idealized section through a lateritic nickel orebody, indicating the composition as a function of depth and indicating the proposed type of treatment. (From H. J. Roorda and P. E. Queneau, "Recovery of nickel and cobalt from limonites by aqueous chlorination in sea water," *Transactions,* v. 82, Fig. 1, p. C79, 1973. Copyright © 1974, The Institution of Mining and Metallurgy. Used by permission of the publisher.)

   Placer deposits, alluvial, beach type, and marine, are widespread sources of cassiterite, platinum, rutile, and rare earth minerals. Gold and diamonds, the historic placer minerals, continue to be of importance in certain areas. Artisanal diamond mining is a major factor in the economy of several African nations. It is estimated that 50 percent of Russia's total gold production has come from placers, and alluvial deposits have long been favorite topics among Soviet geologists because of the famous Lena, Yenisey, and Amur gold fields in Siberia. In an English-language article "Geologic Features of Alluvial Placers," Kartashov (1971) discusses the Russian designation of autochthonous ("in place"—generally the richer and larger) and allochthonous (re-worked) placers. Kartashov places special emphasis on zones of clay in river gravels, zones which may act in the same way as actual bedrock. These zones are called "false bedrock" by American placer miners. The effect of "false bedrock" is to collect the gold at a certain level, giving a misleading high-grade sample that should not be applied to the entire gravel body.

The proving ground for American placer geology was, as would be expected, California, and the classic book was Lindgren's (1911) professional paper on the Tertiary gravels of the Sierra Nevada in which he related types of placer deposits to present and past geomorphic cycles. Still good reading. More recent books on alluvial placers in general are those by Griffith (1960) and Wolf (1976, Vol. 3).

Placer deposits are so directly controlled by the geomorphic environment—premineral, contemporaneous, and postmineral—that the specific processes of alluvial and shoreline development must be understood before any further study can be made. For alluvial geology, the recommended reading is *Fluvial Processes in Geomorphology* by Leopold and others (1964); for shoreline geology, *Beaches and Coasts* by King (1959).

Economic placer deposits are formed and preserved by geomorphic reactions to stress. The stress of flood conditions in a stream is met by rearrangement of the bed load and a downward migration of heavy minerals toward bedrock. The stress of regional tectonism is met by stream and shoreline processes that isolate terrace deposits from further erosion or protect channel deposits beneath a new generation of material.

The concept of orebody heritage has an obvious expression in placer deposits: The ore minerals are derived from an earlier petrologic association. The concept can be applied in additional ways. Placer deposits may reach economic levels of concentration only after several generations of earlier and weaker deposits have been progressively enriched by erosional processes. More than one kind of erosional process may be involved; gold in glacial deposits, for example, is not ordinarily of economic value except where the deposits have been sorted by postglacial streams.

The supergene copper sulfide orebodies, also with strong geomorphic controls that require thorough understanding, are of importance in many parts of the world. They are particularly associated with regions that have had a semiarid climate during relatively recent geologic time and have escaped rapid fluvial or glacial erosion. A "blanket" of supergene sulfide mineralization in porphyry copper deposits may be the orebody or it will at least be an attractive zone in which to begin mining. The principles of secondary sulfide enrichment in copper deposits are explained and described in *Geology of the Porphyry Copper Deposits, Southwestern North America,* edited by Titley and Hicks (1966). An interesting illustration of chalcocite enrichment patterns can be found in the description of the La Caridad porphyry copper deposit,

Mexico, by Saegart, Sell, and Kilpatrick (1974). Figure 3-14 is a cross section through La Caridad showing the inverted saucerlike body of enriched ore with a copper content of about 3.5 times that of the underlying hypogene mineralization. The enriched body has an average thickness of 90 m and a maximum thickness of at least 250 m. Most of the enrichment at La Caridad took place in mid-Tertiary time—prior to the deposition of a local conglomerate that contains clasts of leached chalcocite-type capping. The conglomerate is interbedded with rhyolite flows dated at 24 m.y. The present-day leached capping was derived from chalcocite during subsequent erosion cycles, and the chalcocite "blanket" was, in effect, turned down at its margins by further leaching and reprecipitation.

Lateral migration of the enriched ore has taken place at La Caridad into areas not marked by surface evidence of indigenous limonite—a good point to remember when evaluating capping. In most situations, lateral migration of secondary copper sulfide minerals is not so pronounced as at La Caridad. The far travelers are more likely to be oxidized minerals. There are many instances of oxidized copper mineralization occurring a kilometer or more

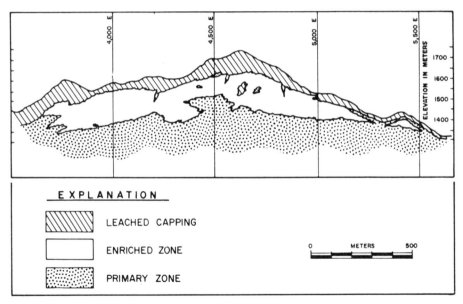

Figure 3-14. Section through La Caridad, showing leached capping, zone of ore-grade chalcocite enrichment, and underlying primary mineralization. (From W. E. Saegart, J. D. Sell, and B. E. Kilpatrick, "Geology and mineralization of La Caridad porphyry copper deposit, Sonora, Mexico," *Econ. Geology,* v. 69, Fig. 5, p. 1070, 1974. Published by Economic Geology Publishing Company. Used by permission of the publisher.)

from their assumed primary source; the Exotica oxidized copper orebody in alluvial gravels at Chuquicamata, Chile, for example, is located 4 km from the exposed sulfide orebody. Note the qualifying words "assumed" and "exposed"; Exotica and other oxidized orebodies of this kind may just as well have migrated vertically from a parent body that has been removed by erosion.

Until recently, supergene sulfide enrichment of nickel and lead orebodies was not considered to be important. Supergene violarite concentrations at Mount Windarra, Western Australia, and supergene galena at Tynagh, Ireland, have now given us some new dimensions to apply wherever geological analogs can be found.

Supergene silver concentrations, both as sulfide and as oxidized minerals, were the "bonanza" ores of mining history. Cerargyrite is the "horn silver" of fabulously rich veins discovered in the early days of the American West.

Uranium minerals are so sensitive to environmental change that it is often difficult to relate a particular orebody to a single present or past geomorphic cycle. Many of the orebodies may have been formed by leaching related to older erosional surfaces, as suggested by Derry (1973) and Barbier (1974). Roll-type uranium deposits (Figure 3-15) may still be mobile, or they may

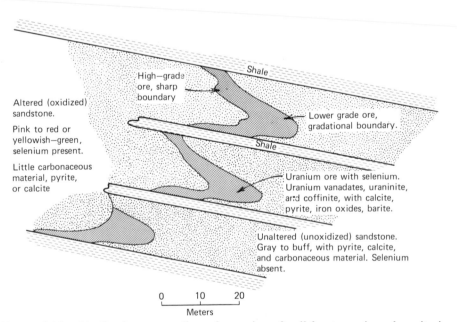

Figure 3-15. Idealized cross section of a series of roll-front uranium deposits in sandstone with discontinuous shale zones, as found in the Colorado Plateau region, Wyoming, and the Texas Gulf Coast.

have resulted from Tertiary weathering processes that have not continued to the present time, as in the Gas Hills and Shirley Basin, Wyoming (Dooley, Harshman, and Rosholt, 1974).

3.4. GEOMORPHIC EXPRESSION AND QUATERNARY COVER

Geomorphology, formerly a deductive science centering about grand evolutionary models, has become increasingly inductive, quantitative, and concerned with tangible problems. As a result, geomorphology is now, more than ever before, one of the exploration sciences. It uses the tools of reconnaissance and provides a framework for reconnaissance; photogeology, radar imagery, and infrared imagery are beginning to provide a wealth of outcrop information and geomorphic detail in jungle, desert, and arctic regions where only fringes and isolated fragments had been previously mapped. The association goes further; interpretation of geochemical data is, in large part, associated with the interpretation of minor landforms, and the recognition of geophysical signals demands an accounting for terrain "noise."

Geomorphic Mapping

Geomorphic mapping, a relatively young branch of geomorphology, is more widely used in Europe and the USSR than in North America. In this kind of mapping, landforms are characterized in great detail with symbols and colors for slope characteristics and for erosional and depositional features. Bedrock geology is not shown in detail, which may be an advantage in that geologists can draw their own interpretations from precise geomorphic observations rather than from someone's prior geologic projections and interpretations. The scale of "detailed" geomorphic maps is generally 1:25,000–1:50,000, which corresponds to the needs of reconnaissance exploration and the magnitude of geologic features having district or regional importance. Some geomorphic maps for engineering site investigations (Fig. 3-16) at scales of 1:5000 and 1:10,000 are more in the orebody and mine planning range. As the techniques of geomorphic mapping develop, they should find increased use in regional exploration, especially as an adjunct to photogeology. In an English-language manual of geomorphic (geomorphological) mapping published in Prague for the International Geographical Union (Demek, 1972), there is a chapter describing applications to mineral prospecting and a detailed chart relating types of mineral deposits to critical landforms and mapping objectives.

In Chapter 2, the point was stressed that a mining geologist must be a structural geologist (as well as somewhat of a mineralogist, petrologist, and stratigrapher). Now with the discussion of weathering, supergene processes,

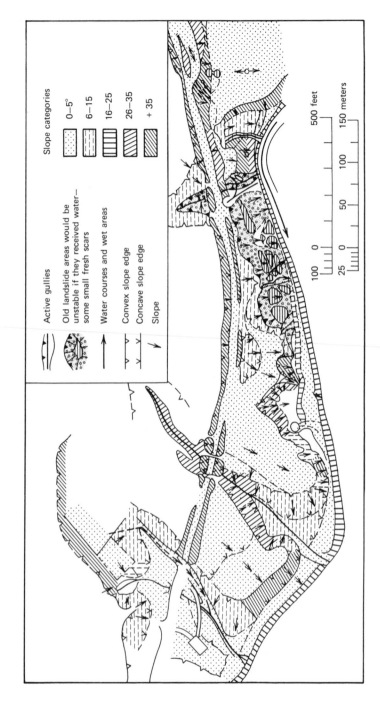

Figure 3-16.   Large-scale geomorphological map of a proposed road alignment in Wales. The lower boundary of the map is a riverside. The narrow zone with steep embankments is an abandoned railway. [From D. Brundsen and others, "Large scale geomorphological mapping and highway engineering design": *Quart.Jour.Eng.Geology*, v. 8, Fig. 10, p. 248, 1975. Copyright © 1975, Geological Society (London). Used by permission of the publisher.]

and topography add "geomorphologist." Without a good understanding of geomorphology, a geologist doing reconnaissance for mineral deposits would have nothing better to begin with than a blind statistical grid. The prime reference book for exploration libraries is *Encyclopedia of Geomorphology*, edited by Fairbridge (1968). To keep up with new developments in geomorphology the journal *Earth Surface Processes* (John Wiley & Sons, Ltd.) is recommended.

Topography

Orebodies commonly have a topographic expression, with siliceous gangue material and siliceous alteration zones forming prominences and with leached sulfide zones forming depressions. Breccia pipes, common features of porphyry copper mineralization, may stand out like "red thumbs" in the terrain (Fig. 3-17). The terms "hill" and "mount" are widespread in the names of

Figure 3-17. An isolated prominence at the exposure of a silicified and mineralized breccia pipe, Cerro Creston, Sonora, Mexico.

mining districts, and the terms often refer to the original discovery rather than to a nearby reference feature. Broken Hill and Mount Isa are examples, and if we translate hill, there are such names as Cerro Bolivar.

The outcrop of the huge iron orebody at Cerro Bolivar, Venezuela, was first noted on aerial photographs as two hills rising above the savanna. Ertsberg (ore mountain), Indonesia, one of the world's great copper mines, was found as a magnificent outcropping knob of hypogene copper mineralization in calc-silicate magnetite skarn.

Siliceous outcrops of veins or bedded ore sometimes extend for considerable distance as narrow ridges. In the Montagne Noire region, France, a thin bed of siliceous dolomite with galena mineralization can be traced for more than 10 km as a prominence bordered by less resistant dolomites and shales (Routhier, 1963, p. 1062).

Even though the most striking topographic expression of ore mineralization is a hill or a ridge, the actual orebodies may have an associated but less direct expression. At Oatman, Arizona, for example, gold-quartz veins are as much as 15 m wide, and some of them crop out conspicuously; but the largest and richest orebodies have been found in the parts of veins that are not so prominent at the surface. The explanation: Gold ore shoots occur where the gangue is composed of calcite as well as quartz. At Goldfield, Nevada, rust-colored craggy quartz ledges are formed from a siliceous alteration zone *alongside* the veins.

In many places, orebodies are expressed as topographic basins and saddles without the benefit of adjacent siliceous zones. It is the linear or arcuate pattern of aligned depressions that is often significant—something that may be overlooked if there are too many landforms of a similar type that are governed by other petrologic contrasts. And, the more complex the drainage pattern, the less likely that an alignment of abrupt bends or branchings will be noticed. The term "noise" can be applied to geomorphology.

Significant topographic depressions are likely to be filled with soil or recent sediments, thus masking their presence. However, the material filling a depression may give some clue that an out-of-the-ordinary erosional feature (a geomorphic anomaly) exists. Patterns of local saline crusts, siliceous and calcareous duricrust, and laterite may be apparent even though the original depressions are not. Harrison Schmitt (1939) has described this kind of occurrence in the San Juan district, Chihuahua, where weathered ore veins and mantos were found by breaking through ribbonlike outcrops of caliche (calcareous duricrust).

To carry the point regarding filled depressions a step further: ridges of caliche that formed in an older and perhaps more significant drainage pattern sometimes stand out in bas-relief in desert areas (Chapman, 1974). The older drainage system with its linear features and geochemical characteristics might have a closer relationship to immediate postmineralization conditions than a newer system formed in less consolidated sediments.

If oxidation of sulfide ore has been particularly active, collapse depressions may occur in the surface, as at Bisbee, Arizona, where subsidence troughs with marginal cracks indicate the presence of oxidized ore at depths as great as 200 m (Wisser, 1927). At Shingle Canyon, near Hanover, New Mexico, slumped blocks of ore-bearing limestone reflect the oxidation and leaching of an underlying higher grade zone of sphalerite mineralization (Fig. 3-18).

Coal beds that have burned in the zone of oxidation are often reflected in a surficial zone of slumping. Where coal is overlain by shale, the overlying rock may adjust by flexure so that the thickness of the original seam is not apparent, but where a seam is overlain by massive sandstone, as in some of the Cretaceous and Tertiary fields of the western United States, the thickness of the burned coal is reflected in the relative displacement of slump blocks.

Postmineralization Cover

With the increasing interest in geochemical and geophysical methods of prospecting for hidden orebodies, the relationship of unconsolidated postore sediments to underlying landforms has become an important topic. "Ghost" structural features in drainage patterns may be inherited from deeper topography; low-amplitude relief features of the unconsolidated cover itself will sometimes reflect differential compaction above an older surface.

The Tertiary and Quaternary conglomerates that cover a large part of the southwestern United States Basin and Range province have come under special scrutiny. In a direct search for indications of underlying ore, the conglomerates—many of them old fanglomerates that contain near-source material—are combed for pebbles and cobbles of capping or gossan. But the search goes further and into more subtle indications. The Locomotive fanglomerate at Ajo, Arizona, contains some unique volcanic clasts that are thought to represent a very localized, late magmatic stage flow associated with copper mineralization, and there are a few small hematitic silver-lead

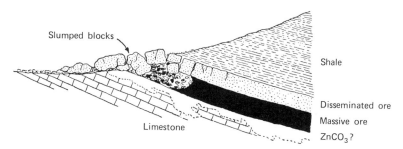

Figure 3-18.   Oxidation subsidence at Shingle Canyon, New Mexico. (From W. C. Kelley, *Topical study of lead-zinc gossans,* New Mexico Bureau of Mines and Minerals Resources Bulletin 46, Fig. 7, p. 35, 1958. Used by permission of the New Mexico Bureau of Mines and Mineral Resources.)

veins in the fanglomerate that may represent a dying stage of mineralization. Other fanglomerates in the vicinity of other porphyry copper mines have also been found to have localized and uncommon volcanic clasts. If the unique volcanic rocks can be related to mineralization in a particular area, then Locomotive-type fanglomerates may provide a reconnaissance guide to ore (Lukanuski, 1975). The guide can be extended to other situations; as pointed out by Gilmour (1971), specific kinds of volcanic rocks are associated with specific types of massive sulfide deposits. Pebbles and boulders of the diagnostic volcanic rocks can serve as guides to favorable conditions beneath postmineralization cover.

The tracing of ore boulders has been practiced in Scandinavia for two centuries, but as a long-distance trail to ore rather than as a direct guide to underlying mineralization. Glacial overburden and glacial soils were long considered too heterogeneous for detailed prospecting and too unrelated to the immediate area. A combination of geochemical and geomorphologic research in recent years, however, has placed glaciated terrain in a more favorable light. Secondary metal dispersion patterns are recognizable in glacial cover and may even be accentuated in organic-rich glacial lake sediments and in ground water (Fortescue, 1972). Methods have also been developed for geochemical prospecting in the permafrost terrain that occupies large areas in Alaska, Canada, and the Soviet Union. These methods combine geochemistry, geomorphology, and the knowledge that cryolithic process will promote the formation of diagnostic sulfate minerals and characteristic water dispersion trains (Shvartsev, 1972).

Certain types of Cenozoic postore cover provide effective screens against the recognition of buried orebodies. As mentioned in reference to "false gossan," hard, iron-rich lateritic cuirasse obscures most leached outcrops of ore and it has notorious "noise" effects in geophysical and geochemical prospecting. Deeply leached lateritic soils, even without cuirasse, have lost nearly all of their identification with the parent rock and its mineralization. Alluvial cover is not normally related to the immediate area, but it can be penetrated by geophysical methods and may contain secondary geochemical patterns formed by ground water. Even so, the geophysical and geochemical advantages are often diminished by the presence of duricrust (caliche), organic material, and chemically active clays, as will be discussed in Chapters 13 and 14.

Assume that we now have a picture of our objective—a conceptual model—and we know how the model should appear in surface exposure. We still have to know *why* this particular combination of geologic circumstances should make an orebody, and we have to know which additional combinations and variations to take into account in our field work. This calls for some of the theorizing that will be discussed in Chapter 4.

CHAPTER 4
METALLOGENIC PROCESSES AND PROVINCES

We have been dealing in empirical models: "This is the association, now go look for more of the same." But imitation is restrictive—and rather dull. The only place to find another Broken Hill orebody is *at* Broken Hill, unless we can put together a flexible model with enough inner circuitry to work somewhere else—perhaps in Mesozoic rocks, in glaciated terrain, or in Africa.

It will have to be a dynamic model, because geologic data, like engineering measurements, are always on the move. Engineers recognize the hazard. An expert in metallurgical plant design once considered the heterogeneity of ore coming to an overworked mill from different parts of a mine and said, "We could lick this problem if it would only *stand still* for a couple of hours."

The porphyry copper model of 20 years ago did not have a barren core at depth; we would have called drill core from the deep center of a deposit exactly like the Bingham Canyon deposit "discouraging." If we had been in Australia and looking for something analogous to Mount Isa, while it was still a perfectly good example of a hydrothermal vein deposit, we would have paid little attention to the Mount Isa shale. We cannot even depend on *rocks* to stand still; mineralization at Rio Tinto and throughout the Iberian pyrite belt is associated with rhyolitic volcanic rocks that were formerly identified by most geologists as intrusive. Once the rock classification changed, the conceptual model had to change.

This is all very stimulating, almost to the limit of endurance, and it has reopened a great many areas that were comfortably filled under "reject" on the basis of simpler models. As mentioned in Chapter 2, ore-forming processes must now be accepted as part of a broader earth drama in which magmatism, sedimentation, diagenesis, metamorphism, and ground-water circulation all play some role. We have to decide where the scenes go. The Homestake orebody was not part of the Tertiary finale; it was part of the opening act in the Precambrian.

Opposition to the conventional hydrothermal ore deposit model came in large part from field geologists. Why were so many areas with all of the favorable structural controls and pluton associations still barren of known mineral deposits after so much investigation? Why did theorists have to invoke fluids from deeply buried plutons to explain ore deposits that followed sedimentary units for several kilometers without displaying any "roots" or "plumbing" from depth? In order to get on with their work, many exploration geologists adopted a policy of "shoot first and then ask questions." Mine geologists, the people who had detailed knowledge of large ore deposits, had doubts about the ruling theory of ore genesis; in the Witwatersrand gold field they found additional orebodies on the basis of placer geometry in spite of all the literature advocating a hydrothermal origin. Resident geologists in the Zambian copper mines based their successful ore search on sedimentary guides, while libraries accumulated articles on a Zambian hydrothermal model. At White Pine, Michigan, mine geologists recognized that the characteristics of the orebody could be squeezed into several models, and they asked visiting colleagues, "Do you want to take the syngenetic or the epigenetic mine tour?"

There were disturbing views from the laboratory as well. Once some of the difficulties with improper sampling were ironed out, isotopic age dating showed that many deposits were of an entirely different age than their associated plutons. Other isotope studies served to point out some additional differences between orebodies and their supposed source rocks. Sulfur isotope ratios in ore directly associated with intrusive rock were relatively uniform. Fine. However, in samples of ore from bedded zones in sedimentary rock, the situation was often just the opposite: there was a wide range in the values and ratios—evidence that isotopic fractionation (segregation) must have taken place during deposition. In some cases, the dissimilarities in sulfur isotope ratios were greater between adjacent beds than with distance along a single bedding. Puzzling, especially if ore fluids were supposed to have spread into the entire system from a certain crosscutting conduit. Ore in metamorphic rocks, in volcanic rocks, and in veins of indirect igneous association had complex sulfur isotope ratios; disappointing, but at least it suggested that single-stage magmatic-hydrothermal processes had not done the entire job by themselves.

Then there were the long-known but unappreciated observations that sulfate-reducing bacteria are H_2S generators, that organic sediments can accumulate metal compounds, and that metals are mobile in bacteria-generated humic acid. Had biochemistry and paleoecology taken a back seat to high-pressure and high-temperature chemistry for too long? Laboratory data said yes. "Light" sulfur and carbon isotopes in sulfides and carbonates could well be bacteriogenic, and there were microscopic ore mineral textures that could hardly be called anything other than biogenetic.

The granitization controversy reached an interesting stage while geologists were searching for alternatives to the magmatic-hydrothermal theory. The timing was right for an integration of the two problems, and the integration was provided by Sullivan (1948) with this thought that ore minerals and rock-forming minerals could both have been flushed out of a metal-rich sedimentary-volcanic sequence by granitization. The "source bed concept" expressed by Knight (1957, 1958, 1959) was a related but more comprehensive proposition in which syngenetic sulfide accumulations in sedimentary rocks were thought to have been mobilized and carried by metamorphic fluids to newer sites of deposition.

A trend had been set. Emphasis shifted from the cooling history of deep-lying silicate melts to the sedimentary and metamorphic history of paleo-basins. The deposits at Broken Hill, Mount Isa, and Rammelsberg now had a good reason for their strong associations with favorable strata: The ore belonged there in the first place. Two prime locations for testing syngenetic ideas (the ideas were not really new in either location) were in the Zambian copper belt and in the ore-bearing stratigraphic unit that extended from the Kupferschiefer in Poland to the Marl slate beds in northern England; these areas were mapped and investigated in more detail than ever before. Dunham (1964) took new evidence from both areas into account in his now-famous lecture on neptunist concepts in ore genesis delivered before the Society of Economic Geologists. He brought Werner's eighteenth-century idea back to life but with a completely new array of processes involving submarine hydrothermal fluids, lagoonal sedimentation, and remobilization. Dunham's view has since been improved upon in detail but not in overall clarity.

Submarine volcanism had to fit somewhere. It was a part of the neptunist model, and it became a dominant environmental factor in explaining the Kuroko ores of Japan and the pyritic ores of Cyprus, where crosscutting epigenetic roots in subvolcanic rocks and syngenetic stratiform orebodies in later sedimentary and volcanic rocks made a believable circuit-and-exit picture. The similarity between these relatively young orebodies and Precambrian massive sulfide orebodies intrigued field geologists who had been working in the shield areas, and the connection between oceanic crustal rocks and metals intrigued research geologists who had been working with the plate tectonics theory.

When the 50-year-old idea of continental drift became the plate tectonics concept in the 1960s, it nearly submerged a great many of the previous notions of the geosynclinal cycle. Plate tectonics had an effect, mostly beneficial, on the notion of a eugeosyncline at the outer edge of a craton and on the sedimentary-volcanic model for ore deposits. Evidence of ore mineralization along subducting (sinking) plate boundaries allowed some sweeping generalizations to be made in regard to the ultimate source of metal concentrations in oceanic material and the stage-by-stage formation of mineral deposits by melting and mobilization.

In a brief review of plate tectonics and related "commotion in the ocean," Chase and others (1975) made the point that geologic theories can never again neglect the two-thirds of the earth that lie outside of dry land. The same can be said for theories of ore genesis, especially since the discovery of metallogenetic processes in action at oceanic spreading centers in the Red Sea, along the Mid-Atlantic ridge, and elsewhere. Syntheses have been offered for the relationship between ore and plate tectonics in journal articles by Sillitoe (1972), Mitchell and Garson (1976), and Sawkins (1972). For more detailed study, entire books on the subject edited by Petrascheck (1973a) and Strong (1976) are recommended. Most of the explanations have the basic elements shown in Figure 4-1. The geologic elements are consuming plate margins where oceanic crust and continental crust are swallowed, accreting margins where new crust is being formed, and transform margins or "faults" that offset some of the consuming and accreting plate margins. The general idea is that ore deposits are formed as a result of plates colliding or separating, with the most favorable sites involving collisions of oceanic crust with continents

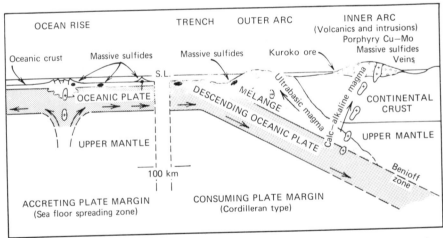

Figure 4-1. Plate tectonic setting of base-metal deposits. (From data by Sillitoe, 1972; Mitchell and Garson, 1976; and Sawkins, 1972.)

or with island arcs. The least favorable collision site for generating ore is thought to be of the continent-continent type where new oceanic material is not involved. Some major ore deposits in the interior of continental plates are thought to be related to hot spots (mantle plume activity) at places of incipient plate separation.

Some of the plate tectonics–ore genesis relationships appear to be quite distinct, as with an envisioned subduction of a metal-rich oceanic plate boundary beneath the Andean and North American Cordilleran provinces. In the Alpine orogen from central Europe to the Himalayas, on the other hand, an ancient sea (Tethys) contained so many continental fragments and island arcs that several alternative explanations have had to be considered for the pattern of mineral deposits. Explanations by Petrascheck (1973b, 1976), Dixon and Pereira (1974), and Evans (1975) make especially good reading; they deal with mining's classic ground in Europe and mining geology's newest ideas.

As with many theories, a good fit from one point of view is a strained or even ludicrous fit from another; Lowell (1974) has pointed out that specific belts of copper mineralization in Arizona and molybdenum mineralization in Colorado can be related to events ranging in age from Precambrian to Tertiary and that such a long-term association would make any connection with metals arising from a moving oceanic plate boundary difficult to believe.

Another aspect of conceptual models, the controlling paleogeographic association of stratabound lead-zinc-fluorite deposits, came back into contention from a nearly discarded early twentieth-century idea. Opposing views were held by European geologists familiar with Alpine-type (syngenetic?) ore and American geologists familiar with Mississippi Valley (epigenetic?) ore. A collection of symposium papers on these stratabound deposits (Brown, 1967, 1970) continues to be the most often quoted reference in arguments about lead-zinc mineralization in limestone.

The stratabound view of orebodies is only one of several views, but it is by far the most productive of new literature. The most imposing treatise, a seven-volume work edited by Wolf (1976), devotes nearly 3000 pages to the geologic characteristics of stratabound and stratiform mineral deposits.

The revival of Werner's old-school neptunist ideas has provided a special place for one of the most profound views of his eighteenth-century opponent, James Hutton: That the present is the key to the past. Most of the current theories of ore genesis are supported by some aspect of present-day geologic processes involving volcanism, ground-water geochemistry, and sedimentation. One particular observation, that metals accumulate in the algal mats of coastal sebkhas—evaporite flats that form along the margins of regressive seas—has become part of an important model for sedimentary ore deposits (Renfro, 1974). The most immediate step from the sebkha model to a field exploration program is the same as that needed for almost any search for stratabound ore deposits: paleogeographic study.

In clawing through the available syngenetic and epigenetic guidelines for exploration, one is led to agree with Stanton (1972, p. 34) "that *perhaps all of these theories are correct*: that no single idea has been universally established simply because there is no universal—or even dominating—mechanism of ore formation." There is little comfort in this view of systematizers, but there is no end of stimulation for exploration geologists who must look for opportunity in every new bit of data: "those barren quarzites between Alamo and Red Hill—if they could be metavolcanics rather than quartzites, we had better look again—maybe they are not so barren after all." Try a new model if you can get one to stand still for long enough.

4.1. THE GEOCHEMICAL BASE

The information in Table 4-1 is fairly well known to physical scientists and even to social scientists. It shows with distressing simplicity that we have everything in the earth's crust that we need. Hence the question: "With all of *this* at hand, what is so difficult about finding an orebody?" An answer, if we ever find one, will have to relate to barren rock as well as to ore rocks (ore petrology) and to the physical and chemical forces that result in limestones and gabbros as well as tin deposits. Economic mineral deposits—those with an acceptable grade—are rare, but the processes that form mineral deposits are common processes.

The geochemical environment of rock and ore deposition is the sum of all

Table 4-1.   Some Metals in the United States Portion of the Earth's Continental Crust.[a]

Element	Abundance (g/mt)	Amount, to Depth of 1 km (mt × 10⁹)
Aluminum	83,000	2,000,000
Iron	48,000	1,200,000
Manganese	1,000	24,900
Zinc	81	2,000
Chromium	77	1,920
Nickel	61	1,500
Copper	50	1,230
Lead	13	330
Uranium	2.2	55
Tin	1.6	38
Molybdenum	1.1	27
Silver	.065	1.6
Gold	.0035	.085

[a] Data from Erickson (1973, p. 23).

the physical and chemical forces that have acted in a particular area, and it determines which minerals will have formed and remained stable. The controlling conditions are temperature, pressure, the relative concentration of the participating chemical entities, and, of course, time. The geochemical cycle, Figure 4-2, can be found in most textbooks in physical geology. Still it is shown here once more because it is a cycle of mineral deposition as well. The geochemical cycle operates in two major environments, the subsurface (solid arrows) and the zone of weathering (dashed arrows), and in the important transition zone of diagenesis between the two. When the components of a rock or a mineral deposit enter a new environment (new temperature, pressure, or chemical conditions), they come into a state of disequilibrium and begin to adjust. The reaction is both a rock-forming and an ore-forming process. The geochemical cycle begins with deep-born magma entering a shallower environment where it reacts and crystallizes, liberating juvenile fluids. Most of the newly formed rock, liquid, and gas undergo weathering and are transported to collection sites in sediments; however,

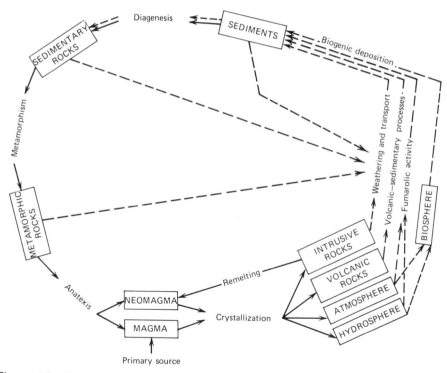

Figure 4-2.   The geochemical cycle. Subsurface environment shown by solid arrows, surface and near-surface environment shown by dashed arrows.

some of the volcanic and fumarolic material is added directly to accumulating sediments and some is added indirectly as biogenic deposits. Sediments and sedimentary rocks are affected by weathering and transportation; so are metamorphic rocks. And the process is repeated in a sequence of subcycles. The magmatic phase is repeated as well by anatexis (palingenesis) of metamorphic rocks and by remelting of intrusive rocks.

As in all abstractions, interpretation is in the mind of the viewer. The actual short cuts and deviations taken by nature are not shown, nor is the fact that we should be dealing with an open cycle. Material and energy are received from several sources outside the system, especially from the primary source beneath the crust.

There are no special places in the geochemical cycle for the formation of mineral deposits; all of the processes are involved and all rock types act as hosts. The choices are nearly infinite; hydrothermal fluids may enter any part of the cycle, and they will contain some product of every part of the cycle in which they have participated. Thus, hydrothermal fluids may comprise juvenile fluids, ground water, sea water, and connate water. Sedimentary ore deposits may be formed, destroyed, and regenerated in almost any part of the cycle, and they may contribute a metal-rich phase to an engulfing magma. Some early-formed magmatic deposits have evidently been cannibalized by later magmas and redeposited.

Generalizations such as these tell us what *may* have happened during the genesis of a particular mineral deposit. But this is of little direct help to a geologist combing through a particular area, whose conceptual model of an orebody must be more specific. What happened in *this* target area of Precambrian terrain in California? In the unseen part of *this* copper-zinc orebody in Colorado?

The more specific indications are somewhere among the characteristics of all possible orebody models. They may, for example, be drawn from evidence that striations formed in cubanite ($CuFe_2S_3$) at a high temperature can be inherited in later forming chalcopyrite; striations of this kind could indicate that a certain chalcopyrite-bearing ore has undergone thermometamorphism. Mineralogic relationships reported for the present-day Red Sea geothermal deposits could be used as supporting evidence for the idea that a certain outcropping massive sulfide deposit is of sea-floor origin. Studies of mineral assemblages—actual ones from other mines or artificial ones made in closed systems that have reached equilibrium in the laboratory—could indicate that the mineralization in a certain length of drill core must be near a carbonatite body or perhaps near a pegmatitic or ultrabasic intrusion.

Mississippi Valley-type Deposits

Conceptual models are drawn in large part from the observation of potential ore-forming conditions in nature and from studies in major mines. Mines and mining districts are the great laboratories of mining geology, but laboratory geochemistry contributes to each model as well. It is too much to discuss here, but at least a single illustration can be afforded to show how geochemical evidence was used in formulating a regional model for ore deposits. This illustration involves Mississippi Valley-type deposits in the central United States, shown in Figure 4-3. A general interest in stratabound lead-zinc fluorite associations was mentioned above, and a paragenetic sequence in part of this area was noted in Figure 2-2.

The deposits are simple in mineralogy (overall, *not* in minor mineral content); galena, sphalerite, and barite are the dominant minerals. The southern Illinois-Kentucky fluorite district, with lesser amounts of galena, sphalerite, and barite, is included. There are faint regional zoning patterns in the silver content of galena and in a few other metal-mineral associations. The orebodies are in Paleozoic carbonate rocks and are stratabound only insofar as they occupy certain beds by local preference rather than by absolute association. Most orebodies are structurally as well as stratigraphically

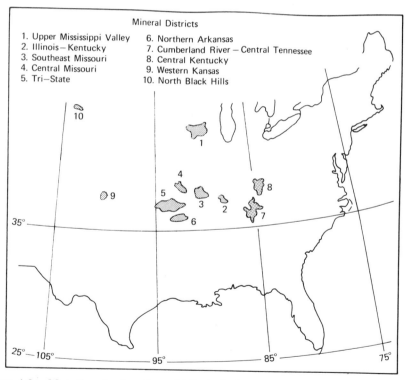

Mineral Districts

1. Upper Mississippi Valley
2. Illinois—Kentucky
3. Southeast Missouri
4. Central Missouri
5. Tri—State
6. Northern Arkansas
7. Cumberland River — Central Tennessee
8. Central Kentucky
9. Western Kansas
10. North Black Hills

Figure 4-3.   Map showing location of Mississippi Valley-type deposits in the central United States.

controlled, there is widespread evidence of open-space filling, and there are occasional breccia pipe "roots" beneath the orebodies. Regional metamorphism is not apparent, and there are only rare occurrences of igneous rock.

Most geologists with long-term experience in Mississippi Valley districts believe that the deposits are epigenetic and that they are unique in many respects. Thus, these geologists do not feel comfortable in extending their models and their ideas on ore genesis to the other lead-zinc districts with which the Mississippi Valley deposits are sometimes grouped in classifications. The Appalachian zinc orebodies, the Pine Point lead-zinc ores of western Canada, and the Alpine lead-zinc ores of central Europe are thought to have nearly as many unique characteristics of their own as they have similarities to the Mississippi Valley deposits. A more flexible and comfortable term is "Mississippi Valley-type" deposits.

The Mississippi Valley ores have long been called telethermal. Yet the deep and unspecified magmatic source called upon for this classification has always seemed rather whimsical. Recent evidence, as reported and evaluated by Heyl, Landis, and Zartman (1974), points instead to a genetic association with basin brines (oil-field brines, connate water), which were heated during deep convection and by contact with a few local intrusive "hot spots." The ore-bringing brines, mixed with ground water, are now believed to have leached their metals from the underlying Precambrian basement and from sandstone aquifers through which they passed on the way to shallower sites of deposition in fractures and beds. Hydrogen sulfide concentrations in the host rocks may have contributed the sulfur for galena and sphalerite. Here we have the skeleton of a conceptual model. Few geologists would argue for a simple or single-stage process; the deposits have been too well studied to permit that. The following evidence has contributed to the basin brine-paleoaquifer theory that now guides most exploration work.

1. *Microscopy.* Fluid inclusions, studied under a microscope with a heated stage, homogenize at temperatures that indicate that they must have formed at less than 200°C. The composition is that of oil-field brines, and in fact, some inclusions contain droplets of petroleum. Deuterium-hydrogen ratios (D/H) in the fluid inclusions suggest that several stages in mineralization must have taken place; δD values (deviation from standard ocean water) indicate a stage-wise decrease from a high deuterium content in the earliest forming fluids to a lower content in the latest forming fluids.

2. *Lead isotopes.* ^{206}Pb-^{204}Pb ratios and other lead ratios indicate that the ores are enriched in radiogenic lead. They are anomalous leads of the "J" (Joplin) type, with "future" age characteristics—more radiogenic than modern lead—and they show isotopic zoning within districts. The

leads were evidently contaminated (in a sense, reactivated) during a complex or multistage traverse from a shallow crustal source.

3. *Sulfur isotopes.* $\delta^{34}S$ values (deviation from standard magmatic or meteoritic sulfur) show a wide range (Fig. 4-4). Like the lead isotopes, sulfur isotopes are zoned within mining districts.

4. *Oxygen and carbon isotopes.* Figure 4-5 shows the variation in $\delta^{18}O$ (deviation from standard ocean water) and $\delta^{18}C$ (deviation from a carbonate standard) in samples taken from a Wisconsin mine. The values decrease—become "lighter"—toward ore mineralization, indicating a heat flow and a partial exchange between the pore fluids in the unaltered limestone and mineralizing fluids in the vein system.

5. *Strontium isotopes.* Little has been done with these except in the few occurrences of igneous rock. The $^{87}Sr/^{88}Sr$ values indicate a mantle source for these rocks—they are related to carbonatites—and indicate that they have no particular connection with the ore deposits. The intrusions, known or inferred, could have provided "hot spots" for localized fluid convection and zoning patterns within mining districts.

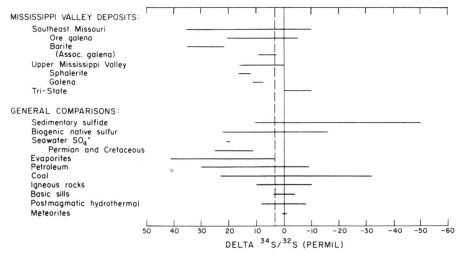

Figure 4-4. Delta 34 S range of Mississippi Valley-type deposits in southeast Missouri, Upper Mississippi Valley, the Tri-State districts. Approximate ranges are shown for possible sources of sulfur that might have contributed to the deposits. (From A. V. Heyl, G. P. Landis, and R. E. Zartman, "Isotopic evidence for the origin of Mississippi Valley type mineral deposits—a review", *Econ. Geology,* v. 69, Fig. 6, p. 1000, 1974. Published by Economic Geology Publishing Company. Used by permission of the publisher.)

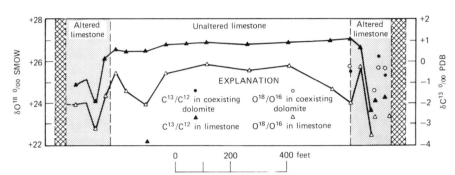

Figure 4-5.   Variation in carbon and oxygen isotopic compositions in carbonate host rock versus distance from ore. Samples are from level workings between two orebodies, Thompson-Temperly mine, New Diggings, Wisconsin. (From W. E. Hall and I. Friedman, *Oxygen and Carbon Isotopic Composition of Ore and Host Rock of Selected Mississippi Valley Deposits*, U.S. Geol. Survey Prof. Paper 650-C Fig. 3, 1969. Courtesy U.S. Geological Survey, U.S. Department of the Interior.

In the Mississippi Valley example we have only scratched the surface of isotopic geochemistry; for some background reading on the use of isotopes in studies of ore genesis an entire issue of *Economic Geology* (v. 69, no. 6, 1974) devoted to this topic is recommended.

Raw theories for the origin of a specific mineral deposit are filtered and concentrated by detailed work in geochemistry. Isotope geology narrows the choice in time, source materials, lithologic associations, and temperature. Fluid-inclusion studies place some further restrictions on temperature and give an indication of the ore-fluid composition. Laboratory simulation places thermodynamic limits on processes and assemblages. All this is for individual deposits, and all of the geochemical relationships must be supported by observed field relationships within the deposit if they are to be of use in mining geology. But the evidence from geochemistry has had a broadening rather than a narrowing effect on the consideration of mineral deposits in general. The need for magma-derived hydrothermal fluids has been brushed aside and the entire range of magmatism, sedimentation, and metamorphism can be dealt with. The array of ore-carrying fluids leaves almost nothing outside the realm of possibility.

4.2. SYSTEMS OF CLASSIFICATION

Mineral deposits may be classified to satisfy a particular need, providing we recognize that we are doing this for convenience rather than to explain some truth in nature that we may never fathom. The residual sulfide "mud" ore and the enigmatic multistage hypogene orebody at Tynagh, Ireland, would not have been accommodated in any classification scheme in use prior to its discovery in 1961.

The need for classification and the design of an appropriate system of classification depend heavily on *communication*—between geologists, between geologists and managers, and between geologists and electronic computers. Geologists must somehow manage the great number of analogies they project from the known into the unknown: "red-bed copper" signifies a certain environment and a certain mineral association even though geologists must furnish their own individual detail and their own exceptions as they gather field information. A manager can visualize a "porphyry copper" deposit, and can attach a general grade, size, and mining cost to it; the manager appreciates its economic importance, and knows that it may have quite a range in geologic characteristics. The computer is the most helpful and the most dangerous of the lot because it adheres to classifications much more faithfully than do geologists or managers. It recognizes what a geologist has told it to recognize, and it delivers only the data asked for. No interpretations—no exceptions. GRBN *is* carbonatite.

We can classify *things* or *processes*. Management and computers prefer the former; most geologists (especially those with academic tendencies) prefer the latter but not being able to agree on the processes generally use some combination of the two. We therefore have to translate "our" language into "their" language at frequent intervals. *Things* are classified according to commodity, as they were for convenience in Table 2-1, or they may be classified according to mineralogy, elements, morphology, or environmental association. *Processes* are classified according to theories of genesis.

There are communication problems with both of these approaches. "Copper deposits" can refer to many things that are only faintly related; the bedded Zambian deposits have little in common with vein deposits at Butte, Montana, and not much relation to the copper-bearing stockwork in volcanics at Rio Tinto, Spain. When we classify things by dominant mineralogy, we can have such misleading labels as "pyrite–pyrrhotite–magnetite–chalcopyrite" deposits in the Val d'Or district, Quebec, that are in reality mined for their gold and silver content. Classify deposits by their metallic elements and we are saddled with "iron-copper-gold" deposits that might contain hematite and azurite in a sandstone bed or might just as well contain pyrite and chalcopyrite in a granite stockwork. Classification by morphology would be relatively clear if the descriptive terms were clear. A "disseminated" ore has an entirely different meaning to a geologist in New Mexico than to a geologist in the Sudbury district, Ontario; a European geologist would call the New Mexico orebody an "impregnation" and the Sudbury ore a "stockwork."

Classification by environmental association is relatively safe, except that "intrusions" may later be reclassified as volcanics (at Rio Tinto) or as metamorphic rocks (as they have been in many Precambrian areas). To emphasize the obvious once more. Which environment? The one we can see at the surface? The one we have not yet seen at depth? The one that metamorphism has nearly obliterated? There is only one foolproof environmental statement: "Gold is where you find it!"

Genetic classification schemes, the ones geologists seem to like, get into trouble as soon as they are applied to a specific orebody. Orebodies in the Colorado Plateau uranium district, the Witwatersrand, and the Zambian copper belt have been thrown back and forth by classification-oriented geologists; orebodies in lesser known districts tend to keep their genetic classifications for longer periods of time, that is, until mine workings become extensive enough and investigations thorough enough to shatter some of the illusions.

The principles and philosophies of mineral deposit classification are fascinating, but they must be left at this point in order to consider the classification systems that are most likely to be encountered. For some depth in the "why" and "how" of classification (and this is important to have) there are the chapters by Graton and Ridge in *Ore Deposits of the United States, 1933–1967* (Ridge, 1968). Park and MacDiarmid (1975), Routhier (1963), and Gilmour (1971) have written very good critiques on ore deposit classifications and their use in exploration work.

Morphologic Classification Systems

Morphologic classifications of mineral deposits differ somewhat between countries, but a preliminary categorization of veins, stratiform or stratabound deposits, and irregular deposits appears to be agreed on at any rate. The "Metallogenic Map of North America" (U.S. Geological Survey, in preparation) expands this terminology to "veins and shear zones," "stratabound-disseminated," "stratabound-massive sulfide," and "stockworks, pipes, and irregular." In North America, the terms "stratiform" and "stratabound" are both in use, with the understanding that the latter type of deposit is confined to a particular stratigraphic zone but is not necessarily stratiform in itself; thus, "stratabound" is the broader term.

Europeans, evidently more concerned with stratiform characteristics than American geologists, often use "stratiformity" as an index to morphologic classification. Pelissonier (1972) describes copper deposits in this way:

Discordant
   (1) Purely vein type
   (2) Multifissured
Subconcordant-discordant
Concordant
   (3) Subconcordant
   (4) Concordant.

"Stratabound" would fit Pelissonier's "concordant" category and would include the intermediate "subconcordant-discordant" category, which he describes as a dense network of mineralized fissures belonging to one lithologic zone.

"Stockwork" and "pipe" are very descriptive terms. Still, an intensely fissured stockwork that occurs in a steeply plunging, elongate shear zone or bedding zone could be called a pipe or it could be called a chimney or a column.

A few additional terms are often found in morphologic classifications. A "manto," such as at Gilman, Colorado (Fig. 2-7), is relatively flat lying and may or may not be stratabound; the European terms "bedded vein," "sheet vein," and "blanket" have a similar meaning. An "impregnation" is understood in North America to mean a filling of pores or interstices in a clastic host rock; in Europe, the term is broader and it is sometimes used to describe mineralization in veinlets. "Pockets" and "lenses" are good descriptive terms. The term "nest" is used in Europe to identify a shallow-seated stockwork.

Morphologic classifications are generally supplemented by reference to other parameters and characteristics. The symbols for mineral deposits to be shown on the "Metallogenic Map of North America," for example, are identified in the following terms:

Principal metals or minerals. Example: Pb, Zn, (Ag)
Size of deposit. Example: large
Geologic environment. Example: predominantly sedimentary rocks of mio-
     geosynclinal type
Geologic class (morphology). Example: veins and shear zones
Age of mineralization. Example: Laramide
General classes of ore minerals. Example: sulfides, sulfosalts, arsenides,
     tellurides

As evidenced in the North American metallogenic map, environmental classifications, like morphologic classifications, are generally used in connection with other parameters.

Genetic Classifications

Genetic classifications are also commonly supplemented by other parameters, and they often turn out to be genetic-environmental classifications. Among these classifications, the most widely recognized is that of Lindgren—with additions by Graton, Buddington, and others. A modified version of the Lindgren classification is shown in Table 4-2.

A condensation of the Schneiderhöhn classification, also widely recognized, especially in Europe, is shown in Table 4-3. This classification, like

Table 4-2.   The Lindgren Classification of Mineral Deposits[a]

I. Deposits produced by mechanical processes of concentration (placer deposits)
II. Deposits produced by chemical processes of concentration
   A. In bodies of surface waters
      1. By interaction of solutions
         a. Inorganic reactions
         b. Organic reactions
      2. By evaporation of solvents
   B. In bodies of rocks
      1. By concentration of substances contained in the geologic body itself
         a. Concentration by rock decay and residual weathering near the surface
         b. Concentration by ground water of deeper circulation; temperature up to 100°C, pressure moderate
         c. Concentration by dynamic and regional metamorphism; temperature up to 500°C, pressure high to very high
      2. Concentration effected by the introduction of substances foreign to the rock (epigenetic deposits)
         a. Origin independent of igneous activity; by circulating atmospheric waters at moderate or slight depth; temperature up to 100°C, pressure moderate
         b. Origin dependent on the eruption of igneous rocks
            i. By hot ascending waters of uncertain origin, charged with igneous emanations (by hydrothermal solutions)
               (1) *Telethermal* deposits (Graton, 1933), formed at the upper limit of the hydrothermal range, at low temperature and pressure
               (2) *Xenothermal* deposits (Buddington, 1935), formed at slight depth, with rapid loss of heat and pressure; temperature low to high, pressure atmospheric to moderate
               (3) *Epithermal* deposits, formed at slight depth; temperature 50–200°C, pressure moderate
               (4) *Leptothermal* deposits (Graton, 1933), formed at intermediate depth but under less intense conditions of pressure and temperature than those associated with mesothermal deposits
               (5) *Mesothermal* deposits, formed at intermediate depth; temperature 200–350°C, pressure high
               (6) *Hypothermal* deposits, formed at great depth or at high temperature and pressure; temperature 300–600°C
            ii. By direct igneous emanations
               (1) From intrusive bodies. Contact *metamorphic* or *pyrometasomatic* deposits; temperature 500–800°C, pressure very high
               (2) From effusive bodies. Sublimates, fumaroles; temperature 100–600°C, pressure atmospheric to moderate
   C. In magmas, by processes of differentiation
      1. Magmatic deposits proper (*magmatic ore deposits* and *magmatic segregation deposits*); temperature 500–1500°C pressure generally very high
      2. Pegmatites; temperature up to 575°C, pressure high to very high

---

[a] Based on the original classification of Lindgren (1933, p. 211–212), on modifications suggested by Graton (1933) and Buddington (1935), and on interpretations by Park and MacDiarmid (1975) and Ridge (1968).

Table 4-3.   An Abbreviated Schneiderhöhn Classification of Ore Deposits[a]

Magmatic Deposits
    Liquid-magmatic deposits
        Chromite deposits
        Platinum deposits in ultrabasic rocks
        Titanomagnetite and ilmenite in gabbros, norites, and anorthosites
        Sulfide liquid segregation
    Liquid-magmatic—pneumatolytic transition deposits
        Intrusive magnetite-apatite deposits
        Intrusive apatite-nepheline deposits
    Pegmatitic-pneumatolytic deposits
        Pegmatite
        Pneumatolytic deposits
        Contact pneumatolytic deposits
    Hydrothermal deposits
        Gold and silver associations
        Pyrite and copper associations
        Lead-silver-zinc associations
        Silver-cobalt-nickel-bismuth-uranium associations
        Tin-silver-tungsten-bismuth associations
        Antimony-mercury-arsenic-selenium associations
        Iron-manganese oxide-magnesium carbonate associations
        Nonmetallic associations
    Exhalation deposits
    Submarine exhalative-sedimentary ore deposits
Sedimentary Deposits
    Weathered zone deposits (oxidation and enrichment)
    Placers (alluvial and eluvial)
    Residual deposits
    Precipitates in lakes and the sea
    Descending ground-water deposits
Metamorphic Deposits
    Contact metamorphosed deposits
    Deposits in metamorphic crystalline rocks formed by dynamic and regional
      metamorphism
    Deposits associated with magmatic and palingenetic mobilization

[a] Adapted from classifications by Schneiderhöhn (1941, 1962) and from the Niggli-Schneiderhöhn classification (Niggli, 1954). In the Schneiderhöhn classification, hydrothermal deposits within specific metal associations are subdivided into plutonic and subvolcanic groups and into epithermal, mesothermal, and katathermal deposits. The term "katathermal" is equivalent to Lindgren's "hypothermal."

Lindgren's, is primarily directed to hydrothermal-magmatic sources of the ore, and it uses some of the same depth-temperature terms. Schneiderhöhn's consideration of ore associations, gangue minerals, and host rock as well as temperature and depth make his classification especially useful as an indicator of geochemical conditions.

The mineral deposit symbols on the "Metallogenic Map of Europe" (UNESCO) are based on a genetic classification with categories representing Lindgren's and Schneiderhöhn's work:

Superficial alteration
Sedimentary
Alluvial
Exhalative-sedimentary
Volcanogenic
Hydrothermal undifferentiated
Telethermal or epithermal
Mesothermal
Catathermal
Catathermal or pneumatolytic
Pneumatolytic
Pegmatitic
Magmatic
Metamorphogenic or metasomatic
Metamorphosed.

In addition, there are symbols to indicate the environment of deposition (volcanogenic, submarine spring, and sedimentary origin), and there are symbols for veins and stratified and irregular deposits.

Simplification—even if it leaves a great many things to individual choice—is needed for perspective. The following arrangement, suggested by Sönge (1974), seems to fill this need. His idea is that the intrinsic features of a deposit, such as mineral association, host rock, and structural framework, should be characterized by a local study rather than by generalization. The terms "exogenetic" and "endogenetic" are more widely used in Europe than in North America. They refer to deposits formed by surface or near-surface processes and those formed by deeper processes.

   I. Exogenetic ore deposits
      A. Formed in the ground-water domain
      B. Formed in surface waters
  II. Endogenetic ore deposits
      A. Related to magmatic intrusions
      B. Related to volcanism
      C. Related to tectonism
      D. Related to metamorphism
III. Organic ore deposits.

4.3. ECONOMIC MINERAL PROVINCES AND METALLOTECTS

First the terms. *Economic mineral province* is an expansion of the term "metallogenic (or metallogenetic) province," expanded so that all mineral deposits, whether metallic or not, can share in the definition. Economic mineral provinces are areas and regions that contain important deposits with enough common characteristics to suggest that the deposits are genetically related. Turneaure (1955) wrote of metallogenetic provinces in this way, and he wrote of metallogenetic epochs because time is as important as space.

Metallotect combines time and space. It has been described by Guild (1968, p. 2) of the U.S. Geological Survey as "a geologic feature believed to have played a role in concentrating one or more elements and to have contributed to the formation of ore deposits." Note that the word "contributed" is used rather than "controlled." The "—tect" in metallotect refers to something that has been built (by geologic processes) rather than to a particular association with tectonic features; thus, a metallotect may be structural, stratigraphic, lithologic, or even geomorphic as long as it fits the user's notion of what has contributed to ore genesis.

A metallotect may be as large as the Andean orogen or as local as one of the half-dozen calderas associated with ore deposits in the San Juan Mountains of Colorado. It may be weak and conjectural or, of course, it may be one of the several strong structural trends that appear to intersect in a vortex, or "crossroads," at a major mining district, as shown in Figure 4-6. A metallotect should normally be a feature that existed during the process of mineralization. But there is the question: How many stages were involved in the mineralization? For example, one metallotect, a major facies change in Mesozoic rocks, could be related to sedimentary ore; another, a series of Laramide intrusions, could be related to a remobilization of sedimentary mineralization, and a third, an unconformity, could be related to supergene enrichment. Metallotects, like individual deposits, have a heritage.

Some economic mineral provinces are so persistent that they cut across several major tectonic domains and stratigraphic associations. They do not coincide with apparent plate margins. In essence, they cut across several metallotects. The Asturian–Pyrenean–Alpine belt of lead-zinc deposits has this characteristic; it extends for 1,500 km from northwestern Spain to northern Yugoslavia and it contains orebodies that have been assigned age dates ranging from Ordovician to Cretaceous. Routhier (1976) and Noble (1976) suggest that such persistent economic mineral provinces may have been born in primitive geochemical provinces in the earth's crust, heterogeneities that have been overprinted by all of the more recent geologic events. Perhaps we should look for evidence of heritage from these paleo-metallotects while using the more easily reconstructed metallotects as guides to regional exploration.

There will be some mention of economic mineral provinces and metallotects in almost every chapter of this book because these are the features that bring geologists and exploration programs to a particular canyon or a specific mountain range in preference to all others. Besides, metallotects are exciting; take a mineral deposit map or a metallogenic map, a geologic map, a series of

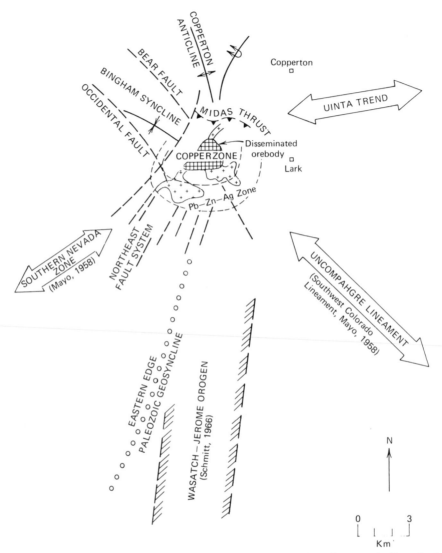

Figure 4-6. Linear elements involved in the "Salt Lake Crossroads," and the Bingham Canyon mining district.

"Landsat" (ERTS) satellite images and go to work. If geophysical or paleogeographic data are available, so much the better. Metallotects will appear.

Lineaments

Major linear features, *lineaments*, have long been used as regional guides to ore. Some are metallotects by virtue of an apparent association between specific geologic trends and orebodies. Others are potential metallotects: "all of these structural features intersecting in Green Basin should have produced *something*" or "there must be some reason why all of these mines line up." Part of the long-standing preoccupation with lineaments derives from the knowledge that the earth's crust tends to fracture in linear zones and from conjecture that these zones could be related to the deep "plumbing" needed for the rise of mantle material or to long-lived paleogeographic features. Part derives from the observation that mining districts are often distributed in a linear or arcuate way. And part is undoubtedly a result of the human urge to perform linear regression analysis on every pattern of more than three data points.

Lineament studies have become increasingly popular with the new wealth of satellite imagery and high-altitude aerial photography. The "new basement tectonics," a study of deep lineaments reflected in overlying geologic features, is reported upon in detail in a symposium volume edited by Hodgson and others (1976).

Some linear metallotects, such as the Colorado mineral belt with its spatial association between mines and porphyry intrusions, are so striking that they must be accounted for in any form of theorizing (Fig. 4-7). Others, such as the Walker line–Texas lineament zone extending from northern California to

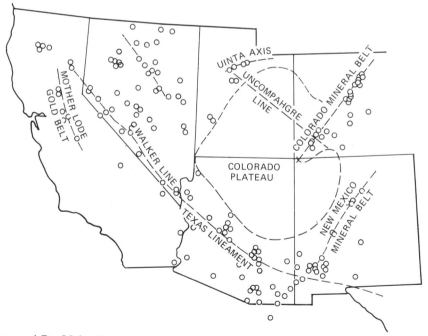

Figure 4-7. Major lineaments and mineral belts recognized in the southwestern United States, with major mines and mining districts.

Texas, are apparent in the regional distribution of mineral deposits and in a few local geologic features, but they do not have an overall association with any identifiable feature.

Obviously, things can be overdone. Once recognized, a local lineament can be extended across a large region by evidence from all kinds of topographic breaks, intrusive boundaries, and basin limits—and there are generally some mineral occurrences that can be used to make the extended lineament into a metallotect. If one lineament will not accommodate all mines and prospects, try two or three; or better yet, try another map projection. If some of the lineaments intersect near major mineral deposits, then the remaining intersections should certainly be investigated. Peter Joralemon (1967) has experienced enough of this to draw the "devil's-advocate" map shown in Figure 4-8. He plotted major regional trends in the southwestern United States and drew circles of 80-km radius around each intersection to represent favorable prospecting terrain. More than three-quarters of the entire area fell within favorable circles, and Joralemon concluded that "the geologist is no better off than he would be without this study."

Metallogeny

Libraries are awash with writings on "metallogeny" (the variously defined and encompassing term for everything that has been discussed in the last few pages). Most of the literature has a direct bearing on regional exploration planning, and most of it makes fascinating reading. But, unfortunately, some of it generates a feeling of skepticism, since metallogeny is a game that can be played by anyone who has a straightedge and a fertile imagination.

Metallogeny is first of all a matter of maps. A good way to begin a study of the subject is to pore over some of the better small-scale metallogenic maps, such as the 1:2,500,000 "Metallogenic Map of Europe," published in a series of area sheets by the United Nations organization. Metallogenic maps (or mineral deposit maps on a geologic background, which are something less) have been published for most countries, and they serve their intended purpose of providing a "jumping-off point" for regional or nationwide exploration ventures.

A group of summary reports on the status of metallogeny and metallogenic maps appears along with several dozen views on metallogeny by North American, European, Asian, and Australian writers in the publications of the 24th International Geological Congress (Gill, 1972). Additional views on metallogeny appear in the symposium volume edited by Petrascheck (1973a) and in the collected papers from a meeting of the International Association of the Genesis of Ore Deposits (Takeuchi, 1971). In a textbook by Ramović (1968) of the University of Sarajevo, metallogeny is treated from the standpoint of geologic age, with an overview of the history of metallogeny and an explanation of the Russian approach to metallogeny in terms of geosynclinal

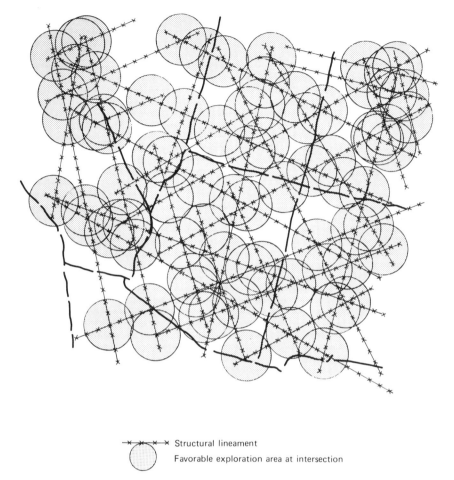

-x—x—x—x Structural lineament

Favorable exploration area at intersection

Figure 4-8. Suggested major structural trends of the southwestern United States. (From P. Joralemon, "The fifth dimension in ore search," SME Preprint 67-I-302, Fig. 1, by permission of the author.)

development. A comprehensive essay on the Russian idea of metallogeny, written by Smirnov (1971), is available in English. The French approach to metallogeny, long associated with detailed studies in paleogeography, is given in detail in Routhier's (1963) French textbook and very briefly in English by Routhier and his associates in the Laboratory of Applied Geology at the University of Paris (1973).

The concept of plate tectonics has been integrated with that of metallogeny in recent years. It is a two-way relationship, with metallogenic studies contributing to ideas about plate tectonic processes as well as receiving ideas from them. Subducting plate margins should be reflected in linear and broadly arcuate features on geologic maps, a welcome arrangement for lineament-oriented geologists. As mentioned in relation to postore faulting in Chapter 2, mineral belts (metallotects), perhaps older plate margins in themselves, that stop at continental margins have all of the disappointing aspects of an orebody that stops at a fault. Now, with the concepts of plate tectonics and metallogenic belts, we can look for the displaced segments.

Metallogeny and geostatistics go together almost as well as metallogeny and plate tectonics. And, as with plate tectonics, one of the immediate applications of geostatistics to mineral exploration should be in the relatively unexplored regions where there is little opportunity to rely upon well-known orebodies and well-mapped geology. If we can produce a deterministic model (a model in which we know all of the parameters quite well), there is little need for the more conjectural techniques of metallogeny or geostatistics. If, however, we are dealing with a thin sample population that may or may not represent the actual parameters (i.e., the real truth), we need a probabilistic model so that we can say, "based on a certain study area or on several such areas, the necessary conditions for ore mineralization in the unexplored area are these: $A, B, \ldots, n$." Here, some problems lie with the sample. Is it a representative sample? Is it really from a similar geologic environment? This takes us back to the comments regarding systems of classification; if we have grouped all "massive sulfides" in one category, our sample may have included deposits ranging from strongly metamorphosed Precambrian gold-pyrite bodies to Cenozoic Kuroko lead-zinc deposits.

Most metallogenic studies used in exploration have both deterministic and probabilistic elements. Normally, we do not know enough about exploration targets to give them specific parameters; yet we do know more than statistical probability would tell us. From this mixed baseline, we formulate a model (a hypothesis), check its salient characteristics in the field, revise the model, and continue with field work until a significant discovery (which may turn out to be an orebody) or a significant discouragement (financial or physical) provides us with enough evidence to make a mine or to quit.

When dealing with geostatistics and metallogeny, there is so much information of potential importance that the use of a computer is essential. A perceptive metallogenic study of copper and lead belts by Laznicka and Wilson (1972) included computer data processing. It had to. The study was based on detailed geologic data from 4,500 ore deposits.

The last few chapters have looked backward to the origins and environments of mineral deposits rather than toward ways of finding them. It has been reported that when a hunting party composed of a geologist, an engineer, and a promoter found a bear track, the geologist went back to see where the bear came from, the engineer went ahead to shoot the bear, and the promoter went into town to sell the hide. The next few chapters will look ahead from the viewpoints of the engineer and the promoter.

PART TWO
THE ENGINEERING FACTORS

Engineering obviously amounts to much more than geotechnics, and geotechnics is certainly not confined to mining. Even so, geotechnics serves as a handy theme for all of the art and science that go into mining engineering. In Chapter 5, a few ideas and terms, the ones most likely to be encountered by a mining geologist, are introduced from geotechnics. In Chapter 6, the ideas are extended to mining and engineering and to a brief description of mining methods. Transportation, metallurgy, and a few other engineering topics are left to be examined in the section on the economic framework. The book will never really finish with engineering until the last page.

Actually, the failure to cram all engineering-related subjects into a single section is a tribute to the engineer. When, as geologists or geological engineers, we advise mining engineers on those subjects that fall within our particular bailiwick, we are talking to people who are geologists to some extent. They are also physicists, chemists, and economists insofar as their job is to turn science into directed action. It would be impossible to put all that under a single label.

CHAPTER 5
SOME ELEMENTS OF GEOTECHNICS

Engineering geology, *geotechnics,* is a part of civil engineering based on knowledge accumulated during the history of mining. Miners and mining engineers have long been applying stress to rock and to soil and learning from experience how the natural materials should respond. There is "good" and "bad" ground. Timber supports and rock pillars hold their loads easily or show some damage when they "take weight." There is easily won ore and there is ore that requires excessive drilling and blasting. In coal mines, the walls "talk" while relieving stress, and miners understand the language well enough to worry when the talking stops or changes tempo.

To civil engineers who design tunnels, highways, dams, or anything requiring an estimate of how soil and rock will respond to stress, geotechnical investigations are a part of the site examination and a basis for site selection. Mining engineers, having now adopted most of the newer ideas in a type of engineering they pioneered in the first place, make as much use of geotechnics as do the civil engineers—and for the same reasons: their projects are bigger and deeper than ever before, tremendous capital costs are involved, and there are competing uses for the same part of the earth's crust.

This is where the similarities end. In civil engineering, geotechnical investigation is a means of choosing the best location for a structure—by recognizing geological hazards and selecting sites where the hazards can be minimized. In mining engineering, site selection is next to impossible. Since gold (or any other mineral) is where you find it, that is where you mine it. The most that can be done is to select the best access to the mineral deposit, the best mining method, and the best sequence of operations.

If the project were a long highway tunnel rather than a mine, there would be some room for choice. The problem might be like one described by Abel (1970), a specialist in *geomechanics,* a field within geotechnics, which deals with the response of natural materials to deformation. A tentative site for an important vehicular tunnel through the Continental Divide west of Denver, Colorado, was selected at Loveland Pass, a location that made use of the existing highway route and the shortest tunneling distance. But mountain passes have an affinity for fractured rock. Fractured rock is generally "bad ground," and a pilot bore—a smaller tunnel—verified this to be the specific situation. An alternate site was found at Straight Creek, 4 km away and in higher terrain. A geotechnical investigation, including another pilot bore, showed that the Straight Creek site would be much better even though it would mean a longer tunnel (3 km versus 1.8 km). The story is oversimplified here; it involved intense geological investigation on the surface and underground at each site. The story has a simple point: we could not move an orebody 4 km.

"Good" rock is not commonly a part of the miner's surroundings. Epigenetic deposits and solid unbroken rock are almost antithetical by definition. Syngenetic mineral deposits are not normally the dominant facies in thick stratigraphic sections, so they too are associated with discontinuities and they are not likely to respond to stress as uniformly as will the rest of the section.

Another point: civil engineering excavations are intended to last, they are not expected to enlarge after they are excavated, and they are designed well within the limits of rock and soil strength. Not so with mines. A mine working that lasts beyond the depletion of its mineral deposit is a scar to be healed unless it can serve as a storage space or an artificial lake. Mine workings are not "installations," and the stress fields around them change constantly as the mine expands. And most mines cannot stay within the limits of rock and soil strength unless a very low percentage of extraction can be accepted; 40 percent final recovery for a mineral body is poor conservation practice; 90 percent is better, but it generally means permitting the rock or soil to fail in some way that can be controlled. In mining by the caving methods to be discussed in Chapter 6, the strength of the rock is purposely exceeded in order to obtain energy from the force of gravity.

5.1. GEOLOGISTS AND GEOTECHNICS

The responsibilities of a resident mine geologist, especially one who is a geological engineer by diploma or by experience, will probably include slope stability investigations in open pits, ground control studies in underground mines, and hydrologic studies. A geologist with a practical "feel" for what is going on in the mine is recognized as a key person whose advice is taken into account. The geologist who is too much of a scientist (with a capital S) to communicate with the people who are faced with geotechnical problems every day is tolerated rather than respected.

The place of geotechnics in an exploration program is not so obvious. An orebody is not yet a mine. But consider this: an occurrence of ore mineralization is not an orebody unless it can be mined. From the very first recognition of a favorable area to the final decision to make a mine, exploration is a process of technical evaluation. Perhaps our model of an orebody is based on a few lengths of drill core with some geophysical anomalies to hold the model together. Is the mineralization too widely dispersed for selective mining? Open pitting or block caving might work. Too deep? Perhaps it could be leached in place. Is there enough water? Too much? Can the minerals be exposed to percolating solutions? Can the solutions be recovered? Suppose we are dealing with a near-surface, bedded deposit. Can it be mined without disturbing the surface occupancy? Suppose it is a deeper bedded deposit. Can we mine a bed this deep and this badly fractured in safety? We must know something about geotechnics before blurting out "We have discovered an orebody!"

A good introduction to geotechnics can be obtained in a textbook by Krynine and Judd (1957); this is one of the very few books that gives some coverage to shoreline and river engineering, permafrost situations, and landslides—all of them mining problems at some time—as well as to the basic engineering characteristics of rock, soil, and ground water. In Krynine and Judd's textbook civil engineering rather than mining is emphasized. Mining aspects of geotechnics are illustrated in an elementary booklet on soil and rock mechanics by Stout (1975). Further background reading in geotechnics is given in an annotated list of recommended reading at the end of this chapter.

The literature of geotechnics can differentiate between rock mechanics, soil mechanics, and hydrology, but the lines are not so easily drawn in nature. A broken mass of rock that has begun to move in a landslide may act like a soil; so may a thoroughly weathered body of rock. Water in pores and fractures is part of soil and rock, and it affects their response to stress. If there has been any indication thus far of a firm boundary between geology and geotechnics, this too should be discounted.

Environmental geology is a relatively new variety of geoscience that is strongly associated with geotechnics and with the use of mineral resources. It has its own journals: *Environmental Geology* and *Geoforum: Journal of Physical, Human, and Regional Geosciences*. Urban geology and the use of the land are fundamental topics in environmental geology and in mining operations. A selected bibliography on environmental geology (Hall, 1975) is available on microfiche from the Geological Society of America. Two textbooks in the field are by Flawn (1970) and Legget (1973).

5.2. ENGINEERING PROPERTIES OF ROCKS: THE SUBSURFACE ENVIRONMENT

As already noted, rock, soil, and ground water are not so distinct as would be indicated by their separation in geomechanics literature. For the time being we can say that rock is "solid," that soil is "uncemented" material, and that soil may include weathered or broken rock. Ground water (or "subsurface water," to use a more flexible term) is a part of both rock and soil. We can generalize further; rocks belong in the subsurface environment, soils belong in the zone of weathering.

Structural Defects

Structural discontinuities in rock have the final say as to whether an underground mine working or an open-pit slope fails or stands. Structural discontinuities are defects because they interrupt what would otherwise be a continuous body of material with a predictable response to stress.

Patterns of intersecting faults, joints, and bedding planes dominate the behavior of rock in open-pit mines because there is little confining force to prevent sliding and because there is likely to be considerable uplift pressure from entrapped surface water within the fractures. The rock above a strong discontinuity that strikes parallel to a pit boundary and dips toward the pit at a low angle may move into the pit even if the overall slope is as low as 15–20 degrees. An open wedge formed by the intersection of two steep fractures striking into a pit face is every bit as bad. Even where the pattern of discontinuities does not promote a simple detachment of wedges and slabs or the toppling of blocks, other and more complex kinds of slope failure may result from a combination of movement along lithologic discontinuities and across fractures. *Circular failures* in which material moves on a curved or spoonlike surface with a near-vertical headwall and a near-horizontal toe is a particularly common type of combination failure in both rock and cohesive soil. In this situation, the characteristics of soil and rock are hard to differentiate.

In underground mines the effect of discontinuities is less direct but still of major importance. Fault surfaces at any depth are places of low cohesion and they are in a sense stress collectors. Joint surfaces and cleavage surfaces are planes of weakness with a low resistance to shearing stress, and they have a tendency to open when disturbed by blasting. Most of these relationships, by one name or another, have been appreciated by miners ever since the introduction of black powder—perhaps even before. A fault surface commonly forms a smooth and natural "hanging wall" to a drift or a stope, something to be used rather than fought. Converging fractures in the back of a mine working are a clue that special support will be needed. A "checked" or well-fractured mine face is easily blasted. Where a major joint plane in a coal seam, the "face cleat," dominates a minor joint plane, the "butt cleat," the geometry of the mine and the direction of advance must conform. The possibility that slab-forming fractures have been opened during blasting or continuous mining is the first thing investigated when a miner enters a new working place. The rock will "ping" (good) or "thud" (bad) when tapped with a pick; a "drummy" roof in a coal mine warns that some open space exists in the overlying strata.

Modern instrumentation in rock mechanics provides better ways of measuring the combined effects of rock strength and structural defects, but many of the basic ideas were already around when miners pinged and thudded and listened to the "talk" of walls and timbers.

A specific advantage to modern rock mechanics studies is that sites and materials can be tested in advance. Mine workings can then be designed in detail. The need for geologic information and careful geologic mapping is obvious, but the actual value of geologic information depends on how well it can be integrated with information on rock strength. Some special techniques are used in this respect.

Detail-line mapping is done by projecting fractures and contacts onto an oriented tape stretched along an outcrop or a mine face (R. D. Call, personal commun., 1975). The mapping is done along several face orientations to avoid overlooking the fracture sets that may run parallel to certain faces. Figure 5-1, a typical data sheet used in this method, shows the type of information collected. Fracture planarity (planar, wavy, or irregular) and fracture roughness (smooth or rough) refer to large-scale and small-scale features. The length of a fracture and the conditions at its terminations (T_1, T_2) express its relative importance and its relationship to other fractures. Fracture-filling material and water-bearing characteristics are important indices to the strength of the entire rock mass.

The fracture attitudes obtained in detail-line mapping are plotted on a stereonet for evaluation. A thorough description of equatorial and polar stereonets, conformable and equal-area stereonets, and their use in geomechanics is given by Goodman (1976).

Another approach to the study of structural detail in mines is *fracture-set mapping* in which all fractures are measured and described in several areas of a mine and then segregated into specific families or groups on the basis of key characteristics. The groups are described and their individual attitudes plotted, as with detail-line mapping, on a Schmidt (equal-area) net. A brief description of the plotting and evaluation of fracture sets on stereonets is provided by Thomas (1973). A special photogrammetric technique is some-

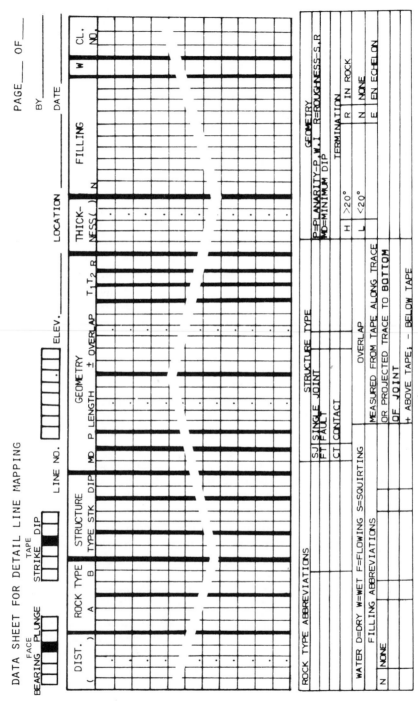

Figure 5-1. Data sheet for detail line mapping of fractures and geologic contacts in open-pit mines. (Courtesy: R. D. Call, Pincock, Allen & Holt, Inc., Tucson, Arizona.)

times used in mapping fractures in pit faces (Ross-Brown, Wickens, and Markland, 1973). This is done on photographs taken from phototheodolite stations in the pit.

A large percentage of the structural information used in mine planning comes from diamond drill core. The spacing of fractures, their attitude relative to the drill hole, and the type of fracture filling must be logged as carefully as possible. Logging of drill core for structural information makes use of a rock-quality designation (RQD): this is based on the percentage of core recovered, counting only the pieces of intact core 4 in. (approximately 10 cm) or longer (Deere and others, 1967). Rock-quality terms in this system are:

RQD, %	Quality
0–25	very poor
25–50	poor
50–75	fair
75–90	good
90–100	excellent

For example:

Total length drilled = 130 cm
Total core recovered = 104 cm
Core recovery = 104/130 = 80%
Summed core lengths greater than 10 cm = 71.5 cm
RQD = 71.5/130 = 55%

RQD can be recorded by geologists while they are logging core from an exploration drill hole. Once the core has been split, such measurements are impossible, structural data are lost and there is no way of performing laboratory tests on the core. This is an important point; some sections of full core must be kept from all drill holes that are likely to be involved in mine planning. Where this is not done, a geomechanics engineer may have to drill new holes right next to holes that could have given the needed information. Ten or twelve pieces of core from each rock type from the ore zone and from each alteration zone will be needed. The length of each piece should be at least $2\frac{1}{2}$ times the diameter of the hole. Some samples of fracture-filling material should also be kept for testing. It is common practice to take color photographs of each box of core before it is split so that the structural pattern can be reviewed.

Seismic investigations are sometimes used in the indirect measurement of "rock soundness." One particular application of the seismic method that has come into wide use is the determination of *rippability*—a measure of the ease with which rock and soil can be removed by bulldozer-rippers and scrapers without blasting. The investigation is commonly done with small hammer-activated refraction seismograph equipment. A comparison between seismic investigations, RQD, and other approaches to designating rock quality is provided by Goodman (1976, Chapter 2).

Table 5-1, taken from a more thorough explanation of geologic information used in rock-slope engineering (Hoek and Bray, 1974), illustrates what is needed in recording structural defects in rock. Note that the faces (the walls) of a discontinuity are important. Smooth, planar walls separated by thick gouge are obviously ripe for shearing movement. Rough or irregular walls in tight contact would have a much greater resistance to shearing stress. The information may be gathered in the course of ordinary geologic mapping or it may be gathered by means of detail-line mapping, and in the logging of core from closely spaced drill holes.

Evaluating Rock Strength

We will be dealing with only a few terms from an entire vocabulary of geomechanics testing; these are defined:

Stress is the force acting on a unit area. It may be *hydrostatic* (equal in all directions), *tensional, compressional, torsional* (twisting), or *shear*. Shear stress, also called rotational or coupled stress, is opposed in direction but not in line. In rock mechanics formulas, σ (sigma) is the symbol for stress normal to the surface on which it is acting; τ (tau) represents the shear stress acting parallel to a surface.

Strain is the response of a material to stress. It is shown in formulas as ϵ

Table 5-1. Geotechnics Information Required for Significant Discontinuities[a]

1. Location in relation to map reference or pit plan
2. Depth below reference datum
3. Dip
4. Dip direction
5. Frequency or spacing between adjacent discontinuities
6. Continuity or extent of discontinuity
7. Width or opening of discontinuity
8. Gouge or infilling between faces of discontinuity
9. Surface roughness of faces of discontinuity
10. Waviness or curvature of discontinuity surface
11. Description and properties of intact rock between discontinuities

[a] From E. Hoek and J. Bray, *Rock Slope Engineering*, 1974, p. 70–71. Copyright ©1974, The Institution of Mining and Metallurgy, London. Used by permission of the publisher.

(epsilon) and is expressed as deformation (shortening or lengthening) per original unit length.

Shear strength is the stress or load at which a material fails in shear. Coulomb's theory of failure states that failure will occur when the maximum shear stress at a point reaches a critical value, the shear strength.

Young's modulus, E, a measure of stiffness, is a constant for each elastic solid. Also called the *modulus of elasticity,* it is the ratio of stress to strain:

$$E = \sigma/\epsilon$$

Poisson's ratio, ν (nu), relates transverse normal strain to longitudinal normal strain under uniaxial stress. It amounts to about -0.2 for many rocks and is expressed as:

$$\nu = -\frac{E\epsilon_y}{\sigma_x} \qquad \text{or} \qquad \nu = -\frac{E\epsilon_z}{\sigma_x}$$

Even though mining is done in a mass of rock with all of its inhomogeneities, we have to start with some rock substance measurements that can be controlled in the laboratory. The laboratory specimens are from surface or underground exposures or, more generally, from lengths of diamond drill core. They are specimens, not samples, because they are carefully selected and they must be as uniform as possible. They give "upper limit" values, and they are indicators, not definers, of the more complex rock mass strength in nature.

Rock tests may involve impact hardness, abrasion hardness, permeability, and a great many other characteristics, but the most important are the *strength* tests that measure the stress needed to rupture a rock and the strain developed during the application of stress (during loading). There are several kinds of rock strength:

Unconfined (uniaxial) compressive strength is tested by loading a cylinder or prism to the point of failure. Figure 5-2 shows the type of test and typical fractures developed upon failure.

Tensile strength is determined by the "Brazilian test" in which a disk is compressed along a diameter or by direct tests involving the actual pulling or bending of a prism of rock. In the bending test, one side of the prism is in tension.

Shear strength is tested directly in a "shear box" or measured as a component of failure in compression.

Triaxial compressive shear strength is tested by placing a jacketed cylinder

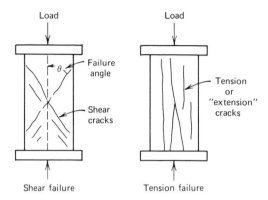

Figure 5-2. Cross-sectional diagram, unconfined compressive strength test on a cylinder or rock. Rocks will fail along "extension" cracks parallel to the direction of loading and along conical surfaces (shear cracks) at a certain angle (θ) to the direction of loading.

of rock in a fluid-filled chamber so that both lateral pressure and axial loading can be applied, as shown in Figure 5-3.

Rock strength can be measured in situ as well as in the laboratory; this is generally done by placing a special jack against a block of rock in a mine and measuring the applied compression or shear stress needed to break the block loose. Rock hardness, an indirect index of strength as well as a direct indication of mining characteristics, is measured by instruments, such as the Schmidt hammer, a spring-loaded plunger that delivers a measured impact.

Measuring Strain. Strain (deformation) is measured in mine workings and then related to stress by referring to laboratory-derived elastic constants. Premining stress, the virgin stress condition, is difficult to calculate, but it is an important mine design parameter. It is commonly estimated and given some quantification by fastening a group of electrical strain gages in a "rosette" on a rock surface, removing some of the adjacent rock, and measuring the response as the original stress is relieved. In one method, "overcoring," a small-diameter hole is drilled in a rock face, instrumented with strain gages, and then encompassed in the core of a larger diameter hole. Deformation of the interior hole is recorded as the existing stress is relieved by the drilling of the outer hole. The sequence is repeated several times as the two holes are alternately deepened, and each incremental piece of core from the small hole is tested in the laboratory to determine its elastic properties. There are additional ways of estimating premining stress by placing hydraulic jacks in freshly cut slots or holes and applying measured pressure until the

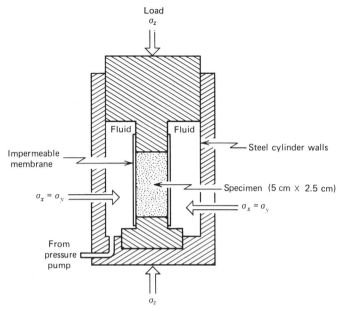

Figure 5-3.   Cross-sectional diagram, triaxial compressive shear test on a cylinder of rock.

opening has resumed its carefully recorded original shape and (presumably) its former stress condition.

Stress conditions developed during mining are of immediate importance in mining operations as well as in mine design. The strain resulting from new stress patterns is measured from time to time or monitored continuously as mining goes on. Electrical strain gages, hydraulic gages, and inclusion (borehole) stressmeters are often involved. Convergence measuring and extension measuring devices are also in common use to show changes in the shape of mine workings; the devices range in complexity from calibrated rods or lengths of wire to automated and remote-recording installations.

Rock Behavior under Stress. Stress-strain relationships are basic to all work in rock mechanics. The descriptive terms for these relationships, *brittle* versus *ductile* and *elastic* versus *plastic,* are familiar to geologists, but they have more specific meanings in rock mechanics work. The relationships, obtained from static (time-dependent) tests, are shown in Figure 5-4. *F* is the point of failure in unconfined uniaxial compression. The line *A* represents a perfectly elastic material where $\epsilon = \sigma/E$. The line *B* represents a perfectly plastic material which will not deform until the stress equals σ_0; this material

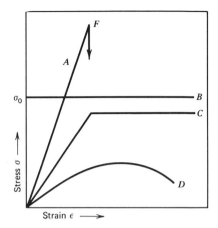

Figure 5-4. Stress-strain diagram for the deformational behavior of four idealized materials.

will not support a load greater than σ_0. The bent line C represents an elastoplastic material. Curve D represents a perfectly ductile material in which strain is not proportional to stress.

There are many more types of behavior, both in theory and in actuality, and some of the behavior is reversible. There are effects of time, temperature, and pore-water pressure, some of which can be approximated in the laboratory. And, of course, there are the effects of discontinuities. Rock mass strength at a discontinuity may be zero. Nevertheless, uniaxial compressive strength tests are widely used as an index to rock behavior, and the results are like those shown in Figure 5-5; the granite in this illustration is almost perfectly elastic and it shatters suddenly upon failure—it is a typical brittle material. The shale approaches ductile behavior. Some other shales would be nearly elastoplastic or plastic in behavior. Marble, in this illustration, approaches elastic-ductile behavior.

Some measured compressive and tensile strength characteristics for common rock types are shown in Table 5-2. In unconfined (uniaxial) compressive strength tests, lateral pressure is zero. The tests are convenient to make, but they do not represent a condition to be expected in a deep mine. At a depth of 1 km, for example, the "dead load" (hydrostatic or, more correctly, lithostatic pressure) of a column of rock with a density of 2.5–3.0 would be 250–300 kg/cm², a sizable portion of its unconfined compressive strength. Since compressive strength increases with confining stress (a fact that will be demonstrated in the next few pages), triaxial stress-strain measurements provide a better index than do uniaxial measurements for what actually happens when the confining stress of a deep environment is added.

Table 5-2. Uniaxial Compressive and Tensile Strength of
Selected Rocks[a]

	Compressive Strength (kg/cm²)	Tensile Strength (kg/cm²)
Intrusive Rocks		
Granite	1000–2800	40–250
Diorite	1800–3000	150–300
Gabbro	1500–3000	50–300
Dolerite	2000–3500	150–350
Extrusive Rocks		
Rhyolite	800–1600	50–90
Dacite	800–1600	30–80
Andesite	400–3200	50–110
Basalt	800–4200	60–300
Volcanic tuff	50–600	5–45
Sedimentary Rocks		
Sandstone	200–1700	40–250
Limestone	300–2500	50–250
Dolomite	800–2500	150–250
Shale	100–1000	20–100
Coal	50–500	20–50
Metamorphic Rocks		
Quartzite	1500–3000	100–300
Gneiss	500–2500	40–200
Marble	1000–2500	70–200
Slate	1000–2000	70–200

[a] Data from Szechy, 1973, and Farmer, 1968.

Figure 5-6 shows the relation of fracturing to stress in a triaxial test. Fractures make an angle θ with the σ_1 axis, σ_θ is the stress normal to the fracture, and τ_θ is the stress parallel to the fracture. Since σ_2 and σ_3 are equal and the specimen has a cylindrical shape, θ should actually be half the apical angle of a fracture cone. The relationship is indicated in an equation expressing Coulomb's theory:

$$\tau = \sigma_\theta \tan\phi + \tau_0$$

where τ is the total shearing resistance, τ_0 is the cohesive strength, and $\tan\phi$ is the coefficient of internal friction.

Test results are related to the Coulomb equation by constructing Mohr's diagram (Fig. 5-7), a widely used representation in geotechnics. The symbols σ_3', σ_3'', σ_3''' represent lateral confining pressures in three consecutive triaxial

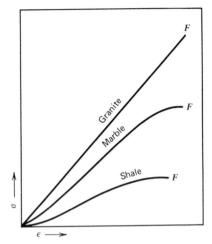

Figure 5-5.   Stress-strain curves for three rocks. (From L. A. Obert, "Rock mechanics," in *SME Mining Engineering Handbook,* A. B. Cummins and I. A. Given, eds., Sec. 6.2, Fig. 6-15, p. 6-19, 1973, by permission American Institute of Mining, Metallurgical and Petroleum Engineers, Inc.)

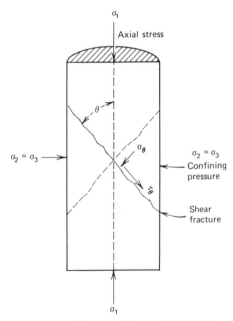

Figure 5-6.   Fracture planes in triaxial test.

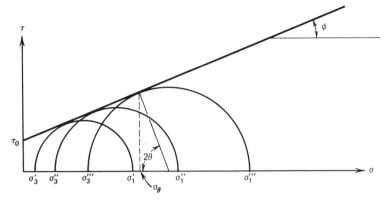

Figure 5-7. Mohr's diagram.

tests on specimens of the same rock. The symbols σ_1', σ_1'', σ_1''' represent the critical axial loads required to break the specimens. The semicircles with σ_3 $-\sigma_1$ as diameters represent Mohr's circles. The common tangent to all three circles, Mohr's envelope, intersects the vertical axis at τ_0, the value of the cohesive strength of the rock, measured in the same units as used for σ. Points of tangency between the envelope and the circles show the values of the shearing stress, τ_θ, and the normal stress, σ_θ, in each of the three tests. The angle between Mohr's envelope and the horizontal axis is ϕ, the angle of internal friction. The angle between the shear surface and the axis of the specimen is represented by θ. Where Mohr's envelope is drawn in the compression direction—to the right of the τ axis—it is essentially a straight line for most rocks; however, it may curve downward in the vicinity of the τ axis for rocks with some degree of ductility.

Some actual numbers: in a selected group of limestones, strengths of 380–3,470 bars (38–347 Mpascals) measured at zero confining pressure became strengths of 3,060–5,990 bars at 1,090 bars confining pressure (Handin, 1966).

The application of Mohr's diagram to mine design? Consider a mine that is planned to occupy a certain range in depth and lithology. For any expected combination of confining pressures and axial loads that will allow Mohr's circle to remain within the tangent envelope, the specific rock should not fail. A circle that goes outside of the envelope, say from a lesser confining pressure at the same loading, indicates that the critical stresses will be exceeded and the rock will fail. Of course, failure may occur along a structural defect regardless of the rock's ultimate strength. The importance of geological mapping and geologic advice in mine design is easy to see.

There are other factors that distort and dominate the picture obtained thus far. Unequal lateral pressures have been measured at depth, and this situation modifies the $\sigma_2 = \sigma_3$ condition assumed in triaxial testing. The tests do not normally take into account the effects of time, temperature change, and pore-water pressure, and all of these tend to weaken a rock in stress. The additional factors are simulated insofar as possible and taken into account in some more complicated laboratory and field tests. The calculations are difficult, and the interpretations provide a challenge for even the most expert geomechanics engineer.

Time-dependent tests are especially important inasmuch as they are related to creep—a change in strain during long-term stress. Many rocks show an initial (transient or primary) high rate of creep that causes localized failure ("popping rock") during the first few minutes or hours after blasting. This brief period is commonly followed by steady-state or secondary creep lasting for days, months, or years. During this second period, localized rock failure with "slabbing" or more violent "rock bursts" may occur as residual strain energy is released, and it may happen without any particular warning. A short period of tertiary creep with a steep acceleration in strain may occur just before a major rupture, giving a warning that can be measured or heard in various ways. Tertiary creep may be so brief in duration that it would go unnoticed without careful instrumental monitoring—or listening.

Rock Mechanics Models

Photoelastic models, plastic sheets and blocks with simulated mine openings, can be stress loaded while the birefringence pattern is observed in a polariscope. This technique has been used to verify theoretical stress calculations for various shapes and groupings of underground workings. Scale models built from rocklike materials and models involving electrical analogs are also used in simulating mine designs and studying the resulting stress patterns.

However, no model technique has been so widely accepted as a type of mathematical modeling called *finite element analysis,* in which large numbers of individual elements are treated as parts of an overall network, a continuum—a fairly close approximation to what must happen in nature. The deformation of each element is related to the expected displacement at nodes where it joins other elements. The summation of all nodal force–displacement relationships gives a set of mathematical expressions that can be varied with alternative patterns of physical behavior. Thus, differences in lithology and structure can be introduced (with certain assumptions) into a computer program and their effects on variations in mine design can be studied. The geological engineering textbook by Goodman (1976) has a chapter on physical models in rock mechanics, a chapter on the finite element method, and an illustrative computer program for finite element analysis.

5.3. ENGINEERING PROPERTIES OF SOIL AND WATER: THE SURFACE ENVIRONMENT

We want to know how steeply a pit wall may stand in soil or rock. How steep does the same material stand in nature where it is undercut by a stream? We plan to pump water from a mine. What has happened to the upland drainage in similar areas where downcutting streams have lowered the water table? Natural erosion and "erosion" by mining (both involve loosening and removal) have enough in common to provide some guidelines for surface mine design. Geomorphology provides the index. So, as in the field of rock mechanics, geologists have an applied science to share with civil engineers.

Soil Mechanics

Most definitions for "soil" begin with "unconsolidated" or "uncemented," they include some mention of "the earth's surface," and then end by citing "rock" as the opposite material. All of these definitions are acceptable as long as we leave some room for the rocklike properties of preconsolidated clays that have been subjected to excess overburden pressure and for "hardpan" accumulations in the soil. Also, we must leave room for the soil-like behavior of thoroughly broken rock.

If a rock-derived material is unconsolidated, it should follow the rules of soil mechanics, rules intended for the materials classified in Figure 5-8. Because soil is of interest to so many scientists in so many ways, there are subcategories and there are more detailed geological, agricultural, and engi-

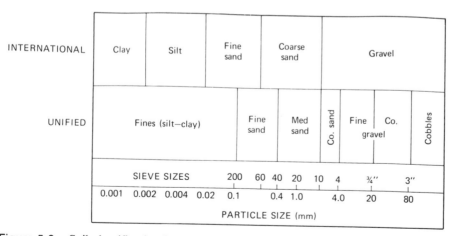

Figure 5-8.   Soil classification based on particle size. The Unified Soil Classification is an American system used by the U.S. Army Corps of Engineers and the U.S. Bureau of Reclamation.

neering classifications, all of which have some bearing on exploration. This point will be noted again in the chapter on geochemical exploration. For a complete explanation of soil classification systems and their engineering significance, there is a good review article by Legget (1967).

Evaluating Soil Characteristics. The behavior patterns of soil and rock are influenced by the presence of water and air; water especially. A stress σ, applied to a saturated soil or rock, is taken up in part by U_w, the pore-water pressure in the voids; therefore, the Coulomb equation can be written as

$$\tau = (\sigma_\theta - U_w) \tan \phi + \tau_0$$

The expression $(\sigma_\theta - U_w)$ can be written as σ_θ', representing the effective normal stress acting on the frictional resistance between soil or rock particles. The equation for a cohesionless soil is

$$\tau = \sigma_\theta' \tan \theta$$

A triaxial compressive shear strength test applied slowly enough to allow the excess (squeezed-out) water to drain from a specimen of cohesionless soil would provide for a Mohr envelope of the type shown in Figure 5-9. The angle of internal friction ϕ, turns out to be 30 degrees for most natural sands; it has larger values, say 40 degrees—indicating better stability—in angular material, in well-graded sediments, and in densely packed material. Remember, these calculations are for soil, a material without structural defects.

If there is no active seepage, a slope in cohesionless soil should remain stable at a value of 1.5 for the *factor of safety* represented by tan ϕ/tan i; i is the slope angle. If there is steady seepage from an exposed water table near the top of the slope, the slope will fail at a lower angle and the factor of safety is often taken empirically as tan $(\phi/2)$/tan i.

If the soils represented in Figure 5-9 were a preconsolidated clay that had recently been stripped of a heavy overburden, there would be an initial or

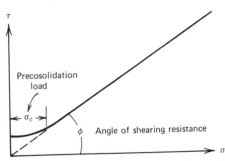

Figure 5-9.   Mohr's envelope for clay soils.

Table 5-3.   Atterberg Limits in Fine-
grained Soils

Stage	Limits between Stages
Liquid	Liquid limit (LL)
Plastic	Plastic limit (PL)
Semi-solid	Shrinkage limit (SL)
Solid	

residual shearing strength and the soil would not follow the characteristics of a cohesionless soil until the *preconsolidation unit load,* σ_c, had been reached.

If $\sigma_\theta = U_w$, then $\sigma_\theta = 0$ and the soil has no shearing strength; it is a flowing mud. A sudden increase in pore-water pressure in very fine silty sands or in mill tailings with this strength characteristic would result in *liquefaction*. This might be brought about by heavy rains or by any vibration that would reduce pore space.

When a flowing mud, a clay suspension, for example, dries out, it passes through the stages shown in Table 5-3. The limits are *Atterberg limits,* terms for changes in the consistency of silt and clay soils. In the plastic stage, the soil moves slowly without any change in volume. The difference in percent water content between the plastic limit and the liquid limit for a particular soil shows the range in moisture within which soil acts with plastic properties. This is its *plasticity index,* an important characteristic in engineering specifications.

A commonly encountered engineering classification for fine-grained soils involves a *plasticity chart* (Fig. 5-10) in which the liquid limit is plotted against the plasticity index. The A line on the chart is an empirical boundary with inorganic clays above and with silts and organic clays below. Solid organic clays, with their low plasticity index and high liquid limit, become flowing mud rather quickly.

In addition to the use of *triaxial compression* testing apparatus, laboratory testing of soils involves *consolidometers* for measuring consolidation under loading, and *direct shear boxes. Unconfined compression tests* are performed on cohesive soils. For field in situ testing, the *vane shear test* is widely used; in this, a four-winged rod is inserted in the soil and turned with a measured force to determine shearing strength.

Hydrogeology

Water is to be sought where needed and to be controlled where it is in excess; it generally falls upon the mine geologist to provide guidelines for both efforts. Guidelines require data, and the most useful hydrologic data will have been

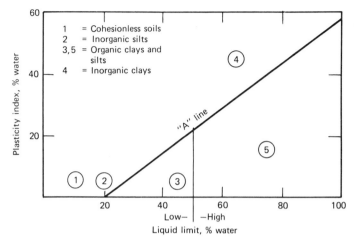

Figure 5-10.   Soil plasticity chart, showing the characteristics of several soil types.

collected over a long enough period of time to understand the trends and the seasonal variations. This is where exploration geologists make their contributions. Surface water flow can be estimated and the location of springs can be plotted during geologic mapping. Measurements can be made during an exploration drilling program. Water-quality samples can be taken and a few simple pumping tests can be made without much extra effort while geologic data are being collected.

Water problems may be acute, and they almost always have social and political aspects. Dewatering a mine may result in dewatering someone's well or stream. Pumped water, often carrying troublesome metal complexes, must be cleaned up and discharged somewhere. Water supply for mining will probably be the same water needed by other people for other purposes. For example, water for mining and processing ore at Akjoujt, Mauritania—barely enough water at that—is piped 120 km, and it must be made available to other users along the way.

A heavy inflow of water may be handled routinely, as it is in the open-cut brown coal (lignite) mines of the Rhine district, West Germany, where 600 wells pump enough water to give a water-to-coal production ratio of 15:1. Or, it may be part of a delicately balanced ground-water system that will not stand intensive pumping. The formation of large sinkholes by ground-water withdrawal in the Far West Rand mining district, South Africa, illustrates the point (Foose, 1968). The water table, originally in unconsolidated debris, was at a depth of 100 m before intensive pumping began. Bedrock, a cavernous dolomite, lay below. A dewatering program at three large mines caused the water table to lower into the dolomite, drying out the unconsolidated cover and allowing it to subside into depressions in the bedrock. Sinkholes began to form within two years after dewatering began, and the sinkholes swallowed large areas of ground, including part of a village and an entire mine crusher installation.

Pumping and drainage of mines are not likely to be the first human disturbance of the water balance in a particular area, but they will certainly have an effect on whatever condition had already been established between recharge and discharge. Natural discharge into streams may decrease and the storage capacity of a pumped aquifer may also decrease unless some provision is made for sealing off the mine workings and letting the water table return to its original level while mining continues. An extensive project in

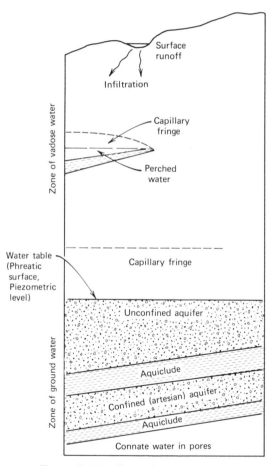

Figure 5-11.   Ground-water terms.

hydrologic control may be needed if a mine is allowed to continue or even to begin operation.

The terminology of ground-water hydrology is rather loose, primarily because water is everyone's business. Figure 5-11 illustrates a few of the more common terms and their synonyms. The "type" condition is that of a continuous ground-water surface or "table" in clastic sedimentary rocks. In crystalline rocks, the water table is a discontinuous reference surface connecting the tops of saturated zones in intersecting joints and faults; perched water occurs in "blind" fissures.

In all types of rock, there are local variations in water level due to the isolation of blocks of ground by barriers of gouge-filled faults and impermeable dikes. This was one reason for the rapid lowering of the local water table in the Far West Rand; the pumped area was a natural compartment closed off by two large syenite dikes, and the water was discharged outside the compartment.

Hydrologic Measurements. Two of the most important measurements in ground-water hydrology are the *coefficient of permeability* (ease of water transmission) and the *storage coefficient,* or "effective porosity."

The coefficient of permeability κ is an element of Darcy's formula: $V = \kappa i$. V is the velocity of flow under laminar (nonturbulent conditions; i is the hydraulic gradient, the ratio of loss in hydraulic head (pressure) by frictional resistance to a unit distance in the direction of flow. κ is determined experimentally for specific areas by field pumping tests and in laboratory permeameter tests. In one widely used in situ permeability test, water is forced under pressure into a section of a drill hole that has been isolated by expandable "packers."

Permeability values for certain rock types are shown in Table 5-4. In order to compare different methods of measuring permeability, the coefficient of permeability κ, in centimeters per second, is shown with two other systems of measurement. Darcy (or millidarcy) units are commonly used by petroleum engineers. Meinzer units are used by hydrologists in describing transmissibility. Note the significant decrease from a high permeability in fractured rocks and clean sands to a very low permeability in dense unfractured rocks. Actually, most rocks are relatively impermeable unless they have *secondary permeability* from cracks and solution cavities.

The storage coefficient within an aquifer is expressed as a decimal fraction. It represents the volume of water that can be expected to drain from a unit volume of ground. It relates to the pores, fractures, and solution cavities open for filling by ground water. The storage coefficient is generally calculated from pumping tests in which observation wells are used to monitor differences in the depression curve or piezometric surface around a well or shaft, as shown in Figure 5-12.

Table 5-4.   Approximate Ranges of Permeability in Certain Rock Types

Description	Permeability Unit		
	Darcy	Meinzer	cm/s
Clay shale or dense rock with tight fractures, considered impermeable in most excavations	0.0001	0.0018	9.7×10^{-8}
Dense rock; few tight fractures, approximate lower limit for oil production	0.001	0.018	9.7×10^{-7}
Clay, silt, or fine sand. Few water wells in less permeable ground	1	18	9.7×10^{-4}
Dense rock with high fracture permeability	2	36	19.4×10^{-4}
Clean sand, medium and coarse (0.25 and 1.0 mm)	500	9,100	0.48
Clean gravel (70% larger than 2.0 mm)	1,250	22,750	1.2

After R. L. Loofbourow, "Ground Water and ground-water control," Sec. 26, Table 26-1, p. 26-15, in *SME Mining Engineering Handbook*, A. B. Cummins and I. A. Given, eds., 1973, by permission American Institute of Mining, Metallurgical and Petroleum Engineers, Inc.

The Marine Environment

Beach-sand deposits and marine placers have been mentioned in Chapter 3. These are of critical engineering interest at present because of all the newly recognized effects of man-made shoreline disturbance, whether or not connected with mining. Here, more than in any other mine site, research in geomorphology can be directed toward understanding potential damage to delicately balanced natural systems. As we obtain more information on sediment transport patterns, wave-energy distribution, and littoral drift rates, it will be translated into the terms of geotechnics and mining. For a review of shallow-water geotechnics, a book on nearshore sediment dynamics and sedimentation edited by Hails and Carr (1975) is recommended.

Undersea coal mining and metal mining have been going on for many years, as has offshore oil and sulfur production. A few of these operations will be discussed in Chapter 6. The geotechnics aspects are a combination of rock mechanics (subsea ground control), soil mechanics (control of bottom sediments), and "ground" water (subsea hydrology).

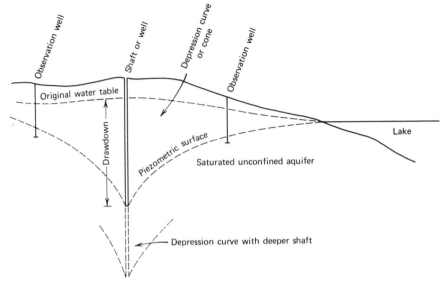

Figure 5-12.   Drawdown by pumping in a mine or well.

Geotechnical problems in the deep-sea marine environment, the potential site of nodular manganese–copper–cobalt–nickel mining, are so unique that they cannot be fully appreciated without considerable background in the physics, chemistry, and geology of oceans. A handbook by Walton Smith (1974) provides a good introduction to the engineering aspects of oceanography. An essential reference book for anyone interested in marine mineral resources is the *Encyclopedia of Oceanography* edited by Fairbridge (1966).

5.4. GEOLOGICAL ASSESSMENT OF ENGINEERING CONDITIONS

In relating geologic information and geologic concepts to the design of mines, we need to provide some special maps and studies. Because the data are collected at the same time as other exploration data, the field procedures will be considered in the appropriate chapters on mapping, geophysics, drilling, and data presentation.

We have some special information already at hand for certain areas in the engineering geology, geotechnical, and land use maps of government geological surveys. A manual published by the United Nations (1976), a special publication of the Geological Society of America (Ferguson, 1974), and a U.S. Geological Survey professional paper (Varnes, 1975) deal specifically with engineering geology maps and their use in environmental planning. In these

maps, emphasis is placed on soil, ground-water, and rock strength characteristics rather than on bedrock geology; bedrock geology is generally shown on
equivalent geologic maps. Figure 5-13, showing the variability of soil-index
properties in a particular clay unit, is taken from a British engineering geology
map.

The scale of most engineering geology quadrangle maps, 1:24,000 or
1:25,000, is too small for mine planning, but the maps provide a starting point
for the detail mapping at 1:1000 or 1:1200 that must eventually be furnished
for a specific project. Detailed geotechnics mapping is an important part of
environment-wise mine design, as explained in a general paper by Rudio
(1974) and in a case history of environmental planning for a new mining
operation in Ontario (Shillabeer and others, 1976).

Something remains to be said (and repeated at intervals in the rest of the
book). It is never too early in the investigation of a potentially economic
mineral body to begin collecting geotechnical data or at least saving the data
for later use. We are dealing with a potential disturbance to nature's
equilibrium, and we must, by all means, learn as much as we can about the

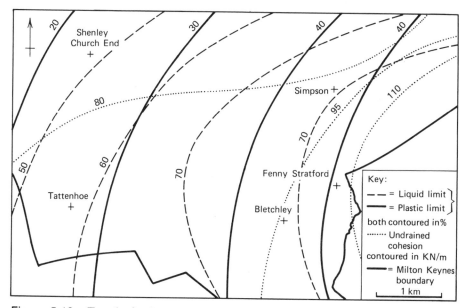

Figure 5-13. Trend of selected engineering index properties in one geological zone
(Upper Athleta zone of the Oxford clay), Milton Keynes map area, England. (From C.
R. Cratchley and B. Denness, "Engineering geology in urban planning with an
example from the city of Milton Keynes," in *Sec. 13, Engineering Geology*, International Geological Congress, 24th, Canada, 1972, Fig. 2, p. 16, 1972, by permission of
the author.)

original near-equilibrium condition. If we wait until we begin stripping overburden, blasting rock, and pumping water, we will have lost the information.

5.5. RECOMMENDED READING*

A. Geomechanics (Rock Mechanics)

Obert and Duval (1967). This is a thorough review of basic concepts, testing methods, and design of underground structures written by two engineers from the U.S. Bureau of Mines. A shorter review, part of a mining engineering handbook, is provided by Obert (1973).

Jaeger and Cook (1969). This is a fundamental treatise, rather mathematical and oriented to civil engineering.

Coates (1970). A widely used textbook, with exercises, this book is specifically oriented to mining.

Farmer (1968). This is the most concise and easily understood of the introductory books on rock strength.

Hoek and Bray (1974). Concerned with open-pit mining for the most part, this book emphasizes the effects of geologic structure. It includes chapters on collecting and representing geologic data. There is also enough coverage of soil mechanics and hydrology to make this a good all-around reference in geotechnics.

Goodman (1976). The chapters on stereographic projection and exploration of rock conditions are especially useful in relating geologic target investigations—mapping and drilling—to problems in geomechanics.

B. Soil Mechanics

Karl Terzaghi (1943) was the pioneer in this field and in many other aspects of geotechnics as well. His book is widely quoted in all subsequent works. A book of Terzaghi and Peck (1967) is available as the revised edition of an earlier work.

Sowers and Sowers (1970). This is a basic book on soil testing, soil properties, and stability.

C. Geohydrology (or Hydrogeology)

Davis and DeWiest (1966). This is the most widely used modern textbook.

Meinzer (1942). This is the classic treatise. Olaf Meinzer of the U.S. Geological Survey was a pioneer researcher in both surface water and ground-water hydrology.

D. Geotechnics Journals

Bulletin of the Association of Engineering Geologists, Dallas, Texas.
Quarterly Journal of Engineering Geology. London, The Geological Society.

---

* Full citations are included in the reference section at the end of the book.

Engineering Geology. Amsterdam, Elsevier Publishing Co.
International Journal of Rock Mechanics and Mining Sciences. London, Pergamon Press.
Australian Geomechanics Journal. Sydney, Australian Institute of Engineers.
Soil Mechanics and Foundation Engineering (USSR). New York, Consultant's Bureau.
Canadian Geotechnical Journal. Ottawa, National Research Council.
Geotechnique. London, Institution of Civil Engineers.
Rock Mechanics. Vienna, International Society for Rock Mechanics. Springer-Verlag.

E. Transactions of Symposia

Society of Mining Engineers, American Institute of Mining, Metallurgical, and Petroleum Engineers; annual Symposia on Rock Mechanics.
24th International Geological Congress (1972) and 25th International Geological Congress (1976) Section 13. Symposium on engineering geology sponsored by the International Association of Engineering Geologists.
International Congresses on Rock Mechanics, Lisbon, 1966; Belgrade, 1970; Denver, 1974.
International Society for Rock Mechanics, Symposia. Example: 16th Symposium, University of Minnesota, Minneapolis, 1975.

F. Case Histories

Engineering Geology Case Histories, Nos. 1–10, 1964–1974. Boulder, Colorado, Geological Society of America.

CHAPTER 6
APPROACHES TO MINING

Mining absorbs human and mechanical energy; in order to be of benefit it must create something of higher value than the energy expended. This is why mines must continue to go deeper into the specific parts of the earth's crust where natural energy has already concentrated the needed material.

How deep? To a depth where the energy cost approaches but never exceeds the value of the mineral. There is no comprehensive formula because something is unique in every mine. Because of highly efficient production methods, mines in the Witwatersrand gold fields of South Africa and the Kolar gold fields of India can go to depths of 3,000 m for ores of lower value and less thickness than those left unmined at 1,000 m elsewhere, but the mining methods used in both of these deep districts are keyed to specific geologic conditions. Coal seams can be mined by surface methods at a stripping ratio (overburden versus coal) of 30:1 in some of the midcontinent United States; in other places, a stripping ratio of 5:1 is excessive. Certain coal fields simply have more economic "leverage" than others, and there are local conditions of rock strength, structure, and hydrology that determine "how deep."

One alternative to fighting the cost of excessive depth is to go wider—into the lower grade material that surrounds certain orebodies with "assay walls" rather than with abrupt physical boundaries. The A/G (arithmetic-geometric ratio) of Lasky (1950) tells us that reserves in an orebody should increase geometrically as the mining grade decreases arithmetically. The reasoning is valid—to a point. Copper ore reserve tonnages have followed the A/G pattern at many large mines, increasing geometrically as the mining grade decreases from 2 percent to 0.8 percent to 0.6 percent, until the cost and availability of energy call a halt to the relationship. Lead-zinc deposits, on the other hand, seldom show this relationship. For any particular mineral deposit geological conditions control the boundaries and geotechnical conditions dictate how much additional energy will be needed to mine and process 10 tons of 0.2 percent ore rather than 1 ton of 2 percent ore.

The energy aspect of mining lower grade ore is illustrated in Table 6-1. Two types of ore are considered, one with 0.5% Cu and one with 4% Pb. The market value of the two ores is certainly not equivalent, but these are typical minimum, or "cutoff," ore grades in many mining districts. Mining, comminution (crushing and grinding), and flotation or other beneficiation (processing) must be done on the total amount of raw ore. Tailings disposal still involves most of the mined ore. These are the bulk stages; filtration and all subsequent steps involve a relatively small amount of concentrate. The proportion of the total energy requirement needed in the actual mining stage is about the same for the two metals, because the lower grade copper ore is mined by larger scale, more energy efficient methods which balance the difference in tonnage handled. Note, however, that the combined mining and treatment of ore (through the stage of tailings disposal) account for a much higher percentage of the energy consumed in making metal from the lower grade material—58 percent for copper as opposed to 34 percent for lead. Comminution and flotation are the high energy consuming stages for copper ore, since every piece of ore—regardless of grade—must be crushed, ground, and processed. If we were to provide for the fact that lead concentrates are much higher in metal content than copper concentrates (therefore less

Table 6-1. Energy Consumption in the Production of Lead and Copper from Low-grade Ores, Direct and Indirect as Electric Power Equivalent[a]

Operation	Crude Lead (kWh per ton)	Cathode Copper (kWh per ton)
Mining	415	1,550
Comminution	415	3,300
Flotation	275	2,200
Tailings disposal	135	
Filtration and drying	135	350
Transportation	135	350
Lead smelting	2,100	
Cathode copper production		4,500
Total	3,610	12,250

[a] Basis: 4% lead ore and 0.5% copper ore. Data from Kihlstedt (1975).

material to filter, dry, transport, and smelt), the contrast would be even more distinct. If we were to add the observation that recovery rates, expressed as a percentage of the total *available* value, tend to drop as the ore grade decreases, the point of the example would be underscored.

This much is apparent: somewhere in the trend toward mining at greater depth and mining lower grade ore there is an energy barrier at which the effort exceeds the benefit. Whatever we can do to keep from coming against this barrier must be done, but it must be done with all possible regard for the safety of miners and it must be done at a larger scale and faster than ever before.

Mining Methods

The "bigger, faster, and deeper" requirements of modern mining are often met by scaling up the mining equipment and giving that equipment enough elbow room. In open-pit mines, there is room to change the scale. Large-scale equipment can also be accommodated in underground mines, but not in the narrow timber-choked passages of a few years ago. Deep, large-scale mine workings are now especially designed to accommodate diesel trucks, trackless load-haul-dump units, conveyor belts, and crawler-mounted drilling jumbos. One of the most important factors in designing this kind of modern underground mine is the rock strength needed to sustain wide roof spans and to permit massive blasting patterns.

A way to obtain the best advantage from whatever energy we have is to use it entirely on the ore rather than wasting it in creating fractures in the wall rock. This is being done wherever possible by using continuous mining methods, by raise boring (Figs. 6-1 and 6-2), and by tunnel boring in place of the conventional drill-blast-support-load cycle. Hydraulic mining (using jets of water to force rock apart along natural fractures and to erode rock) has been used in mining coal seams and uranium-bearing sandstone. These techniques are part of *rapid excavation,* an important field in mining research.

Another way to conserve energy is to use the energy already available in the mineral deposit and in the wall rock. Block caving (see Fig. 6-11) and sublevel caving methods do this by directing the force of gravity against the crushing strength of the rock. Longwall mining methods do this by inducing special stress patterns (see Fig. 6-10) in a carefully controlled system of working faces.

The choice of a mining method for a specific mineral deposit depends on economic needs and geologic conditions—what is desirable and what is possible. Table 6-2, shows what must be taken into account by the engineers who design mining systems and the geologists who advise them.

Physical factors cannot be changed; they represent what we have to work with. The geometry of the deposit may be simple or complex. The geology of

Figure 6-1. Raise drilling machine, capable of boring 1.8 m diameter raises to a distance of 180 m in hard rock. (Courtesy The Robbins Company, Seattle, Washington.)

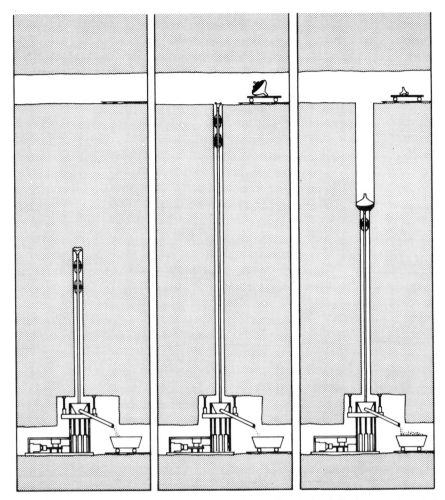

Figure 6-2.   Three stages in raise boring, with a pilot hole drilled upward and then reamed downward to the desired size. (From P. R. Bluekamp and E. G. Beinlich, "Drifting and raising by rotary drilling," in *SME Mining Engineering Handbook,* A. B. Cummins and I. A. Given, eds., Sec. 10.6, Fig. 10-67, p. 10–94, 1973, by permission American Institute of Mining, Metallurgical and Petroleum Engineers, Inc.)

the deposit and its surroundings gives us geotechnical limits. The geography presents some opportunities and restrictions in regard to methods of entry and development.

Technologic factors spell out the conditions under which the mining method must be made to work. An unsafe method is of course unacceptable. A highly

Table 6-2.   Factors in the Choice of a Mining Method

Physical	
Geometry	Size, shape, continuity, and depth of the orebody or group of orebodies to be mined together
	Range and pattern of ore grade
Geology	Physical characteristics of ore, rock, and soil
	Structural conditions
	Geothermal conditions
	Hydrologic conditions
Geography	Topography
	Climate
Technologic	
Safety	Identification of hazards
Human resources	Availability of skilled labor
Flexibility	Selectivity in product and tonnage
Experimental aspects	Existing or new technology
Time aspects	Requirements for keeping various workings open during mining
Energy	Availability of power
Water requirements	
Surface area requirements	
Environment	Means of protecting the surface, water resources, and other mineral resources
Economic	
Cost limits	
Optimum life of mine	
Length of tenure	Prospects of long-term rights to mine

sophisticated method requires that a cadre of skilled miners and mechanics be found locally or brought in—a difficult hurdle in some countries. An inflexible method may be efficient only within a too narrow range of operating conditions. If the method has a well-established precedent, it can be designed full scale; if not, a smaller scale or pilot plant operation may have to be tried first. Continued access through long tunnels or deep shafts requires "good" rock. Subsidence must be carefully controlled, water must not be contaminated, other coal beds and orebodies must not be wrecked, and an oil field using nearby strata may have to be protected.

Economic factors do not normally dictate the specific mining method. However, they do place limits on capital investment and operating costs and on the year-to-year pattern of cash flow. They may indicate that we should start with an open pit while developing an underground mine. And these factors should indicate whether to gear up for a long-term operation.

Geologists cannot advise in regard to anything unless they understand what is being done. Mining is best understood from first-hand experience in mines, but textbooks can provide a good start at any rate. One of the most concise overviews of metal and coal mining, with an introduction to rock mechanics, is a textbook by Thomas (1973). The author and many of his citations of mining practice are Australian, but there are ample illustrations taken from American and European practice. For a detailed introduction to mining engineering, there is the three-volume American work by Woodruff (1966). For a comprehensive reference book, there is the two-volume *SME Mining Engineering Handbook* of the American Institute of Mining, Metallurgical, and Petroleum Engineers, edited by Cummins and Given (1973).

A feel for the terminology of mining is the first step toward understanding the mining profession, and it is a necessary attribute for any geologist who intends to communicate with mining engineers. We would not be surprised to find a mining engineer looking for some of our terms in the American Geological Institute *Glossary of Geology* (1972) immediately after we had explained the importance of a certain nappe structure with imbricated faulting. We must have something similar to pore through immediately after an engineer explains in detail that the best way to mine a certain orebody will be by shrinkage stoping unless there will be too much dilution from a weak hanging wall. The book we need is the U.S. Bureau of Mines *A Dictionary of Mining, Mineral, and Related Terms* (1968). For international work, a useful book is the *World Mining Glossary of Mining, Processing, and Geological Terms* (Wyllie and Argall, 1975), with entries in English, Swedish, German, French, and Spanish.

6.1. SUBSURFACE MINING

The words "mining" and "underground" are a natural couplet (Fig. 6-3). In a dipping tabular body, whether in a bed or a vein, only the narrowest dimension is ordinarily exposed at the surface; therefore, a small pit that furnishes early production will soon reach a practical limit in depth because of excessive waste removal. Mining then goes underground. With favorable topography, a horizontal entry or *adit* can be driven into the deeper zones, preferably at the lowest accessible level so that broken material can be collected by gravity rather than having to be hoisted. For still deeper mining, a *shaft* is sunk in the orebody, or more preferably, near the orebody— generally on the *footwall* side so that fractures caused by mining will not damage the shaft. An *incline, inclined shaft,* or *access spiral* may be used to gain entry, with the last method in favor where low-profile rubber-tired trucks can haul *muck* (broken ore and rock) directly from the mine face to the surface without transferring it. At great depth, a vertical shaft is generally more economic because it is more direct.

In underground ore mining, level *drifts* follow the ore, level *crosscuts*

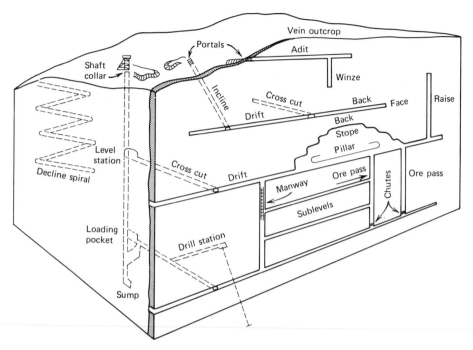

Figure 6-3.   Underground mining terms.

connect drifts, and vertical or inclined *raises* connect the workings from level to level. Raises, designed to serve as ore passes, as manways, or for ventilation, are *driven* upward. Winzes or larger blind shafts are *sunk*. Ore is removed from *stopes*, leaving *pillars* to support the walls. Just before a part of the mine is abandoned, the pillars may be mined out and the walls allowed to collapse.

Whether working in a drift or in a stope, the metal miner's working place is bounded by the *back, ribs,* and *bottom* and by the advancing or limiting *face*. The *hanging wall* of a vein or of a steeply dipping bed may be part of the back, the rib, or both. The footwall is part of the bottom and possibly part of a rib.

The language and elements of hard-rock mining are identified with tradition, but the technology has become a prime subject for research and experimentation. Now that the depletion of shallow high-grade deposits and the economic limitations to surface mining of low-grade deposits have become quite clear, the future source of mineral supplies is also clear: large-scale underground mines. On this basis, a comprehensive report on noncoal underground mining practice and an identification of important directions for research have been supplied to the U.S. Bureau of Mines by Dravo Corporation (1974). The report is recommended as a reference book for all geologists who intend to become involved in the evaluation of deep ore deposits.

Underground coal miners have a few terms of their own. A *drift mine* has a horizontal entry; a *slope mine* has an inclined entry, not within the coal *seam*. Coal seams have *roof* and *floor* rather than hanging wall and footwall or back and bottom. The extraction areas are not stopes, they are *rooms* in room-and-pillar (or bord-and-pillar) mining and *panels* in longwall mining. In longwall coal mining, working *faces* may extend for several hundred meters between two entries.

Another difference between "hard-rock" and coal mining bears special mention: ventilation. Ventilation and dust control are important in all kinds of mining, but they are especially critical factors in coal mining. Entries into a coal seam are commonly driven in multiples so that several air intakes and returns can function simultaneously. A system of *stoppings* and *overcasts* (air bridges) route the air into a coal mine panel; *brattices,* curtains of fire-resistant cloth or plastic, guide the air across working faces.

Underground mine workings may be naturally supported, may require artificial support, or may be allowed to collapse as part of a caving method, depending on their location in strong to weak ore and wall rock. Table 6-3 lists some of the principal methods. Since very few mineral deposits are uniform, most mines make use of more than one method.

Sublevel open stoping in steeply dipping orebodies with strong ore and strong walls begins with a partitioning of the large blocks of ore between main levels and raises into a series of smaller slices or blocks by driving sublevels. The ore is drilled and blasted by miners in the sublevels so that each sublevel retreats en echelon slightly ahead of the next higher level. This pattern allows the broken ore to fall directly to the bottom of the stope. Pillars of ore are left

Table 6-3. Underground Mining Methods

With naturally supported openings	Open stoping
	Sublevel open stoping
	Longhole open stoping
	Room-and-pillar mining
	Shrinkage stoping
With artificially supported openings	Stull stoping
	Cut-and-fill stoping
	Square-set stoping
	Longwall mining
	Shortwall mining
	Top slicing
Caving methods	Block caving
	Sublevel caving

at the top of the stope (crown pillars or floor pillars) to support the next major level and at the ends of the stope (rib pillars) for stability. Mill holes and draw points similar to those shown in Figures 6-4 and 6-6 collect the broken ore in the bottom or sill pillar. One particular advantage to sublevel open stopes as opposed to large single open stopes is that miners can work more safely under low backs in the sublevels than under the less certain condition of a high remote back. Also, access to the working place is easier from the service raises than it would be from the bottom or sides of a large open stope.

Longhole open stoping is a less labor-intensive variety of sublevel stoping in which larger blocks of relatively uniform ore can be taken by fan drilling and massive blasting from locations in the sublevels. As can be seen in Figure

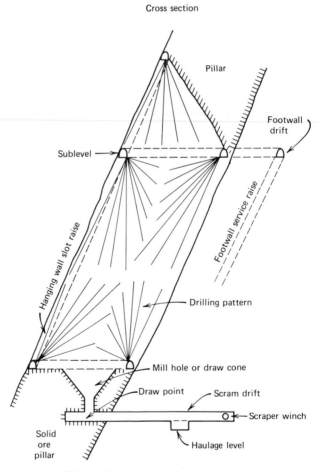

Figure 6-4.   Longhole open stoping.

6-4, the drill-hole fan pattern must be carefully designed and controlled so that unbroken "boulders" of ore will not be left and so that wall rock will not be blasted loose and allowed to dilute the ore. Sublevel longhole stoping methods are used in many large mines, including Mount Isa and Broken Hill, Australia; Sudbury, Ontario; Ducktown, Tennessee; and Mufulira, Zambia. As to dimensions: sublevel stopes at Mufulira are about 60 m in total length and height, with sublevels at about 10-m intervals and with rib pillars 6 m wide. The dip of the bedded ore at Mufulira averages 45 degrees, and the thickness of the ore is 11–14 m.

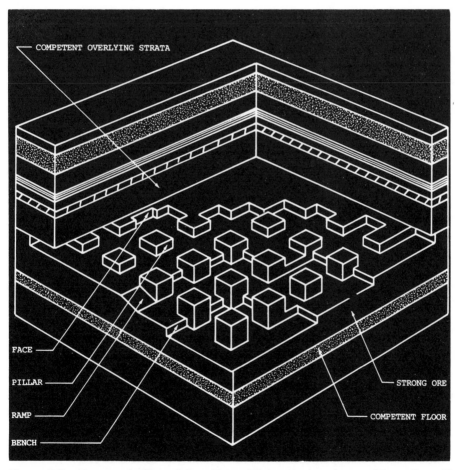

Figure 6-5.   Room-and-pillar mining. (From *Mining Engineering,* v. 27, no. 7, p. 65, 1975. Copyright © 1975, American Institute of Mining, Metallurgical and Petroleum Engineers, Inc. Used by permission of the publisher.)

Room-and-pillar mining, shown in Figure 6-5, is used in coal seams, and it also accounts for 60 percent of the total noncoal mineral production in the United States. It is an especially low cost method where rock bolts (roof bolts) rather than timbers can be used for temporary roof support, allowing fast-moving trackless equipment to operate freely. Thin beds of ore may be mined "full face" in one operation. Thick beds are commonly mined by *benching* after the first full-face advance.

Room-and-pillar mining is well suited to gently dipping and relatively uniform bedded deposits, providing the deposits are not too deep or the rock too weak for the pillars to support the overlying strata. At depths beyond 1,000–2,000 m, hard-rock room-and-pillar mining methods are considered dangerous because of rock bursts and similar results of high stress concentration. The limiting depth in coal mines is much less, on the order of 200–300 m; at this depth, a sudden "bounce" or "bump" is likely to occur when the strength of pillars is exceeded. The result may be an upheaval of the floor in a room, an emission of gas, and a stirring up of explosive coal dust.

As would be expected in mining parallel to bedding and nearly parallel to ground surface, strata control and surface subsidence are major problems in room-and-pillar mining. Stress patterns, methods of destressing "bump" conditions, and the design of pillars are major topics in the literature of rock mechanics. The percentage of mineral extraction is a principal design consideration; it amounts to about 60 percent on the average, but where pillars can be safely extracted ("robbed"), it may be increased to 90 percent or more. A great deal depends on the depth of the back and its ability to hold a wide span. In deep workings under high stress or in shallow mines beneath surface occupancy, large pillars may have to be left and a low percentage of extraction accepted.

The mechanics of stress distribution around pillars is quite complicated, and there are many factors, such as pillar height, structural defects, and creep, to account for. Even so, an empirical formula given by Thomas (1973) can be used to illustrate the problem. The dead weight (total load) of strata overlying the mining area is related to the area of the pillars in the equation:

$$\sigma_p = \frac{\text{(mined area)} + \text{(pillar area)}}{\text{(pillar area)}} \cdot \sigma_v$$

where σ_p is the average unit vertical stress on a pillar and σ_v is the total vertical stress. If the first working takes 40 percent of the coal, the remaining coal is subjected to a vertical stress of 100/60, or 1.67 times the original stress. As extraction increases, the load on the pillars rises rapidly; at 75 percent extraction, for example, pillars would be taking a vertical stress of 100/25, or four times the original load.

Room-and-pillar mining is used in the White Pine, Michigan, copper mine, in many of the Mississippi Valley lead-zinc mines, and in most North American potash and salt mines.

Room widths and pillar widths range from 5 to 20 m. In an experimental oil shale mine in Colorado, an 18-m bedded zone at a depth of 200–260 m was mined in 18-m-wide rooms, with pillars arranged in a pattern to result in an extraction of 75 percent (Agapito, 1974). In a Saskatchewan potash mine, a 2.3-m bed at a depth of 1,000 m is overlain by a high-pressure aquifer that must be kept intact; consequently, room widths are relatively small, 6.5 m, and the extraction is only 30 percent.

Shrinkage stoping, shown in Figure 6-6, is an *overhand, or back stoping* (progressing upward), method in which a major portion of the broken ore accumulates in the stope and helps to support the walls. When ore is broken by blasting, it has a *swell factor* of 1.2–1.7, depending on the thoroughness of fragmentation and the type of ground. The excess volume is "shrunk" from time to time by drawing just enough ore to allow the miners access to the unbroken ore. When the stope is completed, the remainder of the broken ore is drawn. The empty stope may be left open, supported by the pillars, or it may be filled with sand or waste rock (gob) for additional stability. Eventually, the pillars in the empty stope may be mined and the last bit of ore recovered.

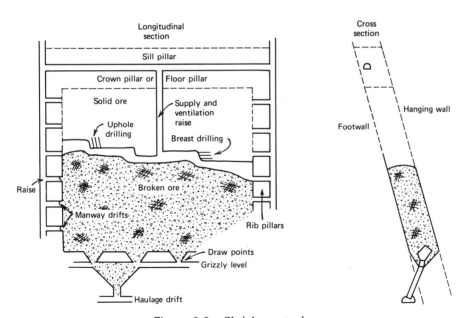

Figure 6-6. Shrinkage stoping.

Shrinkage stoping is best suited to veins and beds dipping at 60 degrees or more, with walls that would slough if left open but which are still strong enough not to squeeze or collapse and dilute the broken ore. The ore itself must have the strength to stand unsupported across the entire back of the stope until the broken ore is removed.

The ore in shrinkage stopes can be regarded as being stockpiled, to be recovered when needed; however, broken ore may oxidize in the stope and generate heat or it may become cemented, causing difficulty in subsequent mining and in mill recovery. Extraction is relatively high, amounting to 75–90 percent, but some waste zones may have to be taken as well, since the process gives little opportunity to leave selected waste pillars. Shrinkage stoping was widely practiced in former times when it had a relatively low cost, but this advantage has diminished because the method is labor intensive and difficult to mechanize.

At the Idarado mine in the San Juan Mountains, Colorado, a zinc-lead-copper vein averaging 1.5–2 m has been mined by shrinkage stoping, with stope heights of 60–100 m and lengths of 120–150 m.

Stull stoping is a variety of open stoping in which short timbers (stulls) are wedged at intervals between the hanging wall and footwall. The maximum width that can be handled in this manner is about 5 m, and both walls must be fairly competent.

Cut-and-fill stoping is used in steeply dipping veins or in beds with weak walls. However, the ore itself must be strong enough to stand across the stope back. Figure 6-7 shows an approach to cut-and-fill stoping in which fill material (sand or mill tailings) is built up as the stoping progresses. The suitability of trackless mining machinery to operations on compacted fill material has made cut-and-fill stoping an important and economical mining

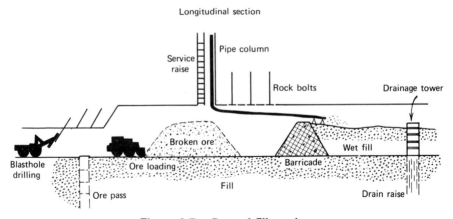

Figure 6-7. Cut-and-fill stoping.

method for wide vein deposits. A less mechanized type of cut-and-fill stoping used in very narrow veins is called *resuing*; in this method, ore and waste rock are broken separately and the waste rock is left in the stope as fill.

Veins at a depth as great as 2,700 m with widths of 3 m have been mined this way. Vein widths of 30 m have been mined at depths of 1,000 m. Stopes are as long as 600 m.

Square-set stoping is the classic approach to mining in heavy ground—classic because it was developed in the 1860s in the famous Comstock lode of Nevada. It is still in use where the ore is rich enough to permit the high cost in labor and timber, but it has been replaced in many mines by caving methods or—where dilution cannot be allowed—by mechanized cut-and-fill methods. A timber set (a skeletal box of interlocking timber on the order of 1.5–3 m wide) is emplaced as each small block of ore is taken, and a stope becomes a network of interlocking timber sets. Completely timbered areas are generally filled with mill tailings or waste rock for additional support. A common practice is to mine the pillars that remain between filled square-set stopes by cut-and-fill or top-slicing methods.

Longwall mining began in the coal fields as a method for extracting seams at depths in excess of 200 m. It is by far the most common method of working in European coal mines, where the shallower seams have been depleted. The method is also used in stratabound ore deposits, including the deep reef gold mines of the Witwatersrand, South Africa.

Longwall mining is best suited to deposits ranging in thickness from 1 to 2.5 m, dipping less than 12 degrees, with relatively incompetent overlying rock and a fairly competent floor. The method lends itself to mechanization, especially in coal mines where fast-moving shearing machines (Fig. 6-8) and ploughs move across a 100–200-m face and load coal into conveyors without any need for blasting. Mobile roof supports with hydraulic legs, adjustable to the load being supported, are moved ahead as the face is mined (Fig. 6-9).

In longwall mining, extraction is nearly complete, caving of roof rock behind the worked face is practically complete, and the resulting surface subsidence is essentially uniform. This uniformity of caving and subsidence is a major reason for the success of longwall mining at depths beyond the reach of room-and-pillar mining. The redistribution of pressures around a longwall face is shown in Figure 6-10. Development of a high-pressure zone ahead of the face in the solid ore or coal and a low-pressure zone just behind the face permits the working area to be held open by supports of very low strength. Normal lithostatic pressure is reestablished in the caved and abandoned waste, or gob, area as the working area moves away. The two sections at a right angle to the direction of face advance in Figure 6-10 show a condition in which there is a time-dependent decay in the sharp flank pressure abutment. Because regularity in the pattern and timing of face movement is so important, geologic information is a critical factor; discontinuities, such as

Figure 6-8. Longwall coal mining with a shearing machine. (Courtesy Mining Progress, Inc., Charleston, W. Va.)

Figure 6-9. Hydraulic roof supports (chocks) at a longwall face. (Courtesy Mining Progress, Inc., Charleston, W. Va.)

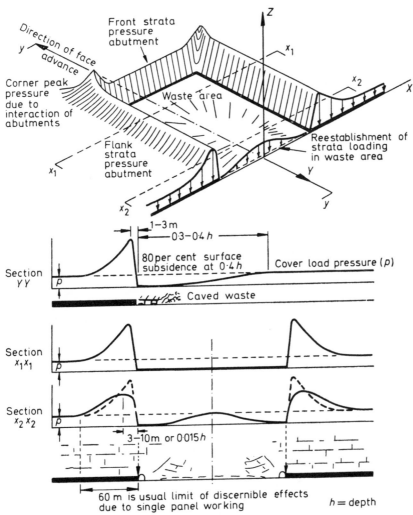

Figure 6-10. Strata pressure redistribution in the plane of the coal seam around a longwall face. [From B. N. Whittaker, "An appraisal of strata control practice," Inst. Mining Metall. (London) *Transactions*, v. 83, Fig. 2, p. A87, 1974. Copyright © 1974, The Institution of Mining and Metallurgy. Used by permission of the publisher.)

faults, washout channels of clastic rock, or splits in the seam, must be located in advance.

Shortwall mining, more flexible than longwall mining and therefore better suited to areas of heavy ground and bad roof, is actually more closely related to room-and-pillar mining. It is done in a conventional mining cycle (undercut, drill, fire, load) or by continuous mining machinery designed either for room-and-pillar or longwall systems. Coal panels up to 50 m wide are removed between entry pillars and the roof is allowed to cave.

Top slicing is a method for mining thick deposits with weak ore and weak walls. It is also used in mining pillars between filled stopes. Top slicing could be called a caving method because a mat of timber is placed on the floor of each horizontal cut, or "slice," and the overlying material is allowed to cave onto it. Extraction rates are high, approaching 100 percent, and it is a relatively safe method despite the disconcerting creaking and groaning sounds in the flexible mat overhead. Like square-set mining, it is labor intensive and requires a supply of cheap timber; therefore, it is not as widely used as in past years.

Block caving, shown in Figure 6-11, is a method with a low cost per ton and a high production rate that is suited to large bodies of weak ore with relatively weak wall rock. While the mining cost is low, the capital cost is the highest of that needed for all underground methods. Closely jointed or fractured ore with low bond strength is needed, yet is must not be so highly altered that it will repack. A block caving mine is developed by driving workings on a haulage level, on a higher *grizzly level,* and finally on an *undercutting level.* Draw-points, or drawholes, on the undercutting level are connected to the grizzly level by *finger raises.* The grizzly level is connected to the haulage level by *transfer raises.* Caving of the block is induced by drilling and blasting on the undercut level. In some mines the edges of the block are further weakened by

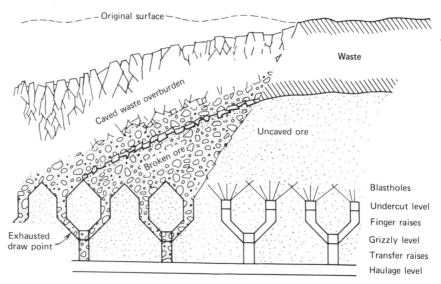

Figure 6-11. Block caving.

cut-off raises, drifts, and even shrinkage stopes. Once the ore block begins to cave, it breaks by tension and continues to break upward through the overburden for as much as 1,000 m to the surface, if all goes well. Progressive breaking and crushing feed the drawpoints. If careful monitoring shows that any part of the caving system hangs up, it must be blasted loose. Extraction is high—sometimes too high. Dilution from wall rock is hard to control and it may add as much as 10–20 percent waste material to the ore recovered in block-caving operations.

Block caving is the underground method used in the porphyry copper mines at San Manuel, Arizona, and El Salvador and El Teniente in Chile. It is the method at the Climax and Urad molybdenum mines in Colorado and the Creighton nickel mine in the Sudbury district, Ontario. Production rates at San Manuel and Climax are both on the order of 40,000 tons of ore per day.

Blocks are designed to be as large as possible so that development work can be minimized, but geotechnics dictates the maximum size of a block that will break and the amount of pressure the development workings must stand. The maximum block size is generally on the order of 50 by 100 m in area and 200 m in height.

Sublevel caving is similar to sublevel open stoping and longhole open stoping. In this method, the walls are weak and are allowed to collapse, following the ore downward. As with block caving, extraction is high and dilution can be a problem. Sublevel caving is a principal mining method in the iron orebody at Kiruna, Sweden, where the mine produces 20 million tons of ore per year.

6.2. SURFACE MINING

If we look at surface mining as geologists, we emphasize the surface and near-surface geologic environment. Temperatures, pressures, and geochemical reactions are similar to those in geomorphic processes. Ground water is surface controlled. Unconsolidated sediments and soil, the surface materials, belong to the picture. Unstable pit slopes are like potential landslides, and over-steepened bench faces are ready for mass wasting.

Looking at surface mining as an engineer, the picture will focus on the low mining cost (which makes the operation feasible) and on the ease of mechanization. Few of the restraints of underground mining apply here. Equipment can be of mammoth size, it can be moved from point A to point B without having to wrestle it through crosscuts and raises, and it can be used under relatively safe operating conditions. The engineer sees an operation that can be designed and controlled with more visible and better known parameters than would be expected underground where everything—rock characteristics, water conditions, ore grade—must be projected from a small amount of information into an unseen block of ground.

Now look at surface mining through the eyes of a passing observer. Ugly. This person may look right past the need for the materials and energy being produced and see a scar on the landscape with too much dust and noise. Underground mining is done in the miner's environment, but surface mining is done in everyone's environment, and many people would rather it were done in someone else's part of that environment.

Surface mines—and they account for 65 percent of the world's solid mineral production—are designed on the basis of how the geologist, the engineer, and the environment-conscious observer think the job should be done. All three approaches are well explained in the handbook on surface mine design edited by Pfleider (1968).

There are fewer distinct methods to consider than with underground mining. This does not mean that surface mining is as simple as the name "dirt moving," the name given to it by underground miners, would indicate. Just for convenience, the methods can be described as open-pit mining, strip mining, quarrying, and placer mining. Where the in situ leaching of orebodies is done entirely from the surface, it could be considered surface mining. In this chapter, in situ leaching is described with its other close affiliate, solution mining, in Section 6.3.

Open-pit mines (Figs. 6-12 and 6-13) range in dimensions from that of the smallest unplanned doghole to that of the world's largest mine at Bingham Canyon, Utah, where an entire mountain has been cut away and replaced by a pit 3.2 km long, 2.4 km wide, and 1 km deep. Local miners still call it "The Hill." Ore production at Bingham Canyon is approximately 100,000 metric tons per day. With a waste-to-ore stripping ratio of 3:1, this calls for drilling, blasting, and removing an average of 400,000 tons of material per day, using power shovels with 5–20 cubic meter capacity dippers, rail cars of 65–80 metric-ton capacity, and trucks designed to haul 60–140 metric tons of ore and rock.

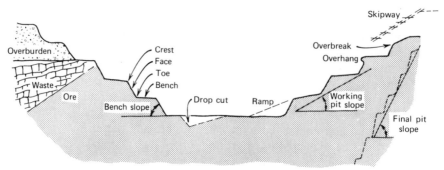

Figure 6-12.   Open-pit (open-cut, open-cast) mining terms.

Figure 6-13.   The Esperanza and Sierrita copper pits, Pima mining district, Arizona. The outcrop of the Esperanza orebody, mentioned in Chapter 3, was recognized during an examination of the New Year's Eve mine, a small underground molybdenum operation near the center of the area shown in the photograph. (Courtesy of Duval Mining Corporation, Tucson, Arizona.)

Open-pit copper mines and iron mines are such huge operations that it is difficult to get a feeling for their relative size, either visually or statistically. The Sierrita and Morenci copper mines in Arizona look equally big to a visitor. Sierrita produces 200,000 tons of ore and waste per day. Morenci produces 160,000 tons of ore and waste. The difference between the two mines takes on some significance if we consider that it would amount to the daily production of several major underground mines. There are well over 100 open-pit mines in the world that produce more than 10,000 tons of ore per day; the yearly ore production from any one of them could be considered an entire orebody in some other mines.

Benches, arranged as spirals in some open-pit mines or arranged as levels with connecting ramps in others, range from 8–40 m wide. The width is designed for safety and for the equipment used. A large-capacity shovel and the trucks that go with it require a bench at least 30 m wide. Heights of bench faces range from 4 to 8 m in weak rock and from 15 to 20 m in moderately strong rock; there are some pits with 60-m faces where the rock is strong and has few structural defects. Bench height, like bench width, depends on the size and reach of the equipment, but it is also designed for adequate slope stability. A *working slope* may have to be kept as low as 20 degrees to avoid rock failure, or it may be able to stand well at 70 degrees. The working pit slope will generally be much flatter in closely fractured areas, in places where bedding surfaces dip into the pit at a low angle, and where fracture-bound wedges "daylight" in the pit wall than it is in the more "solid" parts of the mine. The slope may therefore differ from place to place, but the transition is carefully designed. There are problems deriving from almost any abrupt change in the working slope; a "nose" extending into the pit or a channellike depression in the side of a pit is a prime location for stress concentration and premature failure. Skipways and similar breaks in the pit symmetry fall into this category. The *final pit slope,* a wall in the *ultimate pit,* will generally be higher than the working slope because it is not necessary to maintain benches. Mining to the final pit slope is rather like the final robbing of pillars in an underground mine.

Ore and rock are removed from pit faces in a drill-blast-load cycle or they may sometimes be removed from overburden areas and drop cuts by bulldozer-rippers and tractor-scrapers. Scrapers are the load-haul-dump machines of open-pit mining. At the Twin Buttes copper mine, Arizona, a fleet of 52 scrapers was used in stripping alluvium from the orebody; it took two years of continuous ripper-scraper work at a rate of 225,000 tons per day to expose the orebody.

Haulage in open-pit mines is generally by truck or rail. Some large mines also use skipways or beltways, and some use an underground haulage system as well. Haulage tunnels may be entered directly by pit trains or trucks or they may extend beneath the floor of the pit so that ore can be dropped through ore passes and crushed at that point.

Glory-hole mining is a combination open-pit and underground system. It is similar to open-pit mining with tunnel haulage except that the term refers more precisely to steep-walled, narrow pits in a vein or chimney-type orebody with funnellike mill holes in the bottom of the pit.

Strip mining (Fig. 6-14) is generally done in coal seams or in other flat-lying bedded deposits. There are several more or less comparable terms. Where coal is mined from a cropline on a hillside, it is called strip mining, or *contour mining*; these long, narrow pits may take only three or four faces of coal before the stripping ratio becomes uneconomic. Contour mining may be followed by *auger mining* the finished *highwall*. Augers range from 0.4 to 2.1 m in diameter and can penetrate for 60 m or so. If a larger amount of terrain is involved in a coal strip mine, it may be called an *open cut, open cast,* or simply an open pit.

Small-scale strip coal mining is generally done with bulldozers and front-end loaders. In larger open-pit mines, power shovels operate at the toe of a bank and draglines (Fig. 6-15) operate from an upper bench. Either of these machines may be used to make the initial *box cut* and to strip overburden; the dragline is often preferred for stripping because of its greater reach and its ability to cast spoil. A shovel has a more positive digging action, and is likely to be preferred in mining the coal seam itself. Another machine used in overburden stripping is the bucket wheel excavator, a dinosaurlike affair that can dig and stack as much as 3,000 tons of material per hour in a continuous operation without lifting and swinging; it is used only in soft overburden where blasting is not needed. And there are other monsters. Dragline bucket and shovel dipper capacities may exceed 100 m³. One dragline, "Big Muskie," in the Ohio coal fields, weighs 13,500 tons and delivers a 325-ton bucket load in 1-minute cycles. Heavy equipment, even at a fraction of this weight, demands stable ground.

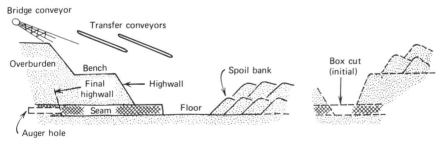

Figure 6-14.   Strip (open-cast) coal mining terms.

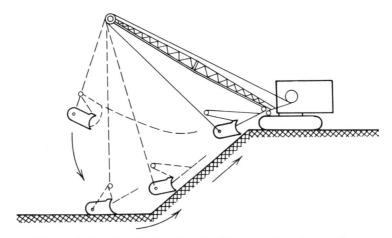

Figure 6-15.   Sketch showing the digging action of a dragline.

Quarrying is done for dimension stone and crushed stone. Dimension stone is generally loosened slowly and selectively by some method other than blasting in order to preserve the strength and shape of the product. Crushed stone quarrying for aggregate and for chemical-grade rock is done with a variety of blasting patterns, depending on the jointing pattern, on the need for a certain size product, and on the proximity to urban areas. The last point—the presence of neighbors (customers)—is one of the distinguishing features of quarry mining. Blasting vibration, dust, and the need for immediate reclamation are of more critical importance than in most other kinds of mining. Geological engineering is especially involved in quarry site selection, environmental monitoring, and multiple-use planning of the area. "Stone" is a much more important resource than its unimpressive name would indicate; it is the most widely used product of the entire mineral industry.

Alluvial and *beach placer mining* is done by dry (land-based) methods and by dredge. Dry mining differs very little from single-bench open-pit mining. The equipment is similar: bulldozers, powered scrapers, and front-end loaders for moderate-sized operations; draglines, power shovels, and bucket wheel excavators in larger mines. Hydraulic caving and removal of gravel banks are practiced in hundreds of tin placer mines in Malaya and Thailand where hydraulic giant or monitor nozzles are directed against banks up to 50 m high. Hydraulic stripping has also been used to remove overburden from some hard-rock deposits of various minerals in other parts of the world, and hydraulic "giants" are sometimes used in underground coal and uranium mining.

Dredging methods are more specifically related to placer deposits; these involve bucket-line dredges and suction dredges. Dredging is done in small lagoons or ponds excavated for the purpose, in rivers, and offshore. Large bucket-line dredges used in Malaysia handle approximately 1,500–2,000 tons of gravel per hour and generally operate in water up to 30 m deep; the deepest digging bucket dredges operate in about 45 m of water—the limitation being the weight of the bucket chain compared with the buoyancy of the dredge. Suction dredges are used in mining onshore and offshore beach sands; they can operate in fairly deep water (70 m) with submerged pumps and "jet assist" boosters. Air-lift, clamshell, and dragline dredges are sometimes used in shallow water, but they are of more interest as potential techniques for deep-sea mining.

Geological problems in placer mining are most commonly associated with the distribution of ore values; placer deposits are complex and are often difficult to sample. There are some major geotechnic problems as well. The size of boulders, the degree of gravel cementation, and the amount of bedrock relief are important. Highly fractured or weathered bedrock provides a difficult situation for the recovery of placer minerals. Arctic placer mining requires the careful thawing of permafrost overburden prior to mining. Perhaps the most critical of all geotechnic problems in placer mining deals with the preservation or restoration of the original environment. Provisions must be made for water clarification, for leveling the hummocky piles of tailings, and for placing the finer tailings on top. For a further introduction to placer mining methods and their geotechnic aspects, there are comprehensive reviews by Griffith (1960) and Daily (1968).

6.3. SOLUTION MINING AND RELATED METHODS

This is where geology, mining, and hydrometallurgy meet in a fascinating group of concepts. There should be some way of simulating nature's leaching, removal, and reconcentration of minerals and metals in the near-surface environment. There should be some way of removing minerals from a mine in the ultimate of small particles: in solution. What works in a vat or in a heap of crushed ore should work, to some extent, in the ground. Some of these things can actually be done in solution mining, and a few of them have been done on an economic basis.

Solution mining or brining of halite is a long-established method. Brine fields consist of wells spaced at intervals of a few tens of meters to several hundred meters and to depths reaching to nearly 3,000 m. The units in the system may be single wells in which fresh water and brine are handled in several strings of concentric tubing in the same borehole or they may be incorporated in a gallery mining system with separate injection and production wells. The gallery system, accounting for most of the salt obtained by brining in the United States and Canada, is generally begun by connecting the wells through a hydraulic fracturing technique in which fluid is pumped into the system at a pressure high enough to overcome the lithostatic load and split the rock along bedding. One unique characteristic of solution mining in salt beds and salt domes is that the resulting cavity may have value as a storage chamber. Better yet, the solution mining of salt may go on while the cavity is being used for storing petroleum products, with a floating hydrocarbon "pad" serving to protect the roof of the cavity and to keep the dissolving action within the main salt bed.

Solution mining of potash in sylvite beds is done in Saskatchewan and Utah. There have been a number of experiments and attempts in other potash fields—at Carlsbad, New Mexico, and in the deep Yorkshire potash in England, for example—but these have run into difficulty because of weak brine concentrations. A particular problem in some areas derives from the hazard caused by the proximity of any solution mining experiments to active conventional mines. To a potash mine superintendent, an uncontrollable ingress of water is the stuff of nightmares.

Frasch process sulfur mining has some similarities to solution mining even though it involves melting rather than dissolving. Some special geologic conditions are needed. The sulfur deposit must be uniformly permeable in itself and yet relatively isolated from other permeable strata. It must be deep enough to provide the pressure for water temperatures to attain 150–170°C without "flashing" to steam and yet not so deep that the cost of wells, about 50 m apart, becomes prohibitive. Because the melted sulfur drains into wells by gravity, the deposit must either be thick enough or inclined enough to allow the establishment of a steep drainage slope. The Frasch process is thermally inefficient, and therefore, only the richest sulfur deposits are mined in this way. Most profitable mines average about 20 percent sulfur content with some zones reaching 50 percent.

Geotechnical aspects of solution mining and Frasch sulfur mining are critical, and they present recurring problems from the design stage through the operating stage. Temperature, pressure, rock strength, and flow pattern must all be accommodated in a delicate balance. Geologists at these mines are faced with some of the same problems that reservoir engineers face in an oil field: wells cease to operate, collapse, or go barren because something is happening at the other end of a long, slim casing. A geologist must read every possible message from the well logs and gages.

Underground gasification of coal has been tried for over a hundred years, but it has not yet reached a clearly economic status. The idea is very attractive. Poor quality and thin coal seams should provide a conservation-wise source of energy even though they cannot actually be mined in competition with better coal seams. The working concept is relatively simple, and the reactions are similar to those once used in surface plants during the nineteenth-century gaslight era. Inject some combination of heated air, oxygen, or steam into a coal seam and permit it to pass through an oxidation (burning) zone and a distillation zone; collect the carbon monoxide-hydrogen gas driven off by the heat of the advancing fire. The carbon dioxide produced in the fire should have been reduced to carbon monoxide en route if there is good contact between gas and coal. Producer gas, suitable for boiler fuel, and a higher quality gas should then be available.

Some of the problems in underground coal gasification are related to a tendency for gas from the reduction zone to work back into the oxidation zone and extinguish the fire. Also, nonuniform permeability in coal seams and collapse of burned areas provide irregular gasification paths and cause a bypassing of parts of the seam. In experiments in a dozen countries, various arrays of underground workings and drill holes and several methods of "linking" the burning paths have been tried. Most of these experiments took place prior to 1960, and most were abandoned after experiencing a poor recovery of the potentially available gas, an inconsistent quality of the gas, and difficulties in controlling the burning process. Experiments in the Soviet Union have been the most persistent, with 1,000 million–1,500 million m^3 of gas produced annually between 1960 and 1970.

Nuclear fracturing and hydraulic fracturing have been proposed as one means of preparing coal for underground gasification. One other approach deserves mention. Commercial-quality gas has been recovered from multipurpose boreholes drilled into coal seams as a safety measure to release methane in advance of underground mining (Fields and others, 1975).

Underground retorting of oil shale, like in situ coal gasification, is an intriguing but difficult process. It has been done on an economic scale in a 17-m-thick deposit of oil shale at Kvarntorp, Sweden, where each gas outlet borehole is surrounded by six electrically heated boreholes in a honeycomb-like pattern. Experiments have been carried out in most of the world's major oil shale deposits. The most recent experimental efforts in the United States have been made in the Green River oil shale; a small plant in western Colorado has pumped 30 barrels of oil per day from a chimney of explosive-crushed oil shale into which burning gas is cycled. For further information on methods of in situ coal gasification and oil shale retorting, there is a review by Wang and others (1973).

In situ leaching (hydrometallurgical mining) of metal ore deposits is a method with considerable promise and some demonstrable success. It grew out of the much older copper recovery practices of heap leaching and dump leaching at the surface. Cement copper, the product obtained by reaction with scrap iron (usually tin cans), has also been recovered for many years from mine drainage water. The most common type of in situ leaching uses the same principles, with solution maintained in a very acid condition, monitored as carefully as possible in routes through old mine workings or stope fill, and circulated through launders or precipitation cones containing detinned and shredded scrap iron.

Sulfuric acid is an inexpensive and widely used leaching agent, and it is assisted—either by circumstance or design—by certain bacteria that oxidize copper and iron sulfides in acid solutions. There are other leaching agents; ammonia, for example, is being used experimentally in leaching copper from the Keeweenaw deposits in Michigan, where it has the advantage of nonreactivity with the calcite gangue.

In situ uranium leaching is done with acid solutions or with ammonium carbonate solutions. Uranium concentrate is recovered from the pregnant solutions by ion exchange. Copper may also be recovered in this way, followed by electrowinning.

Operations at the Big Mike mine, near Winnemucca, Nevada, illustrate a progression from conventional mining to in situ leaching. At the time operations began, the Big Mike deposit was delineated as having 100,000 tons of 10 percent copper ore and 700,000 tons of 2 percent copper ore. The high-grade zone, composed of chalcocite and chalcopyrite, was surrounded by the lower grade zone of mixed oxide-sulfide ore. As a dipping lenticular deposit, it was first mined in a steep-walled open pit, with the high-grade ore marketed directly and some of the low-grade ore stacked on an impermeable pad for heap leaching. After the high-grade sulfide ore was depleted, there was still

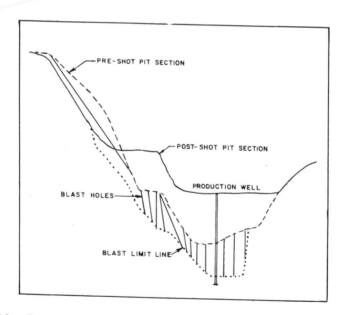

Figure 6-16.   Cross section of the pit before and after blast, Big Mike copper mine, Nevada. (From *Engineering and Mining Journal*, v. 175, no. 7, p. 68, 1974. Copyright © 1974, McGraw-Hill, Inc. Used with permission of McGraw-Hill Book Company.)

about 400,000 tons of lower grade, mixed ore remaining in the walls and bottom of the pit. The stripping ratio, 6.5:1, was unacceptable. A more favorable approach, based on field and laboratory tests, was found to be in situ leaching with an expected recovery of 70 percent at a pH of 2.0 for the leach solution.

The in situ leaching operation at the Big Mike mine is shown in Figures 6-16 and 6-17. A single pit blast with ammonium nitrate-fuel oil (ANFO) explosives prepared the ground in a way that fragmented the ore to an effective size without loosening the relatively impermeable wall rock. During the leaching operation, monitor wells near the pit served to check on any possible loss of solutions.

Figure 6-18 shows methods for preparing larger and deeper deposits for in situ leaching. Escape of leach solutions, always a matter of concern, can be prevented as shown in Figure 6-19, by maintaining a static water table, by chemical grouting, or—if the rock is right—by blasting in a specific pattern to create a compact zone. At the Old Reliable copper mine, a solution mining operation near Mammoth, Arizona, a static water table acts as the barrier to downward flow of pregnant solutions (Ward, 1973).

Figure 6-20 shows the distribution of injection, production, and monitor wells at the Clay West mine, an in situ uranium leaching operation in southeastern Texas. In this mine, the natural permeability of the sandstone ore zone is high enough to permit percolation but yet there are some naturally impermeable clay beds in the section to serve as barriers. Production at the

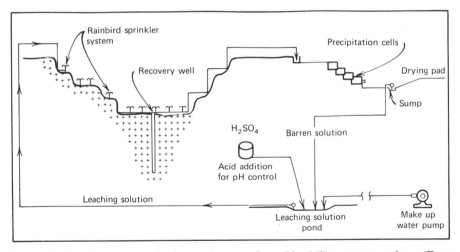

Figure 6-17.   Flowsheet of pit leach operation, Big Mike copper mine. (From *Engineering and Mining Journal*, v. 175, no. 7, Fig. 3, p. 68, 1974. Copyright © 1974, McGraw-Hill, Inc. Used by permission of McGraw-Hill Book Company.)

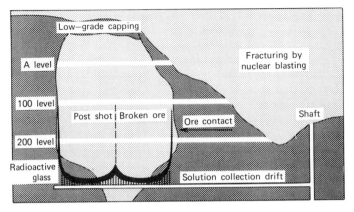

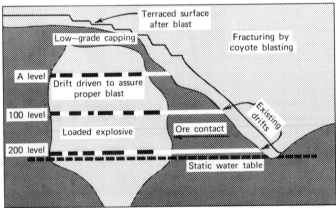

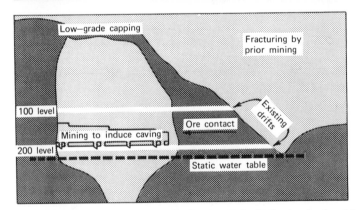

Figure 6-18.   Alternative methods for breaking an orebody for in situ leaching. (From M. H. Ward, "Engineering for in-situ leaching," *Mining Cong. Jour.,* v. 59, no. 1, Fig. 1, p. 22, 1973. Copyright © 1973, American Mining Congress. Used by permission of the publisher.)

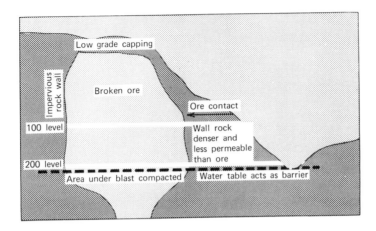

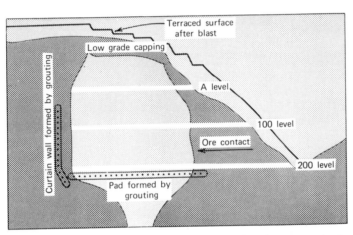

Figure 6-19. Solution control barriers. In the best situation an orebody would be enclosed by naturally impermeable barriers, as in sketch A. It is often necessary, however, to use grout curtains, as shown in sketch B. (From M. H. Ward, "Engineering for in-situ leaching," *Mining Cong. Jour.*, v. 59, no. 1, Figs. 2 and 3, p. 23, 1973. Copyright © 1973, American Mining Congress. Used by permission of the publisher.)

Clay West mine and at several other in situ uranium leaching operations in the region comes from depths of 100–150 m.

In situ leaching is not applicable to most ore deposits. One particular difficulty is similar to that encountered in coal gasification: irregular permeability. A deposit must either be uniformly permeable in its natural state or be

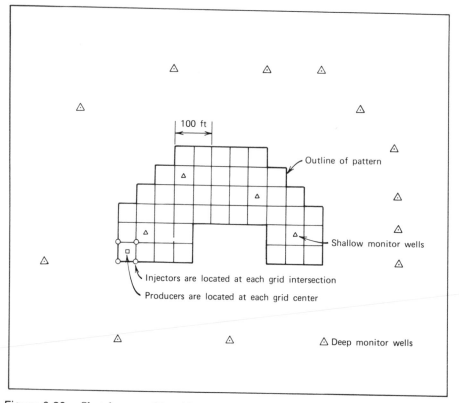

Figure 6-20.   Sketch map of leaching pattern at the Clay West mine, Texas. (From *Engineering and Mining Journal*, v. 176, no. 7, Fig.2, p. 75, 1975. Copyright © 1975, McGraw-Hill, Inc. Used by permission of McGraw-Hill Book Company.)

capable of being made permeable by some method, such as conventional explosive shattering, nuclear explosive shattering, block caving, or hydraulic fracturing. Fragmentation in the shattering of the ore deposit is important; if the fragments are too fine, they may pack together and lower the permeability; if too coarse, they will not expose enough area to the leaching fluids.

Isolation of the orebody from permeable zones in the wall rock must, like the permeability in the orebody, be a natural condition or be capable of being induced. Other conditions to monitor during mining, and to estimate in advance, are ground-water contamination, plugging by iron hydroxide or acid-swollen clay, "piping" or short circuiting of solutions, and the collapse of leached zones.

In situ leaching and related techniques have received special attention in recent years because they create very little surface disturbance, have relatively low capital requirements, and have a low safety hazard. The proceedings of a special symposium on the topic (Aplan and others, 1974) provide a good state-of-the-art review.

Borehole hydraulic mining is an erosional technique that breaks coal and ore into small fragments rather than decomposing them. The work is done by water jets at the bottom of a drill hole, and the fragments are pumped to the surface. Testing of the technique in friable sandstone uranium orebodies has shown it to be especially attractive, but there are many operating parameters yet to be determined.

6.4. MARINE MINING

Beach placer mining is a reality; it has a particular attraction from the standpoint of its low need for energy because the ore has been ground, crushed, and sorted by natural processes. In another kind of marine mining, undersea sulfur has been recovered by the Frasch process from land-based directional drill holes and from offshore drilling platforms. Undersea underground mining from shaft sites on land has been going on as long as mining itself—literally. The ancient mines at Laurium went undersea. Open-pit mining has been done as an undersea operation in at least two places: in the Castle Island barite mine, southeastern Alaska, and at the Cape Rosier copper-zinc mine, Maine. Consideration is being given to "rock site" offshore mining in which shaft sites are located in artificial islands or submarine chambers.

These are shallow-water operations, using land-based technology. The next steps, onto the continental slope at depths beyond 200 m and onto the ocean floor beyond depths of 1,000 m, are still in the realm of feasibility studies. Unexploited reserves of manganese, copper, nickel, and cobalt are interesting because 10 percent of the entire floor of the Pacific Ocean is thought to be covered by manganese nodules containing 1–3 percent of both copper and nickel plus several tenths of a percent cobalt. Metal-bearing muds in the Red Sea and in other areas of hydrothermal activity contain enough zinc, copper, and silver to suggest mining, but the mining problems are however as overwhelming as the possible extent of the resources. In balance, there has been sufficient encouragement to draw nearly every major maritime nation and several large mining companies into marine mining research programs. An experimental ship, the *Glomar Explorer,* has been designed to collect 5,000 tons of nodules per day at depths as great as 5,000 m. A good overview of marine mining technology is provided by Cruickshank (1973). Also, there is a special issue of *Mining Engineering* (April 1975) devoted to ocean and offshore mining.

6.5. ENVIRONMENTAL ENGINEERING

For a detailed account of environmental problems and answers, we do not have to search far; nearly every mining magazine carries frequent articles on the "status of mining's environmental impact." We can read about the same problems from the nonmining end of the telescope in the journals *Environmental Engineering, Environmental Science and Technology,* and *Environmental Geology*. For a broader view (and a geologist certainly needs one) *In Command of Tomorrow* by Brubaker (1975) is recommended.

Where does the geologist fit in? All the way. An environmental impact statement is required by the U.S. government for all major undertakings, and similar detailed statements are required in most of the industrialized countries. Questions must be answered. What will be likely to happen to the surface water and ground water? To the shape of the land? Will there be ground subsidence? Damage to other minerals? What will be the composition and stability of dumps and tailings ponds? There are still more questions, and the answers generally require some quantitative geologic input. The applications of geomorphology are still just as apparent as in all the preceding chapters.

The principal guidebook to environmental impact statements is the U.S. Geological Survey Circular 645 (Leopold and others, 1971). Further guidelines are given by Hyatt (1973) and Turner (1976). A brief commentary on environmental impact statements by two scientists of the West Virginia Geological Survey (Lessing and Smosna, 1975) points out the geological topics that should be included.

The exploration geologist fits into environmental engineering even before a mineral discovery can be called an orebody. As stated in the close of Chapter 5, it is important to establish what the environmental situation *was* before the current mining project came along.

PART THREE
THE ECONOMIC
FRAMEWORK

We mine because someone in the nearby city, in the nation, or on the other side of the world needs the minerals and is willing to pay a reasonable price for them. This is reasonable insofar as the miners are concerned; enough to make their efforts worthwhile. This is also reasonable insofar as the users are concerned; within their limits of monetary, environmental, social, and political cost. But above all, the arrangement must be acceptable to the people who own the minerals.

Unless these factors can be seen in a workable balance when related to a model of mineral deposit, mineral exploration is no more than an interesting academic exercise.

Chapter 7 gives an overall look at the relationship between demand and supply and between goals and capabilities. Comments are included on various approaches to a mineral resource policy, how they have been formulated, how they are changing, and what they may mean to the mining and exploration geologist.

Chapter 8 gets down to specifics—to individual orebodies and ways of expressing their value.

CHAPTER 7
MINES AND MINERAL ECONOMICS

When a poor country gets richer, the quality of life may not improve, but the country certainly develops an insatiable appetite for minerals. And if this cannot be satisfied by its own mines, it will be satisfied by foreign production. National mineral policy and international mineral competition take on all shades of meaning. Since the picture has far greater depth than can be accommodated here, this chapter will concentrate on the aspects most likely to affect a geologist who is outlining an exploration target or keeping a mine in operation.

7.1. MEETING THE DEMAND FOR MINERALS

People emerged from the Stone Age (they became civilized) when they began mining. Yet, they have acted in the most uncivilized ways because of minerals. Nations and tribes have coveted minerals, stolen them, conspired, and fought over them. They still do.

Summarizing the relationship between civilization and minerals: minerals are difficult to live with and impossible to live without.

In balance, miners have provided the needed materials rather well, but not everyone is convinced that they can continue to do so. A relatively new kind of concern for resources, energy, and the environment has forced us to take a good look at what has happened during the last 20 or 30 years. Here is one of the findings: the present doubling time for world population growth is 35 years; the average doubling time for mineral consumption is about 25 years, or by some projections, 12 years (Cloud, 1975).

A supply-demand projection for some major minerals is shown in Figure 7-1. Note that the key term is "reserves," not "resources"; the profound difference between the two terms is recognized by the U.S. Geological

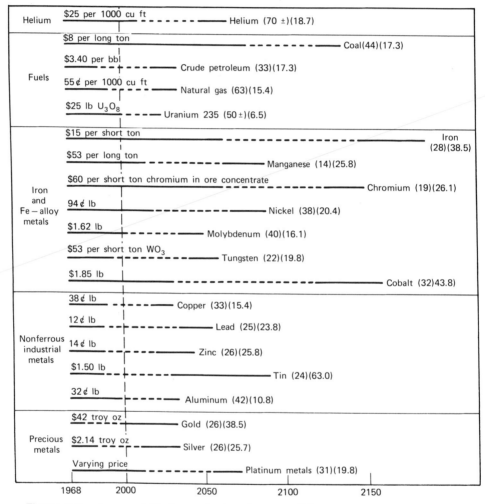

Figure 7-1. Apparent lifetimes of mineral commodities at average estimated global rates of demand to year 2000. Solid line at left is based on reserves at specific market prices. Dashed line is based on five times these reserves. Solid line at right is based on ten times these reserves. The numbers in parentheses are the U.S. percentage of world consumption and the doubling time of world consumption. (From P. Cloud, "Materials, their nature and nurture," *Earth Sci. Newsletter*, p. 3, Dec. 1973, by permission of the author.)

Survey in assessing national reserves and resources, as shown in Figure 7-2, a diagram with two axes, one representing the degree of certainty about the existence or magnitude of the resources and the other representing the economic (technologic) feasibility of resource recovery. The terms "proved," "probable," and "possible" apply to reserves, the portion of the identified resources that can be mined within current technological and economic conditions. The remaining identified resources are *paramarginal* (recoverable at some economic level, such as 1.5 times the present cost) or *submarginal* (conceivably recoverable). The sphalerite fillings in coal cleats in the Illinois Basin, mentioned in Chapter 2, are a paramarginal zinc and cadmium resource. A paramarginal uranium resource is contained in the Phosphoria Formation of Idaho and the adjacent states of the Western Phosphate field:

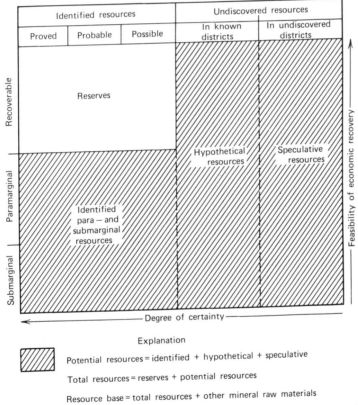

Figure 7-2. Classification of mineral resources being used by the U.S. Geological Survey in assessing total mineral resources in the Unites States. (From V. E. McKelvey, "Mineral potential of the United States," in *The Mineral Position of the United States 1975–2000*, E. N. Cameron, ed., Fig. 4.1, p. 69, 1973. Copyright © 1973, Wisconsin Press. Used by permission of the publisher.)

about 650,000 tons of U_3O_8 in rock with an average grade of at least 0.012% U_3O_8 and with an average grade of at least 31% P_2O_5. If this resource were to become a reserve, uranium would be a byproduct and its level of recovery would be controlled by the market for phosphate rock. A submarginal uranium resource is contained in the Chattanooga Shale in eastern Tennessee: 6 million tons of U_3O_8 in rock with a grade of 0.007% U_3O_8.

Hypothetical resources in Figure 7-2 include those that have escaped discovery in known mining districts (mining geologists would say "in elephant country"). *Speculative* resources are those that could occur in untested places where favorable conditions are known or suspected. Undiscovered resources can only be described in order-of-magnitude figures based on such indices as the amount of unexplored ground and the patterns of already known orebodies.

McKelvey (1973) gives a full explanation of the U.S. Geological Survey classification, with examples of reserves and resources of major minerals. Figure 7-3 is taken from this reference. The estimates differ according to the geologists who made them; this is understandable, and it is good practice to show the range in opinion.

Because most functions performed by one mineral can be performed by another mineral, James Boyd (1975), former director of the National Commission on Materials Policy, suggested that a third dimension be added to the Geological Survey classification, one which would show several use-related minerals in their interchangeability relationships. This third dimension would expand resource and reserve estimates considerably.

Boyd commented on another aspect of reserves and resources. There is an all too prevalent practice of putting inherently limited reserve figures into a statistical analysis and calling them "resources" in order to calculate when mineral supplies will run out. The result is about what we would expect: nonsense.

Much of our feeling for what will happen to mineral demand and supply in the next few decades is related to our assessment of two opposing premises. One premise, the one most often connected with making "resources" from reserves, holds that the world population and the demand for minerals will skyrocket, while technology (*and* geology), interchangeability, and economics wallow along at their present levels. The results: imminent and measurable worldwide limits to growth. A contradicting premise, aptly labeled the "cornucopian view" by Brooks and Andrews (1974), holds that we can never run out of minerals as long as we have such comforting data as the calculations that each cubic kilometer of average crustal rock contains 2×10^8 tons of aluminum, 1×10^8 tons of iron, 8×10^5 tons of zinc, and so on. Technology is expected to come to our rescue. The view from either extreme is ridiculous. But, as with most oversimplified views, both have elements of truth.

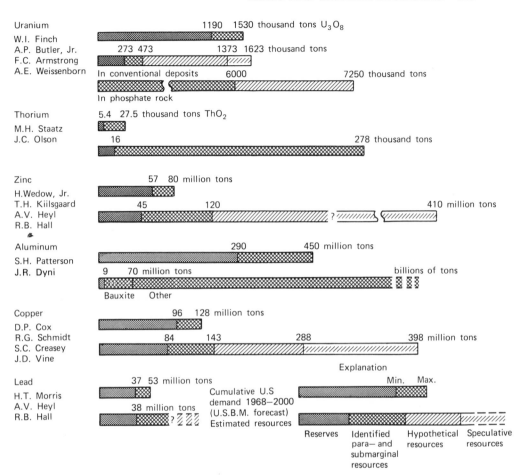

Figure 7-3. United States resources of uranium, thorium, and major nonferrous metals compared with projected demand through year 2000. Names beside each resource bar indicate the U.S. Geological Survey geologists who prepared the individual estimates. (From V. E. McKelvey, "Mineral potential of the United States," in *The Mineral Position of the United States 1975–2000*, E. N. Cameron, ed., Fig. 4.3, p. 79, 1973. Copyright © 1973, University of Wisconsin Press. Used by permission of the publisher.)

We can learn some of the truth by considering this: patterns of mineral resources take on different meanings when they are related in scale to the crust of the entire earth, to countries, to regions, to districts, and to individual mines. The cornucopian view would have its closest approach to reality at the entire earth scale. The limits-to-growth premise could approach reality in an industrial region that, for reasons of politics and economics, cannot replenish its minerals by shipments from elsewhere, by improved technology, or by substitution. For a mine geologist who has "thrown the book" at every possible site for additional ore and has drilled exploration holes in every conceivable direction from a depleted orebody, a local limits-to-growth premise *is* reality.

Earthbound by C. F. Park (1975) provides good reading; it is a geologist's realistic view of the future relationship between man and minerals. "The shrinking world of exploration," an expression coined by Walthier (1976), has been used to describe an apparent trend toward mineral shortages induced by governmental policies rather than resulting from geologic patterns. For more specific mineral resource projections to the year 2000, books edited by Cameron (1973), Marsden (1974), and Govett and Govett (1976) are recommended. Geologists should read the Club of Rome report on the predicament of mankind edited by Meadows and others (1972) and, for balance, an incisive review of the book entitled *The Predicament of the Club of Rome* by Singer (1972). An authoritative, well-condensed, and interesting forecast of mineral demand with reference to Britain has been made by West (1973). A Russian view, based on statistics from outside the Soviet Union and written in English, is furnished by Lukashev (1974). A part of every responsible prediction regarding the grade of ore to be mined in future years is the energy barrier mentioned in Chapter 6. A review of this aspect, with specific examples in regard to metals and existing technology, is provided by Page and Creasey (1975).

7.2. THE LIFE CYCLE OF A MINING OPERATION

The life cycle of a mine or of a mining district (both terms require that something be extracted for profit) is essentially a function of the depletion process. Mines, like people, pass through the stages of youth, maturity, and old age. Unlike people, mines are often resurrected or at least rejuvenated. Figure 7-4 shows two periods of rejuvenation in the life cycle of the San Francisco (Oatman) gold mining district, Arizona, first as a result of technologic improvements and later as a result of a sudden increase in metal price.

The time scale in the life cycle of a mine is indefinite, and we cannot even say that a cycle has been completed as long as there is the possibility of geologic reexamination and the discovery of unsuspected reserves. Some mines have had short, colorful life cycles based on exposed mineralization of bonanza grade but with no apparent roots capable of supporting the cost of deeper development or larger scale operations. Others have had a discontinuous and puzzling history, lasting for decades or for centuries, during which several efforts have failed or have reaped only limited rewards. Engineers and geologists keep trying to unravel the puzzle. Still, some mining operations have supported industrial complexes for 100 years or more despite lowering

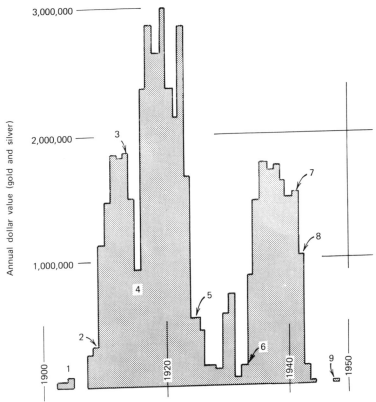

Figure 7-4.   The San Francisco (Oatman) mining district, Arizona. *1*. Early production from exposed orebodies. *2*. Major mine began production from small orebodies (1909). *3*. New company formed to develop deeper ore (1913). *4*. Major mine began production from deeper orebodies (1916). *5*. Major mine closed. Mills continued operating on custom ore (1924). *6*. Price of gold raised from $20.67 to $35 per ounce (January 1934) followed by redefinition of ore reserves and reopening of mines. *7*. Wartime shortages of labor and equipment. *8*. Wartime closing of gold mines (1942). *9*. Postwar mining attempt (1948).

ore grade, increased depth, and tightening economic conditions. In a few remarkable instances, single mines have had centuries of continuous production and are likely to continue for some time; the Almadén mercury mine in Spain, for example, has maintained such a record since 1499.

At the beginning of a representative life cycle, the discovery of mineralization and the opening of some minor high-grade orebodies may antedate the recognition of the main orebodies by many years. At Mount Isa, Australia (Fig. 7-5), the gold mines of the 1870s were opened and exhausted within a

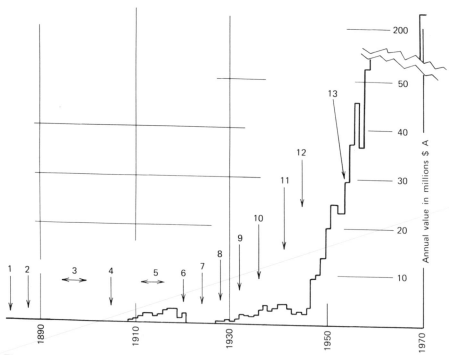

Figure 7-5.   Cloncurry–Mt. Isa mineral field, Queensland, Australia (uranium production excepted). *1*. Early gold production (1867–1900). *2*. Early copper production (1884–1887). *3*. Minor gold-copper-silver production. *4*. Smelters and railway constructed (1908). *5*. Copper and gold production. *6*. Rich copper ore depleted, smelters closed (1920). *7*. Discovery of the Mt. Isa silver-lead-zinc deposit (1923). *8*. Discovery of the Mt. Isa copper deposit (1928). *9*. Silver-lead smelter built (1931). *10*. Zinc production begun (1935). *11*. Copper production replaced silver-lead-zinc production (1943). *12*. Postwar resumption of silver-lead-zinc production (1946). *13*. Copper smelter built (1953).

short time. Copper ore was mined on a moderate scale between 1909 and 1920 and the orebodies were apparently depleted. The major silver-lead-zinc and copper orebodies at Mount Isa were not recognized until several years later, and production from them did not surpass that of the earlier mines until the 1930s.

At the far end of the life cycle, few mines die as suddenly as those in the San Francisco district. The decision to abandon an operation is often made after several years of barely profitable operations in which the producers are simply holding things together in the hope that their geologists (by this time expected to show signs of genius) can find some more ore near the mine. Even after the main operation has been suspended by the company, there may be quite a few years of low-cost scavenging operations carried on by independent lessees who keep mining narrow ore shoots until they have blasted "just one more round" and made their reluctant exit.

Between the beginning of the cycle (sometimes with a bang) and the end (often with a whimper), there are the overall characteristics illustrated in Figure 7-6 in the following pattern:

A. Discovery in the district—prospects and small mines opened and abandoned without significant production.
B. Repeated examination of the district by geologists and engineers.
C. Recognition of a potential major orebody.
D. The preproduction interval.
    1. Preliminary estimates of geologic, technologic, and economic conditions.
    2. Preliminary financing on the basis of high risk.
    3. Delineation and testing of the orebody.
    4. Further financing on the basis of reduced risk.
    5. Development of the mine, supporting plants, and townsite.
    6. Employment and training of labor and technicians.

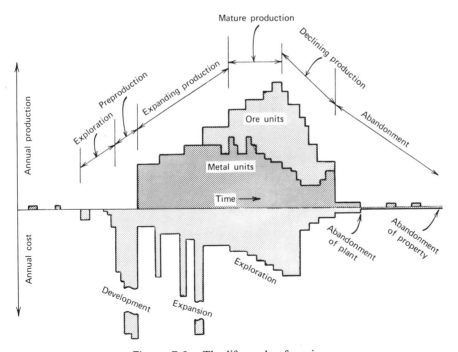

Figure 7-6. The life cycle of a mine.

E. Expanding (youthful) production.
   1. Beginning of dividends to investors.
   2. Addition of new tunnels, shafts, pits, and processing plants.
   3. Vertical growth toward higher value, finished products through smelting, refining, and fabrication.
   4. Horizontal growth toward control of additional materials and facilities, such as
      a. Limestone and coal for smelter.
      b. Phosphate rock for use with smelter acid in manufacturing fertilizer.
      c. Power plants, cement plants, railroad system, explosives and machinery plants within the immediate economic area.
      d. Adjacent mines for access as the mine grows larger.
F. Mature production.
   1. Innovations in mining and processing to offset lowering grade and rising costs.
   2. Verification of the limits of the orebody.
   3. Innovations to extend the life of the orebody.
   4. Increased local exploration for possible extensions and increased "outside" exploration.
   5. Cost reduction and extension of machinery life.
G. Declining (old-age) production.
   1. Sale or lease of assets to a neighboring mine.
   2. Cutbacks in local exploration and development with continuing or increased "outside" exploration.
   3. Blending of ore from pillars with lower grade ore.
   4. Custom processing of ore from other mines.
   5. Cost reduction by concentrating on fewer stopes and working benches.
   6. Mining of shaft pillars.
H. Mine abandonment, with
   1. Salvage of machinery.
   2. Departure of most of labor force.
   3. Milling or leaching of mine dumps.
   4. Custom milling and smelting of ore from other mines.
   5. Lessee operations—sporadic as economic conditions change.

The final phase, abandonment, is not necessarily connected with the physical depletion of the orebody. A mine may be prematurely abandoned by reason of higher taxes, higher operating costs, or a lowering of mineral price. The cost of meeting such emergencies as large inflows of water may be higher than the value of the remaining ore. Prolonged strikes, expropriation, or wartime closing orders may bring an operation to an untimely end. At any rate, the orebody ceases to be an orebody.

Geologists are concerned with every phrase. They enter the scene early, when they are involved in recognition of the orebody. Their work during the preproduction interval is the focus for financing and development. During the youthful phase, they are constantly gathering data and projecting the boundaries of the orebody, while examining nearby ore deposits and nonmetallic deposits for acquisition. Mature production brings the most painstaking local exploration and geotechnics problems. With declining production, the geologists' responsibility changes from an all-out search for the last ore in the mine to a search for a new orebody, first in a nearby location and then at some distance away. Abandonment of a particular mine does not release them. They retain the responsibility for evaluating sources of custom milling and smelting ore, for dealing with the hydrologic problems connected with metal recovery from dump leaching or in situ leaching and for keeping in touch with lessee operations.

If geologic advice and direction are not taken into account at every stage in the life cycle, it may be the fault of the geologist for not learning to "speak the same language" as management. In foreign operations, this can be taken literally as well as figuratively.

7.3. MINERALS AND GOVERNMENTS

Miners, unlike politicians, have no special need for praise; instead, they have a need to be recognized as contributors to a state's economy, and they have a need for profit.

The name of the political system does not matter at this point. The "profit" or "rentability" guideline (value of output minus expenditures incurred in production) is as widely used in collectivized economies as in capitalist or free-enterprise economies. So, in fact, are the terms "capital" and "enterprise."

The largest profit figures and the largest value-added figures may appear in some intermediate product or in the final product made from minerals rather than in the mined minerals themselves. Still, in the final accounting, a certain amount of this is credited to the mine and to the fact that it delivers materials that could not be obtained more cheaply from another source. Stated again: miners need profit. A geologist looks for minerals that can be mined at a profit. No private company and no government will finance an obviously unprofitable mine, nor will they operate a mine for long at a loss. When we read of mines operating in the red, we have to look somewhere "downstream" toward the manufactured product or into future hopes (sometimes desperate) for the rest of the story. Even where a mine is kept in operation by a government subsidy, it is because the value or potential value (social, political, or strategic) exceeds the expenditure, and this relationship can be expressed as a profit.

Why all of this on "profit"? Because the miner's contributions to the economy may be appreciated—even praised—but the miner's best friends sometimes overlook the profit factor in their enthusiasm to benefit from the good work. A commonly cited example relates to the state of Minnesota's tax on iron ore reserves. There came a time when high-grade "direct shipping" ore could no longer be mined at a profit. Low-grade ore, taconite, could not be mined either, unless enough reserves were outlined to justify the huge capital investment. No one could afford to outline such extensive reserves and then pay for simply having them in the ground. Eventually, a more favorable taconite reserve law was passed, but not until a number of mining companies had given up in Minnesota and had sent their exploration teams to Canada and elsewhere. During those years, once-wealthy iron mining communities decayed and some were even classified by the federal government as "depressed areas."

National Mineral Policy

In many countries, the shortsighted practice of encouraging any and all kinds of mining, taxing them out of existence, and then subsidizing their return has been rejected. In its place, national mining policies spell out what is intended. The following statement from the U.S. Mining and Mineral Policy Act of 1970 will serve as an example.

The Congress declares that it is the continuing policy of the Federal Government in the national interest to foster and encourage private enterprise in (1) the development of economically sound and stable domestic mining, minerals, metal and mineral reclamation industries, (2) the orderly and economic development of domestic mineral resources, reserves, and reclamation of metals and minerals to help assure satisfaction of industrial, security and environmental needs, (3) mining, mineral, and metallurgical research, including the use and recycling of scrap to promote the wise and efficient use of our natural and reclaimable mineral resources, and (4) the study and development of methods for the disposal, control, and reclamation of mineral waste products, and the reclamation of mined land, so as to lessen any adverse impact of mineral extraction and processing upon the physical environment that may result from mining or mineral activities (Public Law 91-631).

The words "stable" and "orderly" are significant. In modern-day capital-intensive mining, the exploration effort may take several years, a preproduction period may take several more years, and the payback period may take several additional years. The outlook *must* be for stability. For this reason, the stability of the government itself and evidence of orderly administration are as important as the wording in a national mineral policy. Naturally, there are high-risk investors who will operate in areas and countries with unreliable mineral policies, but their geologists will look for high-grade orebodies that can be brought into production as quickly as possible. They have no option.

The concise wording of most national mineral policies does not do justice to the number of factors that have gone into their formulation and to the complexity of their administration. An interim report on Canada's approach to a mineral policy (Information Canada, 1974) lists the following considerations, all of which have nearly limitless aspects.

I. International factors bearing on mineral policy:
  1. The world's mineral supply base
  2. Ocean mining
  3. Mineral acquisition by consuming nations
  4. Trading blocs
  5. Producer nation arrangements
  6. The general agreement on tariffs and trade (GATT)
  7. International corporations
  8. International monetary arrangements and monetary stability
II. Internal factors bearing on mineral policy:
  1. Social considerations
  2. Stability of communities, employment, and income
  3. Land-use competition and environmental quality
  4. Crude exports or further processing
  5. Increased foreign control of the mineral industry
  6. Increased business uncertainty
  7. An opportunity for improved national unity

Some of the significance of a mineral policy to a nation's economy—its value to the country—can be seen in Figure 7-7 in which the processed materials of mineral origin as well as the minerals themselves are given a value. This analysis still does not place a value on mineral products used in the advanced stages of manufacturing. If this ultimate contribution to the economy could be expressed, it would indicate that very little in the Gross National Product of an industrial nation escapes some dependence on minerals from domestic mines or from foreign mines. Civilized man is a miner.

Producer Nation Agreements

The economic and political power or "leverage" available to organized groups of mineral exporting countries has become especially apparent in

(ESTIMATED VALUES FOR 1975)

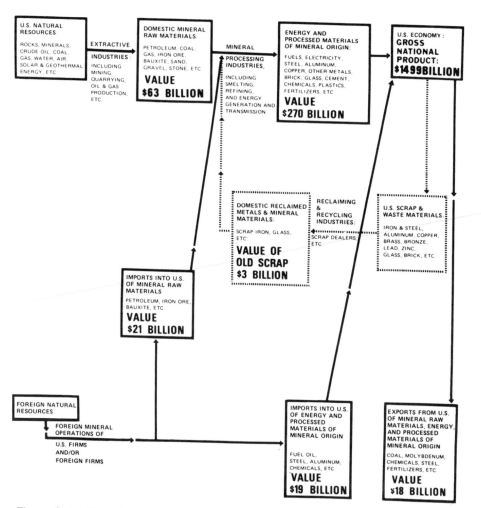

Figure 7-7.   The role of minerals in the U.S. economy. (From "Status of the Mineral Industries, 1976," courtesy U.S. Bureau of Mines, Department of the Interior.)

recent years. The total impact of the oil embargo action taken in the early 1970s by the Organization of Petroleum Exporting Countries (OPEC) on world energy mineral prices, supplies, and exploration patterns was, and is, almost too much to measure in understandable terms. The Council of Copper Exporting Countries (CIPEC) has had a substantial effect on world trade, and there are measurable effects from groups of bauxite, uranium, and mercury producing countries acting in concert. The ideas are to stabilize markets, to obtain a larger share of mining profits, to promote domestic mineral processing, and to use minerals to political advantage. These ideas are attractive to mineral exporting countries, and they add a dimension in raw materials strategy to the considerations that go into selecting exploration targets.

The effects of strategic control by groups of mineral producers (companies as well as countries) depend on a geologic condition: inhomogeneity. A square kilometer of Jamaica is underlain by more mineable aluminum ore than an equivalent area in Norway; the Norwegian area produces more electricity, and a square kilometer of Holland may contain the market for aluminum metal—representing Jamaican bauxite and Norwegian electricity. A "one-world" concept is needed, but since it is not yet a reality, the mineral importing countries respond to geologic inhomogeneity in various ways. One way is to stockpile the scarce commodities; the United States does this in a $700 million mineral and metal stockpile; France stockpiles copper, tungsten, and other critical industrial materials. Another way to overcome inhomogeneity in mineral production is to stimulate domestic exploration for the most critical minerals insofar as permitted by the geologic setting. Britain encourages mineral discovery by subsidizing 35 percent of the cost of approved nonferrous metal exploration projects. Then, there is the sea floor—international terrain—with metal-bearing nodules and sediments that are particularly attractive to industrialized, mineral-deficient countries, such as Germany and Japan. There are many other approaches to assuring a steady mineral supply, including of course, playing the necessary foreign politics.

The position of a geologist in all this is to do the job as well as possible under the provisions of one or more national mineral policies. It is increasingly common to have a company from one country acting as a mining contractor, banks from other countries providing the financing, mineral purchasers from still other countries contracting for the product, and the host country—the owner of the minerals and the taxing authority—setting the operating rules.

Mineral Ownership and Taxation

The literature of mining law—mineral land law—contains much more than a dry enumeration of rights and responsibilities. It is an account of history in the framework of geology, some of which makes fascinating reading, and there are case records that read like adventure stories.

Parr and Ely (1973) give a short summary of North American mining laws. For more detail, there are "legal guides to prospectors" published by nearly every state bureau of mines in the western United States, and there is the five-volume *American Law of Mining,* edited by the Rocky Mountain Mineral

Law Foundation (1966). Foreign mining laws are summarized in a series of U.S. Bureau of Mines information circulars, edited by Ely (1972–1974).

Because mining laws are based as much on precedence as on current mineral policy, they provide insight to national attitudes that have been shaped through the years. Insight is one of the first capabilities needed by geologists working in any country—including their own.

Mexico's centrally administered mining law reflects the nation's trend from small, locally owned operations and large, foreign-owned enterprises serving the export market to the current picture of large-scale "Mexicanized" industries devoted to the production of minerals needed in the country's own economy as well as for export. Even in its most modern form, however, Mexico's mining law contains elements of Spanish colonial mining statutes and of Mexico's geologic-historic association with epithermal silver deposits.

Canadian mining law emphasizes the history of the nation as a federation of relatively independent provinces, each with its own special geologic terrain, and most with an extensive northern frontier. It is only in the territories, Indian reservations, and national parks that the federal government of Canada holds title to minerals.

United States mining law reflects a federal system with more central government control than in Canada and less than in Mexico. It can also be read as a record of egalitarian frontier philosophy, accelerating industrial demand for minerals, and a growing awareness of the need for conservation.

United States Mining Law. Federal, state, and private lands are involved, with the following elements.

Federal lands amount to about one-third of the nation's land. They are located throughout the country, principally in the western states. The right to mine on federal lands is obtained by staking claims under the provisions of the Mining Law of 1872 or by entering into an agreement according to the Leasing Act of 1920. Rights to the surface use of these lands are limited, and they may be held by other people.

Under the 1872 law, a discoverer marks one or more claims on the ground, does a certain amount of "location work," and records the fact with the county authorities. The methods of monumenting claim boundaries vary somewhat from state to state, but the underlying pattern of claim layout and ownership is federal. It is possible to patent claims so that they become real property, or unpatented claims may be maintained by the expenditure of $100 on assessment work, or annual labor, per claim each year. Geological work is a permitted expenditure in some states. If assessment work is not performed, claims are invalid and become open to relocation. An attempt to relocate someone's valid claims falls into the category of "claim jumping," some-times—as in the days of the Wild West—a risky practice.

"Discovery" is a tenuous word in mining law; it may refer to the finding of marketable ore mineralization or it may mean finding evidence ". . . of such a character that a person of ordinary prudence would be justified in the further expenditure of his labor and means, . . ." (*Castle* vs. *Womble,* cited by Parr and Ely, 1973, p. 2–8). The full impact of "marketability" and "the prudent man" is felt when an attempt is made to patent a group of claims. Claims may not only be refused patent but may be declared invalid if, upon federal investigation, the mineralization is found not to be immediately marketable. For this reason, unpatented claims being developed to meet future mineral needs are often left unpatented.

A provision in the Law of 1872 for staking either lode or placer claims gives us some appreciation for the history of the entire American concept of mineral rights. The law was not designed by the government; it came after the fact from district codes put together a few decades earlier by miners occupying federal land in the West without permission. District mining laws, like the miners themselves, brought something from England (the idea of assessment work), from Spain (transfer of claim ownership), and from Saxony (the apex rule). The apex rule (Fig. 7-8) assures the discoverer of rights to the deeper part of the lode or vein even if the dip carries it beyond the claim sidelines; placer claims, of course, do not have this characteristic of extralateral rights.

The idea of extralateral rights was as sound in early-day American vein

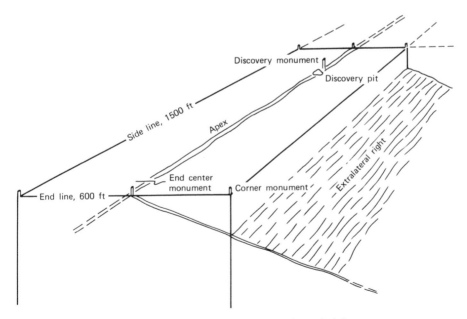

Figure 7-8.   The apex rule—extralateral rights.

mining as it had been in Saxony. Most of the mineable veins were relatively clear-cut tabular bodies exposed in a linear outcrop (the apex). It was only fair that the prospector who recognized the vein and staked the outcrop be given preference over someone who had no valid reason for staking the adjacent land. The apex rule soon came into difficulty, however, with the complexities of newer districts and other kinds of orebodies. Figure 7-9, taken from a typical mining district map, illustrates the point. Veins crossed in outcrop; lode claims overlapped. Veins crossed, branched, and were faulted at depth; some orebodies could not even be defined as veins; lawsuits resulted. Fractional claims were sometimes staked on slivers of open ground between other claims in the hope that some legal complication *might* be used to advantage. Courtroom battles sometimes led to actual underground warfare and, as mentioned in Chapter 1, to the formation of the first resident geological department in the United States. Apex suits are no longer the turbulent affairs they were in the last century, but they are still part of the American mining scene.

The Leasing Act of 1920 related to "coal, phosphate, sodium, oil, oil shale, gas, and certain sulfur deposits." As a conservation law, it provides for the

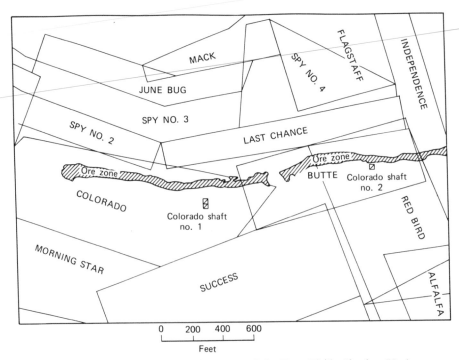

Figure 7-9.   Mining claims in a part of the East Tintic district, Utah.

workings of these substances in large blocks of ground, such as shown in Figure 7-10, rather than in the piecemeal pattern of standard mining claims. As a mineral management law, it specifies performance requirements, rentals, and royalties. Leases are obtained by competitive "bonus" bidding or by selecting ground from larger areas held under prospecting permit. Areas classified as "reserve" by the U.S. Geological Survey are offered to bidders from time to time in specific blocks. For areas without known economic reserves, prospecting permits can be obtained and held for periods of from 2 to 6 years.

Offshore minerals beyond the seaward jurisdiction of the individual states (generally 3 miles), certain uranium reserves, and minerals on "acquired lands" that have come under federal ownership are handled by leasing laws similar to those in the Act of 1920.

Minerals on Indian tribal lands held in trust by the federal government are the property of the tribes and are managed by them. In these areas, which

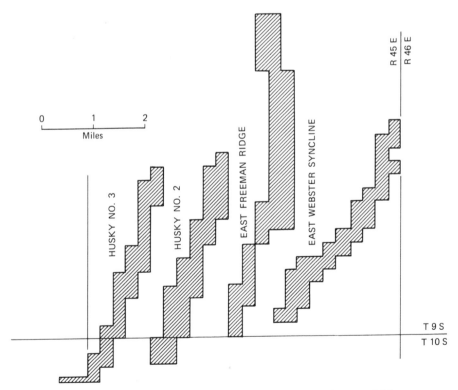

Figure 7-10.   Phosphate leasing in part of Caribou County, Idaho.

comprise large parts of several western states, access to minerals is generally obtained by the purchase of prospecting permits and by competitive bidding.

Another, and growing, category of federal land comprises areas withdrawn from mineral entry for scenic, recreational, military, or other overriding purpose. The most controversial of these are "wilderness areas," which have been designated as containing little mineral potential. Mining geologists have sometimes marveled at the certainty with which the resources in these areas are "written off" in the face of impending mineral shortages.

State lands normally include state ownership of the minerals. While there is no standard formula, most state-owned minerals are subject to lease according to priority of application or through competitive bidding. Many states issue prospecting permits as well, and a few states also provide for staking claims. As with federal lands, some areas have been withdrawn from mineral use. Some state and county governments have excluded mining from specific suburban areas.

Private lands, called "fee" lands, commonly include rights to the minerals, but there are exceptions, especially in the western states, where the federal government has often granted or sold surface ownership without the mineral rights. Where minerals, as part of the public domain, are still "locatable" and "leasable," the surface owners may not be aware of the fact until they find a geologist prowling around their pastures.

Privately owned minerals and valid mining claims are sold or leased in any way the owner feels appropriate. Some of the possible arrangements will be explained in Chapter 19. Where there are extensive private holdings, as in some railroad land grants, private forests, Spanish land grants, and long-established ranches, mining is likely to be done under a leasing arrangement and prospecting done by exclusive permit.

The U.S. system of mineral ownership is complex, but so is the history that provided the pattern. Of course, nothing in history has ever continued for long without change. The American system of mining law will undergo continuing change, especially now that the Mining Law of 1872 is under critical review. What an exploration geologist is mostly likely to find during the next few years is a mineral tenure system in limbo—a good reason for learning where the system came from and where it may be going.

Trends in Mineral Law. As mineral proprietors and landlords, nations permit exploration and mining under two general types of arrangement:

1. *Statutory,* by which discoverers are required to post notices and perform certain work. This is the mining claim arrangement in the United States, Canada, Mexico, and in many countries with a well-established "prospector" tradition.

2. *Contract,* with rights and obligations. This arrangement is represented in

the United States by leasing under the Leasing Act of 1920 and offshore leases. In many countries, this is the only way of obtaining mineral rights.

Obtaining a "land position," a right to explore and mine, is an important step in every exploration program; comments on this step can be found in Chapter 18.

During the 1960s, mineral law in most of the world underwent fundamental changes for various reasons, all connected in some way with national history. In the industrialized countries, the long-term needs of economy and society began to replace a former emphasis on simply putting minerals to work. Mineral laws became conservation laws. In the underdeveloped countries, the changes were directed toward independence and toward ways of sparking their own industrialization. Certain ideas were, and still are, dominant in the underdeveloped countries. Minerals are the quickest source of export credits, and foreign capital is needed to develop them; yet, a foreign-dominated economy is really no improvement over a colonial economy. Since the laws of the former colonial powers could no longer be accepted, some of the changes were experimental and, understandably, confusing to anyone thinking of developing a mine.

Mineral laws in nearly all countries have some element of experimentation. Nations are trying to find a balance between a goal and a capability. An exploration geologist must take the time to understand. The laws came from history.

Taxation of Mines. Taxes on mines have experimental aspects and, like mineral laws, reflect an intended balance between a goal and a capability. The goal is to obtain maximum and steady revenue; the capability is always somewhat lower, its limits are technologic and geologic. With a favorable level of technology and the right geology, tax incentives (reduced rates or initial exemption periods) can bring exploration and industry into a country. But how much incentive is enough? If the balance point is too closely set, an increase in costs or a fall in mineral prices will swallow the incentive and shut down the mines.

The type as well as the amount of taxation levied on mines has a direct bearing on the kind of exploration target to be sought. The three principal types of taxes are the *ad valorem tax,* the *severance tax* or *royalty,* and the *income tax.*

The *ad valorem tax* is based on the assessed value of the plant and the ore reserves. The tax payment is a fixed cost of operation, applied to the entire mine without regard for the level of production. As a cost of operation, the anticipated payment of a high ad valorem tax affects an exploration target by raising the acceptable *cutoff grade*—the grade of the weakest mineralization that can be mined at a profit. This reduces the effective tonnage in an orebody, and if the entire deposit is marginal in grade, it may be completely invalidated as an economic target. Since an ad valorem tax is levied whether the mine is operating or not, it has an advantage to the state in providing a steady, predictable source of revenue. But the advantage has limits. If the tax becomes unrealistically high, the source of revenue goes elsewhere. Minnesota's experience has been mentioned.

Severance tax or *royalty* is levied against each unit of mineral that has been mined and shipped. It is a variable cost rather than a fixed cost of operation. Like the ad valorem tax, it raises the cutoff grade of a potential orebody, but unlike the ad valorem tax, it does not penalize the miner for outlining future ore reserves and it does not apply unless the mine is operating. This kind of tax is sometimes used to encourage local smelting and refining rather than the shipping of ore and concentrates out of the state; the tax is simply reduced for material that has undergone additional local processing.

Royalties are easier for a state to assess and collect than most other kinds of tax, so much so that they are almost too easy to increment past the threshold of what mines can pay. A steep rise in the royalties assessed on minerals by the province of British Columbia in 1975 is a good case in point; it encouraged exploration and new mine financing almost immediately—in Alaska, the Yukon Territory, and almost everywhere else.

The *income tax* is generally levied on a graduated scale, and since it applies only to profits, it is neither a fixed nor a variable mining cost. It has less effect on cutoff grade than the other two types of tax. It does, however, affect the annual cash flow from a mineral body. If the tax rate is too high, it may reduce the present value of a potential orebody to an unacceptable level. Additional information on present value, cash flow, and the depletion allowance—one of the tax credits involved in income taxation—can be found in Chapter 8.

The income tax is by far the most appropriate for conservation of mineral resources. The other forms of taxation may cause lower grade ore to be left in the ground and they may delay capital improvements for mining and processing.

Mineral and Environmental Conservation

There is nothing really new in the ideas that mineral supplies are limited, that population is growing at a dangerous rate, and that the quality of the environment is threatened by the waste products of humans and industries. What is new is our capacity to measure some of these things. The measurements and the projections, like the "doubling times" mentioned at the beginning of this chapter, are frightening.

Even though reactions and remedies range from technophobic "back to nature" prescriptions to science fiction "futurist" formulas, one thing is certain: exploration for minerals will come under increasing control by governments doing what they feel is best (or most attractive politically) for their people.

A key word is "trade-off"—sacrificing one thing, such as electric power, to gain another, such as a cleaner atmosphere. Another key expression is "benefit-cost"—assigning numbers to tangible and intangible benefits to be derived from environmental and conservation controls so that they can be related to the necessary costs.

First, the trade-offs and mineral conservation. Future access to minerals can be enhanced by spending money now or at least by foregoing some of the more immediate benefits. One of the most frequently mentioned ways of implementing this trade-off is to invest in improving mining and mineral processing practices. Others are to reduce the throwaway waste of minerals, to substitute abundant minerals for scarce, and to implement a conservation-wise national mining policy. Criteria for exploration targets enter into most of this, but the more immediate connection is with the concept that known mineral deposits are to be extracted as thoroughly as possible without waste and without impairment of the mineability of nearby deposits. If phosphate rock of marginal milling grade is stripped from higher grade beds in the western United States, there is a requirement that it be stockpiled until recovery is feasible. Permits to mine coal are issued for well-planned extraction of an entire series of seams, not for the mining of the thickest seam in a way that would disturb the stress pattern around the others. A provision for barrier pillars between solution mines and conventional mines is commonly written into saline mining permits; in the deep Yorkshire potash field of England, for example, the stipulation is to leave a pillar width of not less than half the depth from the surface. Britain's long experience with mineral-based industry has prompted some of the world's most stringent conservation and environmental measures. A detailed account of the British experience, edited by Jones (1976b) is well worth reading.

In benefit-cost determinations, two kinds of benefits are measured; direct benefits from the recovery of byproducts or from further use of mine workings or the restored land and indirect benefits related to damages avoided. An example involving both kinds of benefits can be cited. The removal of sulfur from coal would be credited with the value of the recovered sulfur (if marketable) and also with some figure representing the savings by averting pollution damage. Whether or not the benefit balances the cost in the miner's own accounting, compliance demanded by the government will be stated in these terms. Governmental subsidies or taxation allowances for environmental costs are sometimes used to help even the accounts.

The costs of mineral and environmental conservation measures are too widely variable to generalize. The cost of revegetating a tailings berm at an

Arizona copper mine was $1,300 per acre ($3,210 per hectare) (Ludeke, 1973); the cost of restoring mined land to agricultural use in the Ruhr district, Germany, has been reported at $11,000 per hectare (Zurkowski, 1975); and there are all kinds of figures in between. Benefits are even harder to pin down; they may be as modest as the income from a dairy farm on restored land or they may be as great as the saving of the Rhine port at Duisburg, Germany, where a serious silting condition was counteracted by deepening the river through harmonic mining of the underlying coal seams. Whatever the figure, it applies to the cost of mining each ton of mineral, and this, in turn, applies to the evaluation of mineral bodies.

Some costs of environmental conservation are even more directly associated with exploration. In areas of high scenic value, a geologist may be required to bring equipment in by helicopter rather than by road. An access road may be allowed, but only along a specified—and sometimes indirect—route. A proposal for bulldozer trenching may be rejected by the agency evaluating an environmental impact statement. And, the environmental impact paperwork itself can "cost" an exploration team by delaying a project.

CHAPTER 8
ORE VALUE AND THE CONCEPT OF AN OREBODY

No one ever *found* a mine unless it existed at some previous time and was subsequently lost. Ore mineralization is found, orebodies are defined, and mines are *made*.

8.1. THE TIME, THE PLACE, AND THE UNIT VALUE OF MINERALS

A mine, even the ghost of a mine in a remote desert canyon, is evidence that the minerals in that specific location were valuable to someone at some time. The ghost may be brought to life if the minerals in that location have a sufficient value *now*. The value may be expressed in terms of a price in the major trading centers of the world. More likely, "value" and "price" will have connotations that reflect some degree of control by governments and by associations of producers and consumers.

In any event, the value of a mineral is most easily expressed by its price in whatever market is available at a certain time. *Time* is the key word, and exploration is particularly responsive to time trends in mineral value. In a decade when potash prices remain nearly constant and coal prices double, exploration flows from potash to coal—toward the livelier commodity.

Trends in mineral value are so important to geologists' immediate objectives that they must keep as current with them as possible. The "markets" section in *Engineering and Mining Journal* is the most widely read monthly digest of mineral price levels and outlook. *Metals Week* (New York) has more detail. *Coal Week* (New York) contains steam coal and coking coal prices. *Industrial Minerals,* a journal published in London, is a source of international market information for the "non-metallics." For the past records of mineral prices, indications of their stability or volatility, there are the yearbooks of the American Bureau of Metal Statistics and the annual Minerals Yearbooks of the U.S. Bureau of Mines. For the long-term price outlook, most of the "minerals forecast" books mentioned in Chapter 7 have something to say on the subject; also, there are price forecasts in the yearly commodity review issues of most major mining magazines.

Mineral prices respond to market demands, but they do not have such a direct relationship with the general economic cycle as they had in the past. Exponential trends in two of exploration's driving forces, demand and depletion, have created an economy of scarcity with strange new cycles in mineral prices that bridge the periods of economic depression and buoyancy. The reasons (now being studied by practically everyone) are in part a function of geology and in part a function of commodity speculation. Whatever the contributing factors, they have complex origins. International oil embargoes in 1974 affected the long-term price outlook for coal and uranium, but there was also an immediate response in high niobium (columbium) prices because of an expanded market for alloy metals in high-strength pipeline steel. Barite prices increased at the same time as a result of increased oil exploration and the need for heavy drilling mud. Barite miners have recognized this relationship for some time; they say, "When the oil industry catches a chill, the barite industry sneezes."

Mineral commodity prices are especially shaky in their relation to politics. The prices of mercury, a metal with a specialized rather than a broad market, provide one of the very best illustrations (Ryan, 1975, 1976). Figure 8-1 shows the range in price (New York) during 1974 and 1975. Prices began to climb in February on the expectation that a price-fixing meeting of mercury producers was imminent. Then, the news that the Italian government had purchased the old Monte Amiata mercury mine lent some further strength to the market. The mine had been unprofitable; the government might close it down, thus tightening the world supply. In early March, the producers' meeting was postponed and the Italian government announced its intention to keep Monte Amiata operating; mercury prices suddenly turned downward. In April, a new U.S. mercury mine entered production and there was talk of an increase in the U.S. tariff to protect domestic producers. Mixed emotions. Then, the price-fixing meeting of the mercury producers was rescheduled for May and consumers began to think in terms of what had happened to oil, copper, and bauxite when prices were manipulated; mercury prices climbed. In mid-May, the producing countries, Algeria, Italy, Mexico, Spain, Turkey, and Yugoslavia—accounting for 90 percent of the world's mercury production—agreed

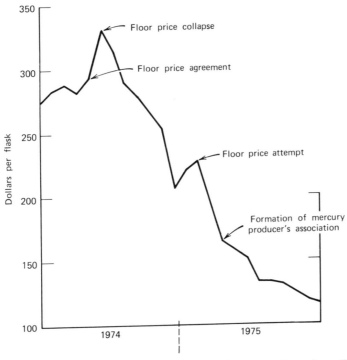

Figure 8-1. New York dealer mercury prices, 1974–1975. (Data from Ryan, 1975, 1976.)

to establish a floor price of $350 per flask (approximately 32 kg); the price climb accelerated.

By mid-July 1974, the producers' agreement was a shambles. In fact, it had never really begun to work; the price speculation had been based on the possibility that it *might* work. Consumers and producers decided to unload their stocks before the price fell—and it fell. The downward trend continued, accelerated near the end of 1974 by the dumping of a large quantity of mercury on the market by a German chemical company that had decided to reduce its inventories.

In early 1975, the producing countries reaffirmed their agreement on a $350 floor price, but the consequent rise in price turned downward again at $228. The formation of a specific cartel, the International Association of Mercury Producers, had little effect on the declining market. Consumers were still hesitant to stock up on mercury, and when the state-owned mines in the cartel countries began undercutting the market in a price war the consumers became even more cautious in their purchasing. Smaller producers and privately owned operations in other countries either closed their mines or stockpiled their mercury. Russian mercury, normally sold in Europe, disappeared from the market. Chinese mercury, reportedly held in large stockpile amounts, remained off the market.

The results: prices ranged from $340 to $118 per flask within two years. The characterization: mercury is as volatile in price as it is in character.

The importance of cyclic changes in mineral demand and mineral price to exploration can be seen in the history of uranium exploration drilling in the western United States during the last few decades (Fig. 8-2). The Nuclear Age created a market for uranium, and the market outstripped the domestic supply during the early 1950s. Government support prices and bonuses led to an unparalleled prospecting excitement and eventually to discoveries, new production, and an apparent uranium oversupply. By 1960, the price had lowered, production declined, and imports of uranium materials—principally

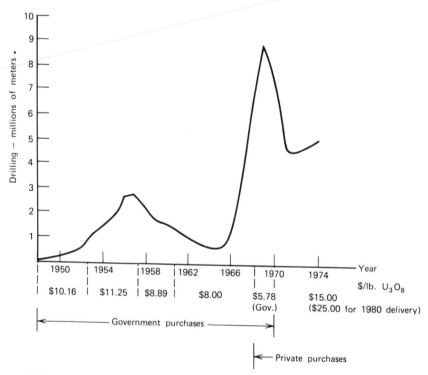

Figure 8-2.   Uranium exploration drilling versus purchase price for U_3O_8 in concentrates, United States, 1948–1974. (Data from U.S. Energy Research and Development Administration reports.)

from Canada—were stopped. Exploration all but collapsed, and mining company offices were inundated by job-seeking geologists whose sole experience had been in uranium. Then, in the late 1960s, private nuclear power industries began to purchase uranium materials and the looming "energy crunch" led to an expectation of a renewed demand for nuclear fuel. Uranium prospecting again became important. In the early 1970s, the price of uranium increased in response to further projections of expanded future needs. The pace of exploration quickened, and mining companies began assigning geologists from other projects to uranium projects. The "boom and bust" pattern is not unique to uranium; it is expressed, even if not so dramatically, in the mining and marketing of most other minerals.

National interest, as well as the international market, places a time value on minerals. War and the threat of war create an economy of urgency, and the needed minerals are produced, imported, and stockpiled with little regard for cost, depletion of reserves, or normal market characteristics. During World War II, new mines were rushed into production in America, while European mines were literally being gutted of ore. In a similar but less frantic way, national programs for economic recovery, rapid industrialization, and regional development take precedence over normal market economics. As in wartime, a need for self-sufficiency is paramount; but there are longer term goals, and the thrust is toward new discoveries rather than toward scraping the roots of known deposits. In some instances, the new discoveries will be protected from premature depletion by restricting mineral exports.

In most nations, even where the price of minerals is not totally controlled by the government, quotas and tariffs on imported minerals and incentives to domestic production are used during certain periods to support and promote domestic industry. Nations heavily dependent on foreign minerals will sometimes use a special price structure to stimulate domestic production. Exploration activity responds to the stimulus.

Individual metallurgical and manufacturing enterprises, as well as governments, will under some circumstances assign a special time value to minerals from small and otherwise unprofitable deposits. A "hungry mill" must be fed if it is to be kept healthy for anticipated future use, and it will accept some rather weak nourishment on a temporary basis as long as the mineralogy of the ore is suitable.

An illustration of special time value for minerals can be cited in one company's step-by-step transfer of phosphate mining operations from southeastern Idaho to southwestern Wyoming, and then to eastern Utah. The Waterloo mine, near Malad, Idaho, was nearing depletion in the late 1940s. A suitable body of phosphate rock was available for open-pit mine development near Leefe, Wyoming, 100 km away, but transition would take time and market contracts could not be interrupted. Smaller deposits in southeastern Idaho met the temporary need. The new mine was brought into production, but a decade later demands on the Leefe deposit outran its capacity and a site for a new operation was chosen near Vernal, Utah, 200 km from Leefe. Time was again needed, and the growing demands of customers had to be supplied, so the company began to use the Leefe mill for processing phosphate rock from high-cost underground mines in the vicinity while the Vernal open-pit mine and mill were being prepared. Eventually, demand was strong enough to keep most of the Vernal and Leefe mines, open pit and underground, in operation.

A market exists at some time and at some place for nearly every reasonably pure mineral. If the mineral has a high *unit* value rather than a high *place* value, it can be sold in hundreds of markets around the world at a price that reflects an international balance in supply and demand. Most metal ores and the less common industrial minerals fit into this category, provided they can be concentrated to economic grades. At the other extreme, the relatively common industrial minerals and rocks have a high *place* value; the market is nearby, and it may in fact exist because the mineral deposit is there. Minerals with high place value, cement rock, for example, seldom enter into international trade except where the mine and the market are both on tidewater and accessible to low-cost marine transport. In practice, most minerals have some aspects of both unit and place value. Coal, salt, gypsum, and sulfur are generally marketed to industries established in the immediate region, but they can be major bulk items in foreign trade where transportation facilities are highly favorable. The significance of this: an industrial minerals geologist tends to develop a "tidewater complex"—the farther inland, the less attractive the mineralization.

8.2. RECOVERABLE VALUE AND MINERAL PROCESSING

Time, place, and unit values are abstract terms. *Recoverable value,* however, is a measurable, intrinsic characteristic of every orebody. Recoverable value pertains only to that portion of the mineralization that can be expected to reach the market; the remainder cannot be paid for because it is left behind or lost during mining and processing.

A story told wherever miners congregate illustrates the notion of recoverable value. A part-time prospector looked up the market quotations for the quantity of metals reported from his sampling of a hard-won initial ore shipment: lead, zinc, antimony, cadmium, arsenic, a lot of sulfur and iron plus a little silver. A good sum! His settlement sheet from the custom smelter: a small credit for lead, a few pennies for the silver, and nothing for the other metals, minus a smelting charge and deductions for the troublesome zinc and arsenic. Balance—zero! He paid the freight bill with the only cash in his pocket and threw his mining tools into the shaft with "Here! You can have it all!" More than a few geologists, well educated in theory but ignorant of mineral processing and extractive metallurgy, have started down the same road until they were halted by an engineer.

In almost no instance can the entire metal content of all of the economic minerals be extracted. The geologist depends on mining engineers and metallurgists to determine which minerals are economically recoverable and to what extent each may be recovered. The geologist must learn to speak "metallurgese."

Weiss (n.d.) and Pryor (1965) have provided comprehensive reference books in mineral processing and extractive metallurgy. The encyclopedic work found in nearly every metallurgical laboratory is by Taggart (1951). In these books, the following steps are described in more detail than can be afforded in this text:

Concentration (processing, milling, beneficiation, mineral dressing)
—Comminution (crushing and grinding)
—Sizing and classification
—Separation
Reduction (extractive metallurgy, chemical extraction)
—Pyrometallurgy (smelting)
—Hydrometallurgy
Refining

The three steps commonly follow in sequence; but some ores cannot be concentrated at all, and for certain ores only one or two steps are needed. Some direct-smelting ores containing oxidized lead-zinc-silver minerals cannot be economically improved by concentrating. Shipments must therefore be inherently high in grade. The in situ leaching of copper and uranium ores bypasses the comminution stage; the product may have to be given additional treatment or it may be recovered in marketable form. In the solution mining of salt and potash, the mining process is a matter of chemical extraction and the surface processes amount to refining.

Where industrial use requires a mineral rather than a derived product, it need only be sized and concentrated to meet market specifications. In no case, however, is preparation a simple matter; the removal of objectionable impurities, such as pyritic sulfur and "ash" from coal or fine-grained silica from fluorite, may require several passes through a comminution, classification, or separation process. In order to be economically practicable, the benefit from each substep in concentration, reduction, and refining must be consistent with the additional energy required. Thus, mill tailings, the refuse from coal preparation plants, and even smelter slags always contain some portion of the economic mineral or metal that cannot be economically recovered. In a few instances, later improvements in technology make it possible to recover some of the originally important minerals as well as newly important byproducts from the waste material.

Where concentration, reduction, and refining follow in sequence, as with the transformation of tetrahedrite-galena mineralization into marketable copper, lead, and silver, the influence of the orebody's geology is apparent throughout. The reduction process is applied to a concentrate of the same minerals as in the original ore but in different ratio; the gangue minerals may have been nearly eliminated, but the concentrate is still a modified image of the orebody. Even at the far end of the sequence, in refining, minor amounts of gold, silver, platinum, and nickel are recovered from impure metals or they may remain as trace amounts in the final product as ghosts of the ancestral orebody. Silver-bearing "lake" copper from Michigan with its special electrical properties is of the latter type.

The influence of geologic conditions is most direct in the concentration step; in fact, the boundary between mining and concentrating is indistinct. As mining becomes more mechanized and less selective, the burden of selectivity is passed on to the concentrator. When ore mining was done in high-grade stopes and narrow open cuts and when coal mining was confined to seams with a minimum of "bone" partings, the "mill" and "tipple" were only of secondary importance. This is no longer true. Processing plants can separate ore and waste at a lower cost than miners can hand-sort the material underground or leave waste zones undisturbed. In the mechanized continuous mining of coal, the entire seam is likely to be mined, and it is nearly impossible to avoid taking intraseam material plus some of the top or bottom rock. The preparation plant does the cleaning, gets rid of the rock and a large portion of the pyritic sulfur that may be present, and delivers a sized product.

Figures 8-3, 8-4, 8-5, and 8-6 show abbreviated flowsheets for a tin concentrator, an industrial mineral concentrator, a copper concentrator, and a coal preparation plant, respectively. A few terms in the illustrations need explanation. We can begin with "crushing."

The comminution process (crushing and grinding) begins at the mine face with the proper use of explosives and mining machines to deliver as uniform a feed as possible to the primary crushers, which may in themselves be installed in the mine. Even where large blocks of ore could possibly be handled in mining, they are avoided because they would require an inordinate amount of energy in crushing. An excess of "fines," especially in coal and industrial minerals, is just as bad, because it may result in wasting of material due to its unmarketable size, in additional treatment to avoid loss in tailings, or in an excessive use of reagents. In distributing the crushing burden between the mine face, the primary crushers, and the secondary crushers, most crushing is handled where it will waste the least energy—and much of the energy requirement depends on geological characteristics. Mineral heterogeneity, weak interfacial bonding, and textural discontinuities promote easy crushing. As the crushed material is reduced in size, the effect of textural discontinuities is progressively diminished, the material gains in strength, and more

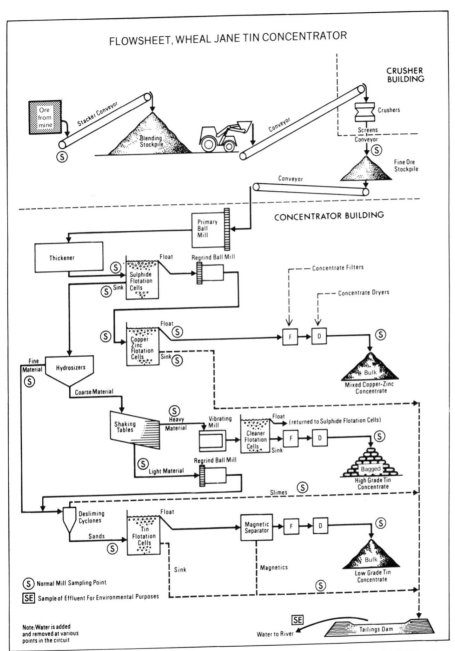

Figure 8-3.   Flow diagram, Wheal Jane tin concentrator, Cornwall, United Kingdom. (From *Engineering and Mining Journal,* v. 174, no. 11, p. 137, 1973. Copyright © 1973, McGraw-Hill, Inc. Used by permission of McGraw-Hill Book Company.)

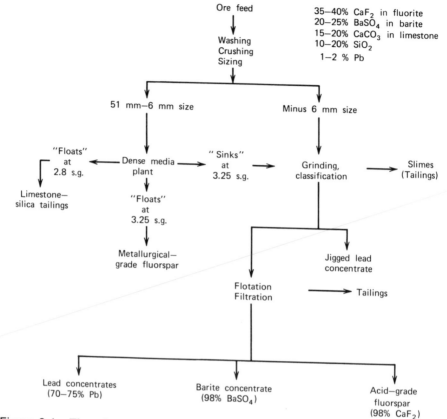

Figure 8-4. Flow diagram, Cavendish fluorspar mill, near Eyam, Derbyshire, United Kingdom. (From *Quarry Manager's Journal,* v. 49, no. 10, p. 417. Copyright © 1900, Quarry Manager's Journal, Ltd. Used by permission.)

energy is required for further size reduction. Consequently, coarse and fine materials are often separated after each crushing, as indicated in Figure 8-5, so that the next step in crushing need involve only the coarser material that actually needs crushing.

Where the separation of minerals for market can take place at a coarse size, as in jigging and sink-float methods (heavy or dense media), the crushed, sized product can be processed without further grinding. Low-sulfur coal, lump barite, bauxite, and metallurgical-grade fluorspar are often handled in this way.

More commonly, grinding is needed. The crushed product is fed to tumbling mills of various types until a *size of liberation*, generally the mineral grain size, is attained through impact crushing, shearing, and attrition. The

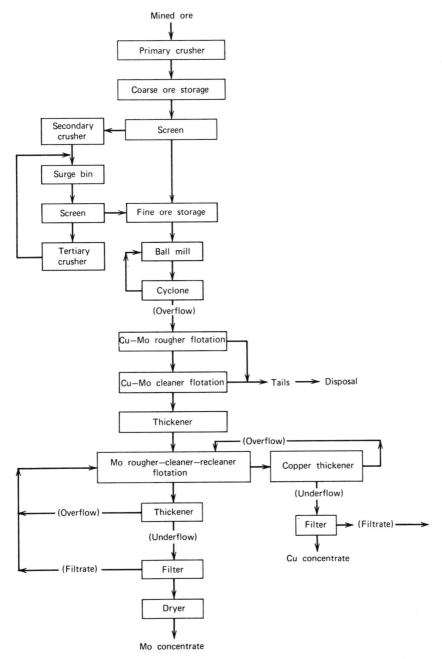

Figure 8-5. Copper-molybdenite concentrator flow diagram, typical of milling practice at porphyry copper mines.

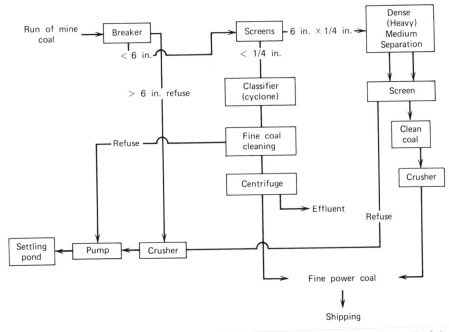

Figure 8-6. Dense medium system for power coal. If pyrite were abundant, all of the coal would be crushed to minus ¼ in. (6.35 mm) before dense medium separation. (From J. W. Leonard, "Mill design—coal," in *SME Mining Engineering Handbook,* A. B. Cummins and I. A. Given, eds., Sec. 28.8, Fig. 28-18, pp. 28–47, 1973, by permission of American Institute of Mining, Metallurgical and Petroleum Engineers, Inc.)

ground product, properly sized (screened and classified), is the feed for various separation methods, among which froth flotation is the most widely used. In Figures 8-3 and 8-4, several methods are used so that cheaper "gravity" processing can be carried as far as possible rather than treating everything by flotation. At Wheal Jane, shaking tables are used to remove the highest grade tin ore before further regrinding and treatment. At the Cavendish fluorspar mill, some of the byproduct lead is recovered by jigging rather than by adding it to the flotation burden. The copper-molybdenite concentrator described in Figure 8-5 delivers two flotation concentrates; in this flowsheet, the "rougher" flotation cells provide the main separation and the "cleaner" cells improve the concentrate.

Whatever the separation method, mineralogy and mineral texture are of prime importance. If the size of liberation is not fully attained, mixed fragments of gangue and ore will be rejected or collected on the basis of only one of the minerals. If too many are rejected, recovery is diminished; if too many are collected, the grade of the concentrate is lowered. Regrinding to achieve a finer size of liberation may be too costly in energy consumption or in excessive handling, and it may result in ultra-fine materials (slimes) that will not respond to classification and concentration.

In nearly all methods of mineral separation, oxidized surfaces and films of supergene minerals, even a micron-thin stain, will cause recovery problems. The effect is similar to that of having mixed fragments. Selective flotation reagents are chosen to coat the surfaces of specific minerals with a water-repellant film that will adhere to a stream of bubbles—and the film will do just that, no matter what the composition of the interior of the mineral grain. Gravity separation in spirals, cyclones, or heavy media vessels has mineral oxidation restrictions; it depends on the density of the pure mineral, not its weathered form. And magnetic and electrostatic separation methods are set for the characteristics of fresh minerals, not weathered minerals.

"Amenable" and "refractory" are key words in mineral processing. A sulfide ore amenable to concentration is typically coarse grained and of relatively simple mineralogy. A refractory sulfide ore, one that is difficult to treat, may be very fine grained and complex, with interlocking grain boundaries and some degree of grain surface coating and may contain microcrystalline impurities and tough or abrasive gangue minerals.

The most common criteria for assessing the recoverable value of a mineral body and the amenability of its minerals to processing are percentage recovery, grade of concentrate, and ratio of concentration. The use of these terms can best be demonstrated in the context of a metallurgical balance sheet for an argentiferous lead-zinc ore (Table 8-1).

The percentage recovery for the silver contained in the original ore was 90.62 in the lead concentrate and 5.65 in the zinc concentrate; 3.72 percent was lost in the tailings. Ideally, all of the lead in the ore would "report to" (be collected in) the lead concentrate. Instead, 1.50 percent of the lead reported to the zinc concentrate where it is not easily recovered in smelting, if recovered at all. Some zinc also remained in the lead concentrate where it could even incur a penalty.

The grade of lead in the lead concentrate is 66.59 percent, as opposed to 6.18 percent in the ore delivered from the mine. The ratio of concentration is the ratio of tons of ore to tons of concentrate. For lead it is 11.33 (100/8.83) and for zinc it is 12.74 (100/7.85). There is no specific silver concentrate.

In addition to technologic reasons for processing mineral shipments into separate concentrates so that they can be treated more economically, there is the aspect of freight costs. In the demonstration just given, freight charges on 83.32 percent of the ore material were avoided. Before jumping to conclusions, we should be aware that freight rates are usually higher for concentrates than for ore and that concentrates require more care in handling. Still, there can be a substantial saving.

Table 8-1.   Metallurgical Balance Sheet for the Flotation Concentration of an Argentiferous Lead-Zinc Ore[a]

	Assays					% Distribution		
	% Weight	Ag oz/ton	Ag g/mt	Pb %	Zn %	Ag	Pb	Zn
Lead concentrate	8.83	40.45	1,386.6	66.59	4.98	90.62	95.16	8.97
Zinc concentrate	7.85	2.83	97.0	1.18	54.21	5.65	1.50	86.78
Tailings	83.32	0.18	6.2	0.25	0.25	3.73	3.34	4.25
Mill feed from mine (ore)	100.00	3.94	135.1	6.18	4.90	100.00	100.00	100.00

[a] Data from Aplan, 1973, and Stather and Prindle, 1970.

8.3. TRANSPORTATION AND MARKETING OF MINERALS

Transportation

Some mining engineers have estimated that the transportation of raw and finished mineral products averages a quarter to a third of the final cost to the customer. The precise fraction is not as important as the overall indication that transportation is a major cost in providing minerals for the market. Most of the transportation cost—for the bulk of the material moved—occurs close to the mining end of the route where it is a large factor in determining whether a mineral deposit can be mined.

Because the impact of transportation cost is greatest in the high place value minerals, cement rock can be used in an illustration modified from an example given by Towse (1970) in a paper on the evaluation of cement raw materials.

Suppose that a cement plant must be located at point A and that we have three possible raw material sources. One source is at the plant site; one source is at B, 50 km away, with only truck transport available; and one is at C, also 50 km away, but with combined truck and water transport. Assume a cement market price of $20 per ton, truck transport at 4 cents per ton-km, and combined water-truck transport 1 cent per ton-km; 1.5 tons of raw material must be used per ton of cement. The situation is:

	Material at A		Material at B		Material at C	
Market price (fixed)	$20.00		$20.00		$20.00	
Raw materials transport	$ 0.00		$ 3.00		$ 0.75	
Production cost (fixed)	13.00		13.00		13.00	
Product transport to market (fixed)	2.00	15.00	2.00	18.00	2.00	15.75
Breakeven raw material mining cost		$ 5.00		$ 2.00		$ 4.25

Even though the mining condition at B may be ideal, the same breakeven cost can be made at A where the mining cost is $2\frac{1}{2}$ times as great. If the mining cost at C could be brought below $4.25 per ton, a pit at C would be the most profitable. The illustration is simple, but it emphasizes the role of transportation costs in establishing a permissible mining cost. Where does the geologist fit in? With geotechnical data—the framework for designing methods and estimating costs in mining.

Truck, train, and ship have been, and still are, the most common means of mineral shipment. Transportation methods have become increasingly expensive, but improvements in materials handling, such as the slurry loading and unloading of ocean vessels and the use of one commodity—one distance "unit trains," have helped keep transportation costs at a manageable level. Long-distance conveyor belts (to 25 km) and aerial tramways (to 50 km) provide transportation for minerals at costs somewhat lower than trucks. Two additional modes of transportation are of growing importance, the pipelining of solids in slurry form and the use of gigantic river barge "tows." The latter is actually a resurrected and modernized method rather than a new one. The distance and the terrain covered by some slurry pipelines are remarkable. Coal and iron ore are pumped 200–400 km in some places. At the Ertsberg copper mine, Indonesia, a 100-km pipeline carries ore concentrates from an elevation of 3,530 m to sea level through 40 km of rugged mountains and 60 km of swamp and rain forest.

Airplanes have been used in a few places to carry high-value mineral concentrates. Air freighting is a concept rather than a means of mineral transportation, but it has received considerable study in relation to Alaskan and Canadian mineral development.

Table 8-2 is included here with the provision that it is intended to show only *relative* costs and with the understanding that there are innumerable factors to be considered for any specific situation. The indication is clear, nevertheless: ocean shipping is by far the most economical.

The "tidewater complex" held by many geologists in their evaluation of high place value industrial mineral deposits is easy to see, and the same attitude carries through to some of the moderately high unit value deposits. Unless cheap transportation is available or capable of being pioneered at a reasonable cost, an interesting occurrence of mineralization 1,000 km from the nearest port or railhead may remain just that—an interesting occurrence.

Marketing

Where crude ores and minerals must be marketed outside a mining organization they may be sold to trading companies, custom mills, or smelters. Some

Table 8-2. Comparison of Mineral Transportation Cost[a]

	Cents per Metric Ton-Km
Ocean shipping	0.02–0.7
Pipeline	0.1–0.7
River barge	0.15–0.3
Rail	0.28–1.1
Truck	3.9–5.0
Air freight	8.4–14.0

[a] Modified from Maddex, 1972.

industrial minerals, hand-sorted "lump" barite, for example, are sold directly to chemical manufacturers.

Trading companies (metal merchants) are the middlemen of mining; they serve as the producers' agents, act as brokers in the metal exchanges, and sometimes arrange financing for mining ventures. About 10 percent of the world's metal trade passes through their hands.

Custom mills or concentrators are the collecting and processing outlets for minerals from small shippers and are sometimes built specifically for this purpose by governments or private companies. But more often they are located at a major mine where they serve smaller nearby mines with whatever excess capacity they may have. Recall that the life cycle of a mining operation may end with custom milling after the main orebody has been depleted. There is one more dimension: a government may keep an older custom mill in operation as a means of supporting a local small mine economy. This arrangement is often managed by a miners' cooperative organization.

Custom smelters will accept certain high-grade ores, but minerals are more commonly marketed in the form of concentrates and the concentrating is most often done by the same company that mines the ore. A large organization may also control the next stage, metallurgical reduction or smelting, and it may carry the product through refining to fabrication and manufacture. Whether the metallurgical reduction of concentrates is handled by a "vertically integrated" company that also does the mining and milling or by a custom smelter, there is an accounting that follows general rules, such as those shown in Table 8-3, and a "settlement" calculation, such as shown in Table 8-4. The units are in short tons, pounds, and troy ounces in keeping with North American metallurgical accounting in 1977.

Table 8-3, a typical "open" smelter schedule, is the kind furnished to independent shippers of ores and concentrates by custom metallurgical plants. Table 8-4 shows a payment that might be received by the shipper or credited to the mining division's account. It is generally called "net smelter return" (NSR) even if it may have involved hydrometallurgical work rather than smelting. Of course, there are many variations depending on negotiated contracts, the minerals, and the design of the metallurgical plant. A regular shipper of concentrates of consistent quality will be in a more favorable position because the plant can gear up for the particular material. Also, there is a limit to the flexibility of metallurgical plants; an uncommon ore assemblage may have to be shipped halfway around the world to reach an appropriately designed smelter.

A shipper of ores and concentrates may have this kind of problem in transportation and marketing: a copper smelter will pay for only the gold, silver, and copper content; a zinc smelter at some greater distance will pay for the same metals plus lead and zinc but will pay less for copper and charge a deduction for arsenic. There are choices:

Table 8-3.   Abbreviated Smelter Schedules—United States (1977)

A. *Copper Concentrates at Copper Smelter*
Payments:

Gold:	Deduct 0.02 troy ounce per dry ton and pay for 92.5% of the remaining gold content at the London gold price
Silver:	Deduct 0.5 troy ounce per dry ton and pay for 95% of the remaining silver content at the Handy and Harman New York quotation, less 4.5 cents per ounce
Copper:	Deduct one unit (1%) and pay for the remaining copper at the domestic refinery quotation for electrolytic wire bars, less 6.75 cents per pound of copper accounted for

Deductions:

Base Charge:	$20 per dry short ton
Zinc:	5 units (5%) free, charge for excess at $0.30 per unit
Arsenic:	0.5 unit free, charge for excess at $1 per unit
Antimony:	0.2 unit free, charge for excess at $1 per unit
Freight:	All railroad freight and delivery charges for account of shipper

B. *Lead Concentrates at Lead Smelter*
Payments:

Gold:	Same as for copper concentrates
Silver:	Deduct 1.0 troy ounce per dry ton and pay for 95% of the remaining silver content at the Handy and Harmon New York quotation, less 3.0 cents per ounce
Copper:	Deduct one unit and pay for the remaining copper at the domestic refinery quotation for electrolytic wire bars less 10.0 cents per pound of copper accounted for
Lead:	Deduct 1.5 units and pay for 95% of the remaining lead at the New York quotation, less 3.0 cents per pound of payable lead

Deductions:

Base Charge:	$17 per dry short ton
Zinc:	7 units free, charge for excess at $0.30 per unit
Arsenic:	
Antimony:	Same as for copper concentrates
Freight:	Same as for copper concentrates

C. *Zinc Concentrates at Zinc Smelter*
Payments:

Gold:	Same as for copper concentrates
Silver:	Deduct 1.0 troy ounce per dry ton and pay for 80% . . . less 2.0 cents per ounce
Copper:	Deduct one unit and pay for 65% of the remaining copper . . . less 6.0 cents per pound . . .
Lead:	Deduct 4.0 units and pay for 80% of the remaining lead . . . less 2.0 cents per pound . . .
Zinc:	Deduct 8 units and pay for remainder at the East St. Louis quotation for Prime Western zinc

Table 8-3. (con'd) Abbreviated Smelter Schedules—United States (1977)

	Deductions:
Base Charge:	$52.50 per dry short ton
Arsenic:	
Antimony:	Same as for copper concentrates
Iron:	7 units free, charge for excess at $0.50 per unit
Freight:	Same as for copper concentrates

1. Produce a bulk flotation concentrate, save on freight, and lose the lead and zinc credits.
2. Produce two selective flotation concentrates and ship them separately to the two smelters.

And there are more problems. How much more will it cost to produce the two concentrates? What is the optimum grade of concentrate? The metallurgical balance sheet is an important guide; the mineralogical zoning within the deposit is another. If the sphalerite-galena ratio increases with depth, this situation will have to be taken into account.

Ferroalloy ores have their own marketing rules. So do industrial minerals, since they are destined for use as minerals rather than as metals. Metallurgical-grade fluorspar (metspar), for example, is marketed on the basis of "effective CaF_2" (CaF_2 content minus $2\frac{1}{2}$ times the silica content). Coal, marketed as an energy source, as metallurgical coke, and as a chemical raw material, has its unique price and quality guidelines.

For a discussion of the many sides of mineral marketing the chapters on metals, nonmetallics, and fuels in *Economics of the Mineral Industries* (Vogely, 1976) are recommended. Smelter schedules are explained by Salsbury and others (1964), and a commodity-by-commodity digest of marketing characteristics is provided by Burgin (1966).

8.4. ROLE OF MINE DEVELOPMENT AND MINE PRODUCTION IN DEFINING OREBODIES

Will the mineralization support a mine? If so, it can be called an orebody. If not, it is nothing more than mineralization—at least for the time being. The answer is sought in a feasibility study, the answer is found by working with mining engineers, and the answer is used to decide whether to continue the project or to abandon it before any more money is wasted.

In most stages of exploration a feasibility study is a deductive matter with a

Table 8-4. Net Smelter Return Calculation[a]

Metal prices (source: *Metals Week* or *Engineering and Mining Journal*) are taken as

Silver	$4.50 per troy ounce
Lead	.24 per pound
Zinc	.39 per pound

"Unit" is 1% of short ton, or 20 lb

I. Assume that the lead concentrate containing 5% moisture is shipped to a lead smelter and that the freight charge is $10 per ton. The equivalent freight on a dry ton is $10.00 × 100/95 = $10.53.

Payments

Silver (40.45 − 1.0) × 0.95 × ($4.50 − 0.03)	= $167.52
Lead (66.59 − 1.5) × 0.95 × 20 × ($0.24 − 0.03)	= 259.71
Value of pay metal	$427.23

Deductions

Base Charge $17.00	$ 17.00
Zinc (no penalty)	
Value f.o.b. smelter	$410.23
Freight	$ 10.53
Value at mine, dry short ton	$399.70

II. Assume that the zinc concentrate, containing 6% moisture is shipped to a zinc smelter and that the freight charge is $12 per ton, equivalent to $12.76 per dry ton

Payments

Silver (2.83 − 1.0) × 0.80 × ($4.50 − 0.02)	= $ 6.56
Lead (no payment; less than 4 units)	
Zinc (54.21 − 8.0) × 20 × $0.39	360.44
Value of pay metal	$367.00

Deductions

Base Charge $52.50 . . .	52.50
Value f.o.b. smelter	$314.50
Freight	12.77
Value at mine, dry short ton	$301.73

III. Net Smelter Return (NSR) on ore mined (value of concentrates divided by concentration ratio):

Lead concentrates $399.70/11.33	= $ 35.28
Zinc concentrates $301.74/12.74	= $ 23.68
NSR per ton of ore	$ 58.96

---

[a] Based on abbreviated smelter schedules shown in Table 8-3 and concentrates described in Table 8-1.

heavy load of assumptions and projections. It *has* to be. A drill hole in ore plus a geophysical anomaly does not define an orebody, but it does provide the basis for a working model of an orebody. Additional investigations, if justified by a preliminary feasibility study, should provide a more exact model, information for a more thorough feasibility study, and the basis for another decision. Each decision calls for a new commitment of money. It may be part of a "stepwise-financing" procedure described by Frohling and McGeorge (1975) in which thousands of dollars in the earliest steps lead to hundreds of thousands of dollars in the intermediate steps, and finally to the main capital investment.

The earliest feasibility study in an exploration program may be part of the basic decision to explore. There are mines at X, Y, and Z. There should be more orebodies in similar geologic terrain in the next mountain range (or even in a matching plate in the next continent, for that matter). If an orebody can be discovered (geologic success), can it be mined as profitably as at X, Y, and Z? Engineering and economic analogies may be just as important as geologic analogies.

The next feasibility study is strongly affected by day-to-day operating decisions made in the field, made in haste perhaps (with a drill rig standing by) and with order-of-magnitude economic figures. The project geologist and the chief geologist examine core fresh from a drill hole, walk back and forth across an iron-stained hill, wave their arms, and sketch on the field map. They have some exciting thoughts. "This could be like the Red Canyon orebody, the one that made the best profit in the Alamo district. We have a similar depth, similar rocks, and probably similar mining conditions. Two more drill holes, here and here, should tell us if we are on the right track."

The ultimate feasibility study, sometimes made years later and at the end of a series of feasibility studies, is more technologic than geologic. It is made after considerable time and money have already been spent. Order-of-magnitude economics have now been replaced by definitive cost figures and by a careful financial analysis. The decision is whether to develop a mine.

A detailed feasibility study made at the Brenda porphyry copper-molybdenum mine site in British Columbia illustrates what may be involved in the final stages before mine development (Chapman, 1970). The study, requiring 15 months time and an expenditure of $3,500,000, had been justified as the result of preliminary feasibility studies showing a huge orebody with a critical and near-marginal ore grade. The detailed study necessitated 72 diamond drill holes, 19 percussion drill holes, extensive drifting, crosscutting, and raising in order to confirm the estimated grade. A laboratory and a pilot mill were constructed to test the ore. Studies of power and water availability, tailings disposal procedures, and markets were made. The mine and concentrator were designed. Throughout the study, economic analyses were made with alternative rates of mining, metallurgical recovery, and metal prices.

Costs in Development, Mining and Processing

Capital costs and production costs—major economic considerations—are part of a mining geologist's "guides to ore." If either geologic considerations or economic considerations are allowed to eclipse the other, waste results. Geologic work may be wasted if it continues on mineralization for which too little of the economic aspect has been considered; the effort might better be spent on more attractive targets. A financial commitment and a superb engineering design may be wasted if they are based on sketchy geologic assumptions. The Ferris-Haggerty copper mine in the once-famous Encampment district in Wyoming was equipped in 1903 with the world's longest tramway, an engineering marvel, but there was no more high-grade ore. Copper prices decreased at the same time. Marginal orebodies became submarginal mineralization, and the entire operation was abandoned within four years. leaving an impressive group of five ghost towns. There are more up-to-date examples of mines designed without an adequate geologic basis, but it would be unkind to cite them while the protagonists are still trying to forget them. Governments have sometimes gone one tragic step further—raising the hopes of citizens in mineral-based regional development schemes for which neither geologic *nor* economic information was adequate.

Happily, geologic and technologic-economic considerations are coming into better balance. Part of this is due to the increased sophistication of investors and governments, part is due to improved methods of economic analysis, and part is certainly due to the appearance of professional geological engineers. It is not that geological engineers are especially capable of estimating mining costs. This is the field of expertise of the mining engineer who has the information and experience for project estimation and evaluation. Rather it is that the geological engineer and the engineering-indoctrinated geologists are able to appreciate the limits of accuracy and complexities in what is being estimated. When we obtain an order-of-magnitude cost figure for use with an incompletely defined exploration target, we have to use it in just that context. If we carry it further without advice from a mining engineer, we will be like children playing with dynamite.

First, some limits of accuracy. Such terms as "order of magnitude" and "preliminary" have definite meanings, as shown in Table 8-5. An order-of-magnitude estimate is a quickie estimate made with a minimum of engineering and geological data. Its usefulness is temporary because the impact of new data—a fresh run of drill core, for example—will be large and immediate. A *definitive* estimate entails, as the name indicates, specific data and considerable time and expense. How do we get from order of magnitude to definitive? By acting on the decisions from a series of feasibility studies. We may never arrive at a definitive estimate since it is not uncommon for most of our original exploration targets to be abandoned somewhere en route.

Table 8-5.   Limits for Cost Estimation[a]

1. *Order-of-magnitude Estimate* (Ratio Estimate)
   based on previous cost data, probable error *over* ±30%
2. *Study Estimate* (Factored Estimate)
   based on knowledge of major items, probable error *up to* 30%
3. *Preliminary Estimate* (Budget Authorization Estimate, Scope Estimate)
   probable error ±20%
4. *Definitive Estimate* (Project Control Estimate)
   based on almost complete data, probable error ±10%
5. *Detailed Estimate* (Firm Estimate, Contractor's Estimate)
   complete with site survey, probable error ±5%

[a] American Society of Cost Engineers. Terms from Zimmerman, 1968.

The most reliable capital and operating cost information comes from "inside" sources, recent data on production methods within the same company, mining district, and geologic surroundings. If the analogy is not close enough, mining engineers will modify the data by using different factors for depth, distance, geomechanical conditions, and so on. If the costs are more than a few years old, they will take inflationary trends into account and apply new values for labor and materials. For definitive cost estimates in a totally new situation, mining engineers will simulate the expected conditions as closely as possible and apply what they know of the capabilities of people and machinery. Obviously, this takes time and requires a well-investigated deposit. It may also require experimental mining and a pilot mill.

For order-of-magnitude and preliminary cost estimates, "outside" sources of information can be obtained from publications of state and national bureaus of mines and from reports of projects in the mining literature. Some of the most frequently used sources of cost information are the *SME Mining Engineering Handbook* (Cummins and Given, 1973), *Mining Engineering* magazine, *Mining Congress Journal, Engineering and Mining Journal,* and the *Canadian Mining and Metallurgical Bulletin.* An annual publication deserving special mention is the *Canadian Mining Journal's Reference Manual and Buyer's Guide*; this contains detailed unit-by-unit data on costs at mines throughout Canada as well as data on exploration, development, and milling costs.

The analogy principle is used. "If room-and-pillar uranium mining costs at a 400 ton-per-day operations near Grants, New Mexico, were reported as ranging from $1.70 per ton at a depth of 90 m to $8 per ton at a depth of 430 m 5 years ago, then our proposed operation would also have been somewhere in this range. Taking the difference in geology into account, our costs would have been in the higher part of this range. With the change in cost index during five years and with the economic conditions of our area, our costs would now be in the $8–$10 range. Now, could our discovery be an orebody?''

In working with this analogy, we obviously have to know something of the geology and engineering conditions in the Grants uranium district as well as in our own. We must have some information on yearly changes in an appropriate cost index and must have some idea of how much the overall costs should differ between the two areas. *Cost Engineering* and *Engineering News Record* magazines contain information on cost updating with periodic comparisons between cost indices (there are several) and with reviews of regional cost differences. Cost indices are generally given in terms of a base year; for example, if one of the cost indices was 1.0 for the year 1968, then it would have been 1.29 in 1974 and 1.33 in 1975. Regional cost differences (actually they refer to building costs) are also based on a certain index area, for example, 1.0 for the eastern United States. On this basis, the cost level might be 0.8 for The Netherlands and 4.0 for Alaska north of Cook Inlet.

The uncertainties in dealing with costs by analogy should now be apparent. And there is the disturbing knowledge that cost accounting practices differ from one mining operation to another. Engineers deal with some of the uncertainties by applying sensitivity analysis (finding the most critical parameters and working on them) and by applying risk analysis (a probabilistic approach in which statistical density functions or ranges are estimated for each cost parameter and then sampled at random to derive an overall cost density function). Neither type of analysis specifies the use of a computer, but neither can go very far without a computer. Sensitivity analysis and risk analysis will be mentioned again in Chapter 19 in relation to the evaluation of prospects.

For the very broadest of capital cost estimates there are *ratio costs* with which a few better known partial costs can be extended toward total costs. The oldest of these is "machinery costs about one dollar per pound"; it is surprising how well this very crude one seems to check out. Another time-honored rule of thumb is "the engineering cost for a new plant is 12 percent of the total plant cost." Then there is the exponential rule for economies of scale: "plants of similar design but different scale are related by the power of 0.6 to 0.7." If a 2000-ton-per-day plant costs $10 million, then a 3000-ton-per-day plant should cost $10 million $\times$ $1.5^{0.7}$ or $10(1.328) = \$13.28$ million. As far-fetched as it may seem the exponential rule is commonly used in order-of-magnitude comparisons between possible levels of production.

Percentage cost distributions—ratio costs—are useful in estimating new mining costs. Assume that the total cost per ton of material handled is the sum of the operating cost (labor plus nonlabor costs) plus the ownership cost. Suppose that in a group of closely related mines the labor cost averages 50 percent of the operating cost per ton and that the ownership cost (distributed capital investment) averages 20 percent of the operating cost per ton. On the basis of new data on mining wages and fringe benefits assume a labor cost of $52 per shift. On the basis of current literature assume a productivity of 120 tons of material per manshift. The estimated mining cost would then be calculated as follows:

Labor cost per ton = $52/120 = $0.43

Operating cost per ton = $0.43/0.50 = 0.86

Total cost per ton = 0.86(1.20) = $1.04

This is the cost of removing material from the mine. If the example were from an open-pit mine in which the productivity is the same for ore and waste, a 2:1 stripping ratio would give a total cost of $3.12 per ton of ore. For additional examples of mining cost calculation from percentage cost figures see Pfleider and Weaton (1973). Tabulations of percentage cost figures for individual mines and groups of mines can be obtained from information circular and reports of investigations of the U.S. Bureau of Mines.

The capital investment in a large new mining venture is measured in tens or hundreds of millions of dollars; some require well over $500 million. The Cuajone copper complex in Peru: $620 million. The Carajas iron ore complex in Brazil: $700 million. The Cerro Colorado copper project in Panama: $800 million. Even relatively small mines that produce a few hundred tons of ore per day require several million dollars in capital investment. A capital investment cannot always be so clearly stated because many mines have grown step by step from early high-grade operations to their present size. And artisanal placer mines and pits operate in many parts of the world with hardly any capital investment. Table 8-6 and Figure 8-7 can be used to make a generalization. "Average" large mining ventures with capacities above 5,000 tons per day will require an investment of $4,000–$8,000 per ton of designed daily capacity. Most smaller mines, especially underground operations, will require $8,000–$14,000 per ton-day capacity. Under some conditions the capital investment may be twice that of the "average" large mine. The effects of isolated sites and difficult climates are illustrated in the capital investment per ton-day capacity for the mines in Greenland, northern Canada, and Indonesia. An item worth special mention is "working capital." This is not part of the fixed investment, but it may amount to 20 percent or more of the total investment in isolated areas where large inventories of supplies must be kept on hand or where the mineral must be stockpiled between shipping seasons.

Tables 8-7 and 8-8 provide some information on the general range of development and mining costs. Even though the spread for each activity is wide there still will be situations with lower and higher costs. In other words, these are order-of-magnitude figures at the very most—not to be strained beyond their intended limit of accuracy.

Table 8-6. Capital Cost of New Mining Projects, 1970–1976[a]

	Cost ($U.S.)
Zeïda, Morocco 2700 t/d open pit, lead-zinc	$2,400
Sierrita, Arizona 75,000 t/d open pit, copper	2,600
Pinto Valley, Arizona 36,000 t/d open pit, copper	2,800
Highmont, B.C., Canada 22,500 t/d open pit, copper	2,800
Brenda, B.C., Canada 21,500 t/d open pit copper-molybdenum	2,900
Tynagh, Ireland 22,500 t/d underground, lead-zinc	3,100
Lornex, B.C., Canada 34,000 t/d open pit, copper	3,700
San Xavier North, Arizona 3500 t/d open pit, copper	3,700
Babine Lake, B.C., Canada 9000 t/d open pit, copper	$3,700
Whitehorse, Y.T., Canada 1800 t/d underground, copper	3,700
Tyrone, New Mexico 26,000 t/d open pit, copper	4,400
Sacaton, Arizona 8000 t/d open pit, copper	4,400
Bougainville, Papua New Guinea 81,000 t/d open pit, copper	4,400
Similkameen, B.C., Canada 13,500 t/d open pit, copper	5,300
Mamut, Malaysia 13,500 t/d open pit, copper	5,900
Las Torres, Guanajuato, Mexico 1800 t/d underground, silver	6,100
Inguaran, Michoacan, Mexico 2000 t/d underground, copper-silver	6,500
Metcalf, Arizona 27,000 t/d open pit, copper	6,700
Ruttan Lake, Manitoba, Canada 9,000 t/d open pit, copper	6,700
Aguas Claras, Brazil 29,000 t/d open pit, iron	7,000
Tilden, Michigan 26,000 t/d open pit, iron	7,200
Twin Buttes, Arizona 27,000 t/d open pit, copper	7,300

Table 8-6 (con'd). Capital Cost of New Mining Projects, 1970–1976[a]

	Cost ($U.S.)
King Island, Australia	7,800
1,800 t/d underground, tungsten	
Paraburdoo, Western Australia	7,800
26,000 t/d open pit, iron	
Ghost Lake, Manitoba, Canada	8,900
220 t/d underground, copper-zinc-silver	
Henderson, Colorado	9,200
26,000 t/d underground, molybdenum	
Sar Cheshmeh, Iran	9,400
41,500 t/d open pit, copper	
Lakeshore, Arizona	10,000
8000 t/d underground, copper	
Wheal Jane, Cornwall, U.K.	11,000
1,000 t/d, underground, tin-copper	
Navan, Ireland	12,600
7,000 t/d underground, zinc-lead	
Mattabi, Ontario, Canada	13,000
2,700 t/d open pit, lead-zinc	
Dorrigo, N.S.W., Australia	13,300
180 t/d underground, antimony	
Davis-Keays, B.C., Canada	14,400
900 t/d underground, copper	
Madrigal, Peru	14,400
450 t/d underground, copper-lead-zinc	
Randfontein Estates, South Africa	15,000
2,300 t/d underground, gold	
Granduc, B.C., Canada	15,000
6,800 t/d underground, copper	
Sturgeon Lake, Ontario, Canada	18,500
1,100 t/d open pit, lead-zinc	
Mount Wellington, Cornwall, U.K.	18,600
540 t/d underground, tin-copper	
Anvil, Y.T., Canada	20,000
6,000 t/d underground, lead-zinc	
Thierry, Ontario, Canada	25,000
3,600 t/d underground and open pit, copper	
Marmorilik, Greenland	25,000
1,800 t/d underground, zinc	
Ertsberg, Indonesia	27,000
7,400 t/d open pit, copper	
Nanisivik, Baffin Island, N.W.T., Canada	40,000
1,500 t/d underground, zinc-lead-silver	

[a] As reported in various mining journals as "cost of project" or "investment." Includes mill and supporting facilities. Production in metric tons of ore; cost in U.S. dollars per metric ton-day design capacity.

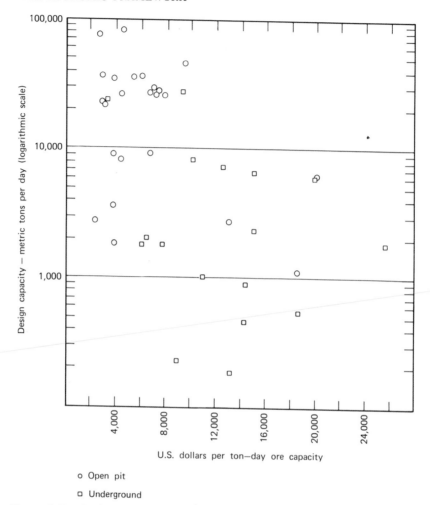

Figure 8-7. Capital cost of mining projects, 1970–1975. Data from Table 8-6.

Table 8-7. Unit Direct Development Costs, including Labor Materials, and Equipment Charges, Underground Mines, 1970–1976

Shafts (without production equipment): $750–$3,000 per meter
Adit Tunnels: $500–$2,000 per meter
Drifts, Crosscuts, and Raises: $200–$400 per meter
General: For 37 Canadian underground mines, report development costs ranged from $0.13–$6.50 per metric ton of ore. The median cost was $1.18.

Source: United States and Canadian mining journals.

Table 8-8.   Direct Mining Costs

Method	Cost per Metric Ton
Square setting	$10–$22
Cut and fill	8–20
Shrinkage	8–18
Sublevel stoping (longhole)	3–15
Open stoping (longhole)	2–8
Room and pillar	2–10
Block caving	1–4
Open pit (ore or waste)	0.40–1.25

Source. United States and Canadian mining journals.
These ranges in cost, including development charges, are
typical of North American mines during the period 1970–
1976. Consider that the open-pit costs are for *material*
(both waste and ore) moved and that the unit cost of
moving waste will be slightly lower than the cost of
mining ore. In most instances, indirect or "overhead"
costs will add 15–30 percent to the direct costs shown.

Processing costs can be generalized, with the same limitations. Typical milling costs at North American flotation concentrators ranged from $1 to $5 per metric ton of ore during 1970–1975. The higher costs were generally associated with smaller mills, those with a capacity on the order of 1,000 tons of ore per day. For still smaller flotation concentrators with capacities on the order of 100–500 tons per day costs of $10–$20 per ton would be more typical. In general, gravity concentration can be done at 25–50 percent of the cost of flotation processing.

The Kaiparowitz project, an underground mining complex planned in southern Utah, capable of producing 10 million tons of coal per year, was estimated in 1975 to cost $300 million. The project was shelved, but the capital cost figure is significant. The Blackwater coal mine, a 4 million ton-per-year surface mine in Queensland, Australia, required a mine investment of $22 million and an infrastructure investment of $23 million (McLeod, 1973). Capital investment and mining costs for the coal industry are difficult to compare with those for hard-rock mining because of profound differences in equipment, methods, operating schedules, and even in accounting. By way of generalization, new underground coal mines require a capital investment similar to that of the "average" large underground metal mine, around $5,000–$8,000 per ton-day capacity. Surface coal mines have capital requirements similar to those of the better situated large open-pit metal mines, on the order of $3,000–$5,000 per ton-day capacity.

Coal mining costs in the United States also include a charge for the union welfare fund and for the cost of coal preparation for market. Total costs for bituminous coal from underground mines were on the order of $15–$20 per ton in 1975. Open-pit coal mining costs, including charges for land reclamation, have been estimated to be 25–30 percent lower than underground costs (Zurkowski, 1975).

Capital and operating costs for solution mines reflect processes that differ considerably from those in conventional mining. Comparisons are difficult because many of the costs are kept confidential by companies that have been successful in their experiments. However, a specific cost comparison between solution mining and conventional mining has been furnished by Hunkin (1975) on the basis of hypothetical uranium orebodies in sandstone. (Table 8-9). The capital cost for solution mining appears to be about one-fourth to one-fifth that of equivalent conventional mining, and the production costs are also considerably lower for solution mining. The penetration of solutions into

Table 8-9.   A Cost Comparison between Solution Mining and Conventional Mining for a Hypothetical Uranium Orebody in Sandstone, Western United States[a]

	Solution Mining	Conventional Mining
Data		
Grade of ore	0.065% U_3O_8	0.15% U_3O_8
Annual capacity	2 million lb U_3O_8	2 million lb U_3O_8
	(700 mt U)	(700 mt U)
Lead time to production		
High	3 years	10 years
Low	2 years	7 years
Date of first U_3O_8 shipment	1977	1982
Cost Estimates		
Capital	(1975 $)	(1975 $)
High	$16 million	$70 million
Low	$ 8 million	$40 million
$/annual lb capacity		
High	$8.00	$35.00
Low	$4.00	$20.00
Overall costs		
3 mine, 1 mill complex		$11.38/lb U_3O_8
4 solution mine operations	$ 8.93/lb U_3O_8	
Land reclamation cost		$ 0.50/lb U_3O_8
Cost of first production	(1977 $)	(1982 $)
shipment	$11.80/lb U_3O_8	$30.64/lb U_3O_8

[a] Data from G. G. Hunkin, "The environmental impact of solution mining for uranium," *Mining Congress Journal,* v. 61, no. 10, Tables 2 and 3, p. 27, 1975. Copyright ©1975 by American Mining Congress. Used by permission of the publisher.

lower grade ore provides for a difference in average mining grade even though the annual capacity in a final uranium product is the same.

Cost estimates for deep-sea nickel-copper-cobalt nodule mining are closely guarded secrets, but the few capital costs that have been quoted indicate a two-step program with $50–$100 million for a pilot operation and $300–$500 million for a commercial operating system (Welling, 1976). Deep-sea mining costs have been estimated at $10–$20 per ton, with additional processing costs at $10–$15 per ton without allowing for manganese recovery (Tinsley, 1975).

8.5. ENGINEERING ECONOMY ANALYSIS IN EXPLORATION AND MINING

The economic basis for exploration work is uncertain—almost by definition. We deal in analogies, probabilities, and order-of-magnitude figures that do not adapt easily to economic analysis. Yet, the need to justify a plan of action arises at the very earliest stage in exploration, and that need grows with each step toward the ultimate mine-making decision. The weightiest decisions will usually be made by someone who is not a practicing geologist, although he or she may have been one before "going administrative." Decisions are made with whatever geologic information is available, and it is the geologist's responsibility to see that as much critical information as possible is taken into account. This responsibility requires that a geologist learn still another language in addition to those of mining, mineral processing, and marketing—the language of engineering economy. There are so many geologic assumptions in financial decisions that the geologist dare not remain on the outside.

So an exploration geologist's personal library must accommodate still another book, a textbook on engineering economy, such as one by Newman (1976), Smith (1973), or Grant and Ireson (1970). Much of the following discussion is a synopsis of these books.

The Time Value of Money

In every basic treatment of engineering economy, the opening theme is that money has a time value. A certain sum of money, one dollar, for example, is worth more now than it would be at some future date, say 5 years from now, because it can be put to work now. If the dollar were to be invested now at an interest rate of 10 percent compounded annually, it would amount to $1.00 × (1 + 0.10) or $1.10 at the end of one year. It would amount to $1.10 × (1 + 0.10), or $1.00(1.10)^2$, or $1.21 at the end of the second year, and finally $1.00(1.10)^5$ or $1.611 at the end of the fifth year. Thus, the 5-year future value of today's investment can be found by multiplying the investment by 1.611, the *single payment compound amount factor* for 10 percent and 5 years, or it can be expressed by the *single payment compound amount formula*:

$$S = P(1 + i)^n$$

where S is the sum of money at the end of n interest periods, P is the original (present) investment, and i is the interest rate. This is compound interest; we let it accumulate without siphoning off any profit along the way.

Or to look at compound interest from another viewpoint. Suppose we want to invest an amount that will grow through 10 percent interest compounded annually to $1 in 5 years. How much would we have to invest? Using the above equation to develop a *single payment present value formula*, we have:

$$P = \frac{S}{(1 + i)^n} = \frac{\$1.00}{(1.10)^5} = \$0.6209$$

Thus the present value of a dollar receivable 5 years from now is $0.6209 under the stated terms of prevailing interest; 0.6209 is the *single payment present value factor* for 10 percent and 5 years. Because the interest rate is now working in reverse, it is called the *discount rate*.

So, today's dollar and the future dollar are not equivalent. The concept deals with interest, not inflation. The future dollar is assumed to have the same purchasing power as the present dollar. Since the effects of inflation on mining investment are so enmeshed with additional factors, such as taxation, the cost of production, and future mineral prices, inflation will be taken up as a separate topic later in this chapter. For now, we will assume constant purchasing power; and we are familiar enough with the use of assumptions in geology to know that some are indispensable, others acceptable, and some downright dangerous. It depends on how far we take them toward specific instances.

The two formulas stated thus far can be used to determine the present value of the expected income from a mine. The expected income from each successive year in a mine's life can be multiplied by a single payment present value factor for a certain rate of interest to give the present value of that year's income. In practice, this certain rate of interest is often called the *hurdle rate*, a threshold below which a company feels that mining investments would be unattractive. A summation of all the yearly present values during the expected life of the mine will then give the present value of all income from the orebody. Remember, at a 10 percent discount rate the present value of $1.00 worth of ore to be mined 5 years from now is only $0.62. From the present value of the entire income, the amount of capital investment is subtracted to give the *net present value* of the orebody, the amount of money in present value dollars that will be received in addition to a minimum acceptable (hurdle rate) return on investment.

The "present" in present value is the date on which some dignitary cuts the ceremonial ribbon and operations begin (and operating problems begin to take shape). But the capital investment will have been made at various times during the preproduction period rather than in one lump sum. Therefore the individual expenditures made during several years must be made equivalent to the starting date of production by applying a single payment compound amount factor to each expenditure at whatever percentage of interest is assigned to capital outlays, that is, the cost of borrowing money.

The net present value term is of special importance to the exploration geologist; it is sometimes called the *acquisition value* of a mining property— the sum of money that can be paid for the mineral rights, for the explored orebody, or for geological investigations under the assumption that all estimated costs, revenues, and reserves are reasonably accurate. Such an analysis is often made on a raw prospect to see if additional work or additional payments are justified and to set a working limit for expenditures. Later, with more information, a more reliable acquisition value can be set as a guideline for additional expenditures.

In Table 8-10 the single payment present value factors give enough information for a demonstration of net present value. Take a capital investment with a starting date value of $4 million and consider that the investment is expected to return $2 million in the first and second years, $1 million in the third year, $0.5 million in the fourth and fifth years, and $1 million in the sixth year. The interest rate—hurdle rate—is 15 percent compounded annually. The situation is this:

Year	Income	15% Discount Factor (from Table 8-10)	Present Value
1	$2.0 million	0.870	$1,740,000
2	2.0 million	.756	1,512,000
3	1.0 million	.658	658,000
4	0.5 million	.572	286,000
5	0.5 million	.497	248,500
6	1.0 million	.432	432,000
Totals	$7.0 million		$4,876,500
Less capital investment			− 4,000,000
Net Present Value (Acquisition Value)			$ 876,500

The factors in Table 8-10 show an important characteristic in the timing of return on investment. At a 6 percent discount rate, a dollar receivable 10 years from now is worth $0.558 in present money and a dollar to be received 20 years from now is worth $0.312. But suppose you might have invested the money elsewhere at 15 percent rather than 6 percent interest. The present

Table 8-10. Single Payment Present Value Factors[a]

n	6%	8%	10%	12%	15%	20%
1	0.943	0.926	0.909	0.893	0.870	0.833
2	0.890	0.857	0.826	0.797	0.756	0.694
3	0.840	0.791	0.751	0.712	0.658	0.578
4	0.792	0.735	0.683	0.636	0.572	0.482
5	0.747	0.681	0.621	0.567	0.497	0.402
6	0.705	0.630	0.564	0.507	0.432	0.335
7	0.665	0.583	0.513	0.452	0.376	0.279
8	0.627	0.540	0.467	0.404	0.326	0.233
9	0.592	0.500	0.424	0.361	0.284	0.194
10	0.558	0.463	0.386	0.322	0.247	0.162
11	0.527	0.429	0.350	0.287	0.215	0.134
12	0.497	0.397	0.319	0.257	0.187	0.112
13	0.469	0.368	0.290	0.229	0.162	0.094
14	0.442	0.340	0.263	0.205	0.141	0.078
15	0.417	0.315	0.239	0.183	0.122	0.065
16	0.394	0.292	0.218	0.163	0.107	0.054
17	0.371	0.270	0.198	0.146	0.093	0.045
18	0.350	0.250	0.180	0.130	0.081	0.038
19	0.331	0.232	0.164	0.116	0.070	0.031
20	0.312	0.215	0.149	0.104	0.061	0.026
25	0.232	0.146	0.092	0.059	0.030	0.011
30	0.174	0.099	0.057	0.033	0.015	0.004

[a] Compounded annually. Present value of $1 to be received at end of nth year.

value of a dollar receivable in 10 years would be $0.247, much less than the 20-year value at 6 percent, and the 20-year value would be only $0.061. The effect of discounting at various rates of interest can be seen in Figure 8-8. Note that a dollar of income falls in value below 10 cents in 9 years at a 30 percent discount rate, in 13 years at a 20 percent discount rate, and not until 24 years at a 10 percent discount rate. A preliminary conclusion, one used by many investors, is that income beyond the tenth year has little effect on the present value of a mining property at the high discount rates that are sometimes set for risky investment. Like most preliminary conclusions, this one has limited merit because it ignores a host of contributing factors. There are many reasons why a mining enterprise would be planned to stay in operation for several decades; some of these will be mentioned in the next section of this chapter.

Thus far each single year's income has been discounted to its present value.

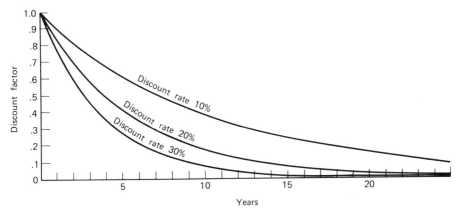

Figure 8-8. Present value of $1.00, $P = S/(1 + i)^n$.

This illustrates the concept, but a more general practice in order-of-magnitude and preliminary economic estimates is to consider income as a uniform flow of money from each year's operation in the life of a mine. The reasoning is sound enough; a mine is designed to operate continuously at a certain capacity and the relation between costs and revenue may be taken as fairly constant until enough detailed information can be gathered to forecast year-by-year changes. A preliminary analysis will at least tell us if we can justify the expense of collecting detailed information. A uniform income during a series of time periods (usually years) is an *annuity,* and the present value of an annuity is expressed in the *uniform series present value* (series discount) *formula*:

$$P = A \; \frac{(1 + i)^n - 1}{i(1 + i)^n}$$

where A is the annual income.

Table 8-11 gives values for the uniform series present worth factor. Consider an acceptable interest rate of 15 percent and that a mine is expected to earn $1 million per year for 10 years. The present value of the income would be: $1,000,000(5.019) = $5,019,000. If the required capital investment were $4 million, then the net present value (acquisition value) of the property would be $1,019,000.

Additional engineering economy factors representing the *capital recovery, sinking fund,* and *Hoskold* formulas can be developed from the formulas and factors given thus far. These factors and their premises will each be discussed in turn.

A present debt multiplied by the capital recovery factor gives the uniform year-end annual payments needed to remove the debt in n years, with the lender receiving an interest rate i on his investment. The capital recovery factor is the reciprocal of the uniform series present worth factor. If we need to know how much money must be allocated each year for 4 years to repay a loan of $100,000 with interest at 8 percent, we can use a capital recovery factor of 0.3019, the reciprocal of 3.312, the uniform series present value factor in Table 8-11 for 8 percent and 4 years. The $30,190 paid at the end of each year will accumulate in the lender's account as follows:

Year	Year End Payment	Single Payment Compound Amount
1	$30,190	$(1 + 0.08)^3 = \$38,031$
2	30,190	$(1 + 0.08)^2 = \$35,214$
3	30,190	$(1 + 0.08) = \$32,605$
4	30,190	$(1.00) = \$30,190$
		Total $136,040

For an investment of $100,000 the lender has obtained a series of payments rather than a single payment, but the series is equivalent in value to the amount that would have been received in a single payment. To check this, multiply $136,048 by 0.735, the single payment present value factor in Table 8-10, and we have a present value of nearly $100,000.

A sinking fund is established to produce a specified amount of money at the end of a certain period of time by making a uniform series of interest-earning investments. To determine the amount that must be invested at the end of each time period, the future amount (the goal) is multiplied by the sinking fund factor. The sinking fund factor is the capital recovery factor minus the interest rate. Therefore, if we want to know how much money must be invested each year at 8 percent to have $100,000 in 10 years, we first find the capital recovery factor (0.1490) as the reciprocal of the series present value factor (6.710) in Table 8-11; we then subtract 0.08 to obtain the sinking fund factor (0.0690); $6,900 must be invested each year.

The capital recovery and sinking fund formulas appear in most forms of engineering economy analysis. The Hoskold formula is, however, unique to mining. Widely used in mine evaluation and mine taxation for a full century, and still in limited use, the Hoskold assumption is that a uniform annual profit should be used in two ways at separate rates of interest. One part of the profit is conservatively placed in a sinking fund at a "safe" rate of return i, where it will yield an amount equivalent to the initial investment when the mine is depleted. The remainder is then available to investors at a higher speculative (risk) rate of return i'. The formula is

$$P = \left[\frac{i}{(1 + i)^n - 1} + i' \right]^{-1}$$

Table 8-11.   Uniform Series Present Value Factors[a]

n	6%	8%	10%	12%	15%	20%
1	0.943	0.926	0.909	0.893	0.870	0.833
2	1.833	1.783	1.736	1.690	1.626	1.528
3	2.673	2.577	2.487	2.402	2.283	2.106
4	3.465	3.312	3.170	3.037	2.855	2.589
5	4.212	3.993	3.791	3.605	3.352	2.991
6	4.917	4.623	4.355	4.111	3.784	3.326
7	5.582	5.206	4.868	4.564	4.160	3.605
8	6.210	5.747	5.335	4.968	4.487	4.837
9	6.802	6.247	5.759	5.328	4.772	4.031
10	7.360	6.710	6.145	5.650	5.019	4.192
11	7.887	7.139	6.495	5.938	5.234	4.327
12	8.384	7.536	6.814	6.194	5.421	4.439
13	8.853	7.904	7.103	6.424	5.583	4.533
14	9.295	8.244	7.367	6.628	5.724	4.611
15	9.712	8.559	7.606	6.811	5.847	4.675
16	10.106	8.851	7.824	6.974	5.954	4.730
17	10.477	9.122	8.022	7.120	6.047	4.775
18	10.828	9.372	8.201	7.250	6.128	4.812
19	11.158	9.604	8.365	7.366	6.198	4.844
20	11.470	9.818	8.514	7.469	6.259	4.870
25	12.783	10.675	9.077	7.843	6.464	4.948
30	13.765	11.258	9.427	8.055	6.566	4.979

[a] Compounded annually. Present value of annuity of $1 to be received at end of each year for n years.

The Hoskold reasoning, that investors will need to buy another mining property when the existing property is depleted, is valid—but the reasoning is not followed in practice. Money is not ordinarily put aside in a sinking fund at a low interest rate when it could be put to work at a higher rate of interest. Also, the amount of the investment for which the speculative (high-risk) rate is calculated actually decreases during the life of the mine. In the last few years of operation not much is being risked.

Calculations using the Hoskold formula tend to undervalue a mining property—a defensible approach if the user is very conservative and *wants* to undervalue the property as a hedge against risk. A comparison between single interest rate and Hoskold present values is shown in the following example. If the expected return from an orebody is $1 million per year for 10 years and if a 15 percent return on the investment is required, the present value from Table 8-10 would be $5,019,000. By the Hoskold method, Table 8-12, using a

Table 8-12. Hoskold Factors[a]

n	12%	15%	20%
1	0.893	0.870	0.838
2	1.652	1.574	1.459
3	2.304	2.155	1.945
4	2.269	2.641	2.333
5	3.363	3.054	2.650
6	3.797	3.409	2.912
7	4.182	3.716	3.134
8	4.524	3.984	3.322
9	4.830	4.219	3.484
10	5.106	4.427	3.625
11	5.353	4.613	3.748
12	5.578	4.778	3.860
13	5.782	4.927	3.953
14	5.967	5.061	4.039
15	6.136	5.182	4.116
16	6.291	5.292	4.185
17	6.433	5.392	4.247
18	6.564	5.484	4.304
19	6.684	5.567	4.355
20	6.794	5.644	4.402
25	7.235	5.944	4.582
30	7.539	6.148	4.703

[a] Compounded annually. Present value of an annuity of
$1 per year for n years, allowing interest to investor at
rate shown, with redemption of capital at 6%.

15 percent risk rate and, for example, a 6 percent safe rate, the present worth
would be $4,427,000. Hoskold factors in Table 8-12 are given for only a few
speculative rates and only one "safe" rate. A much more detailed Hoskold
table is included in the mineral property examination handbook by Parks
(1957).

   All of the illustrations of the time value of money have been based upon
compound interest, annual interest periods, and year-end payments—conventional practice in the preliminary analysis of exploration and mining projects.
There are refinements, however, that come into play as the analysts gain
more data and calculate with "sharper pencils." One refinement is *continuous
compounding*, which is based on the observation that receipts and disburse-
ments do not really occur in lump sums at the end of each year; they are

spread throughout the year in a nearly continuous flow. Thus, when the interest period in a time-value formula is shortened from annually to monthly to weekly, we approach a continuous compounding formula. For an expected income of $1 million per year for 10 years at 15 percent interest the present value figure of $5,019,000 obtained by annual compounding would be $5,386,000 if continuous compounding factors are used. The refinement is justified if geologic, engineering, and economic estimates are comparably accurate and the income is calculated with appropriate variations; otherwise it is incongruous.

Detailed tables of present value, capital recovery, and sinking fund factors can be found in most engineering economy textbooks. These tables are well adapted to occasional use in preliminary economic analyses, but for continued calculations and for iterative runs of figures, a computer or a small electronic calculator is preferred. Pocket-size calculators with a few special function keys and a small storage capacity will in fact do the work of most value calculations directly from formulas without any need for tables.

Estimating Profitability

Regardless of the economic system, funds for capital investment have limits and there is competition for their use. Comparisons must be made, and these are most easily done in terms of profitability. Profitability is certainly not an overriding index; social impact and a host of indirect economic benefits are taken into account, but even in these an attempt is generally made to quantify them and take them out of the "hunch" category. A new mine in an underdeveloped area? It will affect the life-style of the populace: new transportation, communications, schools, sanitary and welfare facilities, and the beginning of a market economy. All have consequences to balance against the required capital expenditure. The index: çost effectiveness. And the units, in socialist and well as capitalist economies: money.

At some point the facts are gathered and an investment decision is made. It may not be the final decision, because there are increments in geologic, engineering, and economic information to be paid for in time and money. Still, even if it is only the first step-wise investment toward a goal, the goal must be identified in the same terms as the cost of getting there.

The most commonly used methods for comparing and screening capital projects in mining—for identifying the goal—are *payback (payout) period, net present value,* and *discounted cash flow return on investment* (DCFROI). The two latter methods take the time value of money into account; the first does not.

The payback period is the number of years required for the earnings to return the original capital outlay. It serves as a coarse screen for retaining desirable projects and rejecting poor ones: the shorter the payout period, the more attractive the project. Thus, for a capital investment of $28 million and an expected annual income (generally, cash flow) of $3.5 million, the payback period is 8.0 years—too long perhaps, compared with alternative projects with payout periods of 4 or 5 years.

Payback period has little real analytical power in selecting between alternatives. It measures the return *of* investment, not the return *on* investment. The life of the mine could extend 2, 10, or 12 years beyond a 3-year payback period, but there is no provision for crediting this fact against an alternative proposition with the same 3-year payback period but with a 4-year life. Payback period has been defended as a method of taking risk into account, since risk is a time function; that is, how long will the investment be exposed to natural and political hazards? Still payback period does not reflect the rate at which the investment is returned in a risky situation. In two projects with equivalent 5-year payback periods, one might return 60 percent of the investment from the mining of a high-grade ore zone in the first 2 years whereas the other returned only 20 percent. Payback period is widely used, but it is regarded as an incomplete selection criterion that must be used in combination with other criteria.

Net present value has been illustrated in the use of single payment and uniform series present worth factors. Net present value, the amount of money receivable in excess of the minimum acceptable rate of return on investment, is widely used in mining, especially for land purchase price guidelines but— like the payout period—it is seldom used alone.

The acceptable interest rate (discount rate) chosen for net present value calculations is determined in various ways. Ideally, it should be the firm's cost of capital, a mixture of the cost of common stock participation, debentures, and loans. In practice, it is also a function of the market value of the entire firm, past experience in mining ventures, and the anticipated risk— especially risk. An 18 percent discount rate might mean "we expect 12 percent on our successful investments, but most of our projects will be abandoned at a loss: use a higher discount rate." Good enough, except that it practically "writes off" the value of ore reserves needed for a long-term operation. There are many reasons for wanting to give credit to ore that will be mined 20 years from now. And, of course, there is the fact that the market value in 20 years may be several times the present level.

The discounted cash flow return on investment (DCFROI) method is essentially a matter of identifying the rate of interest that will be received on the capital investment. To calculate DCFROI, an interest rate is found that will balance the present value of income from the mine with the capital investment; net present value zero. If the earnings are considered on a year-by-year basis, the return on investment can be found by using a single payment present value table. In the following example, a four-place table is used; this is the kind of table found in most engineering economy textbooks.

Year	Earnings	Present Value Factor 15%	Present Value	Present Value Factor 20%	Present Value
1	$50,000	0.8696	$ 43,480	0.8333	$ 41,665
2	70,000	.7561	52,927	.6944	48,608
3	70,000	.6575	46,025	.5787	40,509
4	60,000	.5718	34,308	.4823	28,938
5	50,000	.4972	24,860	.4019	20,095
6	40,000	.4323	17,292	.3349	13,396
7	60,000	.3759	22,554	.2791	16,746
		Total Present Value	$241,446		$209,957
		Capital Investment	230,000		230,000
		Net Present Value (+) $ 11,446			(−) $ 20,043

The rate of return on investment (ROI) is:

$$ROI = 15\% + \frac{11,446}{11,446 + 20,043}(20\% - 15\%)$$

$$= 15\% + 0.363\,(5\%) = 16.8\%$$

Where earnings can be considered as coming in a uniform series the return on investment is more easily calculated by interpolating between the two table factors that are closest in number to the payout period in years, entering the uniform series present value table on the row corresponding to the life of the mine, and reading percentage return figures at the top of the columns. Consider a mining property with annual earnings of $600,000 and a capital investment of $3,200,000, the payout period is 5.33 years. Suppose that the mine is expected to have a 12-year life. As a Table 8-11 factor, 5.33 for 12 years lies between 5.421 (15 percent) and 4.439 (20 percent). The return on investment is 15.5 percent.

Discounted cash flow return on investment is used in two ways. First as a screen—if the rate of return is lower than the agreed-upon hurdle rate, the project is rejected; second, as a method of making a comparison between acceptable projects. The higher the DCFROI, the better the project.

There are several ways of refining the calculations in these three methods of estimating profitability and there are additional criteria for screening and selecting projects. These are presented in engineering economy textbooks and in the literature of finance, such as the books by Bierman and Smidt (1966) and Van Horn (1968). For an explanation of profitability estimates in mining feasibility studies, a brief article by Lewis and Bhappu (1975) is recommended.

A yearly net income and cash flow calculation is the basis for all profitability estimates. While there are many formats, the following is typical:

Revenue	$15,420,000
Operating costs	− 7,340,000
Operating income	8,080,000
Depreciation	− 850,000
Net before depletion	7,230,000
Depletion	− 2,313,000
Taxable income	4,917,000
Income tax	− 2,704,350
Net income	2,212,650
Add: depreciation	850,000
Add: depletion	2,313,000
Cash flow	$ 5,375,650

Revenue is the gross income from sales of the mineral product. Operating costs refer to all of the costs incurred in producing and selling the mineral product, including royalty payments, interest payments, and property taxes. Since development work is done in order to make the ore available for mining, it is an operating cost. Development work may be accounted for in the year it was done or, more commonly, it is accounted for over a longer period of time—even over the entire life of the mine. "Development" and "exploration" are critical accounting terms in the United States because successful exploration is not treated in taxation accounting as an operating cost.

Depreciation is an allowance for the loss in value of the plant and equipment with the passage of time. In the example, a capital investment of $8.5 million was distributed in equal yearly amounts over 10 years of mine life; this is the straight-line method of computing depreciation, the most commonly used method in preliminary mine evaluation. Accelerated depreciation methods, whereby the capital can be recovered more quickly, are taken into account in detailed feasibility studies. Amortization is sometimes used as a loose synonym for depreciation, and it is often listed in cash flow calculations with depreciation. More precisely amortization applies to capital expenditures in such things as research and experimental work which are not actually part of the plant installation. Some capital expenditures do not enter into either depreciation or amortization; these include working capital and cost of land, both of which are considered recoverable at the end of the mine life. Property payments and exploration costs are recovered in various ways in different countries. In the United States they are not normally subject to depreciation or amortization; they are considered recoverable through the depletion allowance.

The *depletion allowance* is a yearly deduction provided to mineral producers so that they may "replace" the orebody (find another one) when it is mined out. In the United States, there are two permissible methods of computing the allowance. *Cost (unit) depletion* allows the total cost of acquiring a mineral property to be distributed among the units (tons of ore) as they are mined. *Percentage depletion,* more widely used, permits the deduction of a specified percentage of the gross income, providing the amount does not exceed 50 percent of the before-depletion net income. In the example, a depletion allowance of $2,313,000, or 15 percent, of the revenue (gross income) was permitted. If the before-depletion net income had been as low as $4 million, no more than $2 million would have been permitted. If royalties had been paid on revenue from the ore, they would have been deducted from the gross income used in calculating depletion. The royalty holder would be expected to claim a depletion allowance on this part. Inasmuch as the depletion allowance is unique to the mineral industries, it is not covered thoroughly in engineering economy textbooks. For additional information on the depletion allowance, two articles by O'Neil (1974a, 1974b) are recommended.

Income tax paid to the federal and state governments is combined and taken as 55 percent of the taxable income in the example. Other taxes, such as state and local sales and property taxes and federal payroll taxes, are included in operating costs.

Because depreciation and depletion are "book" costs rather than actual expenditures, they are part of the *cash flow,* and they are funds that can be used in the same way as net income (the accounting net profit). Most engineering economy calculations use cash flow rather than net income. An alternative approach uses the *accounting rate of return* method, which expresses a ratio of annual net profits to either the total investment or the "average" (midpoint) investment with no provision for the time value of money.

It has been pointed out that engineering economy formulas are based on units of money with constant purchasing power. But inflation has reached such a rate outside of the controlled-economy countries that it must be taken into specific account at some stage in all evaluations.

Costs in many mining operations have risen at rates of as much as 10–15 percent per year and the rise is accelerating—a source of concern in projecting operating costs for more than a year or two. Mineral commodity prices are rising at least in the long term, a cause for optimistic projections. Do the two cancel? A prevalent assumption in mine evaluation is that they do and that mineral prices will rise enough to balance the inflationary erosion of money returned as profit on the original investment. Another assumption is customary. All mining ventures in a similar category will be subject to the same inflationary pressures; therefore the relative merit of one mining venture over another will be unaffected. For the most part the two assumptions are workable and preliminary evaluations are made without a particular allowance for inflation beyond the anticipated startup date.

In a way the specter of intolerable inflation is a variety of risk, and it is sometimes treated in the same way by emphasizing short payback periods, high discount rates, and high hurdle rates. It is also given as a reason for investing in a small or quickly installed mine plant now rather than taking the option of a larger, more efficient plant at some future time. None of this is particularly good mineral conservation practice, but neither is any other method of dealing with economic, political, or geologic risk. This is why the performance record of a national mineral policy is as important as its wording. And this is why a large part of a geologist's job is to reduce the unknown factors in an orebody to the point where a stable, long-term mining operation can be planned.

There is more to be said about profitability estimates. In Chapter 18 profitability estimates will be mentioned in the discussion of exploration objectives, and there will be still more to consider in Chapter 19 in relation to sensitivity analysis and risk analysis as methods of quantifying the chances for a geologic discovery becoming an economic mine. Not that we want to quantify too far; while we are agonizing over the difference between 14 percent and 15 percent DCFROI, we may overlook a geologic factor that could mean zero DCFROI.

PART FOUR
GATHERING AND PRESENTING GEOLOGIC DATA

Up to this point in the book, geologists have been treated as members of a team that includes everyone from geoscience theoreticians to mineral policy makers. Now the work of geologists will be considered in the jobs which they can handle best. Of course, the geologist still is not alone; there is a team of librarians, miners, geophysicists, drillers, and other specialists. Their work will be considered as well.

Because geology has a closely specified purpose when it is applied to mining, field work must be concentrated where it is most needed, and it must involve the data that apply most directly to the objective. The first step is to gather and evaluate the existing data, as explained in Chapter 9, before going into the field for more data. Reconnaissance, the topic of Chapter 10, is done in the field, but the chapter deals to a large extent with aerial photography and remote sensing because these methods are so widely used to indicate the best areas for effective field work.

In Chapter 11, the locale is entirely in the field and on the ground where the geologist does some of the most rewarding work: mapping the terrain and formulating a model for an orebody. Chapter 12 takes the work of mapping into underground mines where it is applied to actual orebodies rather than to conceptual models. Chapters 13 and 14 consider two valuable exploration tools, geophysics and geochemistry.

Conceptual models and the capabilities of the geologists who make them are often tested in the results of drilling and (assuming that a discovery has been made) in the sampling of the mineral deposit, as discussed in Chapters 15 and 16.

None of the field data will provide much help in finding mineral bodies if they remain in a geologist's notebook. Chapter 17 deals with the shaping of field data into information and presenting the information in usable form.

CHAPTER 9
PRELIMINARY STUDIES: OBTAINING AND EVALUATING EXISTING DATA

Mistakes don't come from what you don't know, they come what you know that ain't so.

Josh Billings, *from* Proverb *(1874)*

9.1. DATA, INFORMATION, AND FACTS

Data may or may not have some bearing on the problem at hand, but by selecting a certain amount of the data—"reducing it"—we can synthesize the needed information. Of course, there is a short cut: rely on someone else's synthesis. In doing this we may have to rely on at least a few of their assumptions as well.

A compilation of data published ten years ago on all of the known fluorspar deposits in New Mexico may have most of the information we need to begin a fluorspar exploration program. Most, but not all. What has changed meanwhile? What was the compiler's working hypothesis? How thoroughly was the data base searched? The compiler had no way of knowing how much additional drilling would take place in the Gila district. The compiler may not have considered a few fluorite and barite crystals in the gravels of the Santa Fe Group important enough for comment and, in fact, may not have known of the occurrence. In taking short cuts we must know something of the route.

If geologic information can ever have a central characteristic, it will be that all of the information has an element of subjectivity. Information reflects

281

someone's decision as to what is important; so do data. *Factual* data? Even an aerial photograph represents someone's choice of flight altitude, time of year, camera, and film emulsion. Geophysical and geochemical data are collected at chosen intervals and with chosen parameters; they are not quite facts. We often drown in data, seldom in information, never in facts.

9.2. SOURCES OF PRELIMINARY DATA AND INFORMATION

Appendix C lists some of the sources. The terms "data" and "information" are so completely interwoven that they must be combined for the time being in spite of their separate meanings.

Appendix C lists only the most recent guides to library information sources and only those bibliographic indexes and abstract journals in current use. This does not mean that data from a few decades or even several centuries ago are unimportant. On the contrary, most mineral deposits are discovered in districts where someone has mined before, where an early geologist noted something of possible importance, or where some long-forgotten prospector once filed a mineral claim. The depth to which we sometimes dig into the archives can be called history or even archaeology. Exploration activity is current at the Laurium mine in Greece, the mine mentioned on the first page of Chapter 1, where some of the pertinent information dates from 480 B.C.

One of the most significant mineral discoveries in the Southwest Pacific was made on the basis of a literature search, a careful evaluation of the information from 20 years before, and a subsequent field examination. The presence of copper ore at a high elevation on Mount Carstensz in the rugged backbone of New Guinea was noted in 1936 by two oil geologists on a mountaineering holiday. The occurrence was mentioned in a report published in 1939 by the University of Leiden in The Netherlands, a short time before the German invasion. The report was drowned in the events that followed, but it surfaced again in 1959 during a routine search of library material by geologists of the East Borneo Company. Forbes Wilson, a geologist working in Indonesia on behalf of a joint venture between the Freeport Sulphur Company and the East Borneo Company, learned of the report and verified its legitimacy by contacting one of the Dutch mountaineer-geologists. Having evaluated the information to his satisfaction, Wilson made his way into the Mount Carstensz country in 1960 and re-discovered the magnificent knob of copper ore now known as the Ertsberg. The orebody was eventually estimated to contain 33 million tons of 2.5 percent copper ore. Production began at the Ertsberg mine in 1972 at a rate of 7,400 tons of ore per day.

A great many successful exploration programs have begun with a literature search. This is not a new approach. What is totally new in geological literature search, however, is the practice of data storage and retrieval by computer. Older methods have had to be replaced because the accumulation of paper, microforms, and magnetic tape was getting out of hand. A "doubling time" for scientific literature has been formulated: 10–15 years. More than 26,000 scientific and technical journals are currently being published, and three new journals are born each day (Wood 1973, p. 2).

Specific libraries with collections of geological and engineering literature are not listed in Appendix C because there are so many. Government libraries such as those of the U.S. Geological Survey, engineering society libraries such as that of the Institution of Mining and Metallurgy in London, and university libraries are the principal sources of material. Interlibrary loan services can be used to obtain books, photocopies, microforms, and even maps from these large collections. Among the "guides to library information sources" in Appendix C, the book by Wood (1973) deserves a special comment. This book includes information on discontinued bibliographies and abstract journals that are still of prime importance in a search for data on any particular area. Abstract journals are valuable, but they appear to have a high mortality rate.

Under "sources of abstracts" in Appendix C, there are good reasons for including *Chemical Abstracts* and the geographical bimonthly *Geo-Abstracts*. The chemical industry is just as concerned with sources of minerals as we are, and geographers are even more concerned with cartography, regional landform analyses, and the distribution of mineral resources.

Under "sources of regional information," gazetteers are of special importance. In these books, each identifiable mountain, mine, and town is listed by coordinates. Gazetteers covering foreign countries, compiled by the U.S. Board of Geographic Names, include listings of alternative local names for geographic features and lists of alternative translations of names from other languages; this is especially helpful when ploughing through geologic literature on unfamiliar areas. American state gazetteers can be helpful too; try to find the mines and town of Total Wreck, Arizona, on a current map.

Unedited geologic data on open file in national or state bureau offices may prove to be a veritable bonanza of important maps and reports. Some of these data may be in the form of progress reports made by private companies operating in leased areas, or it may be data relinquished by companies upon the termination of a prospecting permit.

Drill core, well cuttings, and rock samples are available for inspection at many bureau offices.

In some countries, users of open-file data must explain their specific interest rather than just come in to browse. Sometimes, the data can be obtained only upon the issuance of a prospecting permit for the area involved. In most places, however, open-file reports, maps, and available magnetic tapes of geochemical or geophysical data can be bought by anyone for the cost of making the copy.

The geologic and mining information retrieval systems mentioned in Appendix C are normally made available on a subscription or contract basis. These systems provide custom "retrospective" literature searches in answer to specific questions and may also provide current "interest profile" services for selected topics. The information can be received as "hard copy" printout at a remote terminal or by mail; it can also be received as a cathode-ray tube (CRT) display, and in some instances the associated maps can be received by graphic plotter. Microfilm or microfiche cards of the text can often be obtained, and entire tapes can be leased. In practice, some combination of media is used, with the preliminary inquiries being answered by temporary display of titles and a few sample abstracts at a CRT terminal, relevant abstracts received as on-line printout, and full texts of selected material received a few days later by mail.

Several of the bibliographies and abstract journals mentioned in Appendix C deal with published and unpublished translations of geological literature. Many of the translations that have not been published are available in microform or in full-size copy from national and international library services. In the United States, unpublished translations may be obtained from the National Translations Center of the John Crerar Library in Chicago; this material is listed in *Translations Register Index*. Translations available from the British Library, Lending Division (BLL), are listed in the *Commonwealth Index of Unpublished Scientific and Technical Translations*. The European Translations Center (ETC), publisher of the *World Index of Scientific Translations*, collects information on existing translations of scientific work in the languages of western Europe.

The assistance facilities and data reference files mentioned under "sources of aerial photography and spacecraft imagery" are important public services of the EROS data center because of the wide choice in available imagery for most areas. Special processing services and custom-made color composites are available from EROS. Arrangements can also be made with private mapping service companies to select the most appropriate satellite imagery and to compile mosaics for special purposes.

Obtaining information by direct inquiry is a skill developed by many geologists during their careers. Others, the "bull in a china shop" people, never develop the skill. It depends on insight and diplomacy rather than on science and engineering. Insight: most government and research foundation scientists are not specifically assigned to public information services. Some of their work is open to the public and some of their time is available for discussion with geologists who visit during working hours, but we must appreciate that courtesy has limits. Diplomacy: protocol is important, and there are interagency relationships in many countries that may not be apparent to impatient outsiders—especially foreigners. A clumsy request for information from the wrong office may close a great many other doors.

Regional promotion and development organizations are likely to have extensive files of natural resource and industrial information, and they want the files used. One caution: the information on minerals may contain some overly enthusiastic statements.

One important source of information has been omitted in Appendix C. Rumor: someone is said to be exploring here, they seem to have found something there, or they were seen moving in a drill rig with 1,000 m of pipe. This can be valuable information—provided it really *is* information.

9.3. EVALUATING DATA AND INFORMATION

Geophysicists have a saying that applies equally well to the work of geologists: one man's message is another man's noise. Noise is the irrelevant part of the signal or of the data; message is the part we can actually use—the information.

We can begin separating information from noise by skimming the data. Look for key words. "Key-word-in-context" (the KWIK index) is a helpful system used in many abstract journals and the words are compatible with computerized data retrieval. We have some choices, the same ones geophysicists take into account when they separate message from noise. If we call most of the data "information" at first glance, we can be sure that nothing of value has been rejected, but we may have nearly the same unmanageable mess we started with. If we cut right through to a small selection of the most obviously relevant material, we risk throwing away something of value. There are choices in between, all of which take more time.

Thus far, the job has been one of data sorting rather than data evaluation. In some respects, a computer could do just as well. Now, as geologists, we have information and we can evaluate the information by checking it for internal consistency (do maps A, B, and C show compatible structural patterns?), by checking it for credibility (should the breccia pipe really be described as a "blowout"?), and by using the journalist's rule (who? what? where? when?). The journalist's rule needs explanation. Who? A report may have been written by a famous or infamous person. What? The information may have originally been compiled from several sources and for other objectives. Where? The map may have been based entirely on photogeology without anyone setting foot on the ground. When? The map may have been made in 1940.

Ultimately, the place to evaluate geologic data and information is in the field, but field exploration is too expensive to waste on noise.

CHAPTER 10
RECONNAISSANCE

A regional exploration program normally involves the defining of smaller areas of specific interest within a region of more general interest. This is certainly not the only way to find mineral deposits, but it can serve as a theme here.

Reconnaissance is the highest risk stage in exploration; it must therefore be the lowest in cost per square kilometer, and it must be a process of *screening*, not mapping. Geologic mapping is the basic way of gathering information in reconnaissance, but mapping for its own sake wastes time and money that will be needed later.

The step from preliminary study to reconnaissance is a step from office work to a combination of airborne, field, and office work. When does field work begin? Early. Field investigations may begin even during the preliminary data-gathering stage, with orientation visits to key sites of known mineralization and to important geologic features in the region or somewhere nearby. A reconnaissance project can be planned from symbols and words on paper, but the planning will be more realistic if it includes—even briefly—a field experience.

10.1. DESIGNING A RECONNAISSANCE PROJECT

Map Scale

Small-scale maps, such as the 1:5 million metallogenic map of the United States and the 1:2.5 million metallogenic map of Europe, do not show individual mines or closely spaced districts, but they show the "big picture," a basis for making the earliest and most sweeping assumptions. At this scale, gaps between metallogenic provinces can be bridged by geologic projection and the ideas of plate tectonics can be appreciated. Metallogenic maps are of course highly interpretive, a helpful characteristic for cutting through irrelevant detail but at the same time a dangerous invitation to accept theories that were frozen into symbols when the map was made. If we have a new idea, we may have to make our own metallogenic map from mineral maps, geologic maps, and satellite imagery.

For the next step, still in the idea range but involving a more specific region, 1:500,000-scale maps and mineral deposit maps are available for many areas of the world. In some countries, partial metallogenic map coverage is also available at scales of 1:500,000–1:200,000. At any rate, these maps provide a basis for preliminary field visits, for the planning of photogeology studies, and for planning the actual reconnaissance on the ground. Now the ideas are ready to be tested in new map compilations with new field data at whatever next larger scale has good topographic map coverage and some geologic or metallogenic map coverage. The choice of scale depends on the conceptual model of the target; the scale must be large enough to show the anticipated geologic associations without too much exaggeration and yet small enough to emphasize regional patterns rather than intricate geologic detail. This is a *synoptic* view. Detail comes later, when it can be digested in a few well-chosen localities.

In the western United States, reconnaissance often begins at 1:250,000. High-quality topographic map coverage is available at this scale, and there are geologic and geophysical maps that can be photographically enlarged or reduced onto transparent overlay sheets conforming to the scale of the topographic base maps. Landsat (ERTS) imagery is obtainable at standard enlargements to 1:250,000. Field observation sites can be plotted within a spacing of 1 km and related to enough topographic and cultural detail to be located again. After the first-chosen area has been reduced to smaller areas of interest, the appropriate data are transferred to a more detailed set of topographic maps, aerial photographs, and overlay sheets at a scale of 1:24,000 or, in some instances, 1:62,500. At 1:24,000, additional observations can be recorded at intervals of 50 m, airborne geophysics flight lines can be shown, and individual mine workings, local stratigraphic changes, and local structure can be discriminated enough to proceed from reconnaissance investigation into detailed target-area investigation. Detailed target-area work will be described in Chapter 11.

Kreiter (1968), in his handbook on prospecting and exploration, refers to the Russian practice of using 1:50,000-scale and 1:25,000-scale geologic maps as a preparatory step to prospecting. "Prospecting" as such involves the next larger scale, 1:10,000. Kreiter considers a usual mapping area in the prospecting stage to encompass 10–100 km². "Exploration," in the Russian usage of the term, refers to the subsequent outlining of a deposit at still larger mapping scales.

In France, typical of western European countries with a long mining history and with well-mapped geology, the preferred scale for starting mineral exploration projects (exploration in the sense of a search) is 1:50,000. This is the scale of current geologic mapping by the French government bureau of mines (BRGM) and the scale at which a national inventory of mineral deposits is being compiled. Topographic maps at 1:25,000 scale provide the base for the next step. Whether this scale is adequate for selecting small areas in which to do detailed work ("prospecting" in the French use of the term) depends on the coarseness of the mineralization control. Pierre Routhier (1963, p. 1007) of the Paris School of Mines found that some of the basement irregularities associated with ore in the Les Malines district did not begin to resolve well enough for prospecting until mapping was done at 1:10,000 scale. Whatever the choice in terminology, the search for a target area calls for some mapping at scales on the order of 1:50,000–1:10,000. In American parlance, this is part of reconnaissance.

Reducing (Enlarging) the Area

In practice, reconnaissance often departs from an idealized step-by-step sequence because there are various local patterns of information. The earliest field investigations are sometimes directed into unmapped areas lying between areas with better information; in this way a few quick traverses can be made to fill in the geology and to search for fairly obvious targets. Perhaps "Quaternary alluvium" will be found to contain some erosional windows into bedrock.

Field work is sometimes dominated by a specific kind of geophysics, geochemistry, or geology in areas where similar work has already been done with less sophisticated equipment or older ideas. The assumption is that enough new information will be obtained to make the duplication worthwhile. Geophysicists may not have had gamma-ray spectrometer equipment, and geologists may not have been thinking "exhalites and stratabound massive sulfides" when they mapped "volcanics."

Reconnaissance will on occasion work backward from the general large-to-small-area sequence; with good reason. A new idea during a mine examination may inspire a search for similar geologic conditions throughout the region. This was the experience at the Bou Azzer mine in the Anti-Atlas range, Morocco, where an intensive geochemical and geological study of vein-type cobalt orebodies disclosed some intriguing associations between ore, Precambrian ophiolites, and a Precambrian unconformity. The new information at Bou Azzer led to a reassessment of all the previous regional mapping and to a successful districtwide exploration program (Clavel and Leblanc, 1971).

Still another kind of reconnaissance scheme departs from the general pattern. On the basis of existing information and from the impressions received in a few orientation visits by geologists or engineers a company may be attracted to a particular region where "old-time" operators are thoroughly entrenched. The company may gain a foothold by acquiring a mine or by entering into a joint mining venture. Then, from the new vantage point and with local support and information, reconnaissance work and fresh concepts will spread outward. In this kind of situation and in others where there is a "home" district, reconnaissance may be part of a continuing assignment rather than a single program.

The Methods

Reconnaissance, like old-fashioned and classical prospecting, relies on wide-ranging traverses. It is not surprising that traditional methods of prospecting are still part of reconnaissance in spite of modern airborne approaches. Stream courses are still the main traverse lines in areas blanketed by alluvium, soil, or laterite. All of exploration is identified in some way with geomorphology, but in reconnaissance the association has always been especially close.

The tracing of mineralized float and fragments of rock to their source exposures is no less important than it was 100 years ago. A modern aspect of this technique is that the float from ore and from key lithologic units can be identified in the field by geochemical methods or under a binocular microscope as well as with the naked eye. The source of a train of rock fragments may be much more apparent on an aerial photograph than it would be in the field; geomorphology again.

The tracing of ore boulders in glaciated terrain, long practiced in Scandinavia, is more important now than ever because we have geochemistry and new ideas in glaciology to help us. Even the historic patterns of human occupancy have a current-day place in tracing mineralized float; ore boulders in the stone walls of farms served as a mineralogical and geochemical guide to the discovery of the Tynagh orebody, Ireland, in 1961.

The panning of heavy minerals from alluvium, a technique used in the gold rush days, is still a guide to ore, and it provides good regional petrologic information as well. The technique has been modernized and is now used in combination with electrostatic separation of the panned concentrate and with instrumental mineralogical analysis. Critical sites for panning, stream junctions, for example, are often selected on aerial photographs.

The age-old practice of digging pits through thin overburden to obtain geologic information has been made easier by the availability of bulldozers, backhoes, and gasoline-operated blasthole drills. The best sites for pitting and trenching are sometimes determined by recording small hammer-activated shocks by seismic equipment to indicate shallow bedrock.

Drill holes, normally considered to be part of the detailed stages in exploration, are also appropriate to some kinds of reconnaissance work. Light-weight drilling machinery can provide key subsurface information to a depth of several hundred meters, and in situations involving stratigraphic controls to ore mineralization, drill holes are often justified at a very early stage in a regional exploration program. The advantage of early drilling has been demonstrated in the Gas Hills uranium district, Wyoming, where widely spaced holes (10–15 km) were used to provide subsurface stratigraphic, lithologic, alteration, geochemical, and structural information before concentrating the search for sandstone-type orebodies in target areas (J. F. Davis, 1973).

Geophysical and geochemical methods, to be discussed in Chapters 13 and 14, are nearly as important in reconnaissance as in detailed work. Airborne geophysics and stream-sediment geochemistry are especially useful. Because the results are in digital form and thus more comfortably handled than geological data, geophysical and geochemical work sometimes goes one step too far and accounts for an inordinately large portion of a reconnaissance budget. "Inordinately large" because the apparent advantages in quantification and statistical analysis may be offset by not knowing the geologic associations from which the numbers were taken. Reconnaissance information is that tenuous. In an extreme instance of blind quantification, some large areas in the western United States were recently "evaluated" by making impressive isoline maps from geochemical samples taken at wide intervals. And the source of the samples? Anything occurring at a grid point: soil, alluvium, granite, limestone, prospect pits, alkali flats, horse corrals, and so on.

Logistics

Since reconnaissance is a fast-moving affair, campsites are temporary and there is not time for building access roads. Placing exploration crews on site with the essential supplies has always been a major problem and a major expense. Daily weather and the season dictate what can be done. Maintenance facilities and repair services are likely to be far away. Something that must never be far away is ingenuity.

Four-wheel-drive vehicles are the standard means of short-distance travel in relatively open desert and brush country. In remote, wooded, and mountainous terrain and in certain protected wilderness areas, air transportation is more common. Even though the cost of operating helicopters and fixed-wing aircraft is high, the alternative of multiple field camps and slow surface travel would often be much more costly and it would lessen the time spent in doing the most important job: looking at rocks.

Preliminary inspection of an area by helicopter can furnish points of view that even the most detailed aerial photography would not match. More importantly a helicopter pilot can place geologists near significant outcrops, geophysicists on traverse lines, and geochemists at sampling points. In remote areas, a common procedure is to establish a base camp at a site which can be serviced by a fixed-wing aircraft and use a helicopter for access to field sites or to temporary "flycamps" (Fig. 10-1).

In all reconnaissance work, there are heavy demands on the endurance of persons and equipment. Some of these are met by using the knowledge of local surveying and construction people in planning and carrying out operations in a particular climate and terrain. Recognizing that field work in the Arctic has special characteristics, the Geological Survey of Canada has published a booklet, "Tips on Organizing Arctic Geological Field Work" (Kerr, 1974), and Wolff (1969) of the University of Alaska has written a *Handbook for the Alaskan Prospector.*

The treatment and analysis of field samples and raw data from reconnaissance geologic work are as much aided by aircraft service as is the overall logistic element. The turn-around time from collection of samples to interpreted results can be kept to a few days by using the sequence of helicopter–light airplane–airline express–central laboratory–radio communication.

10.2. AIRBORNE AND SATELLITE IMAGERY

Because reconnaissance work begins with a synoptic view, it can begin with small-scale photographs and line-scanning imagery: with remote sensing. In its very broadest definition, remote sensing includes geophysical exploration, instrumental chemical analysis, and all the "hands-off" techniques that use energy from the electromagnetic spectrum shown in Figure 10-2. When the term is used in a more restricted way, as is commonly done, remote sensing refers only to imaging methods that derive data from areas on the earth's surface rather than from depth. Later interpretation of the data—the explaining of a surface thermal anomaly, for example—brings in the depth dimension. Exploration methods that obtain direct signals from depth are more properly called geophysics. The spectral frequencies of remote sensing range from 3×10^8 to 1×10^{15} hertz; this is the range of microwave sensing, radar, thermal scanning, photography, and ultraviolet scanning. Most remote sensing uses *passive* systems that only collect incoming radiation; a few, such as side-looking airborne-radar (SLAR), use *active* systems in which signals are sent as well as received. To explain: bats and dolphins use active remote sensing and so does a flash camera.

Nearly all remote-sensing methods are important or at least potentially

Figure 10-1. Helicopter support for an exploration project in northern British Columbia. (Courtesy Hudson Bay Exploration and Development Co., Ltd.)

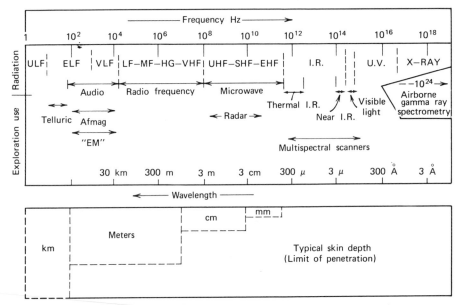

Figure 10-2.   Reconnaissance exploration methods in the electromagnetic spectrum.

important in some specific way to exploration, but none are so widely used as spacecraft imagery and aerial photography in the visible and near-visible wavelengths. Since this chapter deals with reconnaissance in the large-to-small-area context, a good place to begin is by looking at the characteristics of small-scale photographs and photolike images from spacecraft.

Satellite Imagery

Landsat-1 (originally called ERTS-1) was placed in polar and sun synchronous orbit in 1972 and had the capability of covering the entire earth's surface every 18 days. Multispectral scanner (MSS) data were received in four spectral bands. The wavelength range in two of these bands (4 and 5) is from 0.5 to 0.6 and 0.6 to 0.7 μm, corresponding to the green and red portions of the visible spectrum. The other two bands (6 and 7) have wavelength ranges from 0.7 to 0.8 and 0.8 to 1.1 μm, corresponding to near-infrared portions of the spectrum just beyond the visible zone.

The products are computer-compatible tapes as well as derived photographic images, such as shown in Figure 10-3. The tapes have the higher resolution because they are a direct record of the sensor without any loss in detail from subsequent processing. Tapes can be printed photographically or processed in a computer to enhance specific data. The more commonly used

Figure 10-3. Photographic copy of Landsat 1:1 million imagery, southern Arizona. (Courtesy Southwestern Exploration Associates, Tucson, Arizona.)

products, however, are 70 mm black-and-white film negatives and diapositive chips at a scale of 1:3,369,000, covering 185 km on a side. These can be enlarged (or ordered as enlargements from the EROS data center) to 1:1 million, 1:500,000, or 1:250,000-scale.

For geologic reconnaissance study, the more widely used products are black-and-white prints (bands 5 and/or 7) and "false color" composites of several bands (4, 5, and 7). In a false color composite, vegetation appears in shades of orange and red, water in black, and soil and rocks in shades of green, gray, red, orange, and brown. Imagery can be processed in various ways so that the color ratios can be made to enhance certain spectral differences. It can also be processed to deliver more natural-appearing "true color" prints. Color plate examples of processed imagery are included in a U.S. Geological Survey professional paper describing the general features and uses of Landsat (Williams and Carter, 1976). A do-it-yourself color composite can be made by viewing combinations of transparencies and filters in a special chip reader until the best enhancement is found for a particular situation. Geologic features can also be enhanced by selecting imagery from certain times of year when partial snow cover or vegetation bring out subtle differences in structure or lithology. Another way of enhancing terrain features is to process adjacent image frames to make a simulated stereo pair for three-dimensional viewing.

"Shade printing" is a handy method of showing enlarged Landsat data by computer printout. Shade prints can be used as reconnaissance base maps and for precise location of linear or areal patterns of picture elements or *pixels*. The technique consists of representing each pixel (an 80-m square) by overstriking a set of typeface characters. A deep shade of gray, for example, is produced by overstriking B, M, —, and $; lighter shades of gray are produced by overstriking characters such as & and * or by striking a single character.

The Nimbus series of weather satellites provides reflectance and thermal infrared emittance data at lower resolution than Landsat but with the capability of obtaining thermal inertia measurements which show promise for discriminating between surficial geologic units.

The Skylab, an orbiting workshop during 1973, obtained color photography and MSS imagery from scattered test sites. Figure 10-4, a black-and-white copy of a Skylab color photograph, shows a portion of the area covered by the Landsat image in Figure 10-3. The ground resolution (separation of two adjacent points) in Skylab photography is somewhat better than the 80-m resolution in Landsat-1 imagery, but its limited area of flight coverage prevents it from being as widely used in exploration.

Landsat-2, with capabilities similar to Landsat-1, was launched in 1975. Landsat-3, with additional spectral coverage in the thermal infrared range, is a 1977 project. EOS (Earth Observational Satellite) and SEOS (Synchronous EOS), 1978 and 1980 projects, are expected to deliver images with a ground resolution of about 40 m. Geosat, a series of satellite systems designed especially for geologic data collection, has been proposed by a group of geologists representing major oil and mining companies.

Satellite imagery is of demonstrated value in preliminary reconnaissance work, but there are—as with all exploration tools—some serious limitations. Cloud cover and vegetation hide a great deal of terrain and, of course, there are the expanses of soil and Quaternary sediments that mask geologic features from every kind of aerial photography. Still, large structural features sometimes appear as "ghosts" through overburden, and vegetation sometimes

Figure 10-4.   Skylab 1:500,000 photography, southern Arizona. (Courtesy Southwestern Exploration Associates, Tucson, Arizona.)

responds in discernible ways to geochemical patterns. The most useful feature of satellite imagery in comparison with aerial photography is its better synoptic view of large areas—just what is needed in the early stages of reconnaissance. Mosaics pieced together from aerial photographs are not nearly as uniform in tonal quality, they do not normally have the multispectral characteristics and unless they have already been compiled and corrected, are much more expensive.

The low resolution of Landsat imagery prevents it from competing with aerial photography—especially color photography—at any scale larger than that needed for the early stages of reconnaissance. It is, however, an important adjunct to larger scale aerial photography, the long-established "workhorse" of exploration.

An interesting trial of Landsat imagery, subsequent aerial photography, and field work in mineral exploration was made by Stancioff, Pasucci, and Rabchevsky (1973) in a 27,000-km² area of northwestern Sonora, Mexico. As a first step, imagery in band 5 was enlarged to produce 1:1 million-scale diapositives and positive prints. Diapositives from 29 percent of the area were viewed stereoscopically by using the slight overlap in Landsat frames. The remainder of the area was examined monoscopically with and without 5-power magnification. Preliminary lithologic and structural interpretations were recorded on an overlay.

In a second step a photomosaic of Landsat enlargements at 1:250,000 scale was used as a basis for adding more geologic detail to a new overlay. Rock units were differentiated, first by photographic tone and then by more specific textural characteristics. Even though a certain amount of lithologic recognition was possible the rock units were *differentiated* rather than identified; for identification, a few existing geologic maps were used. Eventually, the 1:250,000 overlay became so crowded that two overlays were prepared, one for lithology and one for lineations.

A third step involved comparison between the 1:250,000 overlay information from band 5, additional imagery in bands 4, 6, and 7, and color composites. The additional bands yielded no further information at this point, but a careful examination of the color composites (which had not been available earlier) resulted in additional rock unit differentiation. A cultural feature overlay was prepared; in this work, towns, roads, and existing mines could not be identified without relying on existing topographic maps. On the basis of a thorough literature search, prospect locations were added to the cultural overlay. Finally, areas selected for "ground truth" were photographed in color from low-level flights and inspected on the ground. Alteration zones observed during these investigations were added to the overlays.

As a follow-up step, repetitive Landsat imagery from orbits at later dates was studied and found to be of value in selecting additional linears; in fact, one entire set of linears became apparent for the first time. With the knowledge already gathered about alteration zones the later imagery was studied for further information and some new alteration zones were outlined; in this part of the study, band 4 imagery proved to be of more help than the color composites. During the follow-up step, imagery from nearby areas was also examined in order to improve the understanding of the regional framework.

As a result of the Landsat work, Stancioff and his associates proposed a geologic hypothesis for the occurrence of porphyry copper deposits in northwestern Sonora and they recommended 28 target areas for investigation. In their critique of the project they pointed out that an earlier use of color composites, some preliminary geologic cross sections made from existing map data, and the addition of more overlays for specific structural domains would have been helpful.

Three other evaluation reports on the use of Landsat imagery in mineral exploration are recommended reading. Viljoen and others (1975) give a detailed account of Landsat investigations in southern Africa and western Australia; their paper includes color-photo plates with interpretation overlays for several areas of complex lithology and tectonics. Lathram and others (1973) describe finding previously unrecognized linear zones in Alaska by studying Landsat and Numbus imagery, and they discuss the metallogenic significance of their work—the recognition of a previously unknown trend of copper-molybdenum mineralization. R. G. Schmitt (1976) provides a summary of an investigation for porphyry copper deposits in Pakistan in which computer-processed Landsat imagery was used. A field check of 19 prospecting targets indicated by studies of satellite data revealed 5 sizable areas of hydrothermally altered rock with abundant sulfide mineralization.

Aerial Photography and Photogeology

Exploration geologists often wonder how their work could possibly have been done before the advent of aerial photography. How did the mappers of 50 years ago find those isolated outcrops and how did they find such subtle structural trends? Answer: they used a lot of boot leather and they got their synoptic views of the terrain from hilltops. No hilltop—no synoptic view. They made full use of the best exploration tool ever invented—the geologist's pick. And, naturally, they missed some of the information.

With aerial photography, we can obtain synoptic views at almost any scale, but we will still miss some of the most important information if we take the "inductive leap" from an incomplete observation to a firm conclusion. Fortunately, this hazard is recognized by most exploration geologists and thus they use aerial photographs together with boot leather and the pick.

Photogeologic maps are among the best investments that can be made with the funds available for reconnaissance. The bargain is especially good if the work can be done with government photographs, and the work is especially effective if it is done by geologists who will eventually participate in the field reconnaissance. Photographic coverage from contract flying is still a fairly good bargain, and if large-scale photographs are needed for the field job, it may be the only way to obtain them.

There is hardly any upper limit to the scale at which aerial photographs can serve in exploration, as will be seen in Chapter 11 on geologic mapping; however, most of the available government aerial photographs are taken at reconnaissance scale, on the order of 1:120,000, 1:50,000, and 1:20,000, rather than at a detailed mapping scale. Contact prints, diapositives, and enlargements up to four times the original scale seem to cover most of the demands. Enlargements to five or six times the original scale are satisfactory with some new photography, but with most existing photographic coverage, enlargements beyond four times are too grainy.

Whatever the scale of photography and whatever the intended use, there are certain areas in which it is of little help; weather and low cloud cover may seldom—or never—permit good photography. Photography can be planned for the right season in areas of deciduous trees, but evergreen vegetation may be too dense in all seasons for an effective look at the ground. Topographic

Figure 10-5.   Small-scale 1:125,000 (U-2) aerial photographic coverage for the Globe-Miami district, Arizona, a portion of the area shown in Figures 10-3 and 10-4. (Courtesy Southwestern Exploration Associates, Tucson, Arizona.)

relief may be so great that the photographs are badly distorted unless taken from an extremely high altitude and therefore at a very small scale. Most of these problems can be resolved to some extent. Radar imagery has poorer resolution than photography, but it cuts through cloud cover and even through some light vegetative and light snow cover. Film and camera technology is improving, and this provides for better enlargements from high-altitude photography.

Black-and-White Photography. Government aerial photographic coverage is most generally available in black and white (Figs. 10-5 and 10-6), and most

Figure 10-6.   Larger scale aerial photography from a 1:24,000-scale photograph, Globe-Miami district, Arizona. (Courtesy Southwestern Exploration Associates, Tucson, Arizona.)

photogeologic work is done by interpreting tonal contrast, texture, and geomorphic features in shades of gray. Landforms are commonly quite distinct, and a stereoscopic view emphasizes even the most subtle variations in terrain. *Relief displacement* is the essential characteristic that permits stereoscopic vision. In a vertical photograph of any area, except one that is absolutely flat, the only point that appears in its true map position is the one that lies at the exact center; all other points appear displaced. The higher the terrain and the further its position from the center (principal point) of the photograph, the greater its displacement. Relief displacement for a specific area differs between two successive photographs on a flight line and also between two photographs taken from adjacent flight lines. The usual flight-line overlap is 60 percent and the usual side-lap is 30 percent. In stereovision, we simply imitate the *air base* between photographs with our *eye base* at close range. In most areas, the topography will show vertical exaggeration; hill slopes and dipping-bed exposures will appear steeper than they actually are.

Stereoscopic vision does not actually require a stereoscope; with a little practice, most people can hold up a stereo-pair of photographs and fuse the two images into a single image of three dimensions. It is much easier, however, to use a stereoscope with magnification. Lens-type pocket stereoscopes are standard tools for field work. Most mirror stereoscopes are bulky and better suited to office work, but they cover a larger field of view, as do many other kinds of stereoscopes with combinations of lenses, prisms, and mirrors.

Gray tones on a black-and-white aerial photograph are related to the reflectivity of the outcrops. Quartzite, limestone, and felsic igneous rocks reflect a high percentage of the incident light and have light tones; shale and slate show darker tones; a basalt flow or an amphibolite dike absorbs most of the incident light and appears very dark, as would be expected. Tonal contrast and textural appearance depend on soil and vegetation conditions as well as reflectivity and, of course, upon the shadows which will differ from photograph to photograph as the sun angle changes. Here is a disadvantage that can be turned into an advantage. Low sun angle photography (LSAP) taken very early or very late in the day can be used to accentuate small changes in relief.

Geomorphic relationships—the shape, size, and distribution of topographic features—are the mainstay of aerial photographic interpretation, whether the interpreter is a geologist, a soil scientist, a civil engineer, or any of the dozens of specialists to whom aerial photography is an everyday tool.

Linear geomorphic features are of special interest to exploration geologists, and a strain on their imaginations. In some areas the causes of lines or arcs seen in stream courses, aligned depressions, ridges, and vegetation can be determined well enough to label the linear features as beds, unconformities, or intrusive contacts. However, most linear features are labeled only as "photo linears" or "photo lineaments" until a field investigation can be made; a straight line cutting across several tonal and textural zones may turn out to be a vein, a fault, a dike, or perhaps a fence or an old road. An unexplained lineament may also be the faint expression of a structural feature beneath alluvium or soil that might take a trench or a drill hole to verify. This aspect of photogeology, the use of linear features in surficial deposits as a subsurface exploration technique, is reported upon in some detail by Norman (1976). A "hash" of lineaments in a dozen directions should contain some valuable information on jointing and stress patterns, but it may take intensive data analysis by statistical or optical (diffraction pattern) methods to make much sense from it.

In the past few pages, aerial photographs have been considered only as images of natural features, but this is seldom the complete picture. There are prospect pits, quarries, road cuts, wells, and all sorts of geology-connected cultural features to help in the analysis. One of the most common uses of combined natural and cultural features is in orientation surveys. Suppose we are looking for a uranium deposit in a new area: study some photographs of uranium mining in the nearest "old" area and take the pattern into account. One more step: study the "old" area on the ground as well.

There is much more to photogeology than looking and noting. There is mapping control and there is quantitative measurement. The resulting photogeologic map may look like the one shown in Figure 10-7. Photogeologic map making may call for no more than transferring notes from a photo overlay onto a good base map by drawing ray lines to common topographic features, or it may be so demanding that it belongs to a special field of civil engineering: photogrammetry. The tools and techniques of photogrammetry are matters for the professionals, and we call upon them, through surveying service firms, when a new and accurate topographic map is needed.

Still, a geologist needs to know something of photogrammetry. We use some of the simpler photogrammetric tools, parallax bars and similar "height finders," for example, to find differences in elevation when calculating dip. And we often work with photogrammetrists while they make maps from our data.

A summary of photogrammetric principles and, in fact, an outstanding presentation of photogeologic principles can be found in the chapters on image interpretation, cartographic presentation, and terrain and minerals in the Reeves' (1975) two-volume *Manual of Remote Sensing* published by the American Society of Photogrammetry.

Photogrammetric Engineering and Remote Sensing, the monthly journal of the American Society of Photogrammetry, contains articles on photogeology in almost every issue. Another monthly publication with occasional photogeology articles is *Photogrammetria* (Elsevier-North Holland Publishing Company).

Photo interprétation, a bimonthly publication of Éditions Technip, Paris,

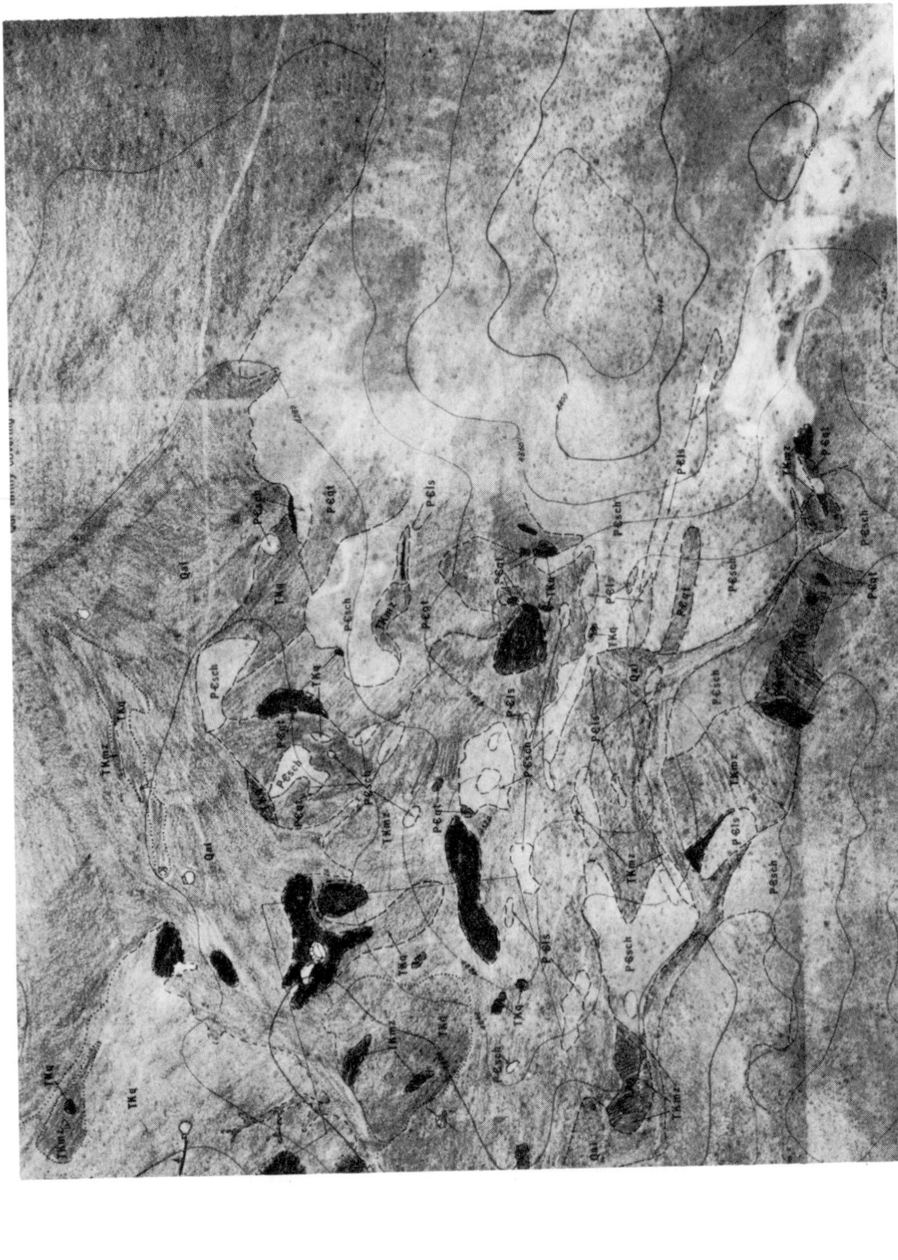

Figure 10-7. A portion of a photogeologic map. (Courtesy Southwestern Exploration Associates, Tucson, Arizona.)

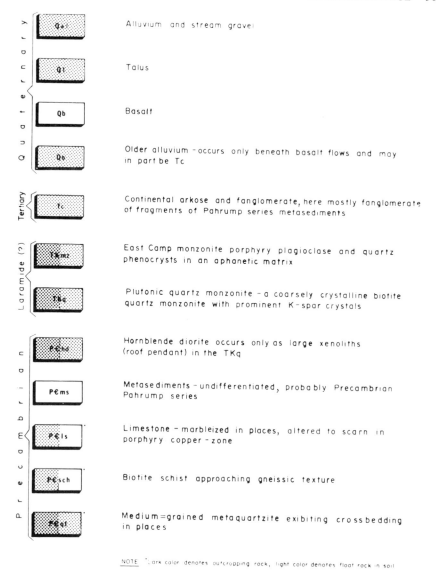

Quaternary	Qal	Alluvium and stream gravel
	Qt	Talus
	Qb	Basalt
	Qo	Older alluvium - occurs only beneath basalt flows and may in part be Tc
Tertiary	Tc	Continental arkose and fanglomerate, here mostly fanglomerate of fragments of Pahrump series metasediments
Laramide (?)	TKmz	East Camp monzonite porphyry plagioclase and quartz phenocrysts in an aphanetic matrix
	TKq	Plutonic quartz monzonite - a coarsely crystalline biotite quartz monzonite with prominent K-spar crystals
Precambrian	P€hd	Hornblende diorite occurs only as large xenoliths (roof pendant) in the TKq
	P€ms	Metasediments - undifferentiated, probably Precambrian Pahrump series
	P€ls	Limestone - marbleized in places, altered to scarn in porphyry copper - zone
	P€sch	Biotite schist approaching gneissic texture
	P€qt	Medium=grained metaquartzite exibiting crossbedding in places

NOTE Dark color denotes outcropping rock, light color denotes float rock in soil

Figure 10-7 (cont'd). Explanation for figure 10-7.

deserves special mention. Each issue includes five or six high-quality stereo-pairs of aerial photographs in black and white or color with interpretation overlays and explanations in French, English, Spanish, or German. The interpretations deal most often with geology, geomorphology, geography, and hydrology.

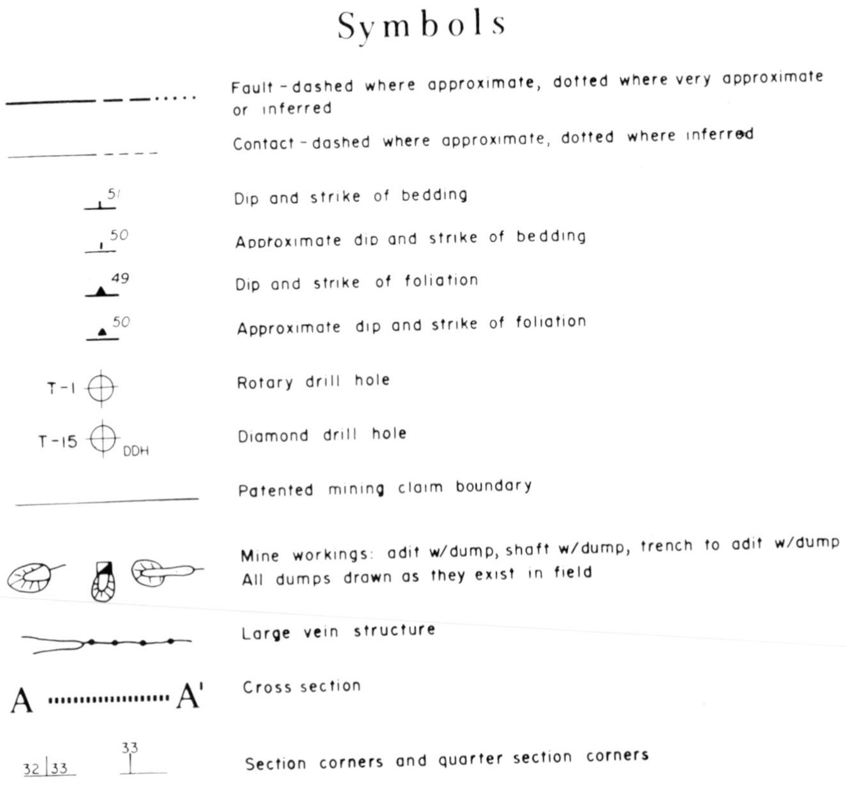

Symbols

———— — — ·····	Fault – dashed where approximate, dotted where very approximate or inferred
———— — — — —	Contact – dashed where approximate, dotted where inferred
⊥ 5'	Dip and strike of bedding
⫪ 50	Approximate dip and strike of bedding
▲ 49	Dip and strike of foliation
▲ 50	Approximate dip and strike of foliation
T-1 ⊕	Rotary drill hole
T-15 ⊕ DDH	Diamond drill hole
————————	Patented mining claim boundary
⊘ ▯ ⊘	Mine workings: adit w/dump, shaft w/dump, trench to adit w/dump. All dumps drawn as they exist in field
———•—•—•—	Large vein structure
A ·············· A'	Cross section
32 | 33 33 ⊤	Section corners and quarter section corners

Figure 10-7 (cont'd). Explanation for figure 10-7.

Color Aerial Photography. Color film has been used for a long time in aerial photography, but it has only recently become competitive in price and in resolution with panchromatic black-and-white film. Since the main costs in aerial photography are in flying and printing rather than in the film, an important application has been found for emulsions that permit color prints, diapositives, and black-and-white prints to be made from the same negative. With this film, preliminary photogeologic work is sometimes done in black and white, followed by work with color prints and diapositives in selected areas.

The advantages of color photography in geologic work are easy to guess. Thousands of tints and shades in color can be separated by the human eye. This gives an entire new dimension to photogeology and is a step toward having a "real" image of the terrain. Still, the procedures and tools of color photogeology and color photogrammetry are essentially the same as those mentioned for black-and-white photography.

Geologists who have used color aerial photography to delineate zones of hydrothermal alteration and to locate reddish-brown areas of capping or gossan now look back on their tones-of-gray work and wonder how much information they missed.

There is an important range beyond visible color in photography. Color infrared film, the camoflage-detection film of military use, provides a direct image without going through the process of artificial color compositing. Multispectral photography—taking photographs by multilens camera—reaches into the near-infrared and (with atmospheric limitations) into the near-ultraviolet spectral zones to obtain several black-and-white images for color compositing. A good description of multispectral imagery use in detecting geobotanical anomalies over orebodies in the Mount Isa-Cloncurry district, Queensland, is provided by Cole and Owen Jones (1974). In that project, flown at scales of 1:15,000 and 1:5,000, ground resolutions on the order of 1.0–0.3 m were obtained.

Scanning Systems

At wavelengths beyond the capabilities of photographic film but still within the field of remote sensing, scanning systems provide optical signals for conversion to electrical signals which are then recorded on tape. Ultimately, a photographic record is obtained from a cathode-ray tube in the same way that photographic images are obtained from the Landsat and Nimbus spacecraft. Two aircraft-borne scanning systems that have found some use in mineral exploration, specifically in reconnaissance, are thermal infrared sensing and radar.

Thermal surveys record the temperature contrast between surface objects or areas. With the Nimbus satellite and aircraft as detector platforms, they have been more successful in monitoring thermal pollution and volcanic activity than in exploration, but their capabilities for using thermal inertia measurements to show evidence of oxidizing orebodies, buried fault zones, and contrasting rock types are beginning to be realized. Side-looking airborne-radar (SLAR) is an active sensor system in which detailed images of terrain can be obtained during day or night and through most cloud cover. An acquisition scale of 1:250,000 is often enlarged to 1:50,000 for use as a base map. Geologic features are not recorded as such, but linears often show up better than with most aerial photography. One technique of aerial photography is analogous: low sun angle photography shows reflections and shadows in a kind of natural "radar" in which the sun acts as a transmitter. The use of SLAR imagery in reducing an 84,000-km^2 area in Nicaragua to a 12,000-km^2 area for priority exploration is described by Martin-Kaye (1974).

For some reading on thermal infrared sensing, SLAR, and other scanning methods, a booklet by Alexander and others (1974) provides a good digest.

More detail is given by Reeves (1975). For continuing reference to new developments there are the *Proceedings of International Symposia on Remote Sensing* published by the Environmental Research Institute (ERIM) at the University of Michigan, Ann Arbor.

10.3. OFFSHORE AND DEEP OCEAN RECONNAISSANCE

Offshore mining is still a relatively small industry in terms of mineral production, but there is more interest in offshore (to 180 m depth) and deep ocean (below 380 m) exploration and reconnaissance than the production statistics would indicate. Some of the interest derives from the recognized possibility of extending current offshore "hard-rock" operations into new areas, some from the projection of iron, tin, and other deposits beyond areas now being dredged, and some—perhaps most—from the potential value of deep-ocean metalliferous (zinc-copper) brines and muds, and manganese (nickel-copper-cobalt) nodules. More than a dozen major mining firms are engaged in deep ocean exploration by dredge sampling, coring, television imagery, and acoustic scanning. Marine seismic, electrical, electromagnetic, and magnetic surveys have been used in exploration programs. Radiometric methods and in situ assay devices have used experimentally.

Even though the techniques and programs of marine exploration are too experimental to cite any common guidelines, a sequence of mapping, geophysical sensing, sampling, "characterization," and evaluation has been described by Cruickshank (1973) in a 200-page state-of-the-art review. A complete book on offshore surveying, including geological and geophysical surveys, has been compiled by Ingham (1975).

The central problems in ocean exploration, and mining as well, are navigation and positioning. Most of the deep-ocean nodule deposits of economic interest are 1,000–2,000 km from the nearest land, within the reach of satellite positioning but well beyond the range of more accurate short-range radar-type positioning systems, and also beyond any simple reference to geodetic control. Even where reconnaissance sites are in continental margin areas and closer to shore control stations the establishing of ocean-floor control points is a complicated matter.

Bathymetric charts, most of which were made prior to the development of sophisticated sonar techniques, are the basic maps for offshore reconnaissance. They may be photographically enlarged to serve as work sheets, but the bathymetric control lines are generally too widely spaced to reflect the small changes in sea-floor relief associated with concentrations of minerals. Cruickshank recommends that target areas at depths of 600 ft (183 m) or less be resurveyed on a grid interval of no more than a quarter-mile (0.4-km) line spacing and at a resolution of at least 1 percent of the water depth. The geologic basis for correlating mineral deposits with sea-floor topography is described by Cruickshank and other contributing authors in a book on the geology of continental margins (Burk and Drake, 1974). A specific case history of offshore exploration for tin deposits in Mount's Bay, Cornwall, with comments on bathymetry and the location of submerged bedrock valleys is given by Tooms (1970).

10.4. RECONNAISSANCE TRAVERSING AND DATA RECORDING

As mentioned at the beginning of the chapter, geologic mapping is one of the prime methods in mineral reconnaissance, but it certainly is not the objective of reconnaissance. Thus, the example of reconnaissance mapping in Figure 10-8 has gaps in the data; these represent areas of minimal interest that were rejected on the basis of smaller scale map studies.

A typical base map for reconnaissance geology is a photogeologic sheet with overlays for linears and for regional geophysics; in this case, field notes are extended "ground-truth" notes. In some instances, the mapping base is a government topographic map, geologic map, or series of aerial photographs at

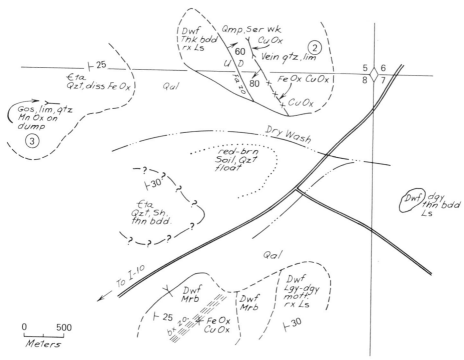

Figure 10-8.   A portion of a reconnaissance geologic map.

STATION NO. _73 APr 47_ (PAGE 1) QUAD: _Anchorage D-6_ AIRPHOTO: _Gul 4-067_
UTH: _E395 700 N6850 500_ ELEV: _3800'- 4100'_ _____ DATE: _June 30_
TEXT TOPIC: (LITH.)(STRUCT.)(GEOMORPH.) MINERALIZ., HAZARDS, OTHER _____
PHOTOS: _Roll 2#'s 6,7 Outcrop of arkose / woody frags_
SKETCH: _Distribution of moraine on S.side of Sheep Valley_

STRUCTURE: _Homoclinal section_ _____

BEDDING: _90 25 S_	CURRENT DIR:
LINEATION:	FOLIATION:
FAULT:	JOINT:
LINEAMENT:	FOLD:

OTHER STRUCTURE: _____

(SAMPLE) OR LITHOLOGY NO: ___ _47 A_ ___ UNIT _Arkose Ridge Fm._ _____

TYPE: REPRES. GRAB OR CHIP,(SPECIAL GRAB,)SSS, SOIL, PAN CONC., SURFICIAL MAT.,
OTHER TYPE:

(PURPOSE,) HAND SPEC., THIN SECT., STAINED SLAB, POL., SECT.,GRAIN MOUNT, MIN.SEP.,
MODE, CHEM., SPEC., K/AR, C–14, X–RAY, NORM, (FOSSIL DETER.)
OTHER PURPOSE:

(DESCRIPTION:)
IGNEOUS, VOLC., HYPA., PLUT., ULTRAMAF., MAFIC, INTERMED.,FELSIC, OTHER IG.:

METAMORPHIC, REGIONAL, CTCT., DYNAMIC, LOW G., MED. G., HIGH G.,PELITIC,
QTS–FELD., CALC., MAFIC, ULTRAMAF., OTHER META.:

(SEDIMENTARY) ARG., SH., MDST., SLTST., (SS.,)CGL., BREC., LS., CHT., COAL, (FOSS.,)
OTHER SED.:

UNCONSOL., CLY., SLT., MD., SD., GVL., CBLS., BLDRS., BLKS., ORG.,
OTHER UNCONSOL.:
OTHER DESCRIPTION:

Figure 10-9. Field notebook entry for computer-assisted processing of geologic data.
(After T. Hudson, G. Askevold, and G. Plafker, "A computer-assisted procedure for
information processing of geologic data," *Journal of Research*, v. 3, no. 3, p. 370,
1975. Courtesy U.S. Geological Survey, U.S. Department of the Interior.)

the original scale or photographically enlarged. Topographic maps are limited
to about three or four times enlargement because, just as some photographs
become grainy, map contour lines become too broad.

In small-scale reconnaissance mapping, field traverse lines and field-note
stations are often plotted and numbered so that the user will know whether

*On Eska Ridge - homoclinical section consisting
mainly of coarse vy hd biot-rich arkose,
orange - weath calc? arkose, and carb. arkose. Carb.
selvages are very bright coal, plant material
commonly is graphitic smear although locally
some woody impressions are present.*

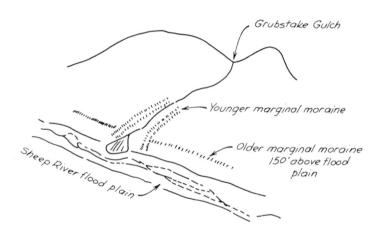

the geologist interpreted something from a distance. Most map symbols will indicate "approximate," "inferred," or "doubtful." A better method of separating field interpretation from direct observation is "isolation of outcrops" in which the outline of each exposure is drawn, but this normally is done in detailed mapping rather than in reconnaissance. Field photographs are of special value in reconnaissance because they can be used to show how closely a feature was observed and they can be compared with images and tonal differences seen on aerial photographs.

In past practice, reconnaissance field notes have been most often kept in a surveyor's pocket-size looseleaf notebook, with location data and comments on one page and a sketch on the facing page. The overall format is generally specified within each organization, but the amount of description is left to the choice of the individual geologist. With this regulated but flexible arrangement, geologists are free to expand whatever detail they feel may be useful

GEOMAP

MAP UNIT :   DATE :

	N-CO-OR	E-CO-OR	ALTITUDE	GEOLOGIST	OUTCROP NO	SERIAL NO
A 1953						
A 1954						
	6 7 8 9	10 11 12 13	14 15 16 17	18	19 20 21 22	23

TYPE OF OUTCROP

124

OUTCROP < 5 m²	A	QUARRY	G
OUTCROP 5-100 m²	B	MINE,PROSPECT	H
OUTCROP 100-2000 m²	C	BORE-HOLE	K
CONTINUOUS OUTCROP	D	KEY-OUTCROP	L
ROAD CUT < 10m	E	UNCERTAIN OUTCROP	M
ROAD CUT 10-50 m	F	BOULDER	N
		PILE OF BOULDERS	P

REFERENCE

25

TECTONICS AS PRECED OBSERVATION	1
PETR. AND TECT. AS PRECED OUTCROP	2
PETROGRAPHY AS PRECED OUTCROP	3
TECTONIC AS PRECED. OUTCROP	4
PETR AND TECT. AS SECOND LAST OUTCR	5
PETROGRAPHY AS SECOND LAST OUTCR	6

OBJECTS

	TYPE I	TYPE II	OBJECT NO
	126	127	28 29 30 31 32 33

TYPE I:
PHOTO	1
SKETCH	2
PHOTO + SKETCH	3
DESCRIPTION	4
PHOTO + DESCRIPT.	5
AREA DESCRIPT.	6
OTHER	7

TYPE II:
SPECIMEN	1
SPECIMENS	2
CHEMICAL ANALYSIS	3
SPEC + CHEM. ANALYSIS	4
MICROSCOPIC ANALYSIS	5
SPEC.+ MICR. ANALYSIS	6

PETROGRAPHY

I	II
34 135	36 137

STRUCTURE

SEE CHART

LAYER THICKNESS

138

< 2 mm	1	2 - 20 m	5
2-20 mm	2	> 20 m	6
2-20 cm	3	VARIABLE	7
2-20 dm	4		

COLOUR

39   140

SEE CHART

GRAIN SIZE

141

mm
< 0.05	1	1.5 - 3.0	5
0.05-0.3	2	3 - 5	6
0.3 - 1.0	3	5 - 30	7
1.0 - 1.5	4	> 30	8

TEXTURE

142

EQUIGRANULAR	1
INEQUIGRANULAR	2
PORPHYRITIC	3
OPHITIC	4
PORPHYROBLASTIC	5
GRAPHIC	6
CLASTIC	7
OTHER	8

MEGAGRAIN

MINERAL

43   144

SEE CHART

HABIT
ANGULAR, ELONGATED	1
ANGULAR, EQUIDIMEN	2
ANGULAR, ROUNDED	3
LENTICULAR	4
ROUNDED	5
VARIABLE	6
OTHER	7

SIZE mm
< 2	1
2 - 5	2
5 - 10	3
10 - 20	4
20 - 40	5
> 40	6

145

146

QUANTITY %
< 1 %	X
1 - 5 %	Y
10 %	1
20 %	2
30 %	3
40 %	4
etc.	

147

MOBILISATE

TYPE

148

PEGM. WHITE - GREY	1
PEGM. WHITE - RED	2
INEQUIGR.WHITE - GREY	3
INEQUIGR.WHITE - RED	4
GRANITIC WHITE - GREY	5
GRANITIC WHITE - RED	6
GRANODIORITIC	7
DIORITIC	8
OTHER	9

MINERALOGY

	I		II		III		IV	
	TYPE	QUANT	TYPE	QUANT	TYPE	QUANT	TYPE	QUANT.
	49	150 151	52	153 154	55	156 157	58	159 160

QUANTITY %

161   (AS 147)

ROCK

STRATIGRAPHY	TYPE	MINERAL, ALTERATION	QUANTITY %
62 63 154	65 66 167	68 169	170 (AS 147)

OTHER OBSERVATIONS

71 172

TECTONICS ETC.

PLANAR STRUCTURES

	TYPE	STRIKE	DIP	PHASE
	224	25 26 27	28 29	230
	233	34 35 36	37 38	239
	241	42 43 44	45 46	247

SS- SURFACE	1
S-SURFACE	2
PLANAR STRUCTURES	3
AXIAL PLANE	4
JOINT	5
SET OF JOINTS	6
FAULT	7
SHEAR ZONE	8
OTHER	9

JOINT FREQUENCY

NUMBER OF JOINTS/5 m
⊥ STRIKE

231

240

248

2 - 3	1
4 - 6	2
7 - 10	3
11 - 20	4
21 - 30	5
> 30	6

STRATIGRAPHICAL SEQUENCE

232

NORMAL	1
INVERTED	2
V { 0°-179°, TOP	3
180°-360°, TOP	4

DIMENSION		SYMMETRY	
0.01-0.05	1	SYMMETRIC	1
0.05-0.1	2	Z < 2/1	2
0.1 - 0.25	3	Z 2/1-4/1	3
0.25-0.50	4	Z 4/1-10/1	4
0.5 - 1.0	5	Z > 10/1	5
1.0 - 2.5	6 ASYM { S < 2/1	6
2.5 - 5.0	7	S 2/1-4/1	7
5.0-10	8	S 4/1-10/1	8
> 10	9	S > 10/1	9

LINEAR STRUCTURES

	TYPE	AZIMUTH	PLUNGE	PHASE
	249	50 51 52	53 54	255
	260	61 62 63	64 65	266

FOLD AXIS, OBSERVED	1
FOLD AXIS, CONSTRUCTED	2
MINERAL LINEATION	3
S-INTERSECTIONS	4
LINEATION UNSPEC.	5
LINEATION+FOLD AXIS	6
LINEAR STRUCTURES	7
SLICKENSIDES	8
OTHER	9

FOLDS

CLASS 1A	A
CLASS 1B (PARALLEL)	B
CLASS 1C	C
CLASS 2 (SIMILAR)	2
CLASS 3	3
CHEVRON	4
PTYGMATIC	5
INTRAFOLIAL	6
FLOW	7
OTHER	8

TYPE	WAVELENGTH	AMPLITUDE	SYMMETRY
256	257	258	259
267	268	269	270

DYKES ETC.

	TYPE	STRIKE	DIP	PHASE
	271	72 73 74	75 76	277

DYKE	1
SILL	2
FILLED JOINT	3
IGNEOUS CONTACT	4
SEGREGATION	5
GLACIAL STRIATION	6
OTHER	7

FILLING
GRANITE GNEISS	A
MASSIVE GRANITE	B
PEGMATITE	C
APLITE	D
AMPHIBOLITE	E
DOLERITE	F
QUARTZ	G
QUARTZ+ORE	H

278

ORE	K
CARBONATE	L
SKARN	M
OTHER	N

THICKNESS

< 2 cm	1
2 -10 cm	2
1 - 2 dm	3
2 -10 dm	4
1 - 2 m	5
> 2 m	6

and they are able to enter their impressions of trends and changes that cannot be measured at any one station. Some geologists prefer to give reconnaissance notes and sketches more room by entering them on $8\frac{1}{2} \times 11$ in. (216 mm $\times$ 279 mm) sheets in an aluminum clipboard in the same way as in detailed mapping (Chapter 11).

Past practice still provides the basic method of keeping reconnaissance field notes, but there have been some significant modifications. The use of small-size tape recorders and the integration of field notes with the needs of computer-based data processing systems deserve special mention. Cartridge and cassette tape recorders are used to lessen the time spent in taking field notes. Traverses and station locations are plotted and sketches are made in the normal way, but the descriptive data are recorded verbally for later transcribing. This procedure is especially handy when dealing with soggy paper and stiff fingers in rain and cold. However, it is rather inconvenient to review field notes in a tape recorder and there is the possibility of accidentally erasing all of a day's field notes. An evaluation of tape recorder methods versus other methods of taking field notes is provided by Klingmueller (1971).

Many computerized information systems are designed to accept geologic field data, providing the data are recorded within certain standards of consistency and reliability. The advantage of integrating field observations and machine processing is readily apparent. So is the main objection voiced by field geologists: that geologic work demands too much individual thinking to have it systematized and squeezed through a narrow pipe designed for some machine's convenience. The objection was most valid in the 1960s when computer specialists wanted to digitize everything whether or not it was subject to quantification. Improved data management concepts and improved computer technology developed in the 1970s, and from these came field data systems more to a geologist's liking.

Most systems are now designed to accept field notes that do not differ so drastically from conventional notebook entries; the form is often like a checklist from which appropriate items can be selected, and there are provisions for summary, generalization, and even field sketching to be entered in subfiles. One U.S. Geological Survey system of this flexible type described by Hadson, Askevold, and Plafker (1975) uses a printed outline page with a facing page of traditional text. Figure 10-9 shows a field notebook entry in this system. Most systems are either project oriented or terrain oriented, and all systems are constantly being revised on the basis of experience and new computer technology.

Figure 10-10 shows a field data sheet from GEOMAP, a system used by the geological surveys of Sweden and Norway and by the Boliden Mining

Figure 10-10. Field data sheet in the GEOMAP format. (From H. Berner and others, "GEOMAP," in *Computer-based Systems for Geological Field Data*, Fig. 1, p. 9, 1975, by permission Geological Survey of Canada.)

Company. The back of this standard field data sheet is used for additional notes in open but still processable wording (Berner and others, 1975).

The tools of a geologist's work in reconnaissance, from notebook to compass and hammer, are nearly the same as those used in detail surface mapping; the techniques are similar as well. The differences are spelled out in the boundary between the two stages. In reconnaissance, the geologist is looking for an area to map; in the detail stage, the geologist has found the area.

CHAPTER 11
MAPPING SURFACE GEOLOGY

Large-scale and detailed geologic mapping is a part of the target investigation or follow-up stage in exploration. Still, a certain amount of detailed geologic mapping will already have been done in areas of critical lithology, stratigraphy, and structure. The need for information from detailed mapping in a part of the reconnaissance area may not become apparent until work has begun in one of the selected target areas. The first few days of large-scale mapping may, for example, show that a supposed intrusive rock actually has volcanic fabric. If the revised lithology has a bearing on the target, then the investigator must, if at all possible, return to the reconnaissance area and to the other "intrusions" that were identified in haste.

Detailed geologic mapping takes place in many locales and for many reasons, but in this chapter the assumption is that detailed work is being done in an area of a few square kilometers to a hundred square kilometers that has been selected through reconnaissance work. This does not mean that detailed geologic mapping is a "one-pass" activity; it will probably be repeated at larger scales and in still smaller areas. Nor is geologic mapping an isolated activity; it is one element in a group of investigations that includes trenching, drilling, geophysics, and geochemistry. A geologic map provides the context for all of the other investigations—they would be of little value without it—and it depends on them as well.

An example of combined geologic and geophysical mapping can be taken from the exploration program that led to the discovery of the Tynagh, Ireland, lead-zinc-silver deposit. This is the deposit mentioned in Chapter 10 in

315

reference to ore boulders in stone walls, and it is the place of supergene galena "mud" mentioned in Chapter 3. It is also an area with very few bedrock exposures. The ore-bearing "reef limestone" at Tynagh, identified in isolated outcrops, was found to have a significantly higher electrical resistivity than other limestones in the same series. The complete pattern was then mapped by tracing a characteristic geophysical expression or "signature" through soil-covered areas (Hallof, 1966).

11.1. RELATION OF MAPPING TO OBJECTIVES

In exploration and mineral property evaluation, time is the overriding constraint. The features that have most direct bearing on a mineral deposit must be recognized and mapped as quickly as possible. If the evidence from mapping is encouraging, more time will be made available for the supporting detail; if not, the additional time can be better spent elsewhere. This is easier to say than to do. The number of potentially important features to map is almost without limit, and a dogged recording of every one of them (even if possible) would be impractical. We must therefore continue the screening process that began with preliminary data gathering and was used in reconnaissance, and we have the same limitations. If the screen is too coarse, nearly everything comes through; if the screen is too fine, we may reject most of the significant features before having a chance to evaluate them. A better approach is to do several "screenings," each at a finer mesh. Statisticians do this with a "decision tree" in which the branches are defined by the presence or absence of a certain feature or by the possible outcomes of a certain event. We can do something similar, even though geologists do not have the advantage of such clearly defined limbs in a firmly stated hierarchical order.

Suppose we are looking for a sandstone-type uranium deposit in an area that has been covered by an airborne gamma-ray spectrometer (radiometric) survey. The first screening or branching uses the most diagnostic features in our conceptual model. In this example, the features are uranium and uranium-thorium radioactivity anomalies. We take this branch; the rejected branch contains potassium anomalies and below-threshold readings. Some of the radioactivity appears to be associated with sandstone beds, some appears to be associated with recognized sources of radioactive "noise," and some is unexplained. The relationships might be determined with no more new mapping than a series of outcrop sketches on aerial photographs or on the "tracking camera" photo strip from the airborne survey. Now we reject the "noise" anomalies and the unexplained anomalies (for the time being) and take the sandstone branch.

In the next screening or branching we map the radioactive sandstone units and other sandstone exposures having obvious and near-obvious affiliations with our model of a uranium deposit. The most interesting sandstone beds should be medium to coarse grained, poorly sorted, and associated with interbedded mudstone or shale; therefore, we pay particular attention to the texture and mineralogy of the sandstone outcrops, especially the shaly and less pronounced ones. We measure stratigraphic sections and project the sequences through covered areas. A proper mapping scale and a special scale for supporting detail should now be apparent. The exposures that agree most closely with our model—those remaining after several more screenings—are mapped in the greatest detail, with special notes on color, weak radioactivity that may have seemed unimportant in the first screening, carbonaceous material, and traces of oxidized uranium minerals.

Other associations and other possibilities for further investigations will probably be revealed. Some of the unexplained anomalies designated for later investigation will eventually be explained; some of the "noise" may even be found to contain subtle messages, but not until the main objective—the mapping of the most direct guides to ore—has been fulfilled.

11.2. SCALE

In North American practice, the smallest scale for detailed mapping is likely to be 1:10,000 or 1:12,000. This is seldom the final scale, but it allows prospect pits 100 m apart to be shown and described separately and it permits a dike or fault zone several meters in width to be represented without exaggeration by a fine pencil line. Good reconnaissance mapping at 1:24,000 will sometimes serve the purpose almost as well. The next step will often be mapping at a 1:2,000 or 1:2,400 scale at which many of the smaller significant features connected with individual mineral deposits can also be shown without exaggeration.

According to Kreiter (1968) the detailed mapping of mineral prospect areas in the Soviet Union is often done at 1:2000 or 1:1000 after reconnaissance or "prospecting" geologic mapping at 1:10,000. The step-up in scale, 5 to 10 times, is about the same as in American practice. A smaller step, say 2 to 3 times, would not be as likely to bring out much new detail.

Once a specific site for drilling or trenching has been located, it is generally mapped at a still larger scale, such as 1:600 (1 in. = 50 ft) or 1:500 (1 cm = 5 m) so that the results of sampling, as well as the structural and stratigraphic detail, can be plotted in their precise locations.

The best mapping scale or the best combination of scales for a particular situation can be determined at the outset of a job by making sketches of the area at two or more trial scales. The minor amount of extra time is well spent. A too small scale may result in an undecipherable clutter when geological, geophysical, and geochemical notes are combined. A too large scale may spread the geologic picture so thin and over so many sheets that the synoptic view is lost. Also, a larger than necessary scale calls for an excessive spread in control points. One mapping sheet with ten control points at 1:10,000 becomes 16 sheets with less than one control point per sheet at 1:2500. Data plotted at a large scale may even be misused, if they outstrip the accuracy of the control and the field methods (compass and pace, for example) and result in a false sense of precision. The accuracy of an enlarged base map is, of course, no better than that of the smaller scale original map.

Transparent overlay sheets for lithology, structure, and mineralization, as shown in Figure 11-1, are often used to avoid clutter, but this practice also requires some advance sketching in order to find the best arrangement.

If there are earlier geologic maps for a part of the area, it may be convenient to use an existing scale for the remainder of the area rather than risk losing some detail (especially in color patterns) in photographic reducing or enlarging.

A multiplicity of map scales adds confusion to a project, but as would be expected, an early-chosen scale may have to be replaced by a better one as more of the geology becomes known. This occurred during the mapping of iron deposits in the Labrador Trough, Canada, where the orebodies are large but the ore-grade pattern is intricate and the orebodies are complicated by small folds and faults. During the mapping of a 40,000-km² area at 1:12,000 scale, a 260-km² area of strong ore mineralization was also mapped at 1:2400. Later, it became apparent that the geologic notes from trenching at intervals of 90 m and drilling at intervals of 60 m could be better shown at 1:1200 than at 1:2400. The larger scale was then adopted for the remainder of the project. Later a special scale of 1:480 was used for mapping ore-control geology in the mine pits (Blais and Stebbins, 1962).

11.3. BASE MAPS AND SURVEY CONTROL

The time available for geologic mapping is best spent in working with rocks and geologic features, not in obtaining survey control. Still, observations must be located accurately enough to show the proper spatial relationships. The ideal way to meet both demands is to plot geologic features as accurately as possible on well-prepared base maps. Topographic features, culture, and survey monuments provide the reference points; geologic data are related to them by occupying the points, by traversing between them, by resection, and by estimation.

Traverses are recorded on base maps by compass direction, with terrain distances measured by pace, range finder, stadia, or tape. Terrain distance is corrected to horizontal map distance and elevation differences are calculated from inclinometer readings. Altimeter readings are sometimes used as an additional means of locating points relative to contour elevations on a topographic base map. New control points, such as mineral claim monuments and fence corners, are added to the map while traversing (although with less

Figure 11-1. Geologic mapping on transparent plastic overlay sheets. Selected control points on the 1:1200 topographic base map are repeated on the lithology-structure overlay and on the mineralization-alteration overlay.

accuracy than primary control points) so that they may be used in additional traverses.

Location by resection is most often done by taking compass bearings or azimuths to three or more map points or aerial photograph points which can be identified on the ground. Accurate direction lines from the identified points will intersect (or at least form a small triangle of error) at the proper map location. Resection locations and traverses are often done together, as shown in Figure 11-2, with the resection locations used as verification for traverse points wherever possible. Minor data points are located by estimation, with compass directions and range-finder distance readings or altimeter readings providing temporary control.

The most common base "maps" for detailed work as well as for reconnaissance geologic work are vertical aerial photographs in black and white or in color. Where enough control points can be taken from planimetric or topographic maps, aerial photographs become *maps* in fact. Data can be plotted on photographic contact prints or on enlargements, with locations estimated from nearby topographic or cultural features, patterns of bushes, outcrops, and irregularities in stream courses. But there are limitations. If the photographs are more than a few years old or if they were taken during a different season, the vegetation patterns will have changed. The forest cover may be so dense in some areas that nearly blind traverses must be made between recognizable clearings. In desert areas of low relief the vegetation and drainage patterns may be so monotonous that reference features are hard to select. It may even be hard to tell which one of a series of photographs to

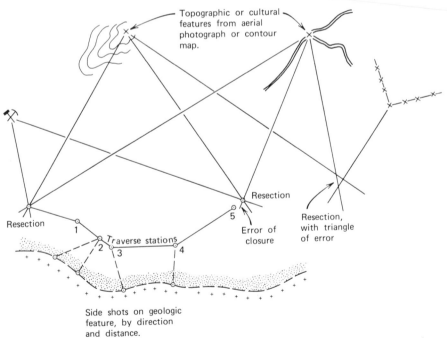

Figure 11-2. Control for geologic mapping by combined traversing and resection.

use. Recognizable terrain features are not, however, without some problems of their own. Topographic distortion—relief displacement—on aerial photographs will affect the accuracy of work in rugged topography, with the distortion generally more troublesome in photographs taken from lower altitudes. Air turbulence, especially common over rugged topography, may cause excessive "tilt" in photographs taken from low-flying aircraft.

Government photographs are widely available and they can be enlarged four times without significant loss in quality. If government photographs are inadequate, large-scale photography and photogrammetric maps can be furnished on short notice (weather permitting) by contractors. Special emulsions and laboratory techniques allow contractors to provide clear enlargements to almost any common mapping scale. By having visible targets at survey control points on the ground prior to flying, the contractors can also provide accurate maps.

Topographic maps, with or without accompanying aerial photographs, can be used as base maps for detailed geologic work if they do not require too much photographic enlargement to meet a chosen scale. The 1:24,000 and 1:25,000 government topographic maps can generally be enlarged to 1:12,000 or 1:10,000 scale without the contour lines becoming too broad. One way to diminish the interference between wide black contour lines and the geologic data plotted across them is to make a "screen print" in which the lines appear in shadowy light gray. Even where field notes are made entirely on aerial photographs, data must be transferred to good topographic maps where they can be related to grid lines, land survey monuments, benchmarks, and triangulation stations. Good tools for this job are a "Sketchmaster" or a zoom transfer scope; both instruments operate on the camera lucida principle in which images on the map and the photograph are made to coincide.

In spite of the wide availability of aerial photographs and topographic maps, it sometimes happens that geologists must make their own base maps. With enough time a triangulation net and control points can be provided by surveyors. A prospect examination of a few days' duration may, however, require the quick location of control points and approximate contours with little or no reference to accurate surveys. It must be done without diverting too much time from geologic work. Which is more acceptable, distortion of the geometric relationships or an incomplete geologic study? Distortions can be rectified, missing data can not. In many situations, a compass-and-pace map or a compass triangulation map with altimeter elevations will suffice until enough encouraging geologic information can be obtained to justify a more refined base map.

Methods and calculations in control surveying beyond the scope of this book are well explained in field geology textbooks by Berkman (1976), Compton (1962), Lahee (1961), and Forrester (1946). The principles are not difficult to learn, but it takes practice to put them to efficient use.

11.4. GEOLOGIC MAPPING EQUIPMENT

Equipment for field geologic mapping includes some special field notebooks or sheet holders, pencils, pen, and scale-protractor. The standard field notebook is a pocket-size surveyor's record book, but many geologists prefer to use a hinged aluminum sheet holder of the type shown in Figures 11-3 and 11-4 so that field notes, base maps, and photographs can be kept together. Sheet holders should measure at least 230 by 280 mm in order to hold aerial photographs and to provide a surface on which to work with a field stereoscope. Special belt pouches, field bags, and vests are available for carrying aluminum notebooks plus extra photographs and mapping equipment. As mentioned in Chapter 10, field notes may be taken by using a small tape recorder and they may be taken in computer-compatible form. Computer coding is especially advantageous when dealing with complex structure and rock fabric in metamorphic terrain, as explained by Burns (1969) in a description of mapping techniques in the Tasmanian Shield and by Roddick and Hutchison (1972) in reference to the Coast Mountains Mapping Project in British Columbia.

2H to 4H pencil leads are widely used for mapping and note taking. Harder leads are likely to gouge into damp note paper; however, 6H leads may be

Figure 11-3. Aluminum sheet holder and plastic scale-protractor.

Figure 11-4.   Leather scabbard for pencils and scale-protractor riveted to the cover of aluminum sheet holder.

used for marking on polyester film overlay sheets. Color pencils, relatively hard so that they will not smear, or "plastic lead" color pencils for marking on film are commonly standardized within an organization so that each rock type or stratigraphic unit is always represented by the exact same color. Ball-point pens give a good copy and will not smear, but a finer and more uniform line can be obtained with Rapidograph-type pens for inking over pencil lines. A transparent plastic scale-protractor with two mapping scales (e.g., 10 and 50 or 1:1000 and 1:400) is used by most geologists for plotting distances and directions on mapping sheets and in notebooks.

For field measurements, a compass and a 30-m tape are basic tools. In addition, there are uses for a range finder, right-angle prism, altimeter, plane table and alidade, pocket stereoscope, and pocket-size electronic calculator. The most popular compass is the Brunton pocket transit, which combines the features of a compass, clinometer, and hand level (Fig. 11-5). The graduated circle in a Brunton compass is marked either in quadrants or in 360-degree

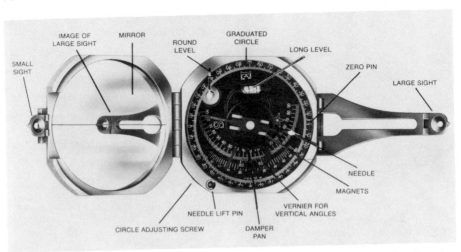

Figure 11-5. Brunton compass or "pocket transit." (Courtesy of the Brunton Company, Riverton, Wyoming.)

azimuth; the latter is preferred because it is more compatible with engineering calculations and computer data handling. Compass direction is measured as shown in Figure 11-6. Vertical angles and level readings are taken as shown in Figure 11-7. A lightweight tripod is often used with the Brunton compass for increased accuracy in traversing. Most compasses designed for geologic use have the same basic compass and clinometer elements. Some European and Japanese models include a magnifying lens for precise directional readings, stadia hairs for distance measurement, and hinged plates for measuring the strike and dip of inclined surfaces in one reading by direct contact. Some models of geologist's compass can be converted to an open-sight alidade by attaching a template or a straightedge.

Most distance measurements are made by pacing, but taping is better for accurate work. Stadia measurements for distances beyond a few tape lengths can be made with a telescopic alidade, with a transit, or with some models of hand level or compass. Electronic tacheometers (distance meters), widely used in engineering surveys, are not yet part of a geologist's normal equipment. Optical range finders have, however, become popular among geologists, in some instances replacing stadia measuring tools. With small range finders, distance measurements can be made with 99 percent accuracy at 100 m and 90 percent accuracy at 900 m.

Pocket-size pentagonal prisms are handy for establishing a series of right-angle lines from a base line and for making quick grid measurements but because there is no magnification, they are best suited to distances of 100 m or less.

Figure 11-6.   Reading compass direction with Brunton compass held at waist height.

Altimeters (aneroid barometers) are used in geologic mapping, especially where mapping is done on contoured base maps and in areas where it is difficult to see reference points. Since precise reading instruments are relatively bulky, they are not in such wide use as are pocket-size reconnaissance instruments with 5- or 10-m divisions.

The plane table, alidade, and stadia rod were the geologist's principal mapping tools for many years. For most purposes, plane-table mapping has now been replaced by mapping on aerial photographs and photogrammetric maps, but the plane table still has a place in mapping open-pit mines and similar areas requiring large-scale plotting of complex geology. Radio transceivers and a new generation of autoreduction alidades in which horizontal distances and vertical differences can be read directly have renewed interest in the mapping of steep terrain and complex geology by plane table.

A pocket stereoscope is an essential tool for all field work with stereo-pairs

Figure 11-7.   Measuring vertical angle with the Brunton compass as a clinometer.

of aerial photographs. Lens-type stereoscopes are most commonly used, but there are small mirror stereoscopes of field size. Stereometers or parallax measuring bars can be attached to lens-type stereoscopes for determining elevation differences in the field.

A pocket-size electronic calculator with trigonometric function keys is handy for all of the surveying, structural, and stratigraphic calculations done in the field. These instruments have, in fact, largely replaced tables and nomograms.

11.5. SURFACE GEOLOGIC MAPPING PROCEDURES

Field mapping is a process of analysis and communication. Geologists put a picture together for themselves and for their associates; this means that most of their observations—even their thoughts—belong on the map where someone can study them, not in a book or side notes. But space on map sheets is too limited for some of the detail unless it can be presented in well-understood symbols and abbreviations.

Typical symbols and abbreviations are given in Appendix A and Appendix B. Most organizations have additional standards and modifications that relate to a particular kind of "home" terrain or to their special group of mineral commodities. Standardization is important. When a geologist remaps an area or extends the coverage he or she uses the most factual data available: another geologist's field sheets. He should not have to puzzle over the meaning of "f gn splt brr 280."

Putting symbols, abbreviations, and ideas on paper follows nearly the same procedure whether the base map is an aerial photograph or a topographic map. A ground location is determined on the base map, the map is oriented, additional map and ground features are matched, and the geologic observations are plotted. If the main control points are too far away or completely off the mapping sheet, a traverse may have to be made from one of them to a convenient secondary control point or it may be necessary to establish the secondary control point by resection. The new control point and subsequent traverse points serve as hubs for plotting geology by estimation. Eventually, all traverses are closed at a primary control point or at another point determined by resection.

Mapping on Aerial Photographs

Even though aerial photographs have parallax distortion, they are generally better than topographic maps for field geologic work because every outcrop is also a potential control point. The ideal geologic mapping base would be a series of accurate and distinct contact prints of color photographs in stereo-pair with a supporting topographic map at the same scale (so ideal that it will seldom be available). In a more attainable situation, mapping is done on enlargements made from aerial photographs taken at a smaller scale. Because the enlargements cannot be viewed with a pocket stereoscope, a set of the original stereo-pairs of photographs should also be taken into the field for concurrent study.

The best way of handling enlarged photographs (and large map sheets) in the field is determined by such things as weather, the density of brush, and the ruggedness of the ground. A two-times enlargement of a typical aerial photograph mounted on lightweight fiberboard measures 46 cm across; this is a common format for mapping, and it is just about the maximum size that a geologist would want to carry in most field situations. While it is good to have an entire photograph for a comprehensive view, the board-mounted enlargements can be cumbersome in a high wind or in thickets. It is therefore often necessary to settle for an enlarged photograph cut into sections of about the same size as contact prints so that they can be carried in an aluminum sheet holder.

Mapping is sometimes done directly on the surface of the photograph, with lines drawn by Rapidograph-type pen and with control stations plotted as tiny needle holes. Unless the photograph is fastened permanently to a board the needle holes can be encircled and numbered on the reverse side. Color lines, generally hard to see on a photograph and quickly smeared by sweat and rain, are drawn *along* the inked lines, not in place of them.

A widely used technique is to plot geologic data on overlay sheets of polyester drafting film. A few gridded lines are drawn on the photograph, and several overlay sheets with matching grid lines are stapled or taped to separate edges of one photograph in a stereo-pair (Fig. 11-8). If the photographs are of sheet-holder size, three overlays can be stapled so that they may be placed one at a time or all together on the photograph. Pencil lines and colors may be used on the drafting film overlays and they will be relatively well protected in the aluminum sheet holder until they can be inked. A large, open board does not give this protection, and three overlays are too many to handle in the wind.

In most practice with overlays, the photograph is left unmarked and the

Figure 11-8. Overlay sheet taped and stapled to one aerial photograph in a stereo pair.

geologic data are plotted on the overlay sheets, as follows:

Overlay 1. Outcrop boundaries, lithologic identifications, geologic contacts, and notes on texture, color, and weathering.

Overlay 2. Structural data, including bedding attitudes, with notes on cleavage, schistosity, and rock fabric.

Overlay 3. Mineralization and alteration, with notes on geochemical sampling and with the locations of explanatory field photographs of outcrops.

The outlines of outcrops are drawn on the first overlay. This is part of a common practice in large-scale geologic work called "outcrop mapping" or "multiple exposure mapping." All geologic mapping is interpretive to some extent, but the geology seen in outcrops or trenches is as close to "factual" data as we can come and it must be identified as such. Areas of abundant float or of a characteristic soil are likely to add some information to the picture, so they are outlined as well. Interpretation between outcrops is also done in the field, but it is kept completely separate on the field map so that someone with new data or new ideas can make new interpretations. Figure 11-9 illustrates

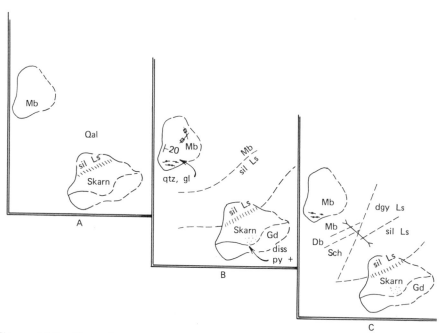

Figure 11-9. Part of an outcrop map. A. As plotted on an overlay sheet showing lithology and geologic contacts. B. Combined data from several overlays, with an interpretation. C. Later interpretation modified by information from a trench.

the outcrop mapping idea and shows what may happen to interpretations when some new data are added.

Rock types are identified on the first overlay sheet. Here again, there is something to be said for factual data versus interpretation. Rocks are described in the field by their appearance, not by their genesis. A rock that appears to be quartzite is mapped as quartzite; this gives the rock a chance to be called a metamorphosed cherty zone in a carbonate sequence, an expression of hydrothermal silicification, or even a felsic metavolcanic rock after laboratory investigation and additional field work. The field description should, of course, be as detailed as time will allow, so that alternative interpretations can be compared. As with the overall aspect of choosing features to be mapped first, there are certain lithologic features to describe first. Color, grain size and habit, fabric (texture and structure), and mineralogy are basic attributes, but they differ too widely among rock types to be

Figure 11-10.   Measuring the dip of a planar surface by direct contact. The strike direction is established by placing the edge of the compass box in a level position on the surface.

adequately discussed in this book. The information is given in petrology textbooks and in Compton's *Manual of Field Geology* (1962), which has four chapters summarizing the field characteristics of sedimentary rocks, volcanic rocks, igneous and igneous-appearing plutonic rocks, and metamorphic rocks.

Planar surfaces, such as bedding, fractures, and foliations, are commonly shown on a second overlay sheet. Attitude is measured by direct contact as shown in Figure 11-10 or by sighting along a level line in the plane of the feature as shown in Figure 11-11. The attitude of linear features, such as ripple marks, metamorphic lineations, and minor fold axes, is measured in direct contact or by sighting the direction from a position above the feature as shown in Figure 11-12 and taking the angle of plunge from a right-angle view to one side. The taking of oriented rock samples, an important step toward studying structural and textural characteristics in the laboratory, is described by Whitten (1966, p. 35). The basic field procedure is to mark the strike

Figure 11-11.   Measuring the dip of a planar surface by sighting along a level line in the direction of strike. The level line is found by using the Brunton compass as a hand level. In the strike direction the surface appears in edge view.

Figure 11-12. Measuring the trend of a linear feature by aligning the Brunton compass and looking vertically from above.

direction and dip reading on the rock with a felt pen or on an adhesive label and to enter the data in a notebook.

Once the field data have been recorded and the type, intensity, and distribution of mineralization carefully noted, they can be transferred from an aerial photograph to a topographic base map in several ways, ranging from estimation by common points of reference to highly accurate photogrammetry. Much of this work is now done in photogrammetric laboratories, where geologic and topographic data can be plotted at the same time. Computerized data processing and plotting are also done in a continuous operation, often in connection with photogrammetric processing.

11.6. GEOLOGIC MAPPING IN SURFACE MINES

Geologic mapping in open-pit mines involves most of the same techniques as detailed mapping of target areas. One of the basic objectives—to describe geology—is also the same. The additional objectives are to provide data for ore-grade control, for geotechnical studies and—ultimately—for mine planning. Patterns in rock type, alteration, mineralization, and weathering can be translated directly from geology to geotechnics, but patterns in structure require some special consideration. Fracture sets and fracture attitudes need more detailed description than in target-area mapping because they are controlling features in slope stability. The fact that a bedding plane or a fault zone dips at a certain inclination and in the same direction as a pit wall may not be of geologic importance, but it may be of considerable importance in designing a safe pit. As shown in Figure 5-1 and Table 5-1, the spacing of discontinuities, their extent, the nature of their surfaces, and the type of filling are to be recorded. Water in fractures and in permeable beds is to be noted and described.

Some unique mapping conditions are present in the fresh new "outcrops" and busy environment of an open-pit mine. Exposures are almost continuous. There is some "cover" in talus piles along bench toes and in roadways, but it is never very far to a well-exposed face in rock. In many pits, only the uppermost faces have been weathered and leached. The deeper faces are likely to be in fresh rock where the original texture and minerals can be seen, an advantage over the disguised appearance of weathered outcrops. There is, however, some disadvantage in fresh rock exposures in that textural contrasts and rock fabric have not had a chance to be etched and accentuated. In a bench face, the most intricate structural detail—every "crack" in the rock—is exposed; too exposed in some respects, because major and minor fracture zones are hard to differentiate where a natural sorting has not been done by processes of weathering and erosion. There is an additional element in dealing with fracture patterns in mine faces: some fractures are emphasized or even created by the blasting pattern.

The atmosphere of thoughtful and uninterrupted investigation in surface mapping does not extend to open-pit mine mapping. Trucks, trains, and shovels have priority, and the geologist must do the work without getting in their way. Blasting schedules are of paramount importance. Freshly broken rock faces are subject to raveling and are less stable than natural cliffs; working along them requires extra caution. A geologist mapping a pit face should not become so deeply engrossed as to ignore vibrations from traffic on the bench above.

Open-pit Scale, Control, and Base Maps

The mapping scale in open-pit mines is not likely to be smaller than 1:2400. This is the scale of mapping at Bingham Canyon, Utah, where the removal of more than 300,000 tons of material per day does not allow much time for great detail. Other mapping scales in common use are 1:1200, 1:600, and in countries with a long-established metric system, 1:2000 and 1:1000.

Mapping control stations are abundant and visible in open-pit mines, but some of them may not last for long. In rail-haulage pits, there are relatively long-lasting reference points in the electrical towers and surveyed "track tag" locations on every few lengths of rail, but in truck pits the survey stations are commonly marked by temporary stakes or laths that are quickly destroyed in the course of mining. Blast holes and sampling sites are normally well surveyed, and they are used as control points for geologic mapping until the bench is blasted. In many open-pit mines, surveyors will establish extra reference stakes for geologic mapping during their periodic traversing of active areas. Some distant reference points are fairly permanent; survey stations and mine installations on the edge of the pit can be used for locating geologic features by resection. Local magnetic disturbance, always a factor to be considered in compass work, is especially common in the midst of rails, heavy equipment, and power cables in open-pit mines.

Aerial photographs and photogrammetric maps are often used as base maps for geologic work in large open-pit mines, many of which are photographed every few months for engineering purposes. Obviously, old photogrammetric maps and old photographs have little bearing on the current pit topography where as much as a million tons of material are removed every few months. The photographs can be "terrestrial" as well as "aerial." At some mines, oriented photographs of the entire pit are taken with telephoto lens from standard stations every few weeks and used to update the geology and base maps.

Recording Open-pit Geology

The traditional methods of pit mapping are by compass and tape, compass and pace, or plane table. Magnetic attraction interferes with compass work; therefore, traverse directions are often obtained by turning angles with the compass. Geologic features are plotted by estimating right-angle distances to critical points at measured distances along a bench traverse or by locating key points on the base map or photograph. Geology and control points are obtained at the same time. Geology cannot wait. It may get blasted away before the mapper can return.

Plane-table mapping is done to good advantage in open-pit mines. Instrument stations on a mine bench are easily located by resection, and a series of stadia rod positions can be plotted quickly along the face and crest. In some instances, a second rod man can obtain data from the crest of the next bench and give locations to the instrument man. Plane-table mapping practice in the Sierrita mine, Arizona, described by Metz (1972), uses radio transceiver communication between the rod man (geologist) and the instrument man. In the Sierrita method, faults and contacts are described by the points where they intersect the toe and crest of the exposure (apparent strike) and by their true attitude at the base of the exposure.

Geologic data are plotted in plan views of each entire face or at a reference elevation near the toe. In the face method (Fig. 11-13), a compact three-dimensional view of the geology is afforded in the steplike appearance of structures on successive faces. The true strike directions of geologic contacts are seen only where they cross the level benches. Geologists mapping by the face method obtain control measurements from a tape along the toe and estimate the geologic picture at higher elevations on the face. The toe method (Fig. 11-14) shows individual mine levels in a record of successive face

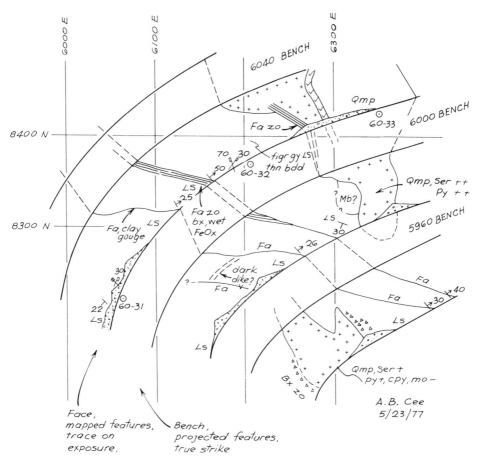

Figure 11-13. Field note sheet, with open-pit mine geology recorded by the face method.

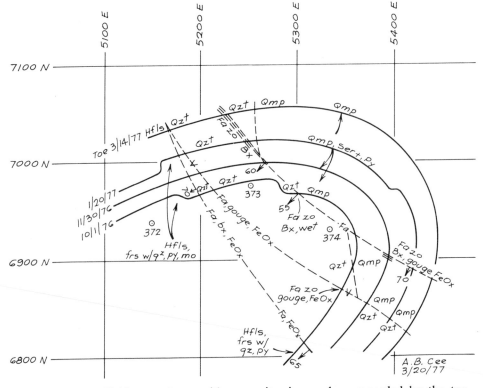

Figure 11-14. Field note sheet, with open-pit mine geology recorded by the toe method.

positions. Because the map represents a single horizontal plane rather than a comprehensive view, all contacts are shown with their true strike direction. In mapping by the toe method, all of the recorded geology is obtained by direct observation; additional sketches are used for situations of special interest higher on the face. Some mine geologists make certain that all information of possible interest is recorded by sketching the entire face in a vertical section view while mapping the toe. The toe method is particularly well suited to the mapping of flat-lying bedded deposits where most of the mineralization occupies one bench.

In most open-pit mines, office maps are eventually prepared in both ways. At Bingham Canyon, Utah, the information from pit mapping by the face method is plotted on a master geologic map and on a simplified "operating" geologic map of the entire mine showing the faces. Then the information is transferred to level-by-level toe maps and vertical cross sections for office study. A geologic map at any open-pit mine is in a sense obsolete by the time it is distributed. It shows only what the situation *was*.

Open-pit geologic mapping can be considered a first step into the complex and unique realm of underground mine mapping. In an open-pit mine, the geologist has the advantage of a perspective view. In an underground mine, the view is gone and any perspective beyond the reach of illumination of a cap lamp and the immediate limits of rock walls must come entirely from a picture in the geologist's mind.

CHAPTER 12
GEOLOGIC MAPPING IN UNDERGROUND MINES

Underground workings are likely to contain the most important information in the most important areas. If the workings are active, they provide a series of fresh geologic exposures with each meter of advance and they provide well-located sites for underground drilling and bulk sampling. Abandoned mine workings are direct guides to a region's most obvious mineralization and provide the most immediate information on ore occurrences. In fact, a few thousand dollars spent in gaining entrance to an abandoned mine can be expected to yield more data from the depth dimension than would be obtained from a new drill hole at a similar cost, and the data should serve as a comparable basis for planning additional drill holes.

In a sense, all geologic projection and analysis are "subsurface geology"; so is the final interpretation of geochemical and geophysical surveys. Underground maps and drill-hole logs simply provide the most direct information. A Colorado School of Mines publication (Le Roy and others, 1977) has served through several decades and revised editions as the standard reference on subsurface geology applied to mining, petroleum, and construction work. Chapters on underground geologic mapping and the representation of mine geologic data are included.

12.1. THE UNDERGROUND MINE MILIEU

Recognition of safety of access into mines, active or abandoned, comes with experience; if this experience has not yet been gained, on-the-spot advice

339

Table 12-1.   Safe Practice in Geologic Mapping of Underground Mines

Potential Hazard or Injury	Safe Practice and Equipment
I. In All Mines	
1. Untreated injury	Inform someone on the surface where you are going and when you expect to return.
2. Injury from flying rock or steel fragments	Safety glasses. Avoid hammering on a rock pick as a makeshift chisel.
3. Foot injury from rock slabs while chipping or washing walls	Safety shoes
4. Head injury	Hard hat
5. Fire and fumes	Self-rescuer (carbon monoxide filter) on belt. Avoid building fires and smoking in timbered areas or "gassy" mines. Caution with carbide lights and candles.
6. Falls	Keep both hands free for climbing ladders. Use safety sling or rope when crossing chutes.
7. Entrapment	Electromagnetic tone transmitter on belt. Learn Bureau of Mines-approved hammer signals for coal mines.
8. Unseen hazards in weak lighting	Test cap lamp in advance. Check lamp brightness periodically. If lamp goes out, call to partner for help or get to the surface by flashlight or candlelight (except in gassy mines).
9. Explosions	In coal mines, use a safety lamp or a methane detector and use Bureau of Mines-approved equipment
10. Falling objects and rock	Avoid standing under shafts, raises, and chutes, and in areas where stability of rock is questionable
II. In Active Mine Workings	
1. Injury from blasting and haulage	Follow general safety rules and check-in and check-out procedures for mine. Check with foreman and miners regarding blasting schedules, active haulage routes, active chutes and raises; inform them of your work location.
2. Injury from machinery	Call out before climbing a manway or entering a stope; establish communication. Obtain permission from foreman before operating any equipment.

Table 12-1 (con'd).   Safe Practice in Geologic Mapping of Underground Mines

Potential Hazard or Injury	Safe Practice and Equipment
3. Electric shock	Wear nonconducting hard hat and use nonconducting tape.
III. In Old or Inactive Mine Workings	
1. Untreated injury or entrapment	Avoid going underground alone
2. Collapse of ground	Approach with caution and test the ground near portals and shaft collars. In marginal shaft conditions, stay clear or use a safety rope—keeping one person at the collar.
3. Injury to others	Unless it is absolutely certain that no one is underground, do not drop rock fragments to test the depth of a shaft
4. Animal and insect bites	Watch for animals, especially snakes and poisonous insects, and especially near the portal or shaft
5. Anoxia	Test for oxygen-deficient air by candle flame or safety lamp (not by carbide lamp). Begin testing at the portal or collar and continue testing to see if conditions change from place to place or during the day.
6. Poisonous gases	Detect by odor (H_2S, SO_2) or by detector tube (CO)
7. Falls and cave-ins	Note condition of timber (test with a knife blade), ladders, landings, and bottom timber lagging. Note appearance of rock and where necessary "sound" with a pick. Do not attempt to repair timber or "bar down" loose rock.
8. Explosions	Avoid abandoned explosives and blasting caps
9. Drowning in water-filled raises or receiving cuts from submerged steel	Probe ahead in "puddles" of water
10. Getting lost	Keep compass directions and distances in mind

must be sought from an experienced miner. Table 12-1 lists the basic safety precautions for underground geologic work. Like all rules, some are too obvious and others are not universally applicable. For example, in most situations a geologist should not work alone but may be justified in doing so in the active part of an operating mine as long as the specific location and schedule are known to the foreman and the miners. In an abandoned mine, however, the rule is firm; there should always be at least two people underground and their schedule must be known to a responsible person on the surface.

It takes practice to recognize geologic features in mine workings and to feel "at home" underground. An experienced mine geologist understands the miner's environment and has learned to read structure and texture in the unique characteristics of freshly broken rock faces. A mine geologist will, for example, use the beam from a cap lamp to obtain highlights, cast shadows, and accentuate subtle rock textures and structures—rather like having a small subsurface version of airborne, low-sun-angle photography. There is no great difficulty in the mining geologist's method of handling a cap lamp, a hand lens, a rock specimen, and a pocket knife at the same time, but it is not an instinctive capability. Much of the disdain for "Professor Noncommittal" and other geological experts who visited early American mining camps was caused by their unfamiliarity with the environment and their failure to recognize even the most elementary rock characteristics through fogged eyeglasses in the dim light of a candle or a carbide lamp while trying to stand on a slippery mine timber.

One of a mine geologist's best assets is an ability to relate observations to the layout of the entire mine. Part of this ability comes from training in geometry, and part comes from daily work with cross sections, block diagrams, and mine models. Even so, constant orientation and reorientation while working underground are necessary because a turning drift or a "corkscrew" raise can be misleading in an environment where there is no opportunity to look at a familiar landmark.

Dust and water are part of the underground environment, and they cause problems for even the most experienced geologists; in addition to making muddy smears on mapping sheets, dust coats and masks the walls of workings. In older workings, a coating of dust, mud, and sulfate efflorescence is common and fresh "outcrops" must be chipped with a hammer. Where water lines are available, a clean rock surface can be obtained by washing the walls with a jet from a hose. In both chipping and washing, over-zealous work is dangerous because loose rocks may be dislodged. In some active mines, as a matter of practice, walls are washed 50 m or so behind the drilling and blasting operation, and they stay relatively clean.

Safety eyeglasses are essential equipment for mine geologists as well as miners, even if corrective lenses are not needed. To prevent fogging when going between cooler workings and warm moist stopes, lenses can be smeared with a silicone lens-cleaning liquid. For experienced (meaning old) geologists, the special upper-and-lower bifocal eyeglasses designed for carpenters and electricians are better than normal bifocals when looking at wall detail while wearing a cap lamp. Contact lenses of all kinds are troublesome and may be dangerous in underground mines because of dust and acid water.

Special Equipment

In addition to the equipment for surface mapping described in Chapter 11, some special equipment is required for underground geologic work. The basics are an electric cap lamp, a special belt for the battery and for a "self-rescuer" device, a nonmetallic hard hat, and a flash-bulb attachment for the camera. Mapping sheets are generally kept in a hinged aluminum sheet holder with a blotter fixed inside the cover. Smaller pocket-size notebooks are sometimes preferred for mapping stopes and development headings where only small increments are plotted at a particular time.

A plastic scale-protractor and an assortment of pencils are kept in a compartmented shirt pocket or, better yet, in a leather pencil-and-scale sheath on the cover of the sheet holder, as shown in Figure 11-4. In order to have both hands free for climbing ladders and for surveying and sampling, a timber cruiser's canvas vest with pockets and pouches can be used to carry the sheet holder, notebooks, camera, and small paraphenalia.

Mapping sheets are of water-resistant paper or frosted plastic film with inked or printed grid lines. Tracing cloth, a photocopy, or any porous paper is unsatisfactory because it will soften and change dimensions when moist. The measuring tape, generally of 30-m length, should have large dark numbers on a light background and it must be made of fiber glass or a similar nonconducting material. Wire-woven cloth tapes, like metallic hard hats, are dangerous around mine electrical systems. For short measurements, a folding rule, a flexible pocket tape, or a wooden stick marked in appropriate divisions may be used.

Compass, range finder, pentagonal prism, and a small plane table with open-sight alidade are used in mine mapping, but in somewhat different ways than on the surface. These special procedures will be discussed in Section 12.4 together with special uses for such common tools as plumb bobs, heavy-duty string, large nails, and lumber crayons. Some instruments unique to underground mine mapping, the hanging compass and hanging clinometer, boxwood rule, and polar protractor, are also mentioned in Section 12.4. A specially designed Brunton compass with a digital readout is sometimes used for underground work; bearings and inclinations taken with this instrument can be read in a three-digit electronic display without any auxiliary lighting.

12.2. Selecting Geologic Data

Recognition of important features to record on a mine geologic map may require several visits to the same underground area—once in an orientation survey for the overall context, once for the basic geologic mapping, and then perhaps several more times for information relating to conditions discovered in another part of the mine or for checking a new hypothesis. The practice of revisiting mapped areas is no different from that done in reconnaissance and detailed mapping in the surface environment, except that the exposures are ephemeral in an operating mine. The stratigraphic detail in today's drift heading or stope face may be blasted away and lost forever when tonight's round is fired.

If the mine is one of several mines to be investigated in a district reconnaissance, the map may be made as a 1:1000 or 1:1200 orientation sketch of features pertinent to the overall problem, with measurements made by pacing on mine levels and counting ladder rungs for depth. Earlier geologic maps may be used as a substitute for new mapping in reconnaissance work, but, as with all use of predigested data, allowance must be made for earlier objectives and theories. The exact date of a mine map is especially important; later workings and drill holes may have changed the entire aspect. In long-lived mining districts, the geology in old and inaccessible workings has sometimes changed classification from "known" on the basis of one undated map in the archives to "questionable" after the discovery of a contradicting, perhaps undated, second map.

A geologist may note a critical feature in several locations during an orientation survey and then concentrate on what appears to be the key to the investigation. At the Westvaco Trona (sodium carbonate) mine, near Green River, Wyoming, mine planning problems called for mapping low-amplitude folds within an evaporite bed. Elevation control stations were available throughout the workings, but the upper and lower surfaces of the bed were seldom exposed. Thin clay layers in the evaporite, representing surfaces that had been horizontal at the time of deposition, were noted in most locations but the number of layers varied from place to place. An orientation survey disclosed three layers in a characteristic series near the middle of the bed. This was the critical feature. The "triplet" was located throughout the mine, and the changes in its elevation afforded a means of rapid structural mapping.

In the first stages of mapping, it is general practice to designate rock types, mineralization, and alteration in broad terms or even as "type A" or "type B" until samples can be examined in daylight and in the laboratory for a more definitive label, such as "quartz diorite." A bleached zone associated with a particular structure would best be mapped first as "bleached" rather than "sericitized."

A mine geologic map is not likely to be a perfectly balanced picture with a fault, vein, or stratigraphic contact every few meters. Most of the pertinent information is likely to be concentrated in a small part of the workings and next to the main ore shoots. It would be misleading to "edit" these features more rigorously than the broadly spaced features in crosscuts and in outlying drifts that have followed weaker mineralization in *search* of ore. A mine geologic map can be objective and yet can emphasize local detail by using auxiliary views and large-scale sketches in critical areas.

12.3. THE MAPPING FORMAT

Most of the rules just mentioned for selecting the data to map in a mine are similar to those of surface mapping. The format—scale, base map, grid, and "topography"—is, however, entirely different.

Scale

The map scale chosen for a specific job reflects the detail and the accuracy required, but these two requirements are not always related. A large-scale sketch of an inaccessible stope face may be made from 2 m below with no opportunity for accurate measurements but still with considerable detail. The accurate measurement of strike and dip in a well-exposed vein structure may be shown even on a relatively small scale map by a solid-line segment in an otherwise dashed line, by definitive numbers rather than rounded-off numbers, or by encircling the symbol.

A 1:1200 scale, the smallest scale at which underground mine workings can be shown without exaggeration, is sometimes used in reconnaissance mapping but it is still too small for most purposes. In North American practice, 1:600 (1 in. = 50 ft) and 1:480 (1 in. = 40 ft) are the most common underground mapping scales. Where more detail is needed, for example where ore is associated with minor fractures, fold axes, and lithologic changes, mapping is done on a scale of 1:240 (1 in. = 20 ft) or at a scale of 1:120 (1 in. = 10 ft).

In countries where the metric system has been long established, 1:500 is a commonly used scale, with a 1:250 scale for greater detail. At the Salsigne gold mine, France, the complex geologic relationship between veins, replacement bodies, and mineralized breccia zones, mentioned in Chapter 2, was determined by first mapping at 1:1000 for the broad context and then once again at 1:250 for detail.

Base Maps and Mine Grids

In operating mines, grid locations and elevations of underground survey stations (wooden plugs with number tags and "spads" or hooks) are shown on accurate engineering maps. In abandoned mine workings, old survey stations may sometimes be found, that may be related to existing maps. There may be some frustration in trying to use these survey stations, however, because older mines are likely to have several sets of stations and several grid systems, each from a different operating company. One system may have been based on the magnetic meridian in a particular year; another may have a special "grid north" representing some long-forgotten convenience in plotting the mine workings. One mine grid in continuous use for 50 years (because of the overwhelming job of recalculating stations to a better system) was established in a peculiar direction so that the general manager could hold a map and look squarely from his front porch at the first surface cut!

Mine maps from abandoned operations have a shortcoming as old as the mining profession itself—there are nearly always more workings than shown on the "final" map. Lessee miners and scavengers may have continued to work on smaller ore shoots for many years after the cessation of company operations.

Inasmuch as a geologist working in an abandoned mine does not generally have the advantage of reliable survey stations, the survey controls must often be provided by less accurate compass-and-tape methods. In this situation, a temporary grid system should be established so that the various mine workings can be related to each other, to drill holes, and to surface geology. The "grid zero" number should be located far enough away from the area of interest so that all measurements will be in one quadrant and stations can be easily plotted by latitude and departure as well as distance and direction. The grid coordinates of the shaft might be, for example, 5000 N, 5000 E.

During an underground exploration project with only a few new workings, a geologist may have to act as a mine surveyor to provide accurate reference points. The fundamentals of mine surveying, not greatly different from those of surface surveying with a transit or theodolite, are given in a standard North American textbook on the subject by Staley (1964). In common with geologic work in mines versus work on the surface, one must adapt to conditions of poor visibility and to very steep traverses in restricted areas with only rare opportunities to close on reference stations. In addition, there is the upside-down experience of setting up *under* a plumb bob and making all elevation measurements downward from overhead stations. A general feeling is that anyone trying to survey a stope should have several extra hands or at least a prehensile tail.

12.4. UNDERGROUND GEOLOGIC MAPPING PROCEDURES

For safety and for the best attention to geologic detail, a two-person team is recommended, with one person plotting data while the partner has both hands free to use a compass, hand lens, or hammer. But since a geologist may have to work alone, comments on one-person improvisations are included in the following paragraphs.

Level Workings: Control Surveys

The basic format is a plan map. Datum is a horizontal plane at waist height; it may be chest height in some systems, and it may be eye height in large workings. Projections are made to this plane even if the projected feature falls slightly outside the workings. In places where a mine working is partly filled with loose rock, datum height is projected through the pile of rock. Gently inclined workings may require the use of an inclined datum. Auxiliary datum planes and auxiliary views are commonly used to illustrate complex or flat-lying features, as shown in Figure 12-2; the views may be wall (rib) sections or they may be cross sections of the working with information projected into the cross section from surrounding walls. In a few localities, such as at Bisbee, Arizona, geologists' notebooks have been arranged to include a continuous view of both ribs as well as a waist-height map.

The position of geologic features relative to control points on the mine level must be known. In an operating mine, the geologist need only hang a plumb bob (or drop a rock fragment) from a survey station in the back in order to establish a point on the bottom of the working. A tape can then be stretched between established points, with the ends of the tape held down by chunks of rock. The tape can be kept out of the mud by fastening the ends to nails or spikes driven into timbers or into rock crevices at measured distances from the control stations. In either example, the priority of mine traffic must be taken into account.

Timber sets or supports, if at a standard interval, can be used as reference locations with distance estimated by counting the number of sets and pacing beyond the last set. This is a particularly handy method in busy mines where there is little opportunity to stretch tapes and make extensive control measurements.

Where no survey stations are available, as in most idle or abandoned mines, underground reference points and an initial direction line must be established at the mine portal or at each level in the shaft. The elevations and positions of levels along an inclined shaft are determined by compass-and-tape traverse. For a station in a vertical shaft, an initial azimuth may be taken by compass and the elevation may be determined by string or by altimeter. Mine surveyors use more precise methods to bring coordinates and directions underground. A gyro-theodolite can be used to establish an azimuth without depending on magnetic readings, or more commonly, "shaft plumbing" is done by hanging wires in a vertical shaft to make a short base line. These methods are essential for engineering work, but they are generally beyond the needs of geological mapping.

The initial underground reference point is established at a nail in timber at the shaft station or at the portal. While one person holds the end of the tape at the initial point, the partner extends the tape to the next chosen point, generally at 30 m or at a junction or bend in the workings. Each person, standing at an end of the tape, takes a bearing on the reflection of the partner's cap lamp in the mirror of a Brunton compass or similar compass in the same way as taking a bearing in surface mapping. The foresight bearing and backsight bearing should match within a few degrees so that an acceptable average can be taken. If the bearings do not agree, the trouble may be from a hammer, a steel belt buckle, or an electronic wristwatch. If there is a local magnetic attraction in stacks of rail, ventilation pipe, or air or water lines, or if the work is being done in a magnetic mineral zone, angles must be turned between a backsight and a foresight direction at each affected station with the Brunton compass on a tripod or, more crudely, by pivoting on one heel. If the initial azimuth or bearing is likely to be affected, a makeshift method of shaft plumbing using the orientation of shaft timbers for direction may have to be accepted for the time being.

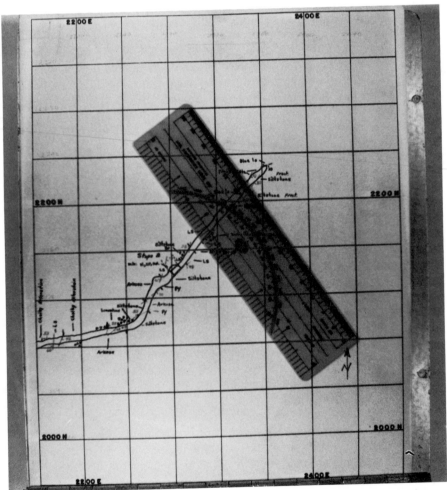

Figure 12-1.   Plastic scale-protractor oriented on a north grid line for plotting an azimuth of 327 degrees.

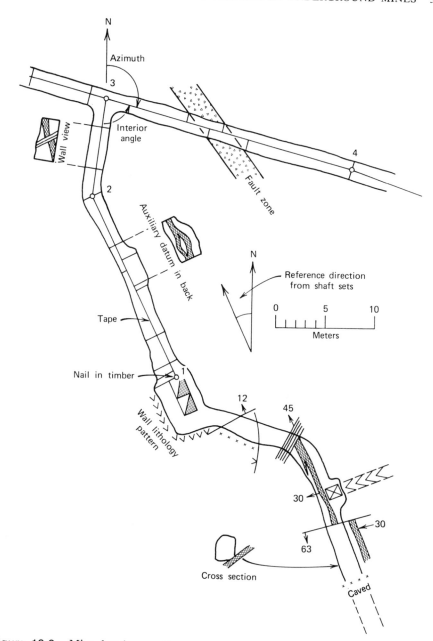

Figure 12-2. Mine level map. Traverse and wall measurements are shown to the north of the shaft.

The bearing or azimuth line to the new station is drawn from the initial point along a plastic scale-protractor oriented on one of the map grid lines (Fig. 12-1). The distance to the new station is plotted and the tape may be left in place until wall topography and geologic detail are obtained. The procedure is then repeated for successive stations. Some geologists prefer to obtain control points for an entire level before concerning themselves with detail; this has merit in that it permits undivided attention to be given to geology and it prevents an interference between early side notes and later mapping of workings. In this level-at-a-time procedure, a small cairn of rock is left at each station for temporary reference (providing there is no traffic) and vertical marks for referencing the geology are made on the walls with lumber crayon or spray paint at 2- or 3-m intervals before the tape is moved to the next station. Whatever the procedure, surveys are closed and errors adjusted as often as possible by making secondary traverses in crosscuts between drifts or in raises between levels.

A geologist working alone may read the direction of a spare lamp at the foresight and backsight or may stretch the tape taut between stations and read the direction by orienting an extended compass over the tape. Another rapid one-person method that can be applied to inactive mines is to stretch a string throughout the level workings and orient the compass on each segment of

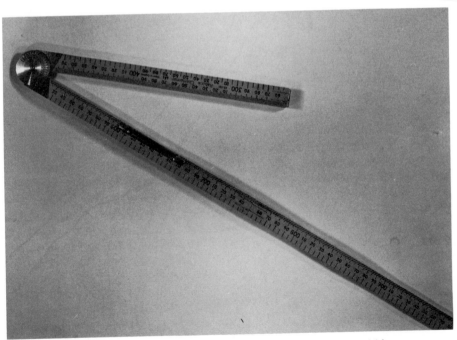

Figure 12-3.   Boxwood rule, with spirit level and graduated hinge.

string between points of contact with the walls; in this makeshift procedure, distance measurements are generally made by pacing between the points of contact.

The wall topography is generally plotted during the control survey, with some minor plotting often done later while measuring geologic features. Measurements are made at a right angle from the traverse tape by extending a pocket tape or measuring stick to significant points on the right and left walls at waist height. Topography is estimated between the measured points. The points need not be closely spaced unless the irregularities in the walls have some geologic significance. Figure 12-2, a mine level map, shows right and left measurements in the workings to the north of the shaft. Topography between stations 1 and 2 was obtained by measuring the distance to the walls at waist height from the tape, thus:

Station	Tape Distance, m	Right, m	Left, m
1	+ 0.0		2.5
	+ 0.6		1.8
	+ 1.5	0.9	1.8
	+ 4.6	0.9	1.2
	+ 9.5	1.2	
	+10.0	2.5	
	+11.8		0.6
	+13.0	1.8	
	+14.5	1.2	0.6
	+20.0	0.9	1.2
2	+20.0		

Level Workings: Geologic Detail

Rock type is most commonly shown by a band of color or by a pattern of symbols along the walls. A less common practice is to show the rock type inside the working as a filling rather than as a border. In either practice, all structures and contacts are shown at waist height or at a chosen datum, such as chest height or shoulder height. Most structures will either cross the workings at waist height or can be projected to waist height from exposures in the back. Gently dipping structures seen near the back or bottom and striking parallel to a working may be projected to a waist-height position beyond the walls, as shown near the south end of Figure 12-2 where a vein is plotted outside of the drift. An auxiliary cross section in the figure illustrates the projection from an exposure in the bottom of the drift.

Planar structures may be plotted with their proper strike direction by recording the waist-height positions on opposite walls and drawing a connecting line, as shown between stations 3 and 4 in Figure 12-2 where the strike of a brecciated zone is obtained by measuring:

3+   8.5 m, left 0.9 m to top of zone

+ 11.0 m, right 0.9 m to top of zone

+ 12.2 m, left 0.9 m to bottom of zone

+ 15.0 m, right 1.2 m to bottom of zone.

The strike of a structure can be verified by standing near one wall and obtaining the compass direction of a point on the same structure at an equivalent height on the opposite wall. Dip can be measured by sighting across the mine working in the strike direction and aligning the extended compass with the trace of the structure while reading the clinometer. Strike and dip may also be taken as an average or a range in measurements made by placing the Brunton compass or a similar compass-clinometer directly on a planar fracture surface. These procedures are the same as in surface geologic mapping, but the sighting method is particularly useful underground because of an enlarged compass shadow cast on the opposite wall.

The angle between intersecting structures and the rake or pitch of linear features on a bedding or fault surface can be measured by Brunton compass in contact with the feature, or the job can be quickly done by using a carpenter's boxwood rule, a folding, hinged measuring ruler with a built-in spirit level and a circle of degrees at the hinge (Fig. 12-3). The boxwood rule can, in fact, be used for a crude, fast survey. Where a local magnetic attraction makes compass surveying difficult, the two arms of the fully extended rule can be aligned with intersecting mine workings and the interior angle read directly at the hinge. Some geologists measure the angle at which structures cross a mine working by leveling the boxwood rule, aligning one arm of the rule with the direction of the workings, and aligning the other arm with the trace of the structure on the walls. This method is speedier than taking right and left measurements to the walls, and it has some advantage where a large number of structural attitudes must be taken in a limited time.

If the boxwood rule seems to be too crude an instrument for mapping geologic structure surfaces in mines, some more sophisticated tools are available. One of the best is an instrument composed of adjustable arms and plates used in U.S. Bureau of Mines geotechnics investigations in places where there is a magnetic influence (Bolstad and Mahtab, 1974).

In most mines, the overall geologic pattern is more important than the precise attitude of a single structure. It is the familiar situation of choosing between a study of the forest or the trees. A mineral deposit is more like the forest. Geologic mapping practice in a typical large-tonnage (block-caving) mine can be taken as an example. At San Manuel, Arizona, where geologic mapping is done at a shoulder-height datum, the scale of mapping is relatively large (1:120) so that the geologists can record abundant side notes on structures, rock types, and alteration, but there is little time or justification for collecting detail on the attitudes of individual structures. The density of fracturing and the characteristics of fracture surfaces are more important. The strike and dip of most structures are estimated rather than measured, and the position is determined by pacing between support sets. While there is no particular attention paid to minor wall topography, there is considerable emphasis on wall sloughing and wet areas, since these are factors affecting rock mechanics.

In a few places, especially on the lower levels of abandoned mines, the attitude of the deepest exposure in a single vein or in a key stratigraphic zone must be measured with greater than normal accuracy. A flattening or steepening in dip might indicate a nearby ore shoot or a major fault. McKinstry (1948, p. 28) describes a way of obtaining precise attitude data by measuring the inclination of a string that has been anchored at two points on a structure in the direction of dip. One point is in the back and the other is the lowest point in an arc. Another of McKinstry's methods involves measuring several offset distances from a plumb line to a dipping surface.

Mapping Notes. In all underground geologic mapping, as many notes and explanatory sketches as can be accommodated should be entered on the mapping work sheet rather than on extra sheets. The notes can be edited later when posting the information, but the work sheet should represent the total factual base for all subsequent interpretation. In some mines it is standard practice to plot a continuous view of the two walls (ribs) as well as the plan map. Abbreviations such as those in Appendix A, and symbols, as shown in Appendix B, should be used. Figure 12-4 shows typical notations on an underground geologic work sheet. If explanations will not fit on the work sheet, they may be made on extra sheets, in a notebook, or with a tape recorder and keyed to the map by letters or numbers.

Underground geologic notes can be taken in a format designed for key-punching and computer treatment in the same way as surface mapping notes. Some computer-oriented field note systems that have been used in underground geologic mapping even though they were originally designed for core logging are GEOLOG (Blanchet and Goodwin, 1972) and GEOMAP and COREMAP (Ekström, Wirkstam, and Larsson, 1975).

Coal Mines. Geologic mapping in coal mines is directed mainly toward geomechanics and the identification of mining problem areas. Faults and fault patterns are of special interest, even faults with very small displacement that may interrupt the stress field and the mining pattern. Faults are often associated with such geological hazards as abnormal gas and water accumulations. Jointing, expressed in the coal seam as a dominant *face cleat* direction and a perpendicular *butt cleat* direction, is an important factor in mine design;

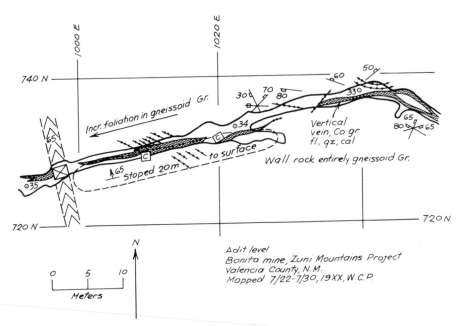

Figure 12-4. Portion of a work sheet for an underground geologic map.

coal ribs tend to spall more readily and roof control is sometimes more difficult where mining in the face cleat direction. "Clay veins" and sedimentary dikes cause instability in the mine roof, and may intersect to form isolated compartments or cells with entrapped gas.

Sandstone and siltstone channels in a coal bed may make it unsuitable for mining with plough-type longwall equipment. Splits in the coal seam and troublesome sequences of roof rock may also restrict the choice of mining method. These features are mapped wherever possible and related to a conceptual model of the mining area. The model is paleogeographic—a coal swamp—with certain stream patterns and sources of sediment. The effects of structural history and present-day topography are added to the model on the basis of information gained in surface mapping and drilling as well as mine mapping. The objectives of mine mapping in the Appalachian coal field are discussed, with case histories, by McCulloch and others (1975). The role of the geologist and of geologic mapping in British coal fields is reviewed by Skipsey (1970).

Stopes, Shafts, and Inclined Workings

The outlines of nearby stopes are shown on level maps by dashed lines; the notation "stoped above" or "stoped below" indicates the position (or former position) of thicker and richer ore that may not appear on the level itself. This does not, of course, give any evidence of the geology in the stope. The mapping of stope geology requires the use of some additional techniques.

In operating mines, stope geology and stope sampling data are generally plotted as each round is blasted. As the stope expands, the geologic map grows as a composite of short increments in plan or in cross section. Survey control is provided by stations in or near the manways. The datum height above a particular mine level is designated by the number of "floors" in square sets or by the vertical distance measured from principal survey stations.

Abandoned stopes are difficult to map because many parts of them may not be safely accessible. Views from a distance and on-the-spot geologic projections are therefore relied upon to a greater extent than in level mapping. Survey control may be sparse, so that a traverse may have to be made from the nearest mine level in the same way as would be done at a cliff in surface mapping. A point of origin on a mine level is obtained at a nail in timber or at a temporary survey station on the bottom, and the traverse is made into the stope through manways or empty ore chutes. Bearings are taken on cap lamps and vertical angles are measured by inclinometer readings along a taut tape. A one-person method of traversing into stopes is to select a convenient compass direction and then measure a series of short horizontal and vertical "step" distances in this direction with a pocket tape or with a measuring stick. This is a low-accuracy method, since some of the steps are in mid-air and they are estimated only as being horizontal and vertical.

Once the location and elevation of a control station in the stope are obtained, the topography and geology may be recorded in plan or in section, depending on the geometry of the mineralization. If most of the geologic detail in a stope can be shown in plan view, as illustrated in Figure 12-5, datum levels are chosen and nearby features are projected to each one, just as if they were waist-high datum in a level working. If detail lends itself better to vertical sections, as in many long and narrow stopes, a series of cross sections or a longitudinal section can be used, as shown in Figures 12-6 and 12-7. In any event, geologic features are projected to the appropriate plane by estimation plus geometric or trigonometric calculations from places where footholds can be gained.

In mapping wide stopes, compass bearings and inclinometer readings are made on two long tapes, with one tape stretched lengthwise as a base line and the second tape extended across it in successive positions corresponding to calculated vertical increments (for a series of plan maps) or horizontal increments (for a series of sections). Uniform increments are preferable, and it is best to place the second tape horizontal and at a right angle to the baseline tape. Yet, the shape of the stope often prevents such thorough access. The second tape and additional tapes may have to be placed wherever

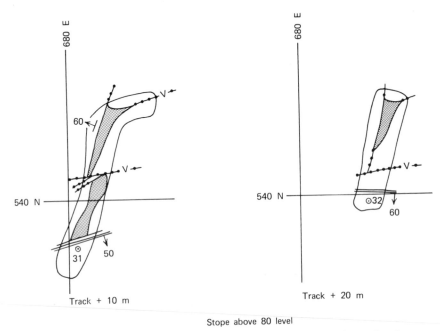

Figure 12-5.   Stope geology plotted in plan view at two datum levels.

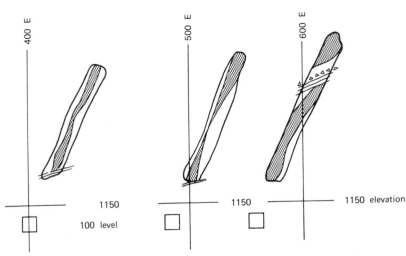

Figure 12-6.   Stope geology plotted in a series of cross sections.

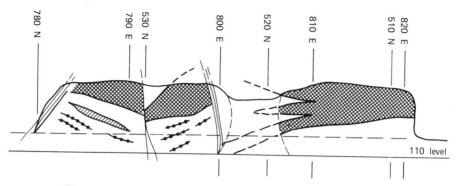

Figure 12-7. Stope geology plotted on a longitudinal section.

possible; the tapes may have to be inclined, they may have to cross at various angles, and they may pass slightly above or below each other. Such modifications are quite acceptable providing the inclination, crossing angle, and the vertical separation can be measured and taken into account. In large open stopes, this calls for considerable ingenuity. Estimating stope topography for geologic control is sometimes aided by sighting spotlighted wall points or extra lamps in a pentagonal prism or in a pocket range finder. Some geologists use a small plane table and open-sight alidade for mapping in wide stopes, with sightings taken on lamps and with distances obtained by triangulation or by estimation.

The reduction of inclined tape measurements to horizontal and vertical projection takes extra time even with a pocket electronic calculator, but it is best done underground so that the relationships can be studied while the geologist is still on site. To save time, a composite sketch-section of an entire stoped area can be substituted or, preferably, a "profile mapping" technique described by James (1946) can be used. In James's method, taped lines are plotted on a composite profile in true length rather than in projection, thereby giving the geologist a preliminary picture of the stoped area as well as data for later reduction in the office.

In the inclined and irregular workings typical of older mines, accurate survey control without a transit or theodolite is sometimes obtained by using a hanging compass and clinometer. In this method, the compass is hung on a strong cord or wire stretched between two accessible points, the clinometer is hung from the cord at two positions equidistant from the end points, and the inclination of the line is taken as the mean of the two clinometer readings. Attachments are available for converting several types of surveying compasses to a hanging compass. In a paper on underground geologic mapping, H. A. Schmitt (1936) describes a method for converting a Brunton compass to a hanging compass.

For detailed stope topography, a polar protractor is sometimes used in connection with a survey by hanging compass. A 360-degree protractor with a slot is set on the survey cord at accessible locations, and distances to the back and walls are measured in a vertical plane at recorded angles. A long stick is used to poke the end of the tape into inaccessible areas. The polar protractor is actually a modification of the "sunflower dial," an instrument formerly used in measuring cross sections of mine workings for ventilation design.

A string survey is an old but still expedient method for plotting irregular workings without the use of a compass or long tape. In this method, an irregular network of strings or wires is stretched from a few survey stations to points in unsurveyed areas so that distances taped in the short and accessible triangles between strings can be used to calculate the distances and angles in the larger triangles by geometry or trigonometry.

Raises and shafts are mapped by methods similar to those described for stopes. The geology can be plotted on successive horizontal plan maps, but under most conditions one wall or two adjacent walls provide the mapping planes. Figure 12-8 shows the geology of an inclined raise plotted on plans and on two vertical wall sections.

12.5. ASSEMBLING MINE GEOLOGIC DATA

At some stage between the mapping of exploration targets and the beginning of mine development, underground geologic maps must be put into a standard system for use with the accumulated data from surface mapping, drill holes, and geophysics. A composite map of all the mine levels and connecting

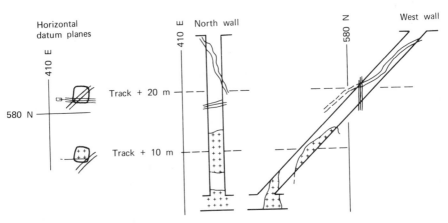

Figure 12-8. Raise geology plotted in a series of horizontal datum planes and in two vertical wall sections. Unless a raise is vertical, one of the two adjacent sections will be foreshortened.

workings can be used for a synoptic picture, but the map will generally be too crowded for detailed geologic information unless the levels happen to be widely spaced on a gently dipping structure, as shown on Figure 17-6. A composite mine geologic map may also be plotted with the levels separated artificially, as in an "exploded" view of intricate machinery. The most common practice, however, is to show all but the most generalized geology on a separate sheet for each level and for each major stope elevation. There still may be too much information to show on a single sheet, and it is always necessary to separate observed data from interpretation; therefore, level maps are often made as a series of overlays.

Each sheet in a series should bear enough identification so that it can be used separately; it should at least have a graphic scale (in consideration of photo-reduction), the mine name, coordinate lines, and an identifying "block" or mine-area designation. The name of the geologist and the *dates* of the basic work and major revisions should be shown.

Large-scale mine geologic maps and interpretive overlay sheets, at 1:240, for example, are used for detailed work, but generally there is a need for smaller scale plans and sections on which the most important features can be emphasized. A large mine may require two sets of smaller scale maps, with one set at 1:1200 showing enough detail for mine planning and one set at 1:6000 for use in district exploration.

Level Plans and Sections

Level maps generally correspond to actual mine levels, and the maps include data from drill holes and isolated workings that penetrate the specific horizon. The level is more often designated by elevation rather than by its depth below a particular shaft collar. Where the horizon intersects a hillside the immediate surface geology is shown in a narrow band of color and symbols. Data located above or below a horizon may be shown; they are clearly labeled ("10 m above") and it is understood to be a vertical projection rather than a projection along a geologic trend if it appears on the "factual" sheet. Information from higher or lower positions may, of course, be projected in a dip direction or along a plunge when drawing interpretive overlay sheets.

The position of a section and the direction it is facing must be shown; this can be done by indicating its bearing or azimuth plus the location of a point through which it passes. A better "fix" for sections is to show the trace of vertical planes containing the north-south and east-west coordinates, as seen in Figure 12-7; this provides enough orientation, but "looking northeast" may also be stated. Mine sections include information on drill holes, workings, and surface geology, but generally with longer projections and fewer control points than on level maps. Forrester (1946) provides thorough instructions for the preparation of sections and describes their use in the complex vein patterns at Butte, Montana, and in the Coeur d'Alene district, Idaho. There is considerable choice in regard to the spacing, orientation, and even the inclination of section planes. Vertical sections are commonly drawn along grid lines at regular intervals of 30–60 m at mines in massive or disseminated orebodies. It is sometimes advantageous, however, to provide sections at less regular intervals, to draw special sections in oblique directions, and to include "bends" in a section for the best view of significant detail. In a tabular or elongate orebody, the preferred orientation for both geologic and ore reserve sections is generally at a right angle to the principal axis, but longitudinal sections, either vertical or on an inclined plane parallel to a vein, are also widely used. Contour lines are drawn on sections to show irregularities in dips and to show isopachs. McKinstry (1947, p. 188–192) has discussed the construction and use of inclined sections, and has also described a convenient method of contouring veins on an inclined datum.

Vertical exaggeration is sometimes used in sections of mine workings, especially where it is necessary to accentuate the intersection of gently dipping structural features. Needless to say, sections with an exaggerated scale are potential troublemakers, so the exaggeration must be clearly stated on the section in order to avoid mistaken assumptions in later use.

Assay Data

Assay maps and sections are supplements to factual geologic sheets. Inasmuch as the values and the coordinates of the samples are reported in digital form, assay maps and sections lend themselves very well to statistical work and to machine plotting. Whether assay patterns are plotted manually or by computer, the metal content and metal ratios provide valuable information for geologic analysis when taken with an underlying geologic map or section. Assay-value intervals on assay overlay maps are often color coded; for example, sulfide copper values in porphyry copper deposits may be colored purple for 0.8 percent and above, red for 0.4–0.8 percent, and yellow for 0.2–0.4 percent, with additional colored overlay maps for molybdenum, gold, and silver values. At Bingham Canyon, Utah, where there are underground as well as open-pit workings, assay overlay maps of this type were used to outline a molybdenite-rich zone between the chalcopyrite-bornite-pyrite orebody and a nearly barren central core.

Supporting Diagrams and Models

The geologic trends and overall relationships in a mine can be shown in a wide variety of statistical diagrams. Ore grade can be related to rock type or to structural domains in histograms and cumulative frequency plots that show the separate populations. Changes in mineralization with depth or distance can be clarified and illustrated by trend surface analysis diagrams. There are many more plots and graphs to use with statistically treated mine data. Koch and Link (1970–1971) show the principal diagrams, explain the derivations, and give detailed examples of their use in specific mines. The calculation of block-caving characteristics from geologic mapping data at San Manuel, Arizona, by Mahtab, Bolstad, and Kendorski (1973) provides a good illustration of statistical data treatment and its presentation in stereonets and histograms.

Large-scale isoline diagrams are widely used to illustrate geologic conditions in mines, and they may be drawn in relation to almost any reference plane—horizontal, vertical, or inclined. Key bed structure contour maps are of special value in studying stratabound orebodies of relatively low dip, where mineralization is associated with gentle folding or with slump structure. In Chapter 2, Figure 2-20, structure contours, isopachs, and assay-thickness isolines are shown in relation to ore controls at the Wheal Jane tin mine in Cornwall.

Among the various types of three-dimensional diagrams described in detail by Lobeck (1958), orthographic diagrams in general and isometric block diagrams in particular are best suited for large-scale illustration of mine geology. The appearance of an isometric diagram, as shown in Figure 17-5, is less natural than that of a perspective diagram, but there is the advantage of being able to measure distances parallel to the three coordinate directions. Isometric grid drafting paper, isometric templates, protractors, lettering guides, and isometric drafting machines are available. Computer graphic techniques are ideally suited to isometric projection because of the uniform geometric relationships. Instructions for making isometric block diagrams and comments on their use in mining geology are given by Badgley (1959). A computer program for constructing isometric diagrams is described by Wray (1970). A general guide to making three-dimensional mine drawings by computer graphics is provided by Notley and Wilson (1975).

Geological scale models are more than exhibits for courtroom use and "Mickey Mouse" illustrations for explaining projects to the board of directors; they are indispensable tools for studying the geology of complex ore deposits in underground mines. Detailed mine geologic models require a lot of effort and display space, but the expense is justified when they become a focus for important discussions involving new exploration workings and drill holes. Scale models may be made from a series of levels or sections, they may be made as "peg" models of drill holes, or they may be of a combination type with plastic plates representing datum planes and plastic rods representing drill holes.

The construction of a mine model often consumes one person's time for at least several months. Effective mine models can however be made on short notice from thick Plexiglas plates with transparent plastic photoprints placed between them. In a mining organization, a stack of Plexiglas plates can be used at meetings of geologists from the various mines; geologists place their own film transparencies of geologic and assay level maps in the makeshift model when it is their turn to describe work at their mine.

A geologic mine model or a few well-edited diagrams can be made to emphasize exploration targets for other geologists, ore-grade patterns for design engineers, stress-field patterns for geomechanics engineers, and litho-logic contrasts for geophysicists. This aspect—communications—deals with whether a geologist's mine maps will be used or ignored. Unless there is some clear way of communicating the summarized information, hundreds of map sheets and dozens of geologic plans and sections may have amounted to nothing more than busy work.

CHAPTER 13
EXPLORATION GEOPHYSICS

Geophysical information is interpreted in relation to patterns in geology, and the patterns are evaluated in respect to known or supposed relationships between rock types, structure, stratigraphic sequence, and ore mineralization. Suppose that an occurrence of ore mineralization, a body of copper and nickel sulfides, for example, is the only kind of geology we are interested in; geophysical work will not actually find the mineralization, but it will show where an electrically conductive zone or a very dense zone of *something* is located at a depth beyond our direct observation. If our interest is broader, as it should be, the same geophysical work might bring out the characteristic "signature" of an ultramafic body or of a major fault zone, geologic features that can be expected to accompany copper and nickel mineralization.

The terms "signal", "message", and "noise" were borrowed from geophysics to illustrate some concepts in data gathering and geologic mapping in other chapters; now the terms can be returned to geophysics. The signal is composed of message (information sought) and noise (extraneous effects). Noise may be inherent in the instrument, it may result from magnetic storms and other transient disturbance fields in the earth and the atmosphere, it may be cultural noise (e.g., from a pipeline) or it may be related to geologic or topographic features in the terrain. Upon reinterpretation, many of these "extraneous" effects, especially the ones called "terrain noise," have often been found to contain something of value. Still, a critical factor in determining the applicability of a particular geophysical method to a given situation is the expected signal-to-noise ratio: too much noise—no information.

Within a geophysical message, there should be *anomalies,* significant departures from the normal pattern of values. Anomalies must be explained in terms of geologic conditions, including a possible occurrence of ore mineralization, and there will generally be a few alternative conditions that could cause similar anomalies. According to Ward and Rogers (1967), several years of searching for massive sulfide deposits by airborne electromagnetic methods resulted in 1 anomaly in 100 being due to massive sulfide mineralization and only one in 5,000 anomalies proving to be an orebody.

By using a combination of the geophysical methods shown in Table 13-1, the number of alternative interpretations can be reduced. For example, a magnetic anomaly appearing in the same position and with approximately the same shape as that of a strong electrical conductor may indicate a body of pyrrhotite or mixed pyrite and magnetite rather than a conductive zone of graphitic schist. If the conductor were not magnetic but were dense enough to cause a gravity "high," it could be a body of pyrite rather than pyrrhotite or magnetite.

The use of geophysics begins in the reconnaissance stage, with airborne methods serving to outline broad geologic features, and it continues into the most detailed stages, where ground methods, drill-hole (downhole) methods, and even underground geophysics are directed toward finding orebodies. In reconnaissance, geophysics is often used as an adjunct to geologic mapping. In the southeast Missouri lead mining district, for example, aeromagnetic anomalies are relied upon to indicate buried hills and ridges of Precambrian rock that are in turn associated with algal reefs and ore deposits in the overlying carbonate rocks. In regions where there is extensive soil cover, electrical, electromagnetic, seismic, and gravity surveys are used in mapping limestone beds of high resistivity, slate beds of low resistivity, and mafic dikes of high density.

The direct application of geophysics to the search for orebodies—radiometric prospecting for uranium ore, magnetic prospecting for iron ore, and electrical prospecting for base-metal deposits—is generally considered to be a part of exploration in virgin areas. But the application is much broader; geophysical prospecting has provided many discoveries in older mining districts. Deep, concealed orebodies in productive districts are tempting geophysical targets because new ideas and new techniques can be more easily applied in searching for orebodies with relatively well-known characteristics. Established mining districts have the special advantage of allowing geophysical access into deep mine workings, but there is an inherent disadvantage as well—that of stray electrical currents and other industrial-connected noise.

13.1. CHARACTERISTICS OF GEOPHYSICAL DATA

The essential requisite for any geophysical method is to have a contrast in some measurable physical property between ore (or whatever is sought) and

Table 13-1. Synopsis of Geophysical Exploration Methods Used in Mining Geology and Mineral Exploration

Method	Unit	Parameter	Physical Property	Compilation	Some Causes of Anomalies	Applications
Magnetic	Gamma	Earth magnetic field	Magnetic susceptibility and remanent magnetization	Contour maps and profiles	Concentrations of magnetite, ilmenite, pyrrhotite, and specular hematite Orebodies Irregularities in basement rock Mafic intrusive and volcanic rock "Black sand" in sediments	A,D,O
Gravity	Milligal	Acceleration of gravity	Density	Contour maps and profiles	Dense orebodies Dense intrusive rock Basement rock irregularities Bedrock irregularities Salt domes	A,D,O
Electrical Self-potential (spontaneous) polarization	Millivolt	Natural potential	Electrochemical action and conductivity	Contour maps and profiles	Conductive orebodies Graphite	D
Resistivity	Ohm per meter	Apparent resistivity with applied current	Resistivity or conductivity	Contour maps, profiles, "sounding" curves	Conductive orebodies Conductive and resistive strata Fissures with conductive fluids	D

A-includes airborne applications; D-includes drill-hole logging; O-offshore applications

Table 13-1 (cont'd).   Synopsis of Geophysical Exploration Methods Used on Mining Geology and Mineral Exploration

Method	Unit	Parameter	Physical Property	Compilation	Some Causes of Anomalies	Applications
Applied potential (mise à la masse)	Millivolt	Potential field with source electrode in ore	Conductivity	Contoured maps and cross sections	Continuation of located mineralization	D
Induced polarization (overvoltage)	Millivolt per volt	Apparent resistivity at two or more frequencies (frequency domain)	Electrochemical effects between electronic (metallic) and ionic (fluid) conductors	Contoured cross section, contour maps, profiles	Conductive orebodies Disseminated mineralization Graphite, serpentine, certain clays and micas	D
		Transient voltage (overvoltage) decay after a pulse of current (time domain)				
Electromagnetic	Mhos per meter (conductivity) and tilt angle of receiver coil	Induced electromagnetic field	Conductivity	Contour maps, profiles, "nested" profiles, vector maps	Conductive orebodies Graphite, certain clays	A,D

Method						
Audio frequency magnetic (AFMAG)	Same as electromagnetic	Natural electromagnetic pulses (thunderstorms)	Conductivity	Same as electromagnetic	Same as electromagnetic	A
Radioactive Radiometric	Counts per time or milliroentgens per time	Natural gamma radiation from uranium, thorium, and potassium minerals	Radioactivity	Contour maps, profiles, "nested profiles, ratio maps	Uranium and thorium orebodies Potash deposits Potassic alteration zones Granitic intrusive rocks	A,D
Nuclear activation		Gamma radiation after neutron bombardment	Radioactivity		Beryllium ore (beryllometer) Hydrogen-bearing minerals (drill-hole logging) Ore minerals (drill-hole logging)	D
Seismic	Distance per time	Velocity of elastic waves	Elasticity	Travel time sections, interpreted depth sections	Bedrock and basement rock irregularities	D,O

A-includes airborne applications; D-includes drill-hole logging; O-offshore applications the adjacent rock. In gravity work, the contrast is in density; in electrical and electromagnetic work, the contrast is in conductivity. Contrast is essential, but there are other things to consider: The measure of an anomaly is heavily dependent on the shape of the body and the ratio of size to depth of burial. An anomaly is composed of a message in the form (Brant, 1965, p. 819):

$$A = \sum \Delta p \cdot F \cdot \frac{V}{r^n}$$

where

A = measure of the anomaly
Δp = the physical property difference involved
F = the acting force, natural or artificially applied
V = the active volume of the body
r = distance of the body from the observation point
n = an integral number, experimentally determined, depending upon the shape of the body, the geophysical method, and the measurement.

A large orebody associated with a strong acting force and in significant physical contrast to its surroundings will yield a large anomaly if it is near the surface and it may still yield a measurable anomaly it is at greater depth or distance (r). The expression V/r^n is, however, an integration over all parts of the body; the attitude of the body is important. The number n is of higher power for a spherical or pipelike body than for an inclined tabular body; thus, an anomaly from a lenticular mass would be much less apparent than the anomaly from a comparable dike or vein at the same depth. The effect of body shape applies to all kinds of geophysics, but it has a unique connotation in the electromagnetic method. Whereas gravity, magnetic, resistivity, and induced-polarization signals depend on the volume of the body, electromagnetic signals depend instead on the area of the body normal to the applied field; in this way, a flat-lying disk can give the same electromagnetic anomaly as a sphere or a thick lens of the same radius.

The depth limit to a direct geophysical search depends on the signal-to-noise ratio as well as on the shape and size of the target body and on the strength of the acting force. A gain in the sensitivity of an instrument or an increase in the applied force will not necessarily help a weak signal from great depth. If, for example, a source of near-surface noise happens to be a conductive zone in the overburden or a magnetic zone in volcanics, the noise will increase as more electrical current is applied or as the sensitivity of the magnetometer is improved. Most geophysical methods have a practical depth limitation of about 100 m as far as response from an orebody is concerned. Some geophysicists suggest that 200 or 300 m can be used as a working limit for certain electrical (induced-polarization) and electromagnetic (AFMAG) surveys. Rules of thumb are occasionally mentioned: induced-polarization methods can obtain a response at depths about twice the minimum dimension of the body sought; and magnetic bodies will be detectable at depths four or five times the minimum dimension of the body. Obviously, these can be taken only as convenience figures. Because so many variables are involved, we should not expect to have a more specific figure for the reach of "geophysics" than for the reach of "geology."

13.2. METHODS AND APPLICATIONS

Table 13-1 shows some of the principal geophysical methods used in mineral exploration. Magnetic, electrical, electromagnetic, and radioactive methods are most popular. Seismic and gravity methods are used to some extent, but not nearly as much as in petroleum exploration. There are literally dozens of approaches within most of the categories; electromagnetic methods, in particular, are used with a great variety of instruments, configurations, and procedures. No more than a few methods and techniques can be mentioned here. For a better overall coverage a textbook on mining geophysics by Parasnis (1973) is recommended. A good "how-to" manual of geophysical exploration for minerals and petroleum is provided by Telford and others (1976).

In addition to the methods shown in Table 13-1 a few other exploration methods are sometimes categorized as geophysics. Remote sensing, discussed in Chapter 10, is particularly hard to separate from geophysics. Vapor sampling and atmospheric particulate sampling methods could be called either geophysics or geochemistry; geophysics because they operate at a distance and geochemistry because they involve sampling. In this textbook vapor and atmospheric sampling are listed with geochemistry even to the extent of admitting "airborne geochemistry." Neutron activation analysis methods, or in situ assaying, are also mentioned in the chapter on geochemical exploration even though they are geophysical methods in the sense that they use a physical field induced by radiation. Instrumental techniques for logging drill holes are dominantly geophysical, but they will be treated in the chapter on drilling for geologic information because their purpose is to enlarge the scope of a drill hole beyond that of a slim cylinder of rock.

Airborne versus Ground Surveys

In general, airborne geophysical methods are used in reconnaissance and ground geophysical methods are used in more detailed investigations. There are, however, many instances in which either airborne or ground methods could be used. In an extended exploration program, combinations and sequences of methods may be appropriate, and there is often a need to weigh their individual advantages.

Airborne surveys have some impressive characteristics. They are fast, they are relatively inexpensive per unit area, they can obtain several kinds of surveys at once, and they can provide a more objective coverage than ground surveys in many kinds of terrain. For example, several hundred line-kilometers of airborne electromagnetic surveying can be done in a day compared with three to five line-kilometers per crew in a ground electromagnetic survey. The cost of an airborne electromagnetic survey, with magnetic and radiometric data included, is likely to be one-fourth to one-fifth the cost of an equivalent ground electromagnetic survey. Airborne survey patterns are reasonably uniform and complete because they do not have the access and traverse problems of ground surveys in swamps, dense brush, and rugged topography.

Airborne methods may sometimes be advantageous because competing exploration groups and mineral land speculators may be lurking in the area. It is easy to locate someone's field camp, trace their newly cut ground survey lines, and "join the crowd." Airborne surveys, on the other hand, can operate in a less conspicuous pattern from supply bases outside the target area.

The airborne advantage in time, cost, and security applies to work in relatively large areas where the cost of aircraft operation can be spread over quite a few line-kilometers of work. Most airborne methods are neither economical nor appropriate in target areas of only a few square kilometers.

Airborne surveys have considerable flexibility, but they have some specific weather and terrain limitations as well. Since many surveys must be flown with a terrain clearance of less than 150 m in order to obtain a suitable signal, days or weeks may be lost because of low clouds. Flight-track recovery, the relating of the finished survey to ground features, is often done by selecting points in a narrow strip of ground photographed during the survey; for this, too, weather must permit some recognizable features to be visible.

An airborne survey will give more accuracy than a ground survey in some areas, but it will seldom provide such detail or such sharp signals as a ground survey. A ground survey can be made with more closely spaced lines, and it can be done with a wider choice of methods and equipment.

Less preliminary work is needed on the actual exploration site for airborne surveys than for ground surveys, but more accurate base maps and photographic coverage may be needed. Ground geophysical surveys have the advantage of being able to tie in to occasional control points and stations, but airborne geophysical surveys are flown so fast and so low that the ground control features must be numerous, accurately plotted, and readily visible. In monotonous terrain where recognizable features are sparse, it may be necessary to follow flight lines by an inertial navigation system or by a doppler (radar) navigation system.

A choice must sometimes be made between helicopter and fixed-wing aircraft for an airborne electromagnetic or radiometric survey. Helicopters have an advantage in being able to maintain a more constant ground clearance above rugged terrain. Also, helicopters have a slow-flying capability which allows for greater accuracy and they can land for a ground check in critical areas. Helicopter geophysical surveys can therefore be used in detailed work as well as in reconnaissance. Still, there are disadvantages. Helicopters are much more expensive to operate than are fixed-wing aircraft, they can cover only a third as many line-kilometers per day at best, they have a relatively short range of operation, and they require more maintenance work per flying hour. The decision to use a helicopter in a geophysical survey is generally based on the assumption that the helicopter will permit an essential level of accuracy or detail that could not be matched in a fixed-wing survey.

13.3. AIRBORNE GEOPHYSICS

The most widely used airborne exploration method is aerial photography, not a geophysical method in itself but a strong accompanying method in all airborne geophysics. Airborne geophysical surveys, in decreasing order of use, are: magnetic, magnetic plus radiometric, magnetic plus electromagnetic, and electromagnetic.

Aeromagnetics

Aeromagnetic surveys are well-established ways of finding indications of lithologic contrast, faults, folds, and concentrations of magnetic ore. Total intensity contour (isogam) maps show distortion of the earth's magnetic field by patterns in crustal rocks. When the regional magnetic field—the more uniform background trend—is subtracted, magnetic anomalies remain. Aeromagnetic anomalies so slight that they rise to only a few gammas (gamma $= 10^{-5}$ gauss) above the regional background may be significant in a mapping program. In a more direct connotation, magnetic anomalies may rise to 10,000 or 50,000 gammas over an iron orebody.

Rock magnetism is a function of magnetic susceptibility, the ease with which the constituent minerals may be magnetized. Among the most common magnetic minerals, magnetite, ilmenite, pyrrhotite, and specular hematite, magnetite has by far the highest magnetic susceptibility and is the most common accessory rock mineral. A strong aeromagnetic anomaly may therefore be associated with a variety of rock conditions, such as a tactite zone or a magnetite-rich mafic intrusion or volcanic flow bordered by felsic intrusions, by rhyolitic volcanics, or by most kinds of sedimentary rocks. Figures 13-1 and 13-2, a surface geologic map and an aeromagnetic map of an area in central Morocco, show the response of a large granite intrusion with a wide zone of contact metamorphism. Some sedimentary rocks, such as

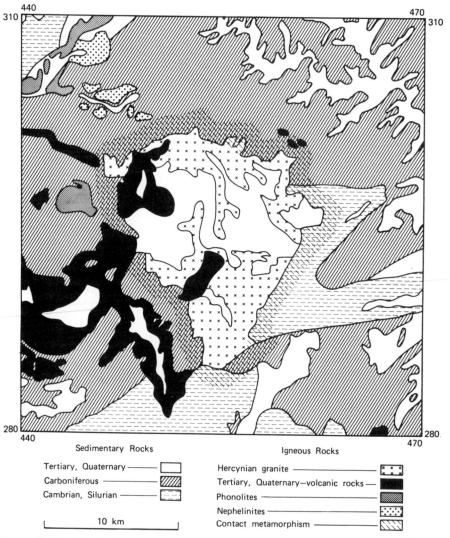

Sedimentary Rocks

Tertiary, Quaternary ——————— ☐

Carboniferous ————————— ▨

Cambrian, Silurian ————————— ▨

|—————— 10 km ——————|

Igneous Rocks

Hercynian granite ——————— ⊡

Tertiary, Quaternary—volcanic rocks — ■

Phonolites ————————————— ▨

Nephelinites ———————————— ⊡

Contact metamorphism ————————— ▨

Figure 13-1.   Geologic map of a Hercynian granitic intrusion with a zone of contact metamorphism in Paleozoic sedimentary rocks, near Oulmès, Morocco. (After Demnati and Naudy, 1975, Fig. 5, p. 336, by permission Society of Exploration Geophysicists.)

ferruginous shale and "ironstone," will of course show a magnetic response. Metamorphic derivatives of ferruginous sedimentary rocks cause some of the strongest magnetic responses. Precambrian banded iron formations have a particularly high magnetic susceptibility.

Remanent (permanent) magnetization, "frozen" in a former condition, is

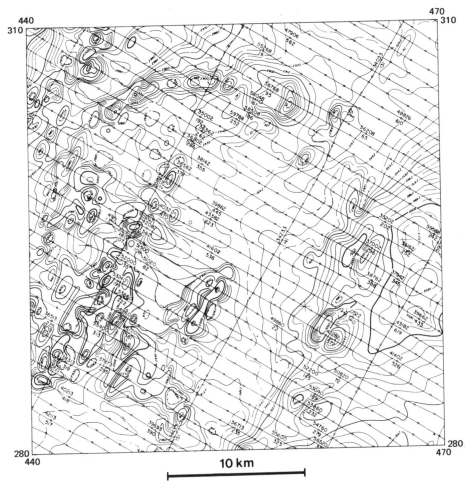

Figure 13-2. Aeromagnetic map of the area shown in Figure 13-1, showing a circular zone of strong anomalies around the granitic intrusion and other anomalies associated with exposures of volcanic rocks. (After Demnati and Naudy, 1975, Fig. 7, p. 338, by permission Society of Exploration Geophysicists.)

related to a special situation in which the intensity and direction of magnetization are independent of the earth's present-day magnetic field. Since aeromagnetic maps are most often interpreted in accordance with the more general condition of *induced magnetization* related to the earth's existing magnetic field, the presence of remanent magnetization (Fig. 13-3) is a disturbing factor—a kind of noise—that must be identified and taken into account.

The most popular instruments for aeromagnetic work are fluxgate magnetometers (resolution on the order of 1 gamma) and proton precession

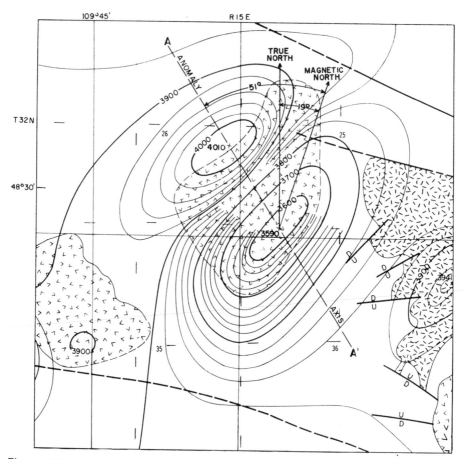

Figure 13-3. Aeromagnetic anomaly associated with Squaw Butte, an exposure of mafic volcanic rocks in Montana. The magnetic high and low closures, representing edges of the volcanic body, appear reversed from the normal situation in the Northern Hemisphere. The reversal is caused by remanent magnetization. (After Zietz and Andreason, 1968, Fig. 12, p. 584, by permission Society of Exploration Geophysicists.)

magnetometers (resolution on the order of 0.1 gamma). The instruments are mounted in a pod on the aircraft wing, trailed behind the aircraft in a bomb-shaped "bird," or placed in a boom or "stinger" extended from the tail section. A third group of instruments, the highly sensitive, optically pumped alkali vapor magnetometers, are more widely used in petroleum exploration than in mining-oriented reconnaissance. In addition to single sensor, airborne magnetometers, vertical gradiometer systems are sometimes used; these consist of sensitive magnetometers trailed at two levels in order to measure the rate of decrease of the earth's magnetic field with height. Gradiometer systems are used to special advantage in resolving closely spaced or steep geologic contacts, distinguishing between shallow and deep magnetic sources and removing the masking effects of regional magnetic gradients.

The fluxgate magnetometer indicates changes in the earth's magnetic field by measuring the time it takes for an alternating current to energize—"saturate"—a permeable ferromagnetic core. Since the pulses from an energizing coil are alternately augmented and opposed by the earth's magnetic field, the voltage from the core received in a secondary coil is affected in even harmonics of the input current frequency. The second harmonic produces a signal that is filtered out and recorded.

The sensing element in a proton magnetometer is a bottle of some hydrogen-rich liquid (water or kerosene) resting within a coil. Spinning protons in the liquid are randomly oriented. When a magnetic field stronger than the earth's field is applied within the coil, the protons realign themselves. The applied field is cut off, and the spinning protons start to precess in phase, oscillating like tiny gyroscopes as they return to their original random alignment. The oscillations induce an alternating voltage in the coil (which now serves as a detector coil) and the frequency of the reference signal, proportional to the earth's magnetic field, is measured.

Since the earth's magnetic field varies with time as well as place, there is disturbance field noise to be removed from the measurements. Many of these variations, such as diurnal variations, can be monitored and the observations corrected. Not so with magnetic storms; these irregular disturbances, most intense in the auroral zones of high latitudes and about the magnetic equator, may last for a few hours to several days. Monitoring of these disturbances is done by a local base-station magnetometer or by a government station. Magnetic surveys, air or ground, are not begun during magnetic storms, and if readings have already been taken, they are discarded.

The data from aeromagnetic surveys are processed and plotted at a map scale that will allow the flight lines to be properly discriminated. At 1:50,000-scale, for example, a flight-line spacing of 1 km is satisfactorily represented by 2 cm on the map. The final maps are often reduced to 1:100,000 or 1:250,000 for matching with regional topographic and geologic maps. There are aspects of data reduction and geophysical modeling that should be appreciated by geologists using aeromagnetic maps, but these are beyond the scope of this book. A review article on airborne geophysics by Richard and Walraven (1975), written for geologists, is recommended for further reading. For additional detail and for a comprehensive treatment of interpretation methods, an article by Grant (1972) and a mathematically oriented book by Grant and West (1965) are suggested.

The aeromagnetic map, generally with total intensity magnetic contours, is

0   5   10

miles

Figure 13-4. Wavelength filtering of an aeromagnetic map to separate shallow and deep anomalies. *Top,* map showing short-wavelength anomalies due to surface volcanic rocks (left of the dashed line) and longer wavelength anomalies due to basement features. *Center,* same map filtered to reveal basement anomalies. *Bottom,* same map filtered to emphasize shallow anomalies due to volcanic rocks. (From E. G. Zurflueh, "Applications of two-dimensional linear wavelength filtering," *Geophysics,* v. 32, Fig. 4, p. 1024, 1967, by permission Society of Exploration Geophysicists.)

interpreted directly or it may be processed further to obtain a filtered map. There are various types of filtered maps, many of which simply assist in discriminating between shallow and deep anomalies, as shown in Figure 13-4. Interpretation is done by referring to geophysical models and by matching whatever geology is known with the more complete aeromagnetic pattern. A fault zone, for example, may be recognized as an anomaly in its own right by comparing the aeromagnetic pattern with models of dipping slabs or it may be recognized by the displacement or truncation of other anomalies. The aeromagnetic signature of a certain lithologic sequence will have characteristics that relate to the magnetite content in its members, and the signature may be traced across the map from places where parts of the sequence are known on the ground. A granitic stock, suspected from preliminary geologic work, may appear as a group of low-amplitude anomalies that stand in contrast to sharper anomalies on its margins. Obviously, the most effective interpretation of aeromagnetic maps, or of any geophysical data for that matter, is done by geologists and geophysicists working together.

Airborne Radiometric Surveys

The principal methods in airborne radiometric surveying are gamma-ray spectrometry and total-radiation radiometrics. Both methods employ the same basic ideas and detectors; because gamma-ray spectrometry is the more versatile, it will be described.

The detecting unit consists of one or more crystals of thallium-activated sodium iodide, a material which emits a flash of light, a scintillation, when struck by a gamma ray. The intensity of the scintillation is directly proportional to the energy of the gamma ray, which is in turn a measurable function of the uranium, thorium, or potassium source. By photomultiplier tube, the scintillation is converted to a voltage, and the pulse height is compared with that of a reference source. Voltage pulses are fed into separate diagnostic channels for uranium, thorium, and radiopotassium and into a total-count channel. The output from each channel is fed to a recorder as a count rate, which shows the number of gamma rays arriving per second within that particular energy range.

The geologic source of gamma rays can be related to the elements detected in the individual channels; uranium possibly from a uranium ore deposit or from a uranium-bearing pegmatite, thorium possibly from a monazite-bearing sand, radiopotassium possibly from a granitic pluton. The choices are many, so are the applications in reconnaissance mapping, and so are the types and sources of radiometric noise.

A type of noise unique to radiometric surveys is *statistical noise* (Richards and Walraven, 1975). Since the gamma-ray count is based on random events, it follows the statistician's Poisson distribution in which the reliability of the data will be very low at a low count rate. In order to obtain a high count rate where its significance will not be cancelled by random background events (statistical noise) the crystal must be as large as can be handled, the flight speed as slow as can be achieved, and the channel "windows" as wide as can be allowed without destroying their selectivity. The limits to crystal size are weight, volume, and cost. Most surveys are flown with crystal assemblies weighing from 15 to 25 kg. An extremely large crystal, such as the one used by the Geological Survey of Canada, weighs 400 kg and cannot be carried in a small aircraft.

A source of noise specifically associated with radiometric surveying is cosmic radiation. This is always present, but like terrain radiation it is randomly distributed and difficult to assess. The cosmic radiation factor is important in trying to obtain a *background* reading that is free from terrain effects. A background reading is often taken by flying at a ground clearance of 600 m where terrain radiation is completely absorbed by the atmosphere. But cosmic radiation coming from space is less absorbed by the atmosphere at 600 m than at the usual terrain clearance of about 150 m; therefore, the background count will still be unduly high. A better method of obtaining a background reading is to fly at normal terrain clearance over a lake, since water absorbs terrain gamma radiation in a depth of approximately 10 cm. But large lakes are seldom where we need them.

The radiometric signal is greatly affected by soil and other kinds of overburden. Gamma radiation is completely absorbed by about 50 cm of soil and rock. Areas of transported soil and alluvium are therefore likely to mask the underlying gamma radiation. Residual soil, on the other hand, may still contain enough of an original radioactive rock component to provide a signal. The problem arising from irregular "cover" is easy to visualize. Bare rock ridges and soil-covered hills will show a higher gamma radiation than stream courses and alluvial plains, unless the alluvium happens to contain transported radioactive minerals. Vegetation and snow are factors as well; they are geophysical "overburden" because they absorb radiation to some extent.

One of the most knotty problems with noise in airborne radiometric work is the change in gamma radiation with differences in ground clearance. Gamma radiation falls off exponentially with distance in the atmosphere so that it is reduced by one-half at a height of about 100 m in the potassium channel and about 130 m in the uranium and thorium channels. As with the irregular overburden factor, and in fact compounding it, the noise effect of high topographic relief is readily apparent. Some of the ground clearance effect can be filtered out of a radiometric survey by using the record of a sensitive radio altimeter.

The flight-line spacing in a radiometric survey is related to geologic features, objective, and cost. More specifically, it is interrelated with the size of the detector crystal, flying height, and air speed. The smaller the crystal, the lower must be the flying height, the slower the speed, and the closer the line spacing. These aspects are covered in more detail, with examples and with guideline formulas, by Darnley (1973). For a general figure, from Darnley, in a gamma-ray spectrometer survey with high-sensitivity equipment at a flying height of 120 m and an airspeed of 100–200 km per hour, a good line spacing for reconnaissance would be 2.5–5 km; for detail it would be 0.5 km.

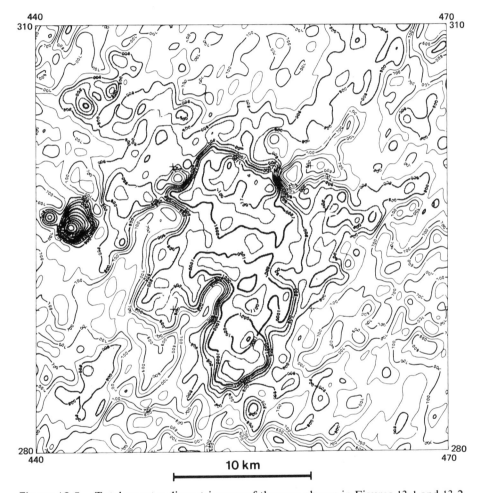

Figure 13-5.   Total count radiometric map of the area shown in Figures 13-1 and 13-2. The exposure of granitic rock is well shown. The magnetic anomalies from deeper portions of the intrusion and from the volcanic rocks do not show up in the radiometric map. A phonolite exposure to the west of the granite causes a strong radiometric anomaly. (After Demnati and Naudy, 1975, Fig. 6, p. 337, by permission Society of Exploration Geophysicists.)

Data from gamma-ray spectrometer surveys are plotted as contour maps and as profiles. Contours are commonly based on the total count rate, on the count rate for each channel, and on the ratio between count rates, as shown in Figures 13-5 and 13-6. The area shown in Figure 13-5 is the same as shown in Figures 13-1 and 13-2 in illustrating an aeromagnetic response. Note that the total gamma-ray count pattern corresponds well to the granite exposure.

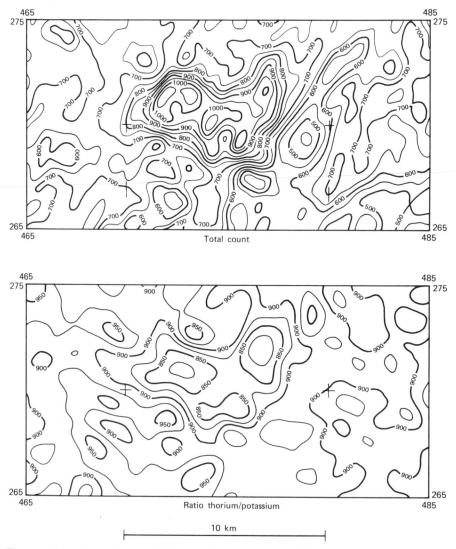

Figure 13-6.   Data from an airborne radiometric survey in Morocco covering an area of Cenozoic rhyolites surrounded by Paleozoic sediments. The rhyolitic volcanics

Interpretation of airborne radiometric data is best done with as much prior knowledge of overburden conditions, rock types, and terrain as can be obtained. Photogeology is therefore usually done just ahead of radiometric work in a reconnaissance program. On the basis of a reconnaissance map, a radiometric anomaly may be explained entirely by the position of a granitic

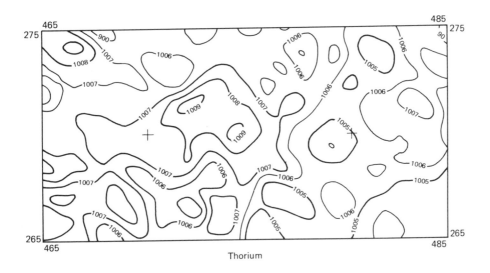

Thorium

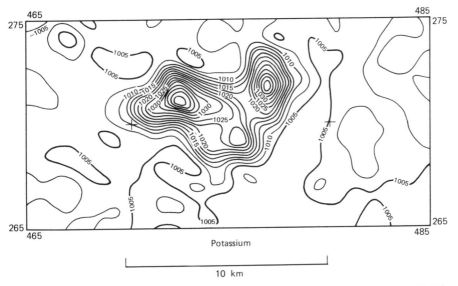

Potassium

10 km

show especially low thorium/potassium values. (After Demnati and Naudy, 1975, Fig. 12, p. 342, and Fig. 13, p. 343, by permission Society of Exploration Geophysicists.)

knob surrounded by alluvium or by a high hill top or it may have a spatial connection with known or suspected pegmatite dikes.

In uranium prospecting the information from several channels can be compared in order to discriminate between weak anomalies that could be associated with a particular kind of uranium deposit and those associated with less favorable or erratic uranium mineralization. Gamma radiation from an arkosic sandstone, for example, should show a potassium anomaly as well as the uranium anomaly, whereas gamma radiation from a phosphatic chert with a small uranium content would not normally show the radiopotassium component.

Airborne Electromagnetic Surveys

Airborne electromagnetic methods provide a means of mapping the electrical conductivity of the uppermost rocks in the earth's crust. There are several methods, all of which have a common principle. An alternating current from a transmitting coil (active system) or from a larger and more remote source, such as atmospheric electricity or radio stations (passive system), generates an electromagnetic field in the earth's crust. Where the field impinges on an anomalously conductive body, eddy currents are induced. The eddy currents generate a secondary electromagnetic field that can be picked up by a detector coil and recorded. In most electromagnetic methods, there is a problem in identifying the conductive body, since graphite zones, conductive overburden, and some clay beds as well as metallic mineralization can cause anomalous signals. One recently developed method uses a three-frequency system in which the response from conductive overburden can be identified and taken into account.

In the active systems, the transmitter generates one or more frequencies, generally in the range of 100–4,000 Hz. The transmitted signals are most often continuous (frequency domain), but they may also be pulsed (time domain). In the frequency domain, it is common to measure the components of the induced or secondary field that are in-phase and out-of-phase (at "quadrature") with the transmitted or primary field (Fig. 13-7). In conventional airborne electromagnetic surveys, the response is shown in parts per million of the primary field or as a ratio of in-phase to out-of-phase components (Fig. 13-8). Transmitting and receiving coils are carried in the aircraft, in a "stinger," in a trailing "bird," or sometimes in separate aircraft.

The TURAIR system, like most conventional and active electromagnetic systems, transmits as well as receives, but the transmitting coil is a ground cable loop several kilometers long. The detector coil is flown across the loop in a helicopter.

The INPUT (induced pulse transient) system measures electromagnetic characteristics in the time domain. It overcomes a difficulty experienced with

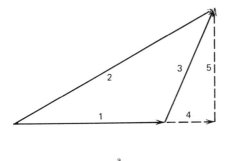
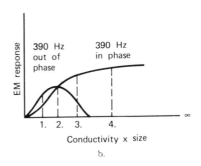

a.   b.

Figure 13-7.   In-phase and out-of-phase relationships in electromagnetic prospecting.
a. Vector diagram; vectors 1 and 2 represent the primary field and the composite effect
of the primary and secondary field at the detector coil, vector 3 is the secondary field,
vectors 4 and 5 are the in-phase and out-of-phase components of the secondary field.
b. In-phase and out-of-phase response at a single low frequency. A body of conductivity 1 will show an in-phase/out-of-phase value of less than unity. Bodies of conductivity 2 and 3 will show ratios of unity and greater than unity. Body 4 would be missed by
the out-of-phase response. (After Pemberton, 1962, Fig. 1, p. 693 and Fig. 4, p. 694, by
permission Society of Exploration Geophysicists.)

most electromagnetic systems where the transmitter is in close proximity to
the receiver, a condition which interferes with the much weaker signal from
the subsurface. The primary signal from the airborne transmitter is actually
terminated in a series of pulses and the response from an anomalously
conductive source in the ground is detectable between pulses as a transient
characteristic—a decay curve—which is identified and recorded.

AFMAG (audio frequency magnetics) and VLF (very low frequency) are
two of the most widely used passive electromagnetic systems in both airborne
and ground surveys. AFMAG depends on spatial variations in the alternating
electromagnetic field imposed on the earth's crust by tropospheric lightning
discharges. Two receiving coils at right angle are used to measure the tilt of
the naturally occurring field where it is distorted from its normally horizontal
plane by conductive rocks and orebodies. VLF systems use the field imposed
by transmissions from special radio navigation stations; as with AFMAG, the
imposed electromagnetic field is ordinarily polarized in a plane parallel to the
surface of the earth except in the vicinity of good conductors.

Airborne electromagnetic surveys have been very successful in locating
anomalies from orebodies, especially from massive sulfide deposits (Fig. 13-
9), but they are considerably more expensive than airborne magnetic and
radiometric surveys. Therefore, electromagnetic surveys are used more often
in a direct search for orebodies than in geologic mapping. Several kinds of
electromagnetic surveys are generally made in the same flight, and electro-

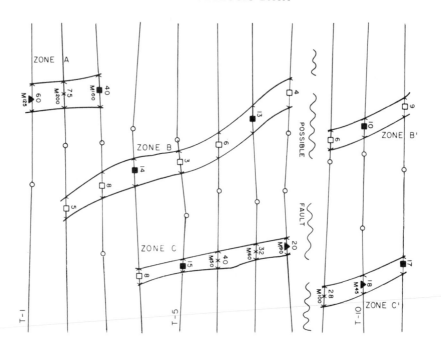

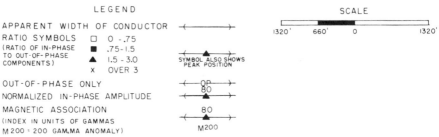

Figure 13-8. Information from a helicopter electromagnetic and magnetic survey across a series of conductive zones. (From S. H. Ward, "The electromagnetic method," in *Mining Geophysics*, Vol. 2, Fig. 82, p. 301, 1967, by permission Society of Exploration Geophysicists.)

magnetic surveys are often combined with other airborne geophysical methods, such as aeromagnetics or radiometrics. The interpretation of airborne electromagnetic surveys, principally in regard to conductive orebodies, is explained in summary by Ward (1967).

13.4. GROUND GEOPHYSICS

One of the principal advantages to ground electrical surveys is their capability of making direct contact with the earth. For this reason, electrical methods

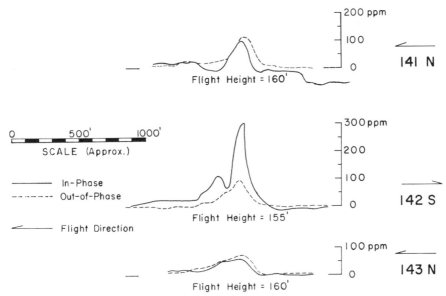

Figure 13-9. Airborne electromagnetic profiles from three flight lines across a zone of pyrite and pyrrhotite stringers in northwestern Quebec. (From G. Podolsky, "An evaluation of an airborne electromagnetic anomaly in northwestern Quebec," in *Mining Geophysics,* Vol. 1, Fig. 4, p. 200, 1966, by permission Society of Exploration Geophysicists.)

are widely used in detailed exploration for orebodies. Ground magnetic methods, like aeromagnetic methods, are used as aids to geologic mapping as well as in searching for orebodies. Ground electromagnetic methods are in general use where massive sulfide deposits are sought; as with airborne electromagnetics, there are dozens of separate techniques. Ground radiometric surveys are not as often applied to geologic mapping as are airborne radiometric surveys; instead, their main use is in searching for uranium orebodies. Gravity methods are applied to regional geologic mapping, and they have a follow-up or supporting function in the interpretation of other geophysical anomalies as shown in Figure 13-12, where gravity and magnetic response from an iron orebody are compared. One use of gravity geophysics, and of refraction seismic methods as well, is in determining the depth and configuration of bedrock in alluvium-covered areas.

Ground Electrical Surveys

An electrical method, induced polarization, is the most popular ground geophysical method used in mineral exploration. Although originally designed for work with disseminated sulfide bodies—porphyry copper deposits, in particular—induced polarization was soon found to give more diagnostic anomalies above massive sulfide and vein deposits than had been obtained with the long-established electrical resistivity method. Electrical methods other than induced polarization are less popular than in past years; they are now used in special circumstances rather than for the overall detection of orebodies.

Electrical resistivity measurements, sometimes used in geophysical methods by themselves, are the basic element in induced-polarization work. Resistivity (ρ), a measure of the difficulty in sending an electrical current through a substance, is measured in ohm-meters. Because conductivity is the reciprocal of resistivity, it is measured in units with a reciprocal name: mhos per meter.

Induced-polarization or "overvoltage" methods use the two modes of electrical conduction that occur in mineralized rocks, ionic (in pore fluids) and electronic (in "metallic" minerals). When a current is applied to a medium containing both types of conductors, an exchange of electrons takes place at the surfaces of the metallic minerals causing (inducing) a polarization and forming an electrochemical barrier.

This barrier provides two useful electrical phenomena. First, an extra voltage, an overvoltage, is needed to send the current across the barrier. When the current is cut off, the overvoltage does not drop to zero immediately; instead, it decays, allowing a current to flow for a short time. Second, the resistivity of a mineralized rock having the electrochemical barrier differs according to the frequency of the applied current, the resistivity decreasing in the higher frequencies. In unmineralized rocks, the applied current is carried only by the ionic solutions in pore spaces and the resistivity is independent of the frequency of the current. Induced-polarization surveys are made by the *time-domain* method, using the decay phenomenon, and by the *frequency-domain* method, using the resistivity contrast phenomenon.

In the time-domain or *pulse transient* method, controlled pulses of direct current are applied to the ground every few seconds through two current electrodes and the overvoltage current is measured between pulses across two potential electrodes. The measure of induced polarization in the time domain—a function of the metallic mineral content—is *chargeability*, symbolized by the letter M and described as the integrated area under the current decay curve normalized by the primary voltage.

In the frequency-domain induced-polarization method, the resistivity of the terrain is measured at two frequencies and the polarization is calculated from the difference between resistivities at the two frequencies. A percent frequency effect (PFE), or frequency effect (FE), related to metallic mineral content, is expressed in larger and more convenient numbers taken from the small numerical difference.

Chargeability and percent frequency effect are equivalent parameters which may be derived from one another by mathematical treatment. Either one may be plotted as shown in Figures 13-10 and 13-11, together with resistivity and sometimes with a derived parameter called the *metal factor* (MF). The metal factor is obtained by dividing either the chargeability or the percent frequency effect by the resistivity and multiplying by a constant. The constant serves to put the metal factor into the range of commonly used numbers.

All common sulfide minerals except sphalerite are electronic conductors; so are most other minerals with a metallic luster, including graphite and some kinds of coal. Thus minerals other than ore minerals will give an induced-polarization response. A response—geologic noise—is also obtained from some clay minerals that are not electronic conductors but have an unbalanced surface charge. Some new work with complex-resistivity spectra offers hope of discriminating between signals from ore minerals and nonore minerals.

Conventional induced polarization requires that current electrodes and potential electrodes be in contact with the ground, but a new technique,

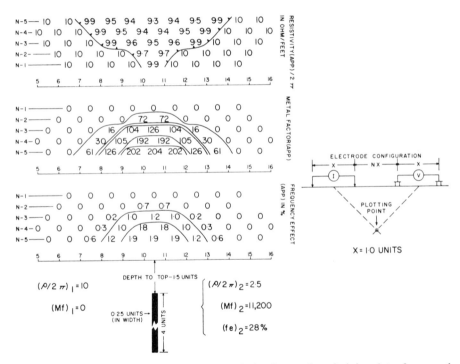

Figure 13-10.   Pseudosections, induced-polarization and resistivity data from scale model experiments, showing the anomalous pattern to be expected in a traverse across a vertical tabular source. Depth (N-1 to N-5) increases downward in the metal factor and frequency effect sections, upward in the resistivity section. (From P. G. Hallof and E. Winniski, "A geophysical case history of the Lakeshore orebody," *Geophysics*, v. 36, Fig. 9, p. 1242, 1971, by permission Society of Exploration Geophysicists.)

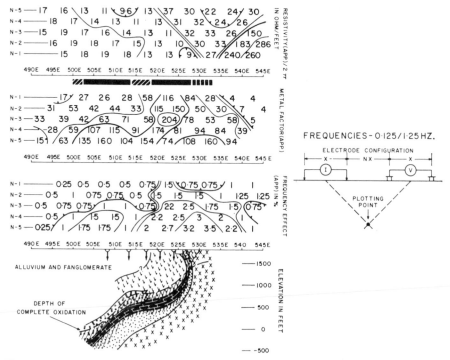

Figure 13-11.   Pseudosections and geologic section from induced-polarization surveying and subsequent drilling, Lakeshore orebody, Arizona. The black band on the pseudosections represents definite (solid), probable (broken), and possible (diagonally broken) anomalies. (From P. G. Hallof and E. Winniski, "A geophysical case history of the Lakeshore orebody," *Geophysics*, v. 36, Fig. 12, p. 1244, 1971, by permission Society of Exploration Geophysicists.)

magnetic induced polarization (MIP), permits a magnetometer to be used in place of the potential electrodes. The MIP method is especially suited to detection of sulfides beneath conductive overburden; it also permits the airborne detection of induced-polarization signals.

Induced-polarization information is shown in profiles, in contoured maps from multiple line surveys, and in contoured "pseudosections" with the format depending on the technique being used and the electrode configuration in the field. Only one illustration will be given here, and only a brief comment on the interpretation of induced-polarization data can be afforded. For further reading, a brief digest of methods, data formats, and interpretation has been prepared for geologists by Hallof (1972). For still more detail, a book by Sumner (1976) is recommended.

Figure 13-10 is a pseudosection with data derived from scale-model experiments in the frequency domain. In this situation, the anomaly is considered to come from a shallow-seated vertical source. The positions of the plotting points at depths of $n = 1$ to $n = 5$ are determined by the separation between electrodes and by their positions in the survey line. Apparent resistivity, plotted as $\rho/2\pi$, shows an increasing effect from the conductive body as the electrodes are given wider spacing for signals from increasing depth. Frequency effect and metal factor anomalies are more definitive than the resistivity anomaly. Figure 13-11 shows the result of an induced-polarization survey over the Lakeshore orebody, Arizona. Note that only the shallower portion of the orebody shows an anomaly at $n = 3$ and that the deeper part of the orebody shows up with wider electrode spacing at $n = 4$ and $n = 5$.

Ground Magnetic Surveys

Ground magnetic surveys have become increasingly popular with field geologists because of the availability of small, easily operated magnetometers. Often the magnetic survey will be run by the geologist as a part of a mapping program. The most common instruments currently in use are fluxgate and proton precession magnetometers. Accuracy is on the order of 0.25–5 gammas. Some still smaller but less sensitive magnetometers with accuracy on the order of 20–100 gammas are in use; in these, the sensing unit is a magnet held in suspension or in balance by a wire or a fiber. Field survey procedures, covered in summary form by Breiner (1973) and in detail by Parasnis (1973), are relatively simple. Magnetic readings, in gammas, can be taken very quickly at intervals of 3–300 m along traverse lines. A trial line with a high density of readings is often surveyed at first in order to determine the best station interval. Changes in the readings from one station to another may indicate a change in magnetic susceptibility in the underlying and adjacent rock, or they may be due to noise factors such as diurnal variations in the magnetic field, magnetic storms, instrument drift, and temperature. Readings must be made at the same stations several times in a day unless a recording magnetometer is maintained at a base station. Cultural noise may be from a fence or from a power line, or it may even be from a pocketknife.

Figure 13-12 shows a magnetic response from an iron orebody and from a dipping series of metamorphic rocks containing a bed of iron formation. Gravity anomalies are shown as well. In this example, the magnetic data are shown in units of vertical magnetic intensity. Depending on the instrument, the data may be gathered in terms of vertical, horizontal, or total magnetic force and may be shown in profile or in isogam contours. Changes in magnetic gradient are sometimes measured or calculated in order to provide an additional index for the interpretation of anomalies (a guide to depth estimation) and for assessing the effect of regional magnetic fields. A thorough

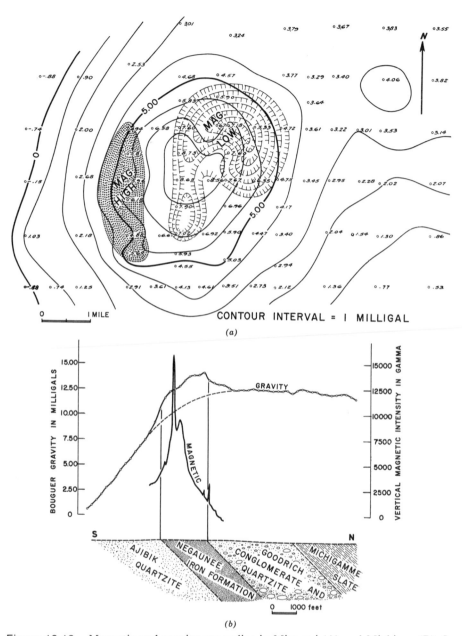

Figure 13-12. Magnetic and gravity anomalies in Missouri (A) and Michigan (B). In the Missouri map the magnetic high and low indicate two edges of the mass, the gravity contours define its shape. Although the magnetic pattern is not quite reversed, as in Figure 13-3, it is distorted by remanent magnetization. In the Michigan cross section the magnetic profile shows the higher grade ore zones within the iron formation; the gravity profile shows the width and dip of the formation. (From G. W. Leney, "Field studies in iron ore geophysics," in *Mining Geophysics*, Vol. 1, Fig. 9, p. 404, 1966, by permission Society of Exploration Geophysicists.)

explanation of the geologic factors involved in ground magnetic interpretation can be found in the book by Grant and West (1965).

Ground Electromagnetic Surveys

Ground electromagnetic systems generally operate in the frequency domain, but there are also time-domain (transient) methods. There is a wide variety of techniques and instruments, including small VLF instruments. Easily portable electromagnetic equipment provides a method for making quick subsurface investigations while mapping or doing reconnaissance work. Too quick and too easy, in one respect; electromagnetic data collected by one or two geologists as a part of general field work do not have the discrimination afforded by crew-operated instruments and methods. More importantly, a geologist working alone with geophysical problems does not normally have the necessary expertise for applying geophysical theory to what is being done.

Figure 13-13 shows a VLF-EM anomaly located above a sulfide zone in Paleozoic volcanic rocks in northern Newfoundland. The edge of the orebody is reflected in an increase in field strength and a sharp change in the inclination (dip angle) of the receiving coil (Heenan, 1973).

The VLF (very low frequency) waves that serve as a source of energy for the geophysical method are transmitted by naval stations for communication with submarines. A new ELF (extremely low frequency) station is expected to provide another source of energy for geophysical use.

Ground Radiometric Surveys

Hand-held radiometric instruments have two types of detectors, Geiger-Müller tubes and scintillators. The Geiger counter, an instrument that became a standard item of equipment for geologists and prospectors during the 1950s, has been replaced by more sensitive scintillation detectors (the scintillometer and the gamma-ray spectrometer) for reconnaissance surveys.

Geiger counters have a relatively weak response to gamma radiation, but they are mechanically robust and have greater electronic stability than scintillation detectors. Geiger counters are therefore preferred for making accurate measurements in areas of strong radioactivity, in uranium mines, and in orebody boreholes.

Scintillometers are used for detecting total gamma radiation, and gamma-ray spectrometers are used for differentiating between radiation from uranium, thorium, and potassium sources, as mentioned earlier in this chapter in regard to airborne radiometric surveys. Special types of detectors for alpha radiation are used in measuring radon gas in soils; in these methods, soil air is either pumped from a shallow hole for a few seconds or is allowed to make contact with the detector for several weeks.

One kind of reconnaissance prospecting for uranium might be considered an intermediate step between ground (on foot) surveys and airborne surveys.

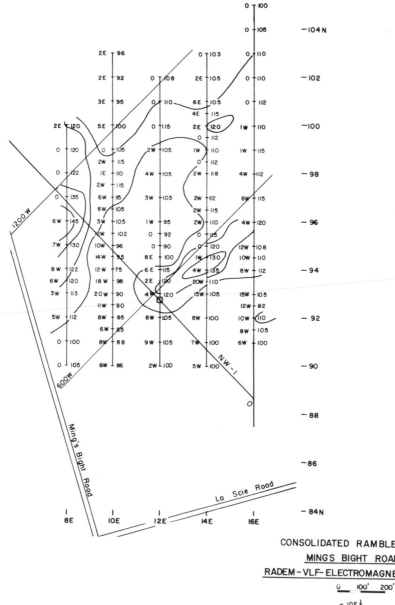

CONSOLIDATED RAMBLER MINES

MINGS BIGHT ROAD AREA

RADEM-VLF-ELECTROMAGNETIC SURVEY

0   100'   200'

DIP ANGLE IN DEGREES

FIELD STRENGTH OF HORIZONTAL COMPONENT OF VLF
FIELD AS A PERCENT OF NORMAL

CONTOUR INTERVAL: 110,120,130,140

VLF STATION CUTLER MAINE   ——— 17·8 K Hz

These are road surveys in which a scintillometer or a gamma-ray spectrometer is mounted on a car; traverses are made at 50–70 km/hr.

In ground reconnaissance surveys on foot, continuous measurements are made in traverses with an instrument with some form of audio speaker or "squealer" so that the operator can watch where he is going rather than watching the galvanometer on the instrument. In detailed uranium prospecting, measurements are taken at grid intervals of a few paces.

Sources of noise in ground radiometric work are similar to those in airborne radiometric surveys. Cosmic radiation and topographic effects acting in various combinations make background readings difficult to obtain. Readings on top of high ridges have a high cosmic ray component, but readings in narrow gulches are also affected from several sides by terrain radiation.

For further information on portable radiometric survey instruments and on ground survey techniques, two books are recommended: the *Uranium Prospecting Handbook* edited by Bowie, Davis, and Ostle (1972) and a panel proceedings volume on uranium exploration methods published by the International Atomic Energy Agency (1973).

13.5. PLANNING AND COORDINATING GEOPHYSICAL WORK

One of an exploration geologist's responsibilities is to coordinate the work of specialists within the organization or from outside contracting firms. There is a special need to coordinate geophysical work with geological investigations because they are so interdependent. A geophysicist chooses field methods and traverses on the basis of interpreted geology, and a geologist uses geophysical information in making an interpretation. The following idealized sequence may be used to prepare for geophysical surveys that are flexible enough to be reoriented while interpretations are being made and while the geophysical crew is still on site.

A. *Preliminary Considerations*
   1. *Geophysical exploration models.* Based on a conceptual model of the orebody and whatever information is available regarding the geology of the area, certain physical contrasts can be expected and a probable range in depth of occurrence can be assigned. One geophysical model

---

Figure 13-13.   VLF–EM response from a stratabound sulfide zone in Newfoundland. The shallow edge of the northeast-plunging ore zone is indicated by high field strength values and significant changes in the coil dip angle. (From P. R. Heenan, "The discovery of the Ming zone, Consolidated Rambler Mines Limited, Baie Verte, Newfoundland," *Canadian Mining Metall. Bull.*, v. 66, no. 829, Fig. 4, p. 83, 1973, by permission Canadian Institute of Mining and Metallurgy.)

may be an ore discovery model, another may be a mapping model directed toward key geophysical signatures in lithology and structure.

2. *Objective*. Limits in cost, time, and scheduling are taken into account. Within these limits and within the framework of the geophysical exploration models, certain sequences of geophysical and geological work can be accommodated. The objective will be to do the work within the best of these sequences.

3. *Procedure*. One or more organizations will be capable of doing the job. In order for them to set up a tentative procedure and offer their services, the status of existing geophysical coverage and the current objective must be known. Also, the following conditions must be spelled out insofar as possible:

   a. Size of the area.

   b. Degree of detail needed.

   c. Indicated orientation of survey lines and spacing of survey stations.

   d. Type of coverage needed—complete or partial.

   e. Sensitivity required in each proposed method.

   f. Required accuracy in survey line control.

   g. Context and the format of the data to be delivered, i.e., raw data, contoured data, interpreted data. If the need is for interpreted data, to what extent?

   h. Intended scheduling of the job.

   i. Kind of terrain involved, seasonal characteristics, and field base facilities.

B. *Preparations for Geophysical Work*. Before the job gets underway, the geologist and geophysicist will design a specific program.

1. Briefing by the geologist.

   a. *Geological conditions*. Using existing geologic maps and—if available—prior geophysical surveys to indicate discontinuities and lithologic contrasts, the geologic pattern will be related in detail to physical properties, such as density, conductivity, and magnetic susceptibility.

   b. *Sources of noise*. Some noise can be anticipated on the basis of existing information. Possible sources of terrain noise (swamps, conductive overburden) may be identified. Sources of cultural noise—mines, pipelines, and abandoned townsites, for example, may be known.

   c. *Access*. The geologist will have some information on roads, terrain, and weather. These are physical conditions of access. In addition, there will be legal conditions of access to be explained—insofar as known. For example, the geophysicist may need formal permission to enter the land, permits to bring geophysical equipment into the country, and work permits for personnel.

  d. *Facilities.* Supplies, campsites, maintenance, and repair will be needed. Aircraft and electronic equipment require special consideration. If the geologist has been working in the area, something will be known of these facilities.
2. *Scheduling.* The season, the time allowable for completion of the job, and possible delays and extensions are taken into account.
  a. *Season.* Weather conditions may dictate the best flying season and the best season for ground access. In tropical areas, monsoon rains may make geophysical (and geological) work all but impossible during certain months. In arctic areas, winter weather and darkness will restrict certain types of geophysical work; for some work, however, ground access and surface traversing may actually be easier across frozen muskeg and lake areas in winter. In forested areas of deciduous trees, aerial photography and ground survey control for geophysical work are most effective in the winter. The best season for geophysics in a particular region will also be the busiest season for contractors and there may be a shortage of available crews; therefore, there may be some advantage in scheduling off-season surveys and accepting a few less than optimum working conditions.
  b. *Delays.* It is almost impossible to avoid some delay due to weather, equipment malfunction, magnetic storms, and unexpected problems with land, government, and people. The scheduling should therefore be flexible enough to permit alternative geophysical methods and traverses while problems are being ironed out. Geologic mapping crews may also need to have some alternative work planned in advance.
  c. *Extensions.* Inasmuch as geophysical work is used in developing the picture of a geologic situation, new dimensions and directions will become apparent while the work is in progress. Extra traverses may be needed. Survey lines may have to be extended into nearby areas—sometimes into areas not yet controlled by permits or claims. The use of additional geophysical methods may be indicated. Geologic mapping may need to be expanded or redirected in order to keep up with new geophysical information. These eventualities cannot be scheduled, but they can be anticipated when schedules are made.
3. *Sampling and orientation.* Samples for the laboratory determination of geophysical parameters and for a simulation of geophysical response can be furnished by the geologist. In addition, the geologist and geophysicist may take an orientation tour of the most significant outcrops. If an orientation survey across a known orebody will be needed, it is the geologist's job to identify a representative orebody in the area or in some analogous area.

4. *Survey control.* Existing map and aerial photograph coverage will need to be studied. Where new topographic mapping or photogrammetry has already been planned, the plans may be reviewed for possible work at a supplementary scale for geophysics and for the location of additional control stations and signals. In places where geological or topographic survey lines must be cut through vegetation, geophysical survey lines can be cut at the same time.

5. *Subsurface information.* Key information from stratigraphic sequences, samples from depth, and dimensions from profiles of known depth are so important to geophysical work as well as to geological work that drill holes may be planned to obtain information of combined significance in the most critical locations. In some instances, a few extra meters of drilling to intersect a significant boundary in physical characteristics or an inexpensive noncore hole to the base of overburden may be worthwhile with respect to geophysics, even if not with respect to the immediate geologic objective. Downhole geophysical information is directly applicable to surface geophysics; certain drill holes may therefore be filled with heavy mud or lined with plastic casing and kept open for geophysical logging.

C. *Coordination work during a geophysical survey.*

1. *Sorting of apparent anomalies.* Some specific geologic work may be needed to strengthen or verify preliminary interpretations.

2. *Key drilling and trenching.* Subsurface information may be needed for depth control points and for index measurements.

3. *Providing for supplementary geophysical investigations.* Where anomalies can be confirmed by additional methods, the work may be done by the geophysical crew that is already on site.

4. *Providing for extended coverage.* Earlier ideas on the limits of an exploration target may be changed by the geophysical data. Additional property access, survey control, and geologic mapping may be needed.

D. *Follow-up work.* After the job has been completed, the geophysicist will interpret the data. The geologist may ask for additional data filtering and treatment to enhance some of the apparent signatures and to bring out more information in specific areas. Additional geologic mapping may be needed to confirm geophysical interpretations. Finally, appropriate targets will be drilled.

Geophysical surveys are keys to the depth dimension in geologic exploration, but they are delicate keys. A geophysical survey is a job for specialists and the interpretation of geophysical data is a job for experts. Both specialists and experts know that their work would be of limited value if it had no geologic guidelines. Geologic specialists and experts know that their work would also be severely limited without geophysical information.

CHAPTER 14
EXPLORATION GEOCHEMISTRY AND GEOBOTANY

Geochemical prospecting and geobotanical prospecting are based on the knowledge that an envelope of primary mineralization is likely to occur around a mineral deposit and a secondary dispersal pattern of chemical elements is often created during the weathering and erosion of the deposit. The primary envelope and the secondary pattern form geochemical anomalies which, if pronounced enough, result in larger guides to mineralization than would be provided by the economic mineral deposit itself.

The primary envelope, called a geochemical aureole or a primary geochemical halo, is an expression of the alteration and zoning conditions discussed in Chapter 2, and it has similar dimensions, centimeters to meters around some orebodies and hundreds of meters to kilometers around large orebodies and mining districts. At Tynagh, Ireland, for example, a zone of anomalously high zinc content extends for 1 km from the orebody and a manganese aureole extends for 7 km (Russell, 1975). A primary geochemical halo does not have to have been formed by epigenetic mineralization. It may represent a weak syngenetic accumulation of ore minerals that strengthens into orebodies where conditions are right. The ultimate size of a syngenetic accumulation amounts to a geochemical province measured in hundreds of kilometers. The ultimate form of a syngenetic accumulation is a function of its original form, but it is further controlled by metamorphism and remobilization.

The secondary dispersal pattern, or secondary geochemical halo, contains remnants of ore mineralization that may be recognizable in rock, soil, sediment, and water samples taken at distances of meters to tens of kilometers from the source. Under certain conditions, dispersed elements from a mineral body may have been so strongly reconcentrated that they form supergene orebodies or more commonly, reconcentrations form spurious anomalies that interfere with the recognition of a more general pattern around the parent orebody.

The use of secondary dispersal patterns is as old as prospecting itself; prospectors have long practiced the tracing of float to its source and panning soil or alluvium for heavy minerals in the resistate. The older practice developed into geochemical prospecting in the 1930s when it began to rely on emission spectroscopy and other sensitive methods of chemical analysis to provide guides where none of the ore and gangue minerals could be recognized.

Geochemistry is now an accepted part of nearly all exploration programs, accepted to the point where 8 million geochemical samples are collected each year in the "Western" nations, and 10 million each year in the Soviet Union (Webb, 1973).

In gaining recognition, exploration geochemistry followed the same route taken by geophysics. A few early successes in the most facile areas led to enthusiastic misuse, disappointment, and, finally, to a renewed application in proper geologic context. The concept is called *landscape geochemistry* in a concise and yet comprehensive review of geochemical environments by Fortescue (1974).

Even though geobotanical associations and Agricola's "small and pale-colored plants" had been used empirically for many years in geologic mapping and for many centuries in prospecting, a specific science of geobotany did not develop until quantitative guidelines became available for studying the detailed geochemical relationships between rock, soil, water, and plants. Geobotanical exploration then became a part of the airborne reconnaissance and through the expression of vegetation in infrared photography, a part of remote sensing. Geobotany is in a sense visual geochemistry in which patterns in plant growth, the presence of indicator plants, and morphological or mutational changes in vegetation are taken as evidence of geochemical anomalies.

Because most exploration programs use several field techniques in combination and in sequence, it is difficult to cite many unqualified geochemical or geobotanical successes in discovering orebodies. Geochemistry deserves credit as the principal guide to the Casino porphyry copper deposit, Yukon Territory (Archer and Main, 1971), to several lead-zinc deposits in New Brunswick, and to the Tanama-Rio Vivi porphyry copper deposits in Puerto Rico. Several uranium orebodies in Utah and New Mexico were found on the basis of geobotanical studies alone (Cannon, 1960). The credit list lengthens when consideration is given to discoveries in which geochemistry or geobotany have had a major contributing role. It has often had an early role as a part of the change in scene from favorable area to target area. Many of the new generation of Irish lead-zinc mines are in orebodies that were first indicated by geochemical studies. Several Zambian copper mining areas were first explored on the basis of the "copper flower," a member of the mint family which served as an indicator plant.

Reconnaissance geochemistry has been particularly helpful in indicating the larger, more attractive targets in areas having small isolated ore mineral occurrences. The target areas for the Panguna porphyry copper discovery on Bougainville Island, Papua New Guinea (Espie, 1971) and La Caridad porphyry copper discovery in Sonora (Coolbaugh, 1971) were outlined in this way. Geochemistry has seldom been the only target index in these areas of prior mining. At Panguna, stream-sediment and soil sampling was accompanied by geologic field mapping, which afforded a comparison with similar conditions of age, rock type, and quartz veining at the Toledo copper deposit in the Philippines. At La Caridad, where copper anomalies were followed 20 km upstream and found to be associated with two abandoned mines, the subsequent geochemical work was accompanied by induced-polarization surveys, geologic mapping, and a thorough study of leached capping.

Theoretical or "pure" geochemistry, the academic parent of applied geochemistry, has a long-standing literature of its own and a representation in the literature of mineralogy and petrology. Exploration geochemistry, part of the applied form, has not had time to develop such an extensive literature.

There are only a few textbooks specifically devoted to exploration geochemistry. Symposia volumes are more numerous, and a great many journal articles are concerned with specific aspects and case histories in geochemical prospecting. The literature of exploration geobotany is sparse, but the geochemistry textbooks give good coverage to *biogeochemistry,* the geochemical method in which plant material is sampled for laboratory analysis. A list of recommended readings is given at the end of this chapter. There is an extensive Russian literature on geochemical prospecting, but the list of recommended readings at the end of this chapter includes only the recent Russian literature that has been translated into English.

14.1. ANOMALIES, BACKGROUND, AND NOISE

The comparison made between the growth of exploration geochemistry and exploration geophysics can be taken a few steps further; they have some more in common. Neither geophysics nor geochemistry locates orebodies; messages and signatures are provided, and these are interpretable in terms of a suspected orebody or in terms of a geologic setting favorable to orebodies. Anomalies are measured in numbers that can be separated from a larger group of numbers constituting the geochemical or geophysical background. There must be a detectable contrast in nature; the body being sought must have an appropriate volume and shape, and it must occur within a detectable depth. Finally, there is noise.

The instrumental noise experienced in geophysics can be restated for geochemistry in terms of variations in sampling practice, laboratory analyses, and field reagents. The disturbance field noise of geophysics can be expressed in geochemistry as irregularities in fluid migration, weathering, and erosion processes that should ordinarily relate dispersion patterns to orebodies. The effects of human occupancy, cultural noise, are of special importance in geochemistry; environmental pollution and contamination are geochemical terrors. Terrain noise in geochemistry may, as in geophysics, be related to the direction a slope is facing, to topographic noise (changes in slope affecting mass wasting and erosion), and to geologic noise (extraneous concentrations of key elements).

14.2. THE EXPLORATION GEOCHEMISTRY SEQUENCE

In a geochemical survey, regardless of the method and its place in reconnaissance or in detailed exploration work, investigations often follow a certain sequence.

1. Selection of methods, elements to be sought, sensitivity and precision to be required, and sampling pattern. Selections are made on the basis of cost, known or suspected geologic conditions, laboratory work on similar material and, most important, an *orientation survey* or equivalent experience in similar terrain and with orebodies similar to those being sought.

2. Preliminary, or first coverage, field sampling program, with occasional check samples and depth (profile) samples to establish a level of reliability and to evaluate noise factors.

3. Sample analysis, in the field (where possible) and in the laboratory, with check analyses made by several methods.

4. Statistical treatment and geologic evaluation of the data, always in connection with available geological and geophysical data.

5. Confirmation of apparent anomalies; follow-up sampling; and analysis and evaluation in smaller areas, using closer sampling intervals and additional geochemical methods.

6. Target investigation, with a provision for resampling and for additional analysis of stored samples.

The provision for resampling and for additional analysis is important. Geochemistry, like geophysics, relates anomalies to a conceptual model of an orebody, and the first few drill holes may change the entire model. An example can be cited. The Mount Pleasant tin-bearing deposit in New Brunswick was discovered after several repetitions of a geochemical sampling, drilling, and evaluation sequence (Hosking, 1963). In the reconnaissance survey leading to the Mount Pleasant discovery, stream sediments were sampled in a 260-km² area; three locations were found to contain anomalous amounts of zinc, lead, and copper. Follow-up sampling of stream sediments and soil provided information on an area of high metal content within the drainage pattern of one of the streams. A more detailed geochemical survey, with soil samples taken on a 30 × 120 m grid resulted in the selection of specific targets for geophysical work and drilling. An electromagnetic survey indicated no additional targets, but the drilling program encountered some weak mineralization. The exploration model was reevaluated and, as a result, the soil samples were reanalyzed for molybdenum and tin. A new group of anomalies became apparent.

Drilling followed, again with inconclusive results, but it gained new geologic information which indicated a need for additional geochemical work. A new try at sampling on a finer grid pattern (6 × 30 m) in a part of the area did not produce the desired information. Then, reanalysis of the original soil samples by more accurate methods provided data for a new group of anomalies; these seemed to fit the growing fund of geological information. Samples were taken once more, this time from bedrock in one of the anomalous zones; as a result of this sampling, a tin-bearing zone was outlined, and it was possible to estimate the extent to which the soil anomalies had been displaced by glaciation. Drill holes were put down again; they were encouraging, and a tunnel was driven to get a better look at the mineralized zone. The result, 11 years after the first reconnaissance survey indicated anomalies in three streams: development work began on a new copper-tin-zinc deposit.

14.3. BASIC CHARACTERISTICS

Most exploration geobotany and geochemistry are applied to the direct search for orebodies. Indirect use as an aid to geologic mapping in soil-covered areas is less common, but it nonetheless deserves mention.

The influence of rock type on plant communities has been recognized for a long time, and the derived contrasts in vegetation have been used in geologic mapping. In the western United States, for example, Douglas fir is often associated with limestone in places where mountain mahogany grows on adjacent quartzite beds. Another characteristic rock-plant association used in photogeology is a dwarfed sparse "serpentine flora" characterized by stunted pine, shrubs, and ferns. Because the ecology of plant communities is not well understood by nonbotanists, geobotanical mapping has generally been done by rule of thumb. The scientific basis for geobotanical mapping is a relatively new development; it is reviewed and explained by Brooks (1972).

An example of geochemical mapping, as opposed to direct geochemical prospecting, is given by Leake and Aucott (1973). In this example, taken from work in England and Scotland, panned concentrates from stream sediment samples were analyzed by X-ray fluorescence; ratios between metals were then used to differentiate between phases of granite, types of metamorphic rock, and areas underlain by volcanic and sedimentary rock.

Geochemical aureoles, or primary halos, were shown in Figures 2-8 and 2-10 in connection with the discussion of zoning patterns around the Altenberg tin deposit in East Germany and the silver-bearing veins at Cobalt, Ontario. Geochemical aureoles are ordinarily represented by anomalously high concentrations of certain elements near the orebodies, but there are exceptions where a key element is significant because it decreases in amount toward ore. In geochemical aureoles associated with fluorite veins in Derbyshire, Durham, and Northumberland, England, the lead, zinc, and fluorine concentrations commonly increase toward the vein, whereas strontium shows a decrease toward the vein. The depletion in strontium is explained by a recrystallization of the limestone wall rock that caused liberation and removal of the strontium originally contained in calcite (Ineson, 1969).

Epigenetic aureoles are referred to as *wall-rock anomalies* if they occur in the host rock and as *leakage anomalies* if they extend upward into barren preore rock. Geologic factors that control the intensity of wall-rock anomalies and leakage anomalies are the same as those that control hydrothermal alteration: mobility of the participating elements, fracture patterns, and rock permeability and reactivity. Hydrothermal alteration and leakage anomalies in surface exposures were used by Morris and Lovering (1952) of the U.S. Geological Survey to locate blind lead-zinc orebodies through 60 m of overlying rock in the East Tintic district, Utah. Their work and the subsequent exploration by private companies are now considered to be an outstanding example of the successful use of integrated field and laboratory methods in exploration (Bush and Cook, 1960; Lovering and Morris, 1960).

The strength of a geochemical anomaly is sometimes expressed in terms of *geochemical relief*, the level to which the anomaly rises above the plateaulike threshold of local background values. Other factors to be evaluated when dealing with a geochemical anomaly are the topographic site of its occurrence and its immediate geologic association. A point to be noted here is that most soil anomalies will have been displaced from their parent bedrock mineralization by creep; actually, the only soil anomaly that would occur immediately above an orebody would be in residual soil, in flat terrain, and over a vertically dipping body. Hydromorphic anomalies result from the precipitation of new mineral material at places where ground water reaches the

Table 14-1. Examples, Indicator and Pathfinder Elements

Ore Association	Indicators	Pathfinders
Porphyry copper	Cu, Mo	Zn, Mn, Au, Rb, Re, Tl, Te
Sulfide ore complexes	Zn, Cu, Ag, Au	Hg, As, S (as SO_4), Sb, Se, Cd
Precious metal veins	Au, Ag	As, Sb, Te, Mn, Hg, I, F, Bi, Co
Skarn deposits	Mo, Zn, Cu	B
Uranium (sandstone)	U	Se, Mo, V, Rn, He
Uranium (vein)	U	Cu, Bi, As, Co, Mo, Ni
Ultramafic orebodies	Pt, Cr, Ni	Cu, Co, Pd
Fluorspar veins	F	Y, Zn, Rb, Hg

surface; hydromorphic anomalies are commonly associated with a seep at a break in slope or stream banks or a bog.

Two additional terms are basic to geochemical prospecting: *indicator element* and *pathfinder element*. An indicator is one of the major ore elements in the body being sought. A pathfinder is associated with the orebody, but it is more easily detected, freer from noise, or more widely dispersed than the indicator elements. The selection of a pathfinder element requires a conceptual model of the expected orebody; for example, arsenic would not be a strong pathfinder to every kind of copper deposit, but it would serve as a pathfinder to copper in massive sulfide deposits. Table 14-1 shows a few of the most common orebody, pathfinder, and indicator element associations.

14.4. METHODS AND APPLICATIONS

Table 14-2 lists the major methods of geochemical prospecting. The most popular reconnaissance method is stream-sediment surveying; for detailed investigations, the most popular method is soil sampling. Rock sampling is also widely used in the western United States and in similar areas where rock outcrops are abundant. Vapor, vegetation, and water sampling are used to advantage in special situations.

Rock Sampling

Rock sampling is the most flexible method insofar as the choice of a sampling site is concerned. It can be done in outcrops, in mines, and in drill core. The surface of the rock is cleaned by washing and scaling, and the sample chips are taken within standardized areas or at standard intervals. A 500-g sample is commonly taken in fine-grained rocks; up to 2 kg are taken in very coarse grained rocks. The sample is larger and the sampling is more time consuming than in sampling soil and stream sediments.

Table 14-2. Major Methods of Geochemical Prospecting

Sample Source	Causes of Anomalies	Case History Reference
Rock	Syngenetic concentration	Birch-Uchi Lake area, Ontario, Davenport and Nichol (1973)
	Wall-rock aureoles	Highland Valley, British Columbia, Olade and Fletcher (1975)
	"Leakage"	Johnson Camp, Arizona, Cooper and Huff (1951)
	Postmineralization dispersion	Troodos Complex, Cyprus, Govett and Pantazis (1971)
Soil	Residual accumulation	Rio Vivi, Puerto Rico Learned and Boissen (1973)
Glacial debris	Dispersion	Tverrefjellet, Norway Mehrtons, Tooms, and Troup (1973)
Stream sediments	Dispersion	Shawinigan, Quebec Felder (1974)
	Heavy mineral accumulation	Cheviot Hills, England Leake and Aucott (1973)
Lake sediments	Accumulation	Manitowadge-Sturgeon Lake, Ontario Cooker and Nichol (1975)
Surface water	Dispersion	Gerona Province, Spain Schwartz and Friedrich (1973)
Ground water	Dispersion	Pima district, Arizona Huff and Marranzino (1961)
Snow	Hydrochemical accumulation	Clyde Forks area, Ontario Jonasson and Allen (1973)
Vapor	Oxidation of ore	Notre Dame Bay, Newfoundland, Meyer and Peters (1973)
	Radioactive decay	Beaverlodge area, Saskatchewan, Dyck (1972)
Vegetation	Selective concentration	Barrytown, New Zealand Quin, Brooks, Boswell, and Painter (1974)
Sea water	Primary dispersion	Atlantis II Deep, Red Sea Tooms (1973)
Marine sediments	Secondary dispersion	Mount's Bay, Cornwall, England Tooms (1970)

Rock geochemistry has several advantages over other methods. The data can be directly related to primary aureoles in detailed sampling and to geochemical provinces in reconnaissance sampling. The geologic context of a rock sample is direct; structure, rock type, mineralization, and alteration can be noted when the sample is taken. Rock samples are not as likely to be contaminated by extraneous material as are soil and stream-sediment samples, and rock samples can be stored for later testing with less risk of their undergoing chemical change. Contamination is relative rather than absolute; even the most well cleaned outcrops have been leached and reconstituted to some extent.

An obvious disadvantage to rock sampling is rarity of sampling sites. Even where exposures are plentiful, they do not generally occur exactly where wanted. The scope of rock samples is relatively narrow, geochemical variations are generally weak, and the effect of inherent rock type is strong. A rock sample represents conditions only where the sample was taken. Stream-sediment samples, in contrast, represent conditions within an entire catchment area. A specific problem arises where visible mineralization is exposed. The rock sample will obviously be high grade and not necessarily representative of a wall-rock aureole. The usual solution is to take two samples, one in the mineralized zone to obtain data on metal ratios and one nearby in unmineralized rock. Prospect pits are often sampled for metal ratio data. In any event, descriptive notes and a field sketch must be made.

There is one additional disadvantage to rock sampling. The samples can be tested only in the laboratory. Soil, water, and stream-sediment samples, on the other hand, need not be crushed and therefore can be tested in the field by colorimetric methods so that the more obvious anomalies can be followed up immediately.

Soil Sampling

Soil sampling is advantageous in the many areas where outcrops are scarce. The target has had a chance to spread out in a "fan" from a smaller zone in the underlying rock. The principal disadvantages to soil sampling in comparison with rock sampling are that a high geochemical noise level and the effects of a complex history must be considered.

We consider soils (plural): residual and transported, mature and juvenile, zonal and azonal. There are in-between types of soil; and there is a matrix of controlling factors, such as topography, time, biologic activity, parent material, and climate. In Chapter 3 there was the question "Which surface environment? Modern? Pliocene?" Now, once more: "Which climate? Which topographic relief? Which parent material?" Conditions for development of lateritic soil may not exist now, but they may have existed when a certain soil profile was formed. A linear soil anomaly may not be associated with a vein at all, it may represent the truncated edge of an earlier metal-rich soil that has been tilted and incorporated into a new soil profile.

Figure 14-1 represents a typical profile in zonal soil and shows some of the profile variations that may occur in four climatic environments. The B horizon, the zone of mineral accumulation, is most commonly sampled in geochemical prospecting. Note that the B horizon is missing in some profiles; other horizons can be used if the same horizon is sampled at every site. In some areas a complete profile series of samples is taken at critical locations; the idea is to locate a characteristic near-the-orebody profile in which metal content increases or stays constant in going from the B to the C horizon. In samples taken at greater distance from the orebody the B horizon will ordinarily be richer.

Transported soils on desert pediments and in glaciated terrain are difficult media for geochemistry. Hydromorphic anomalies can reflect deeper conditions, and deep-rooted plants can add metals from the bedrock to the soil as organic material but, in general, there is a sharp geochemical discordance between a soil and a rock that is not its parent. Some of the first successful

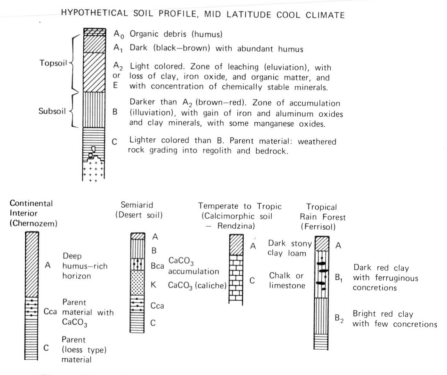

Figure 14-1.   Zonal soil profiles, showing principal horizons (layers).

applications of soil sampling to exploration were in the Canadian Appalachians where the effects of glaciation were light and the preglacial soil was not too disturbed. But when the same "New Brunswick style" prospecting was tried in the strongly glaciated, water-logged muskeg soils of the Canadian Shield, it failed. Then after a few years of experiment, study, and field trials, there came a new understanding of glacial geomorphology and geochemical associations in glaciated and permafrost terrain. Arctic and subarctic areas became suitable places, but not necessarily easy places, for geochemistry. Special considerations for soil geochemistry and other geochemical techniques in glaciated terrain have been reviewed in books edited by Kvalheim (1967), Jones (1973, 1975a), and Boyle, Shaw and Webber (1972). The subject of geochemical prospecting in glaciated terrain has been emphasized during recent years because of the large mineral-rich areas involved in Canada, northern Europe, and the Soviet Union.

Holes for soil sampling are dug by a mattock, post-hole digger, or mechanical trencher and a sample is taken, sieved to minus 80 mesh, and 20–50 g of the fine fraction are collected for analysis. Soil surveys are generally made on a grid pattern, with the sample locations 300–1,500 m apart in reconnaissance work and 15–60 m apart in follow-up surveys.

Stream-sediment Sampling

Stream sediments are a natural composite of all the material upstream from a sample site. They obtain metals by erosion of soil and rock and from inflows of ground water. The metals may be contained in grains of minerals, but they are more commonly held in soil particles or precipitated coatings on rock and mineral fragments. An anomalous drainage "train" may decay rapidly downstream or it may extend for 20 km as at La Caridad. Since many trains are persistent, stream-sediment sampling is effective in reconnaissance work where a single sample location may represent a very large catchment area. In some areas, one stream-sediment sample is taken for each 100 or 200 km² of terrain; more commonly a sample represents only a few square kilometers of terrain, with two to three sample sites per kilometer along a main stream and at points on tributaries just above their junction. In detailed stream-sediment surveys, samples may be taken every 50–100 m along the stream.

Stream-sediment samples are generally easier to collect and easier to process than soil samples. However, cultural noise is a particularly irritating problem because so much of humankind's trash ends up in streams and so much of the work of civilization is sited along streams. About 50 g of minus 80 mesh material are taken from near the middle of the stream, from the active or "live" stream bed, from pools, or from accumulations of fine material beneath boulders. Organic material likely to contain erratic metal concentrations is avoided. Stream banks are not sampled in reconnaissance, but they are often sampled in the follow-up stage to locate the source of an anomaly. Stream banks are in older, more complex, alluvial material and they are likely to have been affected by slumping and hydromorphic anomalies. It is often difficult to relate samples to specific former channels.

Stream sediments do not escape the "which stage" questions posed for outcropping orebodies and soils. Present-day divides between stream valleys may not correspond to those in the past. An anomaly may have to be traced first to a prior stream pattern, then to the source. A particularly interesting situation has been revealed by stream-sediment sampling in Cornwall, where strong tin anomalies occur in isolated positions as much as 10 km from the nearest primary mineralization. The reason: new tin anomalies were formed by streams reworking an earlier train of cassiterite made by longshore currents during Pliocene submergence (Dunlop and Meyer, 1973).

Where enough heavy resistate material can be found, stream sediments are sometimes panned and the concentrates collected for analysis. In still another stream-sediment sampling method, iron oxide and manganese oxide coatings are scraped from cobbles in the stream bed and analyzed for their content of scavenged metals.

Stream-sediment geochemistry is unsatisfactory in the disorganized drainage of glaciated terrain, but lake sediments have been found to provide a good representation of anomalous metal concentrations in their immediate catchment area. In the glaciated terrain of the Canadian Shield lake-sediment samples, at minus 80 mesh or sometimes finer, are taken at least 5 m from the shore in order to avoid vegetation or from a depth of 20 to 30 cm into the bottom material in order to avoid temporary or seasonal geochemical associations.

Water and Vegetation Sampling

Water sampling is one of the oldest methods in geochemistry, yet it is not widely practiced in exploration. Samples are easy to take, but they are likely to be unstable in storage for even a short time. Factors controlling the dissolved metal content in surface water—dilution, pH, temperature, organic complexes, and interfering elements—are difficult to evaluate, and the metal content is relatively low compared with that of stream and lake sediments. In contrast to surface water, sediments accumulate over a long period of time and thus provide a more representative sample of the drainage area. The season has a marked effect on interpretations made from samples of stream water because of the metal salts that are flushed into drainages by heavy rains after a dry season.

Nevertheless, stream-water samples are effective guides to fluorine mineralization (Schwartz and Friedrich, 1973), and stream-water sampling is useful in places where sediments are lacking. Stream waters have been used to advantage as indicators of uranium mineralization through their content of dissolved uranium complexes and radon. Lake-water samples and bog-water samples have been used as guides to underlying or adjacent mineralization. Snow and meltwater are also geochemical sampling media, especially in permafrost areas where natural processes are now known to aid mercury vapor and metal complexes to work up from underlying mineralization. A good review of Canadian and Russian work with snow in hydrogeochemistry has been written by Jonasson and Allan (1973), and a review of several geochemical techniques in permafrost terrain is provided by Horsnail and Fox (1974).

Ground water, generally of low pH and therefore a better carrier of metals than surface water, has been sampled in wells and springs and used as a guide to ore. Well sampling is an old technique, but it presents problems with contamination from the well installation. A potential use of alkaline ground water in the desert areas as a medium for porphyry copper exploration has been demonstrated by Huff (1970). Figure 14-2 shows an anomalous molybdenum area in an Arizona copper mining district where ground water has been sampled directly in wells and indirectly in deep-rooted vegetation.

Vegetation sampling is essentially a means of soil and ground-water sampling for chemical analysis. Roots of trees and bushes in desert areas can penetrate the ground-water level at depths of 50 m or more, and they can provide a sample of chemical conditions that otherwise would be completely masked by transported overburden. Elements that produce good results in biogeochemical surveys include copper, zinc, molybdenum, silver, uranium, gold, lead, and mercury. The idea is simple enough—plants extract elements from depth and transmit them to foliage. The interpretation is however more complex than for any method thus far discussed. Sampling is also a simple process—clipping twigs and leaves—but the choice of what to clip is very difficult.

The processes by which plants accept or reject certain elements differs from species to species. Many plants have an exclusion mechanism by which the intake of a certain toxic element is totally restricted, partially restricted, or held at a certain threshold level. Essential nutrient elements are so readily accepted by some plants that a high background concentration will exist almost everywhere, thus making it hard to recognize anomalous concentrations.

According to Brooks (1973), there are at least 20 categories of variables to be taken into account in biogeochemical prospecting. Some of the more important variables pertain to plant type, plant organ, age of the plant and of the organ, health of the plant, soil characteristics, and drainage. The need for an orientation survey before any biogeochemical sampling program and the need for uniformity in sampling are apparent.

Phreatophytes, deep-rooted plants that draw water from the zone of

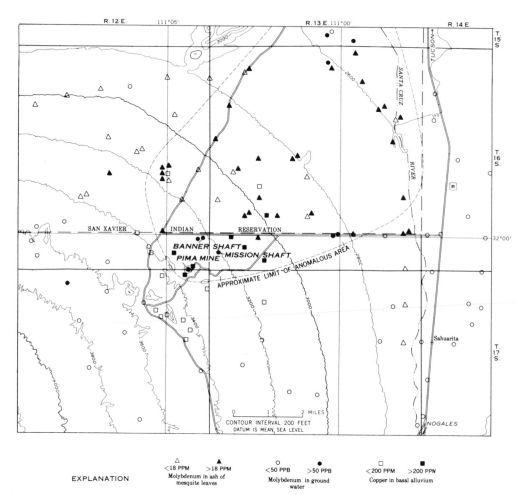

EXPLANATION

△	▲	○	●	□	■
<18 PPM	>18 PPM	<50 PPB	>50 PPB	<200 PPM	>200 PPM

Molybdenum in ash of mesquite leaves   Molybdenum in ground water   Copper in basal alluvium

Figure 14-2.   Geochemical map showing anomalies resulting from the dispersion of metals in ground water, Pima district, Arizona. (From L. C. Huff and A. P. Marranzino, *Geochemical Prospecting for Copper Deposits Hidden beneath Alluvium in the Pima District, Arizona.* U.S. Geological Survey Prof. Paper 424-B, p. 308, 1961. Used by courtesy of U.S. Geological Survey, U.S. Department of the Interior.)

saturation beneath the water table, are sampled in preference to xerophytes, the shallow-rooted plants that depend more directly on rainfall. In the southwestern United States, mesquite, a small tree that grows on desert pediments and sends a long taproot to a depth of 20 m and more, is often sampled in biogeochemical surveys.

Samples of approximately 100 g are taken from the same organ (e.g., leaves or young twigs) in each bush or tree and sent to the laboratory for ashing and analysis; the final ashed sample commonly amounts to 10–30 g. Ideally, vegetation should be sampled on a uniform grid, but some compromise must generally be made with a naturally irregular pattern of occurrence. Mesquite, for example, is often confined to intermittent stream courses across desert pediments, thus leaving broad areas devoid of anything to sample.

Vapor Sampling

Mercury-vapor sampling, used as a guide to sulfide orebodies since about 1950, has been emphasized more recently as a result of new and sensitive methods in mercury detection. Mercury vapor is collected from soil, atmosphere, and water. A portable backpack spectrometer described by Robbins (1973) can be used directly in the field; in the most common application of this equipment, gas is pumped from small-diameter drill holes in soil. The portable spectrometer has also been mounted in trucks, planes, and helicopters, with atmospheric air pumped from a collector to a filter and into the spectrometer. The processes involved in the release of mercury vapor to soil and air by oxidizing orebodies are reviewed by McCarthy (1972). These processes are now well enough understood to provide an increasingly valuable index in prospecting.

Other vapors can be sampled to detect halos of volatile elements and compounds around orebodies. Methods have been developed for the vapor associations shown in Table 14-3. The most effective samples are taken from soil, where gas concentrations are as much as a thousand times greater than in the atmosphere. Radon and helium collected in surface-water and groundwater samples have proved effective guides to uranium mineralization. A special method of gas detection in uranium prospecting makes use of the

Table 14-3.   Examples, Vapor Associations in Geochemical Exploration[a]

Mercury	Hg deposits, U deposits, sulfide assemblages
Helium	U deposits, sulfide assemblages, deep-seated fractures
Carbon dioxide	Hg deposits, deep-seated fractures, and (as $CO_2:O_2$) oxidizing sulfide ore
Radon	U deposits
Sulfur dioxide	sulfide ore
Hydrogen sulfide	sulfide ore
Fluorine, iodine	skarn, greisen, and porphyry copper deposits
Organometallic compounds of Hg, Cu, Pb, Zn, Ag, Ni	sulfide ore

[a] Data from Levinson (1974), Curtin, King, and Mosier (1974), and Ovchinnikov and others (1973).

radon cup, a receptacle containing a radiation-sensitive film that is placed in the soil. Atmospheric sampling, airborne geochemistry, is plagued by problems with wind, rain, vegetation screens, and man-made contamination, but it has been demonstrated as a way of detecting organometallic vapors and it has given significant results over known orebodies. Vegetation can be a source of useful vapor anomalies as well as a source of atmospheric sampling problems. As described by Curtin and others (1974), conifers exude metal-bearing aerosols in detectible quantities.

Most analytical work in vapor geochemistry is done in the laboratory or with portable spectrometers. One additional method for detecting gases from oxidizing sulfide mineralization deserves mention. Dogs have been trained to detect sulfur dioxide by their keen sense of smell. The use of "prospecting dogs" by the Swedish Geological Survey to locate mineralized glacial boulders beneath soil cover is described by Nilsson (1973).

14.5. FIELD AND LABORATORY ANALYTICAL METHODS

Field Tests

Some geochemical analyses are performed at the sampling site without any need for specific sample preparation. The most widely used field method of analysis is based on colorimetric tests, generally made with dithizone (diphenylthiocarbazone), a reagent that forms colored dithizonates with various metals. Separate tests can be performed for individual metals, with the color in solution compared with a group of color standards and translated into grams per ton or parts per million (ppm). A more commonly used test is less specific; this is the cold extractable total heavy metals (THM or cxHM) test for soil and sediment samples, first described by Harold Bloom, a geochemist at the Colorado School of Mines; it is also referred to as the Bloom test. The test, which takes less than a minute, is based on the reaction of dithizone at a particular pH with an entire group of heavy metals, including copper, lead, zinc, cobalt, and tin. The test is most sensitive to zinc, so that a small amount of zinc will often mask the reaction with other metals. Because only the loosely adsorbed metals are released into solution in the THM test, a small portion (5–20 percent) of the total heavy metals present is all that is determined. Even though there are many variables that govern the amount of metal available for cold extraction, the test provides a good way of identifying strong geochemical anomalies. A dithizone test for total heavy metals in water is often made at the same time as the THM test on stream sediments; the approach is well taken because metals may be adsorbed on mineral grains or dissolved in stream water in varying proportions.

On-site geochemical analysis for fluorine in natural waters, using specific ion electrodes in portable equipment that also measures pH and Eh, has been described by Schwartz and Friedrich (1973) with examples from a fluorspar exploration program in northern Spain.

A portable radioisotope-excited X-ray fluorescence spectrometer has been used to determine the metal content of ore-grade concentrations in the field, but the sensitivity is too low for direct use in most geochemical work. The main application for portable X-ray fluorescence analyzers is in the rapid assaying of metals (Ti, Cr, Mn, Fe, Ni, Cu, Zn, Pb, Sn, Mo, Ag, Zr, Co) in samples from orebodies. In Cornwall, some 60,000 m of drill core have been analyzed by portable X-ray fluorescence equipment with a precision and accuracy (against chemical analysis) of better than 1.5% Sn. In drill core at Endako, British Columbia, precision and accuracy on the order of 0.2% Mo have been reported. In situ measurements on the walls of mine workings in Cornwall indicate that zinc and tin can be determined with an accuracy approaching 0.3% metal (Gallagher, 1970). The use of a portable X-ray fluorescence analyzer for investigating mineralized outcrops and stream-sediment concentrates in the Illimaussaq intrusion, Greenland, is described by Kunzendorf (1973).

Nuclear irradiation is applied in a special way to the in situ analysis of beryllium in rock. The field instrument, the beryllometer, has a sensitivity of about 20 ppm for finely ground ore in laboratory tests, but it is much less sensitive on irregular rock surfaces. Another radioisotope method of in situ analysis is neutron activation. It is primarily a laboratory method, but field instruments have been developed and experiments with them indicate that 30 elements in soils and rocks may be detectable. Senftle and others (1971) have described the method and a technique of neutron activation in which nickel ore in simulated deposits has been detected and analyzed, indicating that the method can be used on rock exposures, in drill holes, and for in situ analysis of metals in marine manganese nodules.

Radiometric techniques for direct field analysis and detection have been classified as both geochemistry and geophysics. Gamma-ray spectrometer surveys, discussed in Chapter 13 as a type of airborne radiometrics, are most generally considered to be a part of geophysics, but they have been called "airborne geochemistry." Radon gas determinations made in soils by using a portable radon counter or a radiation-sensitive film are more often called geochemistry.

Laboratory Methods

Exploration geochemistry began in the 1930s with emission spectrography in the laboratory as the key analytical method. Emphasis then changed to field methods of colorimetric analysis. Today, emphasis has returned to the laboratory. New laboratory methods are much more sensitive than field methods and are relatively inexpensive. The quick turnaround time between field and laboratory made possible by modern transportation and communication is a major factor.

Preparation of material for laboratory analysis is relatively simple. Soil and stream-sediment samples are oven dried, sieved, ignited to remove organics, and digested by acid attack or by fusion with an alkaline or acid flux. Rock samples are pulverized, sieved, and digested by acid attack or fusion. Vegetation samples are ashed or digested by acid attack.

The most widely used method of analysis is atomic absorption spectrophotometry (AAS); it has an advantage over most other methods in its capability for determining about 40 elements with good accuracy and precision at a low level of detection. The atomic absorption method is also inexpensive and fast, and it is a relatively simple method to learn. According to Levinson (1974, p. 258), the atomic absorption method is used for about 70 percent of all laboratory determinations of geochemical samples in North America.

Emission spectrography is in substantial use, and it is widely used in the Soviet Union. It is a good method where large numbers of elements must be determined at a wide range in concentration and in various chemical associations. Emission spectrography is particularly well adapted to analyses for boron, beryllium, and molybdenum. A relatively new instrument, the plasma torch emission spectrometer, is so sensitive and economical that it promises to be the successor to atomic absorption spectrophotometry.

Laboratory colorimetric methods of analysis are similar to field colorimetric methods, but they benefit from further sample preparation and more carefully controlled conditions. Although less accurate than most other methods, they are still in relatively common use.

A few additional methods of laboratory geochemical analysis are preferred for certain conditions, but they are not nearly as popular as atomic absorption, colorimetry, and emission spectrography. X-ray fluorescence analysis (XRF) has been supplanted by atomic absorption except for those elements that are beyond the practical reach of atomic absorption and other methods; these elements include the rare earths, columbium, tantalum, tungsten, and zirconium. The portable field version of XRF equipment has been mentioned in relation to its special advantage in the analysis of ore-grade material.

Electron microprobe analysis, a technique in laboratory mineralogy associated with high-magnification microscopy, provides ultra-detailed geochemical data on the composition of ore minerals. The microprobe uses electron-excited X-ray fluorescence rather than X-ray–excited fluorescence. The electron beam can be focused in widths as small as 1 mu on a polished surface of ore so that the grain can be scanned in a series of parallel lines to give a two-dimensional picture of elemental distribution. An overall reference work in electron microprobe analysis is provided by Wenk (1976). An application of microprobe analysis to exploration geochemistry is described by Desbrough (1970) of the U.S. Geological Survey, who investigated micro-scale variations in the silver content of gold grains from placer occurrences to determine the relationship between low-silver rims and the unaltered gold-silver ratios in the interior. The metal ratio in the interior of placer gold grains can be used to discriminate between alternative lode sources for placer gold.

Paper chromatography, specific ion electrodes (for halogen analysis), neutron activation fluorometry (for uranium), and the classic laboratory wet and fire assay methods of laboratory analysis are discussed by Levinson (1974) and compared with each of the more widely used analytical methods, element-by-element and at various levels of concentration. Instrumental methods of analysis, including nuclear activation analysis, are reviewed by Wainerdi and Uken (1971).

Mobile laboratories in trailers, tents, or helicopter-lifted shelters were once simply colorimetric and wet analysis laboratories. Now, they are more likely to contain emission spectrographs, atomic absorption units, and radioisotope devices. Sometimes there are advantages to mobile laboratories despite the rapid service obtainable from the large commercial laboratories. Large laboratories are geared to a wide range in sample material and elemental analysis, not to specific situations at a mine or prospect that might require special treatment. Also, commercial laboratory facilities may be overtaxed during peak seasons or in areas of intense exploration activity. The gain of even a day or of a few hours between sampling and analytical results may be of utmost importance in fast-developing exploration projects.

14.6. FIELD NOTES IN EXPLORATION GEOCHEMISTRY

The collection of geochemical samples is a quick process, in fact, this is one of the principal reasons for geochemistry's importance in exploration. A large number of data points can be studied for patterns and trends. But samples taken without context can be as misleading as samples that are improperly taken or incorrectly analyzed. Even in the common situation where the sampler is not a trained geologist, the field notes are an important part of the sample. On-the-job training by geochemists or group training sessions are often provided for samplers so that field notes will be as uniform as possible.

In a soil survey the sampled horizon and its thickness, color, and texture should be noted. If there is evidence of slumping, organic material, transported soil, and rock debris or possible contamination, it should be noted as well. Field tests for heavy metals or elements should be recorded. Of course the sample site must be accurately located on a map or a photograph. The sample location should be marked at the site so that it can be located again (unless some uninvited observer is too interested in what is being done).

Stream-sediment samples require notes on the location of the sample relative to the active stream bed, stream size and volume, stream profile (steep or gentle), nature of nearby outcrops, presence of organic material, and

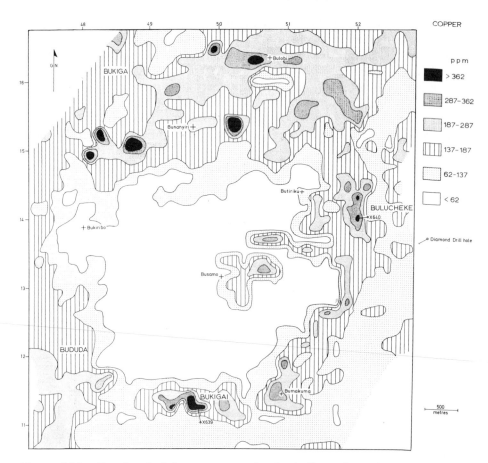

Figure 14-3. Geochemical data representation by isolines. Secondary dispersion of copper in minus 150-mesh fraction of soils over the Butiriku carbonatite complex, Uganda. (From J. H. Reedman, "Residual soil geochemistry in the discovery and evaluation of the Butiriku carbonatite, southeast Uganda," *Transactions,* v. 83, p. B8, 1974. Copyright © 1974, The Institution of Mining and Metallurgy. Used by permission of the publisher.)

possible sources of contamination. A sketch showing the relation of the sample site to stream tributaries is useful.

Rock samples have the most specific geologic context. Therefore, the geochemical sampling notes should include as much information on rock type, alteration, mineralization, and fracturing as time will allow. Time will probably not allow very much detail. To speed note taking, exploration groups often provide special checkmark-type data sheets. Some groups provide computer-compatible cards like those used in geologic mapping.

14.7. TREATMENT OF GEOCHEMICAL DATA

Geochemical data are essentially geologic data, and they must be used in relation to geology. If geophysical data are available, so much the better for interpretation.

Geochemical maps are most often point-plot maps, with the analytical results for individual elements or closely associated groups of elements plotted on separate transparent sheets keyed to a geologic map and topographic map. On some geochemical maps, the sample points with higher geochemical values are shown by circles of larger size; higher values in a stream survey are sometimes indicated by wider drainage lines. Contours of equal metal content, isograds (shown in Fig. 14-3), can be drawn where a suitable density of data points will permit. Ratios between key elements, such as Cu:Mo or Ag:Zn may also be plotted and contoured. Figure 14-4 shows the ratio of zinc to cadmium in soils in the Coeur d'Alene district, Idaho. This was some of the information used by Gott and Botbol (1974) to determine the amount of displacement on the Dobson Pass fault mentioned in Chapter 2.

Geochemical profiles are plotted as shown in Figure 14-5. Finally, maps and profiles are made on the basis of statistical analysis if enough data are

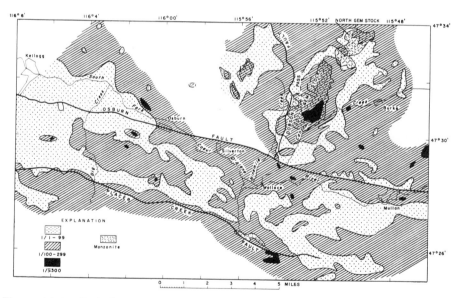

Figure 14-4.   Ratio Zn/Cd in soils of the Coeur d'Alene district, Idaho. (From G. B. Gott and J. M. Botbol, "Zoning of major and minor metals in the Coeur d'Alene mining district, Idaho, U.S.A.," in *Geochemical Exploration 1972,* M. J. Jones, ed., Fig. 11, p. 8, 1973. Copyright © 1973, The Institution of Mining and Metallurgy, London. Used by permission of the publisher.)

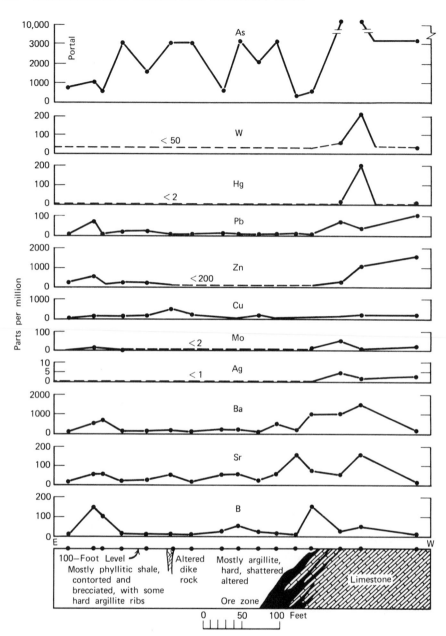

Figure 14-5. Geochemical profile of the 100-ft-level adit, north workings, Getchell mine, Nevada. (From R. L. Erickson and others, *Geochemical Exploration near the Getchell Mine, Humboldt County, Nevada,* Fig. 3, p. A8, 1964. Courtesy of U.S. Geological Survey, Department of the Interior.)

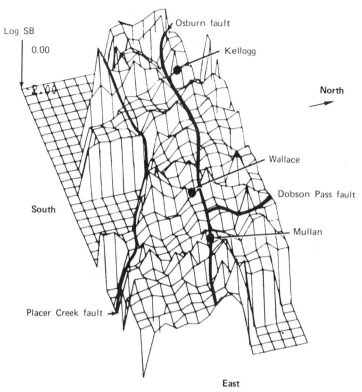

Figure 14-6. Computer-generated perspective diagram. Distribution of antimony in the Coeur d'Alene district, Idaho. The Dobson Pass fault, mentioned in Chapter 2, is shown. (From G. B. Gott and J. M. Botbol, "Zoning of major and minor metals in the Coeur d'Alene mining district, Idaho, U.S.A.," in *Geochemical Exploration 1972*, M. J. Jones, ed., Fig. 17, p. 10, 1973. Copyright © 1973, The Institution of Mining and Metallurgy, London. Used by permission of the publisher.)

available. Computer-generated perspective diagrams, as shown in Figure 14-6, and computer-generated anomaly maps are sometimes used for visualizing a geochemical situation.

Statistical Aspects

Geochemical data are taken in numbers, but the numbers are not information unless they have a geologic context. Even then, the information is only qualitative unless it has been given a statistical context. Geologic and statistical interpretation must be taken together; either approach is blind—or at least dangerously myopic—if taken alone. Ultimately all exploration data will be interpreted in terms of a potential orebody, as will be seen in Chapter 18. There are, however, a few statistical aspects to the interpretation of geochemical data that must be mentioned here.

Geochemical thresholds for rock type, locality, and region can be established by inspection (experience), by rule of thumb (laziness), or by a careful orientation survey. Threshold values are also often determined by a statistical analysis of the accumulated sample data.

In a simple geochemical situation with samples obtained from a single population, the threshold value for an element is often taken as its mean value plus two standard deviations. Two standard deviations from the mean enclose 95 percent of the data in a normal frequency distribution curve; 2.5 percent of the values will exceed the upper limit of this region, meaning that there is only 1 chance in 40 that a background value will rise above this level. Some geologists prefer to take three standard deviations as the threshold, thereby enclosing more than 99 percent of the values in the background population. Any value or group of values above this level would be even more clearly anomalous. In a great many investigated areas, the geochemical population within a rock type is lognormally distributed so that the frequency distribution will show a bell-shaped normal curve only if the elemental concentrations are plotted on a logarithmic scale. In other areas, the population is normally distributed and the elemental concentrations may be plotted on an arithmetic scale. In working with geochemical data, one of the first things to determine is which scale to use. A statistical technique, the chi-square test, can be used to evaluate the fit of any data against a lognormal curve or any specified curve. However, a quick test using arithmetic and geometric mean values can be made on geochemical data to determine if the distribution is more nearly normal or lognormal. The arithmetic mean is calculated by adding the values and dividing by the number of samples. The geometric mean of n numbers is the nth root of their product; the calculation can be made directly on some small electronic calculators or it can be made by adding the logs of the values, dividing by n, and obtaining the geometric mean as the antilog. If the median value and the geometric mean are in agreement, the distribution is lognormal; if the median value agrees with the arithmetic mean, the distribution is normal.

Suppose that data are taken from more than one population, say from two types of geochemical dispersion or from a background population and an anomalous population. The frequency distribution curve should then have two peaks—it should be bimodal. Each of the two major populations would have its own threshold or if there is a major population and an anomalous population, the single threshold should be apparent. However, there is often too much overlap or inequality between populations for the situation to appear so clearly in a bimodal frequency distribution curve. A clearer representation is sought by plotting the cumulative frequency rather than the frequency distribution of the values against the concentration. A normal distribution from a single population should plot as a nearly straight line on logarithmic probability paper. If there are two populations, a break or bend will appear where they join, as shown in Figure 14-7, which illustrates a stream-sediment survey in Guatemala. The bent line for zinc in Figure 14-7 shows an excess of low values in an otherwise lognormal distribution; the situation might have resulted from the inclusion of a lithologic unit especially low in zinc or it might have resulted from the inclusion of too coarse material in one set of samples. In any event the low values are not part of the main population. The bent line for copper shows an excess of high values in an essentially lognormal distribution; the threshold value might be taken at the break. If it were not for the abnormally high copper values (a more interesting situation than where a curve is bent by low values), the lognormal line xO would continued toward z rather than turning toward y. The molybdenum line shows first a positive and then a negative bend, suggesting that there are two distinct populations, corresponding to lines a and b. The elementary distributions a and b can be calculated to represent the sources of the compound molybdenum line. If one of the elementary distributions represents the

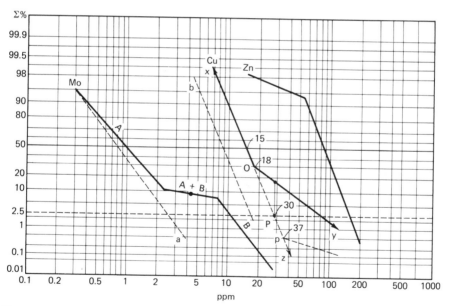

Figure 14-7. Cumulative frequency distribution of copper, zinc, and molybdenum in complex populations, stream sediment geochemistry, Guatemala. (From C. Lepeltier, "A simplified statistical treatment of geochemical data by graphical representation," Econ. Geology, v. 64, Fig. 5, p. 545, 1969. Published by Economic Geology Publishing Company. Used by permission of the publisher.)

anomalous population, the threshold is in the middle of segment $(A + B)$. Sinclair (1976) gives a full explanation of cumulative probability plots, such as that used in Figure 14-7; he also includes a clear explanation of several additional statistical methods used in exploration geochemistry where two or more populations are involved.

Statistical Interpretation

Statistical interpretation in geochemistry may deal with one value at a time for each specific location (*univariate analysis*) or with several values at once (*multivariate analysis*). The statistical principles are covered by Agterberg (1974) and Davis (1973). The role of statistical interpretation in geochemistry with examples of univariate and multivariate procedures are reviewed by Nichol (1973) in more detail than can be afforded here.

Univariate Analysis. Two of the most commonly used methods in univariate analysis are *trend surface analysis* and *moving average analysis*. Both methods, and in fact most statistical methods applied to large quantities of geochemical data, require computer processing. Trend-surface analysis involves the fitting of mathematical surfaces to actual data. The surfaces are expressed by polynomial models such as shown in Figure 14-8 or by Fourier models, which are expressions of periodic wave-type functions. The difference between the actual data and a computer trend surface is a *residual* (Fig. 14-9). A residual map isolates local deviations from a regional trend and, ultimately, anomalies from background. The higher the degree or order of a polynomial trend surface, the more complex it is, the closer it comes to fitting the data, and the fewer the residuals.

Moving average analysis shows trends in geochemical data by computing a mean value for the point values falling within a search area (or standard "window") that is moved from position to position across a map. The larger the window, the greater the smoothing effect. A smaller window has the same effect as a higher order polynomial surface or a more complex waveform in Fourier models—it reduces the number of residuals. Trend surface analysis and moving average analysis have separate advantages, but they will often point out similar trends. Nichol, Garrett, and Webb (1969) found that both methods of treating minor element data from rock, soil, and stream-sediment samples taken in a regional survey in Sierra Leone resulted in comparable patterns. Figure 14-10 shows two nickel patterns derived from the same sample data.

Multivariate Analysis. Multivariate methods of analysis include *discriminant, factor, cluster,* and *regression* analysis. Discriminant analysis provides a means of measuring the extent to which an unknown sample belongs to either

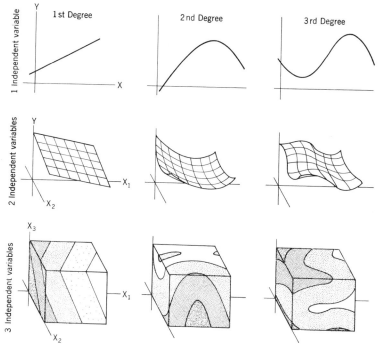

Figure 14-8. Typical trends from curvilinear regression, trend surface analysis, and hypersurface analysis, involving one, two, and three independent variables. First, second, and third degree trends are shown. (After J. W. Harbaugh, Kansas Geological Survey Bulletin 171, Fig. 4, p. 8, 1964. Courtesy of the Kansas Geological Survey.)

of two distinct populations, even if each population has a complex geochemical behavior. In this method, the existence of two populations must be known or at least suspected in advance; they might be two rock types or two drainage systems known to exist in the same area.

Factor analysis is a widely used method for identifying common underlying factors among a certain number of independent variables. From a data matrix containing information for rows of samples (entities) and columns of variables (attributes), correlation factors or principal components are extracted, characterized as data points, and expressed in terms of an encompassing envelope. The envelope, an ellipsoid around the swarm of points, has three perpendicular axes (eigenvectors), and the data can be re-expressed in terms of the axes. The largest variance (largest eigenvalue) is then apparent. The second and third largest variances are apparent in the remaining eigenvalues and eigenvectors. In essence, a large and complicated data matrix is reduced in this way to a model in which correlations can be seen and measured. If the

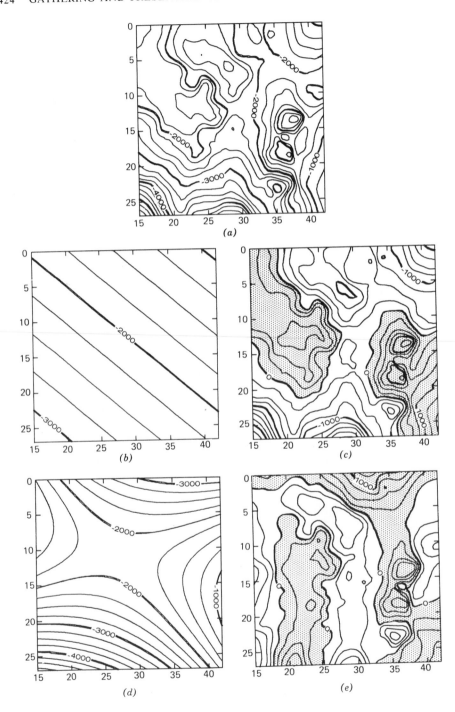

purpose of a factor analysis is to determine and evaluate correlations between samples, the analysis is referred to as *Q*-mode; if the purpose is to correlate variables, element concentrations, for example, the analysis is *R*-mode. The principles of factor analysis are described in condensed form for geologists by Klovan (1975). An example of factor analysis in exploration geochemistry, a study of stream-sediment samples from the Mount Nansen area, Yukon Territory, is provided by Saager and Sinclair (1974). In the Mount Nansen study, the geological significance in a large number of initial variables was found by reducing them to a three-factor *R*-mode model that indicated a correlation between one group of metals and local volcanic rocks and a correlation between another group of metals and porphyritic intrusive rocks.

Cluster analysis is a method of classifying multielement data into groups or subpopulations on the basis of interelement correlations. The procedure is to arrange the variables in a treelike hierarchical dendrogram in which the finest branches represent individual elements (*R*-mode), samples (*Q*-mode), or a sequence of samples (*P*-mode). The branches are linked into limbs and eventually into a statistically adequate trunk. An extensive literature is available on cluster analysis in geology. A basic explanation is given by McCammon (1968). A specific demonstration of cluster analysis used as an aid to the interpretation of multielement stream-sediment sample data is provided by Obial and James (1973). In this example, an *R*-mode analysis showed three main groupings of elements in samples taken from the Derbyshire lead-zinc mining field in England; one of the groups was related to clay minerals and resistates, another group was related to manganese and iron oxides, and a third group was related to the dominant type of ore mineralization. A dendrogram used in the Derbyshire study is shown in Figure 14-11. Culbert (1976) explains the use of *P*-mode analysis with an example of its application to copper exploration in British Columbia.

Multiple regression analysis, the statistical method used in trend surface analysis, can also be used to identify and measure relationships between individual elements in an area covered by a geochemical survey. This type of analysis provides a good index for recognizing pathfinder elements by their association with ore mineralization and for recognizing troublesome "noise" anomalies, such as those caused by scavenged copper or zinc on precipitated crusts of manganese or iron oxide. As with trend surface analysis, a

---

Figure 14-9. Examples of trend surfaces and residuals. (*a*) Isoline (contour) map of raw data. (*b*) First-degree trend surface. (*c*) Contoured residuals from first-degree trend. (*d*) Second-degree trend. (*e*) Contoured residuals from second-degree trend. Areas of positive residuals shown by shading. (Adapted from maps of elevation data by J. C. Davis, *Statistics and Data Analysis in Geology*, Fig. 6.14, pp. 336, 337, 1973. Copyright © 1973, John Wiley & Sons, Inc. Used by permission of the publisher.)

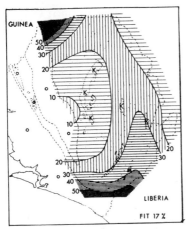

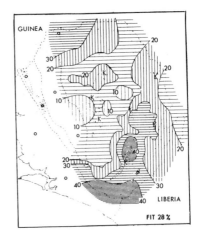

Stream sediment

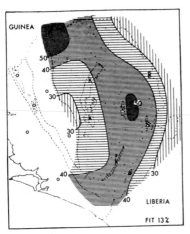

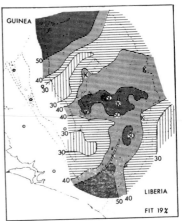

Soil

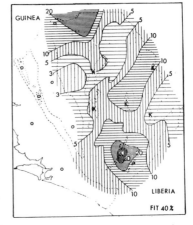

Rock

426

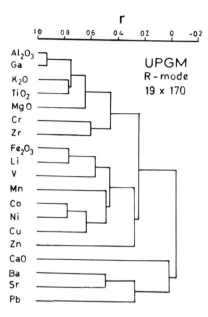

Figure 14-11.   R-mode cluster analysis by the unweighted pair-group average method (UPGM), from 170 samples collected in a stream sediment survey in Derbyshire, England. The 19 elements appear to be divided into a clay mineral-resistate group, a manganese-iron group, and an ore mineralization (lead) group. (R. C. Obial and C. H. James, "Use of cluster analysis in geochemical prospecting, with particular reference to Southern Derbyshire, England," Fig. 7, p. 246, in *Geochemical Exploration 1972*. Copyright 1973, The Institution of Mining and Metallurgy, London. Used by permission of the publisher.)

mathematical model is fitted to the data; in this approach a number of independent variables (geochemical values) are related to a chosen dependent variable (also a geochemical value). Some of the independent variables will then appear to be effective predictors of the dependent variable, according to the line or model best fitted to the data.

Geochemical work provides a part of the data needed for mineral discovery. The data may be interesting, but data generally have an equivocal pattern

Figure 14-10.   Nickel distribution patterns indicated by trend surface analysis (cubic or third-degree surface) and rolling mean analysis, basement complex of Sierra Leone. (After I. Nichol, R. G. Garrett, and J. C. Webb, Fig. 4, p. 207, and Fig. 7, p. 211, 1969, "The role of some statistical and mathematical methods in the interpretation of regional geochemical data," *Econ. Geology*, v. 64. Published by Economic Geology Publishing Company. Used by permission of the publisher.)

until a meaningful pattern can be developed by geological and statistical interpretation. The pattern and its residuals, bends, and correlations will be even more interesting, but our target—the hidden deposit—still has no real substance. A drill hole or a series of drill holes can provide the substance or can shatter the conceptual model that was so carefully assembled.

14.8. RECOMMENDED READING

A. Basic Books in Geochemistry

Fairbridge (1972). This encyclopedia of geochemistry contains information on the behavior of each element in the subsurface and surface environments; summary descriptions are also given in regard to geochemical cycles, soil chemistry, and chemical aspects of hydrology. Two other volumes of geochemistry interest in the same *Encyclopedia of Earth Sciences* series deal with geomorphology and with mineralogy and economic geology.

Perel'man (1967). This is a good background book on the behavior of minerals and elements in the surface environment. It is concerned with the geochemical basis for prospecting.

Wedepohl (1958—). This is an encyclopedia in loose-leaf form, with periodic revisions and additions. Each element is treated in a separate section, with its geochemical characteristics, mineral associations, and ore deposits.

B. Textbooks on Geochemical Prospecting

Levinson (1974). This is a thorough, practical, introductory text with information on the geochemical behavior of elements in primary and secondary halos and on field methods, exploration programs, and analytical methods.

Siegel (1974). This book is shorter than Levinson's, but it provides a good explanation of geochemical and geobotanical methods and interpretations in prospecting and environmental monitoring.

Granier (1973). In French, this condensed textbook covers basic principles and the methods and programs of "strategic" and "tactical" geochemical prospecting, with illustrations drawn principally from France and Africa.

Hawks and Webb (1962). This is still a good introductory text and a valuable reference book, even though the methods have advanced considerably since its publication.

Ginzburg (1960). This is the classic work on geochemical prospecting. It covers the transition from theoretical geochemistry to exploration geochemistry, and it records some of the pioneering work done in the USSR.

C. Textbooks on Geobotanical Prospecting

Brooks (1972). This is the only recent textbook specifically devoted to geobotany and biogeochemistry, and it contains one of the few available detailed lists of characteristic flora associated with mineral deposits.

D. Short Summary Works on Geochemical Prospecting

Bradshaw (1975)
Bradshaw, Clews, and Walker (1971)
Boyle and Garrett (1970)
Andrews-Jones (1968)

These are well-written overviews concerned principally with field techniques and immediate exploration objectives.

E. International Geochemical Symposia Proceedings

First—Ottawa. Cameron (1967)
Second—Denver. Canney (1969)
Third—Toronto. Boyle (1971)
Fourth—London. Jones (1973b)
Fifth—Vancouver. Elliott and Fletcher (1975)
Sixth—Sydney (in connection with the 25th International Geological Congress 1976)

F. Journals Dealing with Exploration Geochemistry

Journal of Geochemical Exploration: Amsterdam, Association of Exploration Geochemists.

Geochemistry International (USSR): Washington, American Geological Institute.

Some other journals with frequent articles on geochemical prospecting are *Geochimica et Cosmochimica Acta, Chemical Geology, Economic Geology, Geoexploration, Bulletin of Canadian Institute of Mining and Metallurgy, Transactions of the Institution of Mining and Metallurgy* (London), and *Mineralium Deposita.*

CHAPTER 15
DRILLING FOR GEOLOGIC INFORMATION

The Peerless Mining Company has obtained encouraging results in a drill hole on its Eagle Mountain property. A significant interval of mineralization was encountered, with an average assay value of 3.8% zinc and 6.1% lead. Other sections of the drill hole were reported to contain appreciable values in silver and copper. A second drill hole is in progress.

Such a quotation could be from any mining journal. It creates an image. Someone's ideas and someone's interpretations have identified a target; a decision has been made and a risk accepted. Now, the stream of drill cuttings changes color; the change is slight, but it is at the right depth. A small audience gathers for the opening of the core barrel: the Moment of Truth.

The image fits a real-life drama, but it shows only a small part of the relationship between geologists and exploration drilling. Drill holes (boreholes) enter an exploration program long before the Moment of Truth, and they stay until the last ore is mined—sometimes even longer. Drill-hole information can be related to stages in exploration as follows:

1. *Orientation.* Drilling may have already been done for minerals, oil, or water somewhere in the region. In an old mining district there may be some drill holes in or near the actual target area. If the core, cuttings, and logs are available, they provide background information for a new program.

431

2. *Reconnaissance.* Some drilling is often done for regional stratigraphic or lithologic information, especially in areas where stratabound orebodies are sought.

3. *Target Area Investigation.* Subsurface information is obtained on key structure and stratigraphy and zoning and for use as reference points in geophysical interpretation.

4. *Target Testing.* Drill holes show a presence or absence (or a tantalizing hint) of mineralization. With an encouraging discovery in one or more drill holes, the target becomes a prospect.

5. *Evaluation.* The mineralization is outlined and sampled to determine its size and grade and, ultimately, whether it can be called an orebody.

6. *Preproduction.* The prospect is now on the way to becoming a mine. Drilling is done for further delimitation of the orebody, detailed ore-reserve calculations, geotechnical and metallurgical investigations, and mine development guidelines.

7. *Mining.* Drilling continues, generally under the direction of mine engineers and resident geologists. The objectives are to block out additional ore reserves and to provide information for mine planning.

8. *Imminent Depletion.* Drilling takes on an air of desperation. If no new ore can be found in the immediate area, the mine is finished.

The first drill-hole information in a new area is likely to be from holes put down by someone else on the basis of prior objectives and concepts. Some of the holes may still be open for deepening or for geophysical logging, but this is rare. More often, the only information will be in the form of written geologic logs, assays, and geophysical logs. Sometimes the core will be available, and the obvious task will be to inspect it, verify it as being authentic, and relog it.

A great many successful exploration programs have begun with the relogging and reinterpretation of old core and drill cuttings. The discovery of the Kalamazoo orebody in Arizona, mentioned in Chapter 2 in the context of postore faulting, is an illustration. Using published information from drill holes at the San Manuel mine, Arizona, Lowell (1968) made a new interpretation of the movement on a large fault that had truncated the copper orebody. Briefly stated, interpretations held that a small portion of the orebody had been displaced upward and had been eroded away, but Lowell believed it had moved downward and was still nearby. He considered the attitude of the San Manuel orebody and its alteration pattern; the orebody had evidently been sliced in half rather than simply nipped on one end, therefore the displaced portion should be another large orebody rather than just a small segment. Seven holes had already been drilled in the most likely target area without finding ore-grade mineralization. What the drill holes had encountered was seen in a reexamination of the cuttings, and it sharpened Lowell's interest.

Four of the holes had gone from a zone of propylitic alteration into a zone of quartz-sericite alteration, and one hole had penetrated weak copper mineralization; all of this suggested that the drilling might have stopped just short of an orebody. The next drill hole was Lowell's; at 750 m it passed from weak pyritic mineralization into the Kalamazoo orebody, now known to be one of North America's major copper deposits.

When extensively explored prospects or large mines are being evaluated, there may be more than 50 km of drill core to examine within a few months, enough to call for temporary help from nearly every geologist in an entire organization. "Drop everything and go log core at Cerro Grande" is not an uncommon assignment where joint ventures between major companies are involved.

Reconnaissance drilling has been practiced for many years in the search for Mississippi Valley-type lead-zinc deposits. Regional changes in stratigraphy, unconformities, and regional reef structures are important enough to make an early view of the third dimension worthwhile. With increasing interest in stratabound orebodies and in areas covered by alluvium and glacial debris, reconnaissance drilling has become even more common.

Target area investigation, target testing, and prospect evaluation provide the themes for most exploration drilling and the theme for this chapter. The use of drill-hole data as control points for surface geophysics, mentioned in Chapter 13, will be carried a step further in this chapter, into drill-hole geophysics. Logging of drill holes for geotechnical information during pre-production work and mining will be covered here; ore sampling by drill hole and ore reserve calculations will be covered briefly in Chapter 16.

The final contribution of drilling in the life cycle of a mine is carried out in the hope that it will *not* be a final contribution. It is accompanied by as much drama as in target testing; more in some ways, because an anxious mining community may join the geologist in praying that the next drill hole will give it a new lease on life. The right drill hole can even make the geologist a local hero.

15.1. METHODS OF DRILLING AND RECOVERING SAMPLES

Among the available methods in exploration drilling, three are by far the most popular: diamond core drilling, rotary drilling, and percussion drilling. The principal features of these and several other methods are given in Table 15-1. Detailed descriptions of the methods are provided in a book by McGregor (1967).

Diamond Core Drilling

Diamond core drilling (Fig. 15-1) is the most versatile of all methods, and it is designed specifically for mineral exploration. *The Diamond Drill Handbook*

Table 15-1.   Exploration Drilling Methods and Normal Characteristics

	Diamond Core	Rotary	Continuous Coring	Downhole Rotary	Downhole Hammer	Percussion	Churn
Geologic information	good	poor	fair	poor	poor	poor	poor
Sample volume	small	large	small	large	large	small	large
Minimum hole diameter	30 mm	50 mm	120 mm	50 mm	100 mm	40 mm	130 mm
Depth limit	3000 m	3000 m	1000 m	3000 m	300 m	100 m	1500 m
Speed	low	variable	low	high	high	variable	low
Wall contamination	low	variable	variable	variable	variable	variable	variable
Penetration—broken or irregular ground	Poor	fair	fair	good	good	good	good
Site, surface and underground	S + U	S	S	S + U	S + U	S + U	S
Collar inclination, range from vertical and down	180°	30°	0°	30°	180°	180°	0°
Deflection capability	moderate	moderate	moderate	high	none	none	none
Deviation from course	high	high	high	little	little	high	little
Drilling medium, air or liquid	L	A + L	L	A + L	A	A + L	L
Cost per unit depth	high	low	moderate	high	low	low	high
Mobilization cost	low	low	variable	variable	low	low	variable
Site preparation cost	low	low	variable	variable	low	low	high

Figure 15-1.   Diamond drilling at a prospect in northern Manitoba. (Courtesy Hudson Bay Exploration and Development Co., Ltd.)

(Cumming and Wicklund, 1975), used through several editions and during several decades, is a standard reference to the basic techniques and applications. Diamond drilling is relatively expensive, but it can be done in most surface and underground locations and holes can be directed at any angle. It is the only method capable of providing a complete record of geologic structure and rock texture. It is also the only commonly used method that will deliver samples for geomechanics testing.

Even though diamond drilling has broad capabilities, it has limitations. Some kinds of broken and abrasive rock are nearly impossible to core at a reasonable cost. There are special methods for recovery of core in soft rock (protective sheaths and tubes), but recovery is generally poor in soft or thoroughly sheared zones. Careful use of drilling muds aids core recovery, but if recovery is still too low, it may be necessary to collect sludge (bit cuttings) as well as core. Sludge makes a less reliable sample than core, but it is often saved as a matter of insurance until the core is recovered.

In diamond drilling the sample is cut by a diamond-armored bit, recovered in the inner tube of the core barrel, and brought to the surface. In wire-line diamond drilling, the most widely used method, the inner tube is hoisted through the drill rods without removing the rods from the hole. The circulating medium may be diesel fuel where a delicate permafrost condition or water-soluble rocks are involved, but in most areas it is water with various combinations of mud and other additives, which lubricate the bit and core, stabilize and seal the hole walls, and carry the cuttings to the surface. Mud engineers are important consultants to drilling projects, and their work is much more sophisticated than the title indicates.

Although geologists would prefer to have the largest possible core for study, a small-diameter core is generally accepted since the cost of diamond coring increases with hole diameter as well as with depth. Also, the smaller the core diameter, the greater the attainable depth with small, less-expensive, portable drill rigs. In continental Europe, core size is expressed directly in millimeters; in English-speaking countries, the letter code designation below is more common.

Size Code	Core Diameter (mm)	Hole Diameter (mm)
XR	18.3	30
EX	21.4	36
EXT	23.8	36
AX	29.4	47
AXT	32.5	47
BX	42.1	59
NX	54.8	75

Larger core sizes are described in inches or millimeters without a letter code. The first letter in the code refers to the hole size, the second letter indicates the core barrel series. The third letter, T, indicates a thin-wall core barrel, lighter in weight and giving a slightly larger core. The X series of core barrels was standard for many years, and the terms EX, AX, and BX are still general labels for core sizes, but a W (wire-line) series of core barrels and rods is now more common. The wire-line hole diameter is the same as the corresponding X hole, but the core is slightly smaller because of the removable inner tube. BW core, for example, has a diameter of 33.3 mm rather than 42.1 mm. The use of a protective split tube inside the wire-line inner tube promotes good core recovery in coal, clay, and soft ground but reduces the core size by a small amount. Where casing must be inserted through bad ground, the hole can continue with the next smaller size diamond bit. With two strings of casing, an NX hole will be reduced to an AX hole unless the cost of reaming the hole to accept larger size casing can be justified.

Rotary Drilling

Rotary drilling is ordinarily faster and cheaper than diamond drilling. A rotary drill hole is good for geophysical logging, but it does not deliver a core sample except where a special bit is used in place of the standard roller-type rock bit or fishtail (drag bit), for soft rock, in short and very expensive core runs. Wire-line coring can be done with heavy oil field rotary equipment, but it cannot normally be done with the equipment designed for shallower and smaller diameter mining exploration holes. Instead of core, rotary drilling provides rock chips or cuttings of a few millimeters in diameter that come to the surface in a stream of air, foam, or mud. Rock type and mineralogy can be identified to a fair extent from the cuttings, but all of the larger textural and structural features are lost. Dry cuttings are preferred, but the depth of drilling with air circulation may be limited by water inflow and by the pressure and capacity of the air compressor. Wet zones can sometimes be drilled with air circulation by adding a foaming reagent. Dry cuttings are collected and separated from finer dust in a cyclone sample chamber. Wet cuttings are collected in a settling tank or in various kinds of cyclones, vibrating screens (shale shakers), or filters.

Rotary drill rigs are usually mounted on trucks. They are heavier than most diamond drill rigs, less flexible in where they can be placed, and limited in their ability to drill inclined holes. Small rotary bits can be handled by diamond drilling equipment, generally for going through overburden or weathered rock, but the equipment is not designed for extensive rotary drilling or for penetrating hard and abrasive rock. The relative advantages of large rotary rigs, diamond drill rigs, and combined drilling methods in deep exploration projects are discussed by Eyde (1974).

Large truck-mounted *continuous-coring drills* and dual-tube type drilling systems have some of the advantages of both diamond drilling and rotary drilling equipment. Still, most are very heavy and can only drill near-vertical holes. They are more expensive to operate than conventional rotary drills, but they deliver larger pieces of sample. Continuous-coring drills are designed to recover cylindrical pieces of core in a fluid stream rising in the center of a double-wall drill pipe—the reverse of the normal pumping circuit in drilling. A more common technique, however, is to use the reverse circulation stream to bring up large chips of rock broken from the bottom of the hole by a special rock bit. An envelope of near-static fluid around the outer drill pipe keeps the sample from being contaminated by wall-rock sloughing, and it protects the walls from contamination from material from elsewhere in the hole—an important consideration where radiometric logging must be done in a "clean" hole.

A *downhole rotary drill* uses a fluid-driven or air-driven motor located near the bottom of the drill string. Because the rods do not turn, they can be fitted with a deflecting "sub" which may be oriented in a chosen direction to make a hole deviate as much as 15 degrees per 100 m from its original path. As with other rotary drilling equipment, a diamond bit can be substituted for the rock bit in order to take core at critical locations.

Percussion Drilling

Percussion drilling is done with a sectional string of drill steel on a conventional blasthole drill or on a downhole hammer drill from surface locations (Figs. 15-2 and 15-3), or from underground stations. Conventional blasthole drilling equipment is relatively light, except for the air compressor that is needed in surface sites, and it is capable of drilling at nearly any angle. This popular method for outlining shallow orebodies and for probing out from mine workings provides quick, inexpensive samples of finely broken rock chips in a stream of air or water. With conventional percussion drills, the holes are small and the maximum hole depth is limited to about 100 m. The downhole hammer drill (Fig. 15-4) operates in the drill hole and transmits an impact directly to the bit without losing energy through the drill string. It can drill larger holes and penetrate to greater depths than a conventional percussion drill. In percussion drilling rock chips of about the same size as those from a rotary drill are blown out of the hole and collected in a cyclone sample chamber. Percussion drills are generally better for penetrating hard or abrasive ground than are diamond drills and rotary drills.

The churn drill is a venerable percussion machine that has served in mining as well as in oil and water drilling since drilling began. Even though it has

Figure 15-2. Percussion drill, crawler-mounted for surface sites. (Courtesy Ingersoll-Rand Company.)

given way to faster and cheaper methods in most exploration work, churn drilling still has some unique advantages. Given time, a churn drill can obtain sample fragments by chopping its way through almost any kind of ground—unconsolidated, heterogeneous, hard, or completely fractured. Casing can be driven into the hole a short distance behind the bit, so that walls in loose rock will not cave. Churn drilling, also called cable-tool drilling, is a long-standing method for sampling placer deposits and mine dumps, and it is still a principal

Figure 15-3.   Downhole hammer drill on a truck-mounted rig. (Courtesy Ingersoll-Rand Company.)

method of water-well drilling. The sample is collected in a bailer whenever enough broken rock accumulates at the bottom of the hole. Some churn drills designed specifically for placer exploration are light and compact; most, however, are cumbersome. One more point about churn drills: in whatever part of the world the need arises for an exploration drill hole, a churn driller and a machine can be found if water is being pumped.

Other Methods

Jet drilling is a placer sampling method in which a casing and chisel-pointed bit are advanced by percussion while water is forced ahead to loosen the

Figure 15-4.   Downhole hammer drill with a tungsten carbide "button" bit. (Courtesy Ingersoll-Rand Company.)

material and to flush out and bring the sample to the surface. Jet drilling is replacing churn drilling in many placer mining districts where minerals other than gold are involved.

Auger drills are of limited use, but they are important in soil sampling, beach placer sampling, and in evaluating clay deposits. Most are machine powered, but geochemical soil samples are often taken by muscle power. In any event, auger drilling stops at the first boulder.

One more drilling method uses muscle power; in fact, it is the epitome of labor-intensive drilling. The Banka drill or Empire drill is still used in some parts of the world for placer exploration. In this method a large-diameter casing is rotated by a man- or animal-powered sweep, while workers stand on an attached platform and remove material from inside the sinking casing by tools on long rods.

15.2. DRILL-HOLE PATTERNS AND SEQUENCES

The patterns and methods in an exploration drilling program depend primarily on the intended use of the data. In reconnaissance, drill holes are likely to be isolated and are often drilled to investigate a stratigraphic sequence or to probe beneath an unconformity or thrust fault. Topographic relief and accessibility may dictate the site and type of equipment. If information is needed on a certain geologic contact in gently dipping beds, the site may have to be the topographically lowest site available.

Even though reconnaissance drill holes are located for geologic orientation rather than for target investigation, they still are part of a sequence in drilling that may eventually make them part of a target-oriented pattern. In the Gas Hills uranium exploration program in Wyoming, mentioned in Chapter 10, Davis (1973) followed the sequence and pattern that he recommends for similar programs involving a search for sandstone-type orebodies. The first holes, reconnaissance holes, are drilled 10–15 km apart; they are located to penetrate formations of potential interest and to provide generalized structural data. If favorable lithology is found in a part of the region, additional holes are drilled at a spacing of 2–5 km. At this stage trends are sought in favorable host rock, alteration zones, and accompanying geophysical and geochemical patterns. "Fences," or lines of holes, are then drilled across possible ore trends, with the fence holes at a relatively close spacing (100–200 m) and with the fence lines at intervals of 1–2 km. Ultimately, when uranium orebodies are to be outlined, fences of drill holes may be at intervals of only 30 m, with holes 10 m apart.

A reconnaissance-to-detail case history in which drilling was used to find stratigraphic guides to sandstone-type uranium ore has been described by Bigotte and Molinas (1973). The French atomic energy commission (CEA) began subsurface investigations in the Arlit area, Niger, by drilling fence lines at intervals of 3.2 km with holes spaced at 800 m. On the basis of a paleogeographic map made from the reconnaissance drilling information and indications of uranium mineralization in three of the holes, CEA geologists narrowed the search to a target area. Further drilling was done on an oriented grid with holes concentrated in a smaller area and at decreasing intervals of 800, 400, 200, and 100 m. A more detailed paleogeographic map (1:1000) was then prepared and used to develop the emerging picture of a major orebody underlying 2.3 km^2 of barren rock. Detailed drilling within the orebody was eventually done at a grid spacing of 25 m.

A general practice in following the trends of geologic ore control or of geophysical or geochemical anomalies from reconnaissance drilling to target drilling is to place the first fence line across the trend, using inclined holes to intersect planar zones, such as veins, at a high angle. Vertical holes might miss a steeply dipping tabular orebody. A vertical hole might also follow a thin, steeply dipping vein without indicating its actual thickness. With some encouragement in the first fence line additional holes are drilled to determine the width of mineralization; these are offset holes, still in the fence line but drilled a short distance from the best of the original holes. The next fence line might be 100 m or a kilometer away at a distance indicated by the apparent geologic control or length of anomaly. The hole spacing and inclination are based on the width and attitude of the mineralization found in the first fence line. More fence lines are drilled at greater distance until the entire length of the favorable zone is delimited. Finally, short fence lines of holes are drilled between the best of the preliminary fences and more holes are drilled between the best of the holes in the fence lines until the target area can be confirmed or a zone of weak mineralization can be soundly damned for having caused so much trouble.

Target-area Drilling Patterns

A target area may be drilled in a continuation of reconnaissance drilling, as in the Gas Hills and Niger uranium examples, or it may be drilled as the direct test of a favorable geologic interpretation or strong geochemical or geophysical anomaly. In the direct test, a pattern for additional holes develops around the first few drill holes.

The pattern is likely to be simple at first. A certain orientation fits the geologic control or the trend of the anomaly, and an orebody of a certain size fits the exploration model. A common practice is to drill at a spacing that will allow two neighboring holes to penetrate the minimum-size ore zone. At this stage the pattern is flexible and drill sites are as often occupied on a basis of accessibility as they are on a basis of a uniform spacing. New trends and new models may develop as the holes are drilled, and the pattern is adjusted until a satisfactory coverage seems to have been made. This step-by-step approach is generally completed with the first dozen or so holes in a target area.

The preliminary pattern is abandoned in some projects before it serves its purpose. Some of the early holes may be so exciting that nearby offset holes are drilled immediately and boundaries of the target area are left open for later investigation or for the ultimate insult of having someone who has been attracted by all the activity find a better orebody.

In a well-planned and adequately financed program, drilling on a more systematic grid pattern comes next. A tight grid pattern in the area of greatest interest and a compatible but more widely spaced grid of drill holes in the remainder of the target area will afford adequate coverage without wasting money in needless drilling. Earlier drill holes and existing mine workings are incorporated into the new pattern insofar as possible, but the new grid is otherwise quite specific in that the locations are to be occupied as closely as a reasonable cost for site preparation will permit. A typical drill-hole grid incorporating earlier drill holes is shown in Figure 15-5.

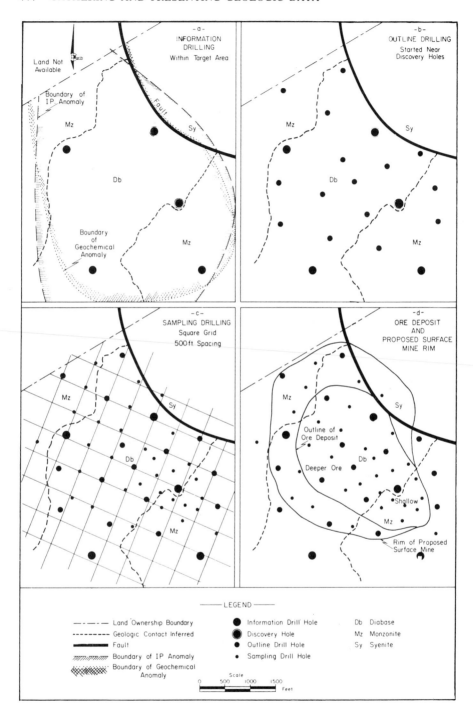

Statistical Patterns

Statistical approaches to widely spaced geological drilling and the more closely spaced grid drilling are well reported in the literature of operations research and geostatistics. The two-volume textbook on statistical analysis of geologic data by Koch and Link (1970–1971) contains several chapters devoted to grid systems in exploration, target investigation, and ore-reserve calculation.

Statistical patterns in exploration drilling have gained substantial acceptance among geologists since these techniques left the blind "punchboard" approach of a decade or two ago. In current practice a geologic model of the expected orebody is the dominant consideration. Other considerations include a justifiable expenditure, a stated probability that the orebody actually exists, and a stated probability that the orebody can be detected by available methods.

An increasingly used approach to drill-hole planning in exploration is to devise a mathematical simulation model that can be tested in a series of computer runs until an optimum pattern is found. In the simulation approach certain stochastic or natural variables—the number of orebodies and their size, shape, orientation, location, and value—are assumed. Deterministic or controllable variables—drill-hole locations, azimuths, inclinations, lengths, and costs—are then assigned. Alternative assumptions about the stochastic variables and alternative patterns for the controllable variables are considered in all of their most probable combinations until the most cost-effective layout is determined. Koch, Schuenemeyer, and Link (1974) describe the use of such a simulation model as a guide to drilling for concealed orebodies from deep mine workings in the Coeur d'Alene district, Idaho. Singer and Drew (1976) use the term "physically exhausted" for areas that have been drilled to the point where no room is left for simulated orebodies.

A statistical drilling grid or pattern is similar in one way to the data screens that a geologist uses in locating a target area. If the mesh is too coarse, most of the orebodies (the information) will fall through; if the mesh is too fine, the cost of a series of target-area investigations will exceed the value of the ultimate discovery. When used properly, drilling grids and data screens are used a step at a time rather than all at once and the results of each step are evaluated before going to the next step.

All statistically derived patterns for drilling employ a number of assump-

---

Figure 15-5. Successive steps in an exploration drilling program to detect, outline, and sample a disseminated copper-molybdenum sulfide deposit. (From P. A. Bailly, "Exploration methods and requirements," in *Surface Mining*, E. P. Pfleider, ed., Fig. 2.1–2, p. 31, 1968, by permission American Institute of Mining, Metallurgical, and Petroleum Engineers, Inc.)

tions. The range in target size, depth, orientation, and shape estimated by the geologist is assumed to be correct. The existence of one, two, or more targets is assumed. A certain concentration of geologic signals is assumed around each orebody, and the means of differentiating between message and noise is assumed.

Some simplifications, both helpful and dangerous, are also likely to be accepted for the sake of statistics. Drill holes are often planned to be "hits" or "misses" even though this is seldom the way it will happen; actually each drill hole will yield some information and some will show marginal mineralization. Simplified geologic patterns that lead to a statistically sound drill-hole grid may ignore some basic structural or lithologic control. For example, the spacing and orientation of the grid may coincide with a periodicity in folding, faulting, or sedimentation that places the orebodies exactly between drill holes. Statistical experts can design their analyses to cope with weak assumptions, oversimplifications, and subtle geologic controls, but they have no way of knowing what the problems are unless a geologist advises them.

It is easy to find flaws in any attempt to quantify and grid the elusive conditions of ore discovery. Most mining geologists find the flaws almost immediately. Still an attempt at quantification is generally made. Premature

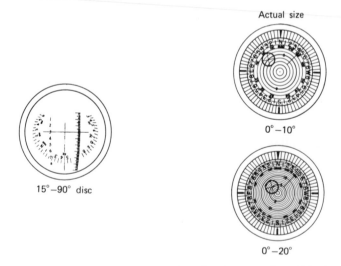

Figure 15-6. Instrument readings from a directional survey in a EW-size drill hole. For the 0°–10° and 0°–20° film discs, the inclination of the hole is 5° from the vertical; the magnetic direction of the hole is N. 45° E. For the 15°–90° disc, the inclination (30°) is read from the intersection of the horizontal line with the calibrated scale; the direction (N. 45° E.) is read from the intersection of the vertical line and the graduations at the lower edge of the disc. (Courtesy Eastman Whipstock Inc., Houston, Texas.)

abandonment of a project or the costly overdrilling of a poor target area is more likely if quantification is ignored.

Borehole Surveying

Quantification is wasted if the positions of the data points are not accurately known. A drill-hole collar can be easily located, but the location of a sample point at a depth of a few hundred meters requires the use of borehole surveying tools. Oil-well drilling uses special straight-hole monitoring equipment, but mineral exploration drill holes are seldom as carefully engineered, and the bottom of almost any mining drill hole is likely to be somewhere other than its intended location. Diamond, rotary, and percussion holes are notorious wanderers, and the amount of wandering is almost impossible to predict. Low-angle drill holes tend to droop; many drill holes tend to become oriented normal to bedding planes and to spiral toward the right; but two drill holes in two similar situations may take entirely different paths. Holes at depths near 1,000 m have been found to be 500 m off course. Even at depths of a few hundred meters, sample positions may be displaced so far from their supposed location that geologic projections are seriously affected. Borehole surveys indicate the direction and inclination of the hole in a series of instrument readings, as shown in Figure 15-6, or in a continuous record. The directional readings are taken by a tiny magnetic or gyroscopic compass.

Deflections: Multiple Hole Drilling

The inherent flexibility in a string of rotating drill rods is used to advantage where part of a hole must be deflected and redrilled to avoid drill steel that has twisted off deeper in the hole or to sample several places from a parent drill hole. The multiple drill-hole technique, originally developed as "whipstock" drilling in the petroleum industry, is most often used in diamond coring where a thin ore zone is overlain by a great thickness of barren rock, as in the Witwatersrand gold fields of South Africa and the Elliot Lake–Blind River uranium region of Canada. The drill bit is deflected by placing a thin wedge in the hole to obtain one or two degrees of change or by "arc cutting" with a multijointed drill rod to produce a uniform change of about 1 degree per 10 m advance. The branch holes are surveyed after being drilled. Where the additional sample intersections must be carefully placed, the wedges can be oriented and the new holes can be surveyed and realigned every few meters. Where a deflection must begin at more than 1 or 2 degrees, the new hole will generally have to be drilled at a smaller diameter so that smaller drill rods can flex in the parent hole. As an alternative to recovering core in deflected diamond drill holes, cuttings from multiple branch holes can be obtained by using a downhole rotary drill, which is easier to orient and can build up a larger deflection angle.

15.3. LOGGING OF DRILL-HOLE DATA

A part of the data comes to the surface as core and cuttings, part is obtained from instruments inside the hole, and part is read from the performance of the drilling machinery. All drill-hole data gathering is called logging.

Geologic Logging

In noncore drilling, the cuttings are separated at depth intervals of about 1 or 2 m, split into smaller samples, and bagged for shipment to the base camp or laboratory. A geologist will often inspect the sacks or piles of cuttings at the drill site. Where an important zone is expected, the geologist will take small grab samples almost continuously from the stream of drill cuttings and pan them for heavy minerals or look for index minerals. Drilling with air circulation brings cuttings to the surface almost immediately; when drilling with water, however, allowance is made for "sample lag," the time for cuttings to reach the drill collar. At a depth of 1,000 m, the lag might be as much as half an hour and by the time the cuttings are sampled the drill might have penetrated 3 or 4 m into a critical zone. Dry cuttings are relatively easy to sample; they are wetted and examined under a hand lens or a binocular microscope. Wet cuttings, somewhat more difficult to sample and often mixed with drilling mud, may have to be washed before examination. During the logging of drill cuttings in the laboratory or in camp, it is common practice to save some rock chips and a panned concentrate of heavy minerals from each sample interval and glue them to a "sludge board" or "chip board" for later inspection.

Core is often taken for assay, metallurgical testing, and geomechanics information as well as for geology. The core is broken into convenient lengths and divided or "split" lengthwise into two sections, with half for assaying and metallurgy and half for the geologic record. Unbroken pieces of core with lengths about $2\frac{1}{2}$ times the diameter are often saved for geomechanics testing. Because each geomechanics determination involves successive tests on the same material, 10 or 12 samples are required for each ore type and each important rock type. No breaking, splitting, or separating is done until geologic logging has been completed.

Geologic logging of core begins at the drill site, where decisions must be made regarding a large number of geologic and drill-hole engineering problems. Delay for a more thorough logging of core might result in expensive useless drilling beyond the zone of interest, premature abandonment of a hole, loss of sludge that should have been taken, or insertion of casing in a critical section of the hole that should have been left temporarily uncased for geophysical or directional surveys. Drill-site logging is not normally recorded; it is simply a quick examination of the wetted core with a hand lens or a binocular microscope.

In the field camp or laboratory, a new sequence of boxed core is first examined or "scouted" for major lithologic changes and structural zones, and it is then logged in detail. How much detail? As in geologic mapping, the potential information is almost without limit and time is a constraint. Unlike geologic mapping, some of the information—half of the core—is kept so that it can be reviewed for more detail. Logging is therefore most often done quickly and for the main objective of the moment. The main objective of the drilling is to find or to outline an orebody, not to log core. Structural information, such as fracture spacing, orientation, and filling material, must be recorded in any event because it will be lost when the core is split. Of course, *all* information will be lost if the core is thrown away or poorly stored. Adequate and long-term core storage requires a certain amount of time, room, and expense, but the value of the information is important.

Most organizations have a standard format for core logs. Two computer-oriented formats, COREMAP (Ekström, Wirstam, and Larsson, 1975) and GEOLOG (Blanchet and Goodwin, 1972) have come into use; with these much of the logging data and the drill-hole directional survey can be processed and machine-plotted on maps and sections. In most core logging the essential petrographic description includes the color, fabric, texture, diagnostic mineralogy, and rock name. Other essential data pertain to alteration, mineralization, and structural discontinuities. Percentage core recovery is noted, with special attention given to intervals of lost or badly broken core; these may represent possible fault zones and zones of mineralization. The logging of drill holes in placer deposits demands special consideration and a special logging format; Wells (1973) provides a thorough description of what is involved in this specialized field of investigation.

Geotechnics Logging

In addition to the data given in geologic logs, geotechnical logs require more detail on discontinuities and they may require an estimate of rock strength. The angle between a discontinuity and the axis of the core is measured in a special rock mechanics goniometer (Fig. 15-7) or more crudely with a simple clinometer made from a celluloid protractor. The absolute orientation of a series of discontinuities can sometimes be calculated from their attitudes in two or three nonparallel drill holes or by taking an oriented core from one drill hole; all of these techniques are explained and illustrated in a book on geological engineering by Goodman (1976). Discontinuities can sometimes be oriented from geophysical logs in a single hole, or they can be "seen" in an oriented borehole camera picture. The orientation of drill-core information is especially important in mine design; inasmuch as none of the existing techniques are entirely satisfactory, there is considerable emphasis on the development of better tools.

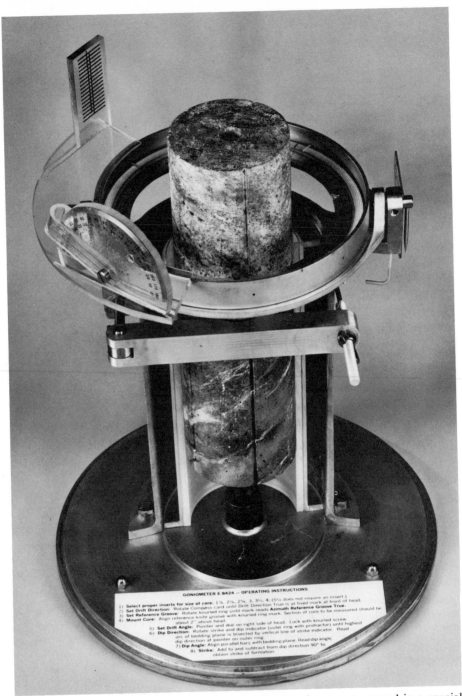

Figure 15-7. Goniometer for orienting planar features in core recovered in a special marking (grooving) core barrel. (Courtesy Christensen Diamond Products Co., Salt Lake City, Utah.)

The overall pattern of discontinuities and the location of broken zones are often recorded by taking color photographs of each box of core before the core is split. Mine development engineers are especially glad to have this record; the record from a hole that missed the orebody cannot be ignored: it deals with the wall rock.

Rock strength information is obtained from unsplit pieces of core. It is also expressed in core logs by rock quality designation (RQD) or impact hammer test indices mentioned in Chapter 5. In situ stress has been measured in experimental drill holes by stress relief and rock fracturing methods, but the techniques need further refinement.

Other data for geotechnical logs may be obtained from the driller's daily report (zones of water loss and gain, bit wear) and from a drilling rate or penetration rate recorder (hardness and "drillability"). Relationships between drilling parameters and mine design (things to keep in mind for the anticipated orebody) are given by Wilderman (1973). A format for geotechnical logs is shown in Figure 15-8; this log form is taken from a comprehensive report on the logging of rock cores for engineering purposes in the *Quarterly Journal of Engineering Geology*, 3, p. 1–24 (1970).

Special hydrologic tests can be made by keeping exploration drill holes open for use as observation wells or by pumping water into open sections of exploration drill holes that have been sealed off by "packers." The overall procedures are reviewed by Greenslade, Brittain, and Baski (1975), pumping test calculations are described by Dick (1975).

Caliper (hole-diameter) logs and certain geophysical logs are used in geotechnics as well as in geology. The geotechnics-geophysical logs most commonly involve nuclear, resistivity, self-potential, and sonic (seismic) measurements, all of which deal in some way with rock density and permeability.

Geophysical Logging

As indicated in Table 13-1, most surface geophysical methods have an equivalent drill-hole (borehole) method. The techniques of drill-hole geophysics did not however develop from the surface methods; they came from geophysical well logging, well-established techniques for stratigraphic correlation and subsurface mapping used in the petroleum industry. The mining geophysics textbooks and the serial literature mentioned in Chapter 13 contain basic information and case-history reports on drill-hole methods. However, much of the up-to-date literature on developments in drill-hole geophysics is located in petroleum-oriented serials, such as *The Log Analyst*, the journal of the Society of Professional Well Log Analysts. Transactions of the society's annual logging symposia provide state-of-the-art summaries that can be taken as forecasts of what will be tried with smaller tools in mining drill holes.

DRILLING METHOD		GROUND LEVEL	CO–ORDINATES	BOREHOLE NO.
Shell and auger to 4.80 m Rotary core drilling, water flush to 25.00 m		+43.63m O.D.	7268/5423	30

MACHINE	CORE BARREL AND BIT DESIGN	ORIENTATION	SITE
Pilcon '20' and B.B.S. 10, truck mounted	F design barrel, diamond bit	Vertical	CASTLECARY DEVELOPMENT 'C', GLASGOW

SOIL SAMPLES DEPTH AND TYPE	DRILLING AND CASING PROGRESS	WATER RECOV. (%) & A.M. LEVEL 20 40 60 80	R.Q.D. 20 40 60 80	CORE RECOV. % AND SIZE 20 40 60 80	DESCRIPTION OF STRATA	O.D. LEVEL	SYMBOLIC LOG
0.50–0.96 U(10)							
0.96 D					Stiff, becoming hard, brown silty CLAY with occasional cobbles and boulders (Till)		
2.00–2.46 U(10)							
2.46 D							
3.50–3.96 U(10)							
3.90 W		22 ▽					
3.96 D							
4.50 D					4.80	38.83	
PERMEABILITY cm/sec × 10⁻⁵ 10 20 30 40	22.3.67	23 ▼24 ▼		SF ⑨	Thick bedded pale grey and brown coarse strong SANDSTONE with fine pebbles and conglomerate bands. Steep clay lined joints 6.00 to 7.50 m. Dark brown fine con— glomerate 9.05 to 9.60m. Mudstone flakes at 10.15m. Very coarse at base (KIRKHILL SANDSTONE)		
13.2				HWF ⑫			
	22.3.67				11.00	32.63	
3.5				⑧	Thin bedded grey moderately weak MUDSTONE, sandy to 12.40m, with ironstone nodules throughout		
					14.00	29.63	
				⑮	Medium bedded grey fine strong SANDSTONE becoming laminated 17.50 to 18.70m		
30.2	23.3.67						
					(Borehole continued to 25.00m)		

KEY: U(10) – 0.1m dia. undisturbed sample
D – disturbed sample   Casing depth
W – water sample   Borehole depth
2 – day
▽ – ground–water depth first encountered
▼ – morning water level
② – rate of penetration (mm/min)

REMARKS

Borehole chiselled 1.05 to 1.90m, 4.0 to 4.45m.

LOGGED BY: M. Jones	SCALE 1/100		
BLOGGS BROS. INC.		CLIENT STRATHCLYDE CITY CORPORATION	REF. MJ/7964/30 FIG. 1

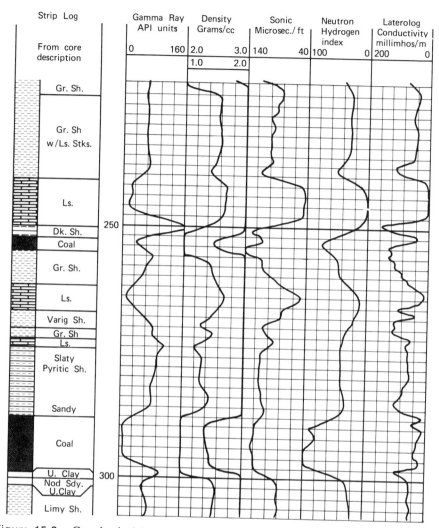

Figure 15-9.   Geophysical logs from a coal field exploration drill hole. (From J. C. Jenkins, "Practical applications of well logging to mine design," Fig. 2, p. 11, 1969. Preprint 69-F-73. Used by permission of the American Institute of Mining, Metallurgical and Petroleum Engineers, Inc.)

Figure 15-8.   Format for the geotechnical logging of rock core. [From Geological Society (London), "Working party report, the logging of rock cores for engineering purposes," Quart. Jour. Eng. Geology, v. 3, p. 16, 1970. Copyright © 1970, Geological Society, London. Used by permission of the publisher.]

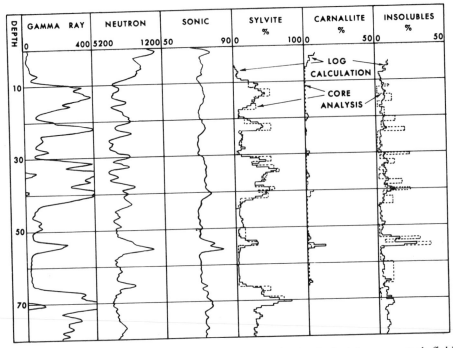

Figure 15-10. Geophysical logs from a drill hole in the Saskatchewan potash field. (From M. P. Tixier and R. P. Alger, "Log evaluation of nonmetallic mineral deposits," *Geophysics*, v. 35, no. 1, p. 137, 1970, by permission Society of Exploration Geophysicists.)

For a review of drill-hole geophysics techniques used in mineral exploration, a publication of the Geological Survey of Canada by Dyck (1975) is recommended reading. A bibliography of geophysical logging techniques for mineral deposit evaluation—composition and grade—is provided in a U.S. Bureau of Mines Information Circular by Scott and Tibbetts (1974); in this publication the techniques are classified by method and type of mineral deposit in which they are best applied. Case-history references are also given, and there are citations to foreign literature.

Drill-hole geophysical surveys are used in several ways. They provide indirect lithologic, stratigraphic, and structural information, indications of the mineralogy and grade of ore zones, index measurements for surface geophysical work, and, most importantly, a means of enlarging the radius of the slender cylinder of drill-hole data. A few techniques can be used to gain the ultimate benefit from drill-hole geophysics: to detect ore mineralization between drill holes and between points at depth and on the surface.

For the overall geophysical signature of a drill hole, nuclear, seismic, and electrical logs are often used as shown in Figure 15-9. The natural gamma-ray log shows a relatively high response from the dark shale zone and a low response from limestone and coal. In the density log (gamma-gamma log), which records the backscattered rays from a source of gamma radiation in the logging tool ("sonde"), the contrasts between relatively dense limestone, low-density coal, and shale can be seen. The sonic log or acoustic velocity (seismic) log shows a contrast between the more elastic limestone beds and the less elastic shale and coal. The neutron log uses a neutron-emitting source and a neutron detector in the logging tool; it shows differences in hydrogen density, as seen in the high "hydrogen index" response from the coal beds and the low response from the dense, relatively nonporous (low water content) limestone. Laterolog, the Schlumberger Well Services trade name for

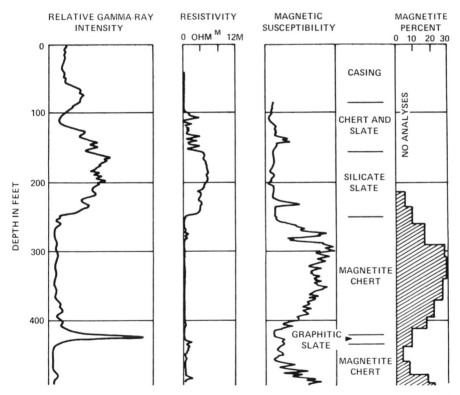

Figure 15-11.   Geophysical logs, showing correlation between magnetic susceptibility and magnetite content. (After C. J. Zablocki, "Some applications of geophysical logging methods in mineral exploration drill holes," 1966, SPWLA Seventh Annual Well Logging Trans., Paper U, courtesy Society of Professional Well Log Analysts.)

a focused resistivity log, shows the lower conductivity of limestone and coal relative to shale.

Figure 15-10, from the log of a drill hole in the Saskatchewan potash field, shows the response of sylvite (KCl), carnallite ($KMgCl_3 \cdot 6H_2O$), and insoluble (shale) zones in a thicker halite zone. The potassium-bearing intervals are apparent in the natural gamma-ray log. The hydrogen-rich carnallite and shale zones are apparent in the neutron log. The sonic log helps to differentiate between the shale zone and the more elastic saline beds.

The use of nuclear, electric, and magnetic logging in iron ore exploration is illustrated in Figure 15-11. The magnetite chert (taconite) is expressed by low radioactivity, low resistivity, and high magnetic susceptibility. The graphitic slate zone is identified by its high radioactivity. Note that the gamma-ray log

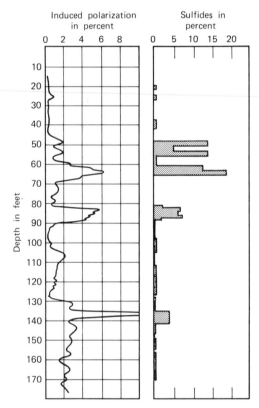

Figure 15-12. Induced-polarization log in sulfide-mineralized gabbro. (After Zablocki, 1966, SPWLA Seventh Annual Well Logging Trans., Paper U, courtesy Society of Professional Well Log Analysts.)

was taken in the cased as well as the uncased part of the hole. Most geophysical logs, other than nuclear logs, cannot be taken inside the steel casing. Where the walls of the hole are firm enough, a plastic casing is sometimes substituted but this does not permit the kinds of logging in which electrodes must touch the rock walls.

Induced-polarization logging (Fig. 15-12) is the most popular type of drill-hole geophysics in exploration for sulfide minerals. Induced-polarization and many other electrical and electromagnetic methods of drill-hole logging can provide information on the direction as well as the presence of nearby mineralization. In most of these techniques, electrodes or coils are placed at

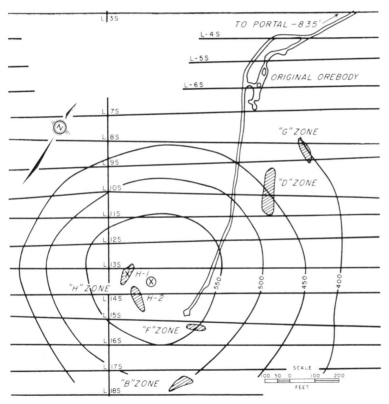

Figure 15-13.   Applied potential pattern from a current source in the H-1 mineralized zone, York Harbour, Newfoundland. (From W. H. Pelton and P. G. Hallof, "Applied potential me.hod in the search for massive sulfides at York Harbour, Newfoundland, " *Transactions,* v. 252, Fig. 6, p. 123, 1972. Copyright © 1972, American Institute of Mining, Metallurgical and Petroleum Engineers, Inc. Used by permission of the publisher.)

Table 15-2. Planning Factors in Exploration Drilling

A. *Preliminary Considerations*
  1. Objective: reconnaissance information (lithology, stratigraphy, structure), target area information (guides to ore), or target testing (discovery).
  2. Target geometry and depth, according to geologic, geophysical, and geochemical interpretations. Alternative exploration models. Key information needed for alternative interpretations.
  3. Budget limitations and maximum acceptable cost (current exploration limit).
  4. Site conditions: accessibility, seasonal limitations, facilities.
  5. Drill-hole engineering conditions: lithology, structure, existing mine workings, possible problems with sample recovery.
  6. Minimum acceptable information: major geologic contacts versus detail—in and out of the ore zone.
  7. Need for additional information.
      a. Borehole geophysics
      b. Borehole directional surveys
      c. Drilling time and caliper logs
      d. Geomechanics data
      e. Hydrologic data
      f. Metallurgical samples
      g. Geological laboratory samples
      h. Photographs of drill core.
B. *Program Design Considerations*
  1. Minimum core size and percent recovery or minimum sample volume for:
      a. Geologic information
      b. Geologic sample
      c. Representative grade sample
      d. Rock mechanics testing
      e. Metallurigical testing.
  2. Minimum hole size for:
      a. Borehole geophysics
      b. Hydrologic tests
      c. Casing and reducing in bad ground.
  3. Surface versus underground drill sites; potential value of an underground drilling base.
  4. Vertical versus inclined holes: distance to target, length of sample in ore zone, drill-hole engineering problems.
  5. Available drilling methods: compatiable combinations of methods (noncore drilling to top of critical zone), cost, contamination of sample or of walls.
  6. Multiple hole drilling from parent hole, by wedge, whipstock, arc-cutting tools, or dyna-drill.
  7. Casing program and hole abandonment: provisions for returning to deepen holes, special casing for geophysics, protection of ground water from contamination.
  8. Core and sample treatment and storage.

Table 15-2 (con'd).   Planning Factors in Exploration Drilling

   9. Contractor's quotations:
- a. Cost per unit depth for core and noncore drilling, with hole diameter and depth increments
- b. Core recovery provisions; cost of special core-handling or sample-handling methods and equipment
- c. Diamond loss
- d. Reaming
- e. Casing installation
- f. Cost of lost drill pipe and of casing lost or left in place
- g. Cement and mud
- h. Cementing; drilling cement
- i. Delay time and downhole survey time
- j. Water haul
- k. Mobilization
- l. Hole-to-hole moving time
- m. Site preparation and access costs
- n. Number of shifts per day or week.

C. *Program Evaluation During Drilling*
1. Changes in design with new information.
2. Completeness and decisiveness of the program: need for extended pattern, deepened holes, additional downhole information.

measured distances and directions on the surface as well as at measured depths in the drill hole. This directional capability is important in that a "near-miss" or an edge intercept of an orebody can be recognized.

In the applied potential, or *mise à la masse*, technique the range of drill-hole information can be extended by placing a current electrode in or near an orebody and mapping the response pattern on the surface or in another drill hole. Figure 15-13 shows a map taken from an exploration case history in Newfoundland where applied potential geophysics was used to outline a group of massive sulfide orebodies (Pelton and Hallof, 1972). The drill-hole electrode was placed in the H-1 mineralized zone at a depth of approximately 200 m. The potential pattern, measured in the surface traverse lines shown on the map, appeared to arise from a source at the circled "X" to the east of the electrode. A second survey with the electrode in the H-2 mineralized zone showed nearly the same pattern, indicating that the orebodies were joined electrically and the mineralization extended to the east. This interpretation was verified when mine workings were extended to the new orebody.

Hole-to-hole geophysical measurements are discussed by Scott and others (1975). In a field investigation of roll-front deposits in the South Texas uranium district, Scott and coworkers found that induced-polarization and resistivity measurements were anomalous between barren holes that straddled an ore zone. Log data from the individual drill holes did not show nearly so well which holes were near ore.

15.4. PLANNING EXPLORATION DRILL HOLES

Since exploration drilling is done according to geologists' needs, it is done according to their specifications and tempered by their budgetary constraints. Table 15-2 lists what is likely to be involved, but it does not do justice to the detailed planning, the uncomfortable geologic projections, and the sudden decisions that must be made. In spite of all planning, every drill hole is an adventure.

CHAPTER 16
SAMPLING OREBODIES AND ESTIMATING RESERVES

Assuming that our measurements and interpretations have led to a significant discovery of ore-grade mineralization, measurement and interpretation have a new purpose: to see if the technical success is an unqualified success.

Theories and practices in sampling and ore reserve estimation are so important to every phase of mining that they go far beyond the capability of this book. The summary given here can be extended into more detail by readings in two volumes, one on geological, mining, and metallurgical sampling published by the Institution of Mining and Metallurgy (1974) and one on ore reserve estimation and grade control published by the Canadian Institute of Mining and Metallurgy (1968). Textbooks by McKinstry (1948), Parks (1957), and Forrester (1946) describe basic principles of ore sampling and ore reserve calculation that continue to be valid even in today's atmosphere of instrumental ore analysis, geostatistics, and computers. Statistical approaches to ore reserve calculation are discussed in a two-volume textbook on statistical analysis of geologic data by Koch and Link (1970–1971).

16.1. SAMPLING SURFACE EXPOSURES, MINE WORKINGS, AND DRILL HOLES

Any sampling—even the most simple—brings with it a cascade of possible errors, some of which are related to the structure of the ore, and

461

some to its distribution and its texture, with still others resulting from the particular sampling technique used, from the way the technique is applied or from the sampling apparatus used (Gy, 1968, p. 5).[1]

An orebody will eventually become a stream of broken ore from which a metallurgist must recover the values that geologists and engineers said were there. As a consulting metallurgist, Pierre Gy saw broken ore and all the miscalculations associated with it coming into processing plants—and he provided the cascade simile for geologists to reflect upon.

All definitions of "sample," whether from statistics ". . . the part of a population used to estimate parameters . . ." or from mining textbooks, ". . . representative of an orebody . . .," have two stated or inferred elements: typicality and smallness. A collection of samples should be typical of an orebody; otherwise it will be a collection of specimens. Each sample is an infinitely small portion of the parent body; otherwise sampling would amount to mining—the very thing we are seeking to justify or reject by making an infinitely small expenditure.

The word "sample" ordinarily denotes something that has been physically removed from its natural location to be tested in the laboratory. The visual estimation of ore grade and ore boundaries, part of a geologist's everyday work, can be called sampling in a broad sense. In some mines geologists actually measure and record the width and mineral composition of orebodies at intervals between the physical sampling locations in order to complete the picture of ore grade and ore distribution. With practice the resident geologist becomes the resident "eyeball assayer."

Bias in Sampling

A collection of samples may be representative of the entire orebody or only of some accessible portion of the orebody. More often it will be the latter. Because we seldom have equal access to every geologic zone in an orebody, we cannot expect to take an unbiased set of samples of the entire deposit. Consider the zoned zinc-lead-copper orebody at Gilman, Colorado, and the zoned barite-lead-copper orebody at Walton, Nova Scotia, mentioned in Chapter 2. Representative samples of these orebodies could not have been taken from the shallower workings.

If sampling begins near the surface, it begins under a particular cloud of suspicion. The exposure represents the oxidized zone, not the orebody at depth; as discussed in Chapter 3, *something* has been leached, oxidized, or rearranged.

The effects of surface leaching and associated secondary enrichment are

---

[1] Pierre Gy, "Theory and Practice of Sampling Broken Ores," CIM Special Volume No. 9, Ore Reserve Estimation and Grade Control, p. 5, 1969. Copyright ©1968 by The Canadian Institute of Mining and Metallurgy. Reprinted by permission of the publisher.

most often identified with sulfide ore deposits, but the effects actually apply to nearly every kind of mineral deposit. Iron ores are made and modified in the zone of weathering, coal oxidizes, and phosphate rock is enriched. Even the most prosaic minerals and rocks are affected. Sand and gravel exposures may be winnowed, resorted, leached, and indurated. Limestone outcrops may lose objectionable impurities. These impurities can appear at depth to change the cement-grade or chemical-grade material into worthless rock.

Underground mine sampling sites have not been exposed to leaching and oxidation for such a long time, but they have been exposed to contamination, and the sites are not well distributed. Dust and mud on mine walls and in cracks contain an abnormally high percentage of brittle or soft minerals, including the ore minerals. In old mines sulfate efflorescence of copper, zinc, and other metals may coat the walls. The only accessible locations for sampling in many old mine workings are in relatively stable rock, not in the more intensely fractured—and more mineralized—stoped and caved areas. Pillars in stopes, convenient locations for taking samples, may have been left behind in an irregular pattern for good reason: they contain the lowest grade ore.

Drill-core samples would be relatively free from bias if recovery were 100 percent. At 95 percent recovery, however, the small amount of missing material may have been soft, friable, or soluble minerals (e.g., veinlets of molybdenite or of soft silver halides and sulfosalts) representing significant economic values. Cuttings from non core drilling and the sludge sometimes recovered in diamond core drilling purport to represent specific increments in depth. Or do they? The cuttings that arrive in a sample at the drill-hole collar are composed of minerals with a wide range in density, and therefore with a range in travel time from the bottom of the hole. Some of the material may have come from the walls of the drill hole; some may have been lost in fractures; some may have accumulated in open fissures and vugs until it was regurgitated into a later sample. Where drilling fluid is recycled, as in permafrost and desert areas, some of the finer mineral particles are recycled as well. In the most unreliable situation of all, a sample from an unsurveyed part of a deep drill hole may have been taken far from its assumed location.

Whatever the sampling site, if a biased or unrepresentative sample can be recognized, the source of the problem may also be apparent and the resulting data can be treated accordingly. Natural geologic bias can be taken into account. If the orebody is known to have a zonal pattern, samples from each zone can be considered as belonging to a subpopulation. If leaching or secondary enrichment is recognized in some samples, these too can be classed in a separate population. Obviously, the best way to deal with contamination in sampling is to avoid it. A contaminated surface at a sample site can be removed before the sample is taken by washing and chipping until a fresh surface is exposed.

One of the most difficult kinds of bias to avoid or to explain is that associated with human factors. Where samples are taken by more than one person or by more than one procedure the samples may have to be treated as being from more than one population. The "salting" of samples, incorporating too much high-grade or waste material, may occur when crushing equipment or sample sacks have not been cleaned. Due to fatigue or carelessness softer material is sometimes taken in greater amount than harder material. Salting may be the subconscious action of a sampler who hopes that the orebody is rich or one who is trying to compensate for what appears to be a nonrepresentative exposure. These are matters of psychology rather than geology. The intentional salting of ore samples and mine faces is a matter of criminology. There is an impressive history of frauds, ranging from the crude mixing of high-grade ore with lower grade material to such sophisticated practices as exploding gold dust into rock crevices or injecting silver chloride into sealed sample sacks.

Channel Samples

The basic and time-honored sampling method, but not the one in most common use, is to cut a linear channel or slot at a uniform width and depth of several centimeters. The tools are a hammer and moil (a pointed stub of drill steel) or a pneumatic hammer with a pointed or chisel bit. In some mines a hand-held diamond saw has been used to cut the sides of a uniform channel. While the sampler cuts a channel a second person collects the fragments in a clean box or sack or on a canvas sheet.

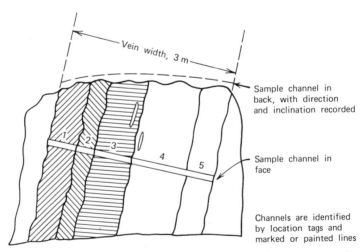

Figure 16-1. Sampling channel in a zoned vein exposure at a mine face. Five samples are taken, one from each distinct zone.

Even though a precise sample can be taken from a carefully cut and measured channel the procedure is more expensive and tedious than most situations warrant. A more common practice, therefore, is for a sampler to take chips from a chip channel (a somewhat broader band of about 0.5 m) with a field pick, collecting the fragments in a sack. Chip sampling is not a haphazard technique; it is done with as much attention to uniformity as is the deeper and more time-consuming channel sampling.

Channel and chip-channel mine samples, amounting to about 2 or 3 kg from each separate zone or from each meter of sample length, are taken from the back, face, or ribs of an underground working (Fig. 16-1). Samples are never (unless absolutely unavoidable) taken in the bottom of a working because of probable contamination. The channel is cut at a right angle to the ore zone, if possible. In any event the attitude of the ore zone and the location of the sample are recorded.

Provided sample recovery is acceptable, the most uniform channel sample is given by a drill hole. At both surface and underground exposures it is often possible to drill short holes with a light machine.

Drill Core and Cuttings

Drill core and drill cuttings furnish most of the samples used to evaluate prospects, and they furnish most of the samples for testing and extending ore reserves at operating mines. In addition, cuttings from blastholes are often sampled for grade control and detailed mine planning. At many large mines, several exploration drilling sites and dozens of blast holes are sampled on a daily basis and this work is commonly a responsibility of the geology department.

Since a major objective in sampling at prospects and mines is to obtain enough material for accurate ore-grade calculations and metallurgical testing as well as delineation of the orebody, A to N size core (30–55 mm diameter) are taken in preference to the less expensive E size core (21 mm). The larger core sizes are especially preferred in irregularly mineralized and low-grade deposits. After geologic logging has been done, the core is split lengthwise and half of the core is taken for assay. As mentioned in Chapter 14 in relation to field geochemical tests, X-ray fluorescence analyzers have been used to directly assay the entire drill core. At some mines a diamond saw is used to remove a thin slab from the core for the geologic record, thus allowing a larger portion of the core to be sent to the laboratory for assay. Where core recovery is poor, sludge is sampled and the sludge assays are combined with core assays, but with an air of suspicion if the combined weights add up to other than 100 percent of the weight of material that should have come from the hole.

Diamond drill coring and rotary noncore drilling are used to sample coal beds, with coring preferred for geotechnical data in the roof rock and information on crushed or friable zones in the seam, since these zones will be associated with a loss of coal in the washing plant. Coring is especially important where zones of oxidized coal must be outlined and impurities, such as sulfur, must be identified. In coal sampling large amounts of material are needed for testing; in some mines this has been accomplished by blasting a cavity at the end of a large-diameter drill hole and hydraulically removing the broken sample.

In sampling sandstone-type uranium deposits core holes are drilled only where geologic detail is needed. Less expensive noncore drill holes are often used for closely spaced sampling, since they are sufficient for radiometric logging and provide cuttings for assay.

Samples are collected from noncore drill holes at depth intervals of 1 m or more, depending on the variability of the ore. The advantages over core drilling in lower cost and sometimes in larger sample volume are offset in part by the weakened ability to recover geologic data, contamination from the hole walls, and the mixing of cuttings from adjoining sample intervals. Where air is used as a circulating medium mixed samples can be avoided to some extent by blowing the hole clear at the end of each sampled interval.

The accurate sampling of blastholes is difficult because these are busy sites and sampling is not their main purpose. Samples are generally collected in small scoopfuls from a pile of cuttings at the collar of the hole. A more thorough job is sometimes done by recovering all the dry cuttings in a cyclone collector or all the wet cuttings in special funnels or drill fittings. The entire volume from each selected interval can then be carefully mixed and sampled.

Placer Sampling

Sampling of placer deposits poses some unique problems because of the loose material involved. Sinking of shafts and trenching are preferred because a large sample can be recovered, but drilling is more common because it is cheaper, especially below the water table. In drilling placer deposits, casing is driven into the deposit as the bit advances so that gravel and sand from the hole walls will not contaminate the bottom sample. In churn drilling, the most common method for sampling alluvial gold placers, gravel is lifted by bailing; boulders can be penetrated, although large boulders must first be broken by repeated chopping or sometimes by explosives. In jet drilling and rotary drilling, more common in nongold placers, sand is flushed out of the hole. In auger drilling and clamshell drilling, sand is lifted from inside a casing and collected at specified depth intervals. Since the value of a placer deposit depends on the recoverability of the minerals by gravity methods, the sample is ordinarily processed in a sluice box or rocker and panned for a preliminary estimate of the grade. A major part of the sample is then concentrated by scaled-down production techniques. For further information on drilling and sampling placer deposits, the book by Wells (1973), the circular by Romanowitz, Bennett, and Dare (1970), and the proceedings of a conference on alluvial and open-pit mining published by the Institution of Mining and Metallurgy (1965) are recommended.

Bulk Sampling

Bulk samples, which may consist of large-diameter drill core, the contents of a trench or mine working, or entire train load shipments, are needed at some time in the evaluation of most large prospects. Trench samples are taken as though they were giant channel samples, with attention to the width and depth of the trench or the mine working and with separate samples taken from each identifiable zone. To evaluate an Idaho phosphate rock deposit, the entire phosphatic section, 60 m thick, was exposed in a long bulldozer trench and cleaned to bedrock, a narrow slot was then excavated in the bottom of the trench, and each individual bed was sampled separately. Enough material was taken from each bed so that 50 kg were available for metallurgical testing.

Broken ore in cars or on conveyor belts is commonly sampled by mechanical devices. Mechanical sampling may, of course, introduce systematic errors in place of human errors, but systematic errors are generally easier to identify and correct. The sampling of broken ore, an important subject with special problems, is described by Gy (1968), the metallurgist whose comments on cascading errors were previously cited.

Taking samples for pilot plant metallurgical testing may be a veritable mining operation in itself. At the Carol iron mining complex, Labrador, the pilot plant treated 75,000 tons of ore. Even where ore characteristics are well-known, bulk sampling may be needed. Evaluating and planning for the Berkeley pit in the midst of the intensely investigated area at Butte required the extraction and testing of more than 20,000 tons of rock (Waterman and Hazen, 1968).

Sampling Records

Except where mining follows immediately and destroys the sites, sample locations and drill-hole collars are marked so that they can be revisited. A record is kept by the sampler, generally in a book that has numbered tear-off tags to place in plastic envelopes or tubes and include in the sample container. The tag generally shows nothing more than the sample number, the date, the sampler's name, and instructions: "assay for Cu, Mo, Au, Ag." The sampler's notebook (Fig. 16-2), usually a standardized form, includes data on the exact location, the relation of the sample to the ore pattern, and where possible, comments on the rock type, weathering, and alteration. A sketch of the site is added if several zones have been sampled.

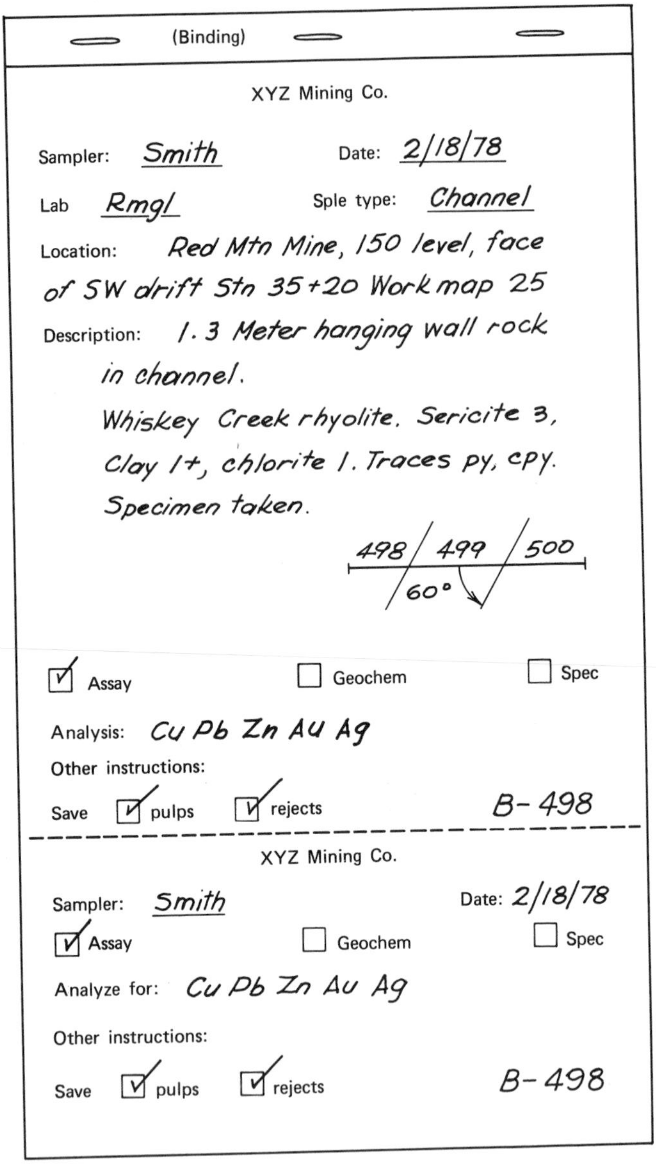
(Binding)

XYZ Mining Co.

Sampler: Smith Date: 2/18/78

Lab Rmgl Sple type: Channel

Location: Red Mtn Mine, 150 level, face of SW drift Stn 35+20 Workmap 25

Description: 1.3 Meter hanging wall rock in channel.
Whiskey Creek rhyolite. Sericite 3, Clay 1+, chlorite 1. Traces py, cpy. Specimen taken.

498 499 500
60°

☑ Assay ☐ Geochem ☐ Spec

Analysis: Cu Pb Zn Au Ag

Other instructions:

Save ☑ pulps ☑ rejects B-498

XYZ Mining Co.

Sampler: Smith Date: 2/18/78

☑ Assay ☐ Geochem ☐ Spec

Analyze for: Cu Pb Zn Au Ag

Other instructions:

Save ☑ pulps ☑ rejects B-498

Figure 16-2. Representative page from a sampler's notebook.

16.2. SAMPLING PATTERN AND SPACING

The pattern of sampling, like the pattern of target-area drilling, is a function of the sampling method, the accessibility of the site, the geologic situation, the objectives of the project, and the requirements for statistical analysis.

Uniform grid sampling is preferred for deposits of any appreciable size so that an optimal statistical coverage can be obtained. In practice the final pattern is generally a compromise between what is preferable and what is convenient or economical. Where sampling is done by drilling, for example, access to a particular site may be difficult but once the site is prepared it can be used for a fan of holes into other parts of the orebody, as shown in Figure 16-3. Branch holes can also be drilled from a parent drill hole, as explained in Chapter 15.

To sample relatively uniform, large copper deposits drill holes are commonly spaced 50–100 m apart. In the porphyry copper orebody at Bingham Canyon, vertical diamond drill holes were first spaced on a square grid 250 m apart and fill-in holes were then drilled at a 125-m spacing. Closely spaced blastholes are sampled at most porphyry copper mines for grade control during mining.

In irregularly mineralized deposits preliminary, or "scout," samples are often taken at intervals of 20 m or more with subsequent fill-in sampling at closer intervals—down to a spacing of 2 or 3 m if necessary. Uranium-bearing

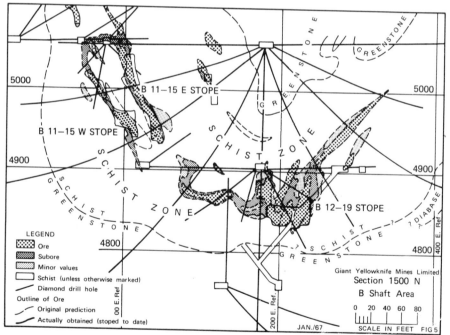

Figure 16-3. Fan drilling from a single site in mine workings, Giant Yellowknife mine, N.W.T. (From A. S. Dadson and D. J. Emery, "Ore estimation and grade control at the Giant Yellowknife mine," in *Ore Reserve Estimation and Grade Control*, Fig. 5, p. 221, 1968, by permission Canadian Institute of Mining and Metallurgy.)

conglomerate beds in the Elliot Lake district, Ontario, were first sampled by drilling from the surface on 300-m centers; one or two branch holes were deflected from each parent drill hole. Later chip-channel samples were taken from underground mine exposures at approximately 3-m intervals and sample holes were percussion drilled into the back at 6-m intervals. Where the ore zones were not yet exposed they were sampled from nearby workings by drill holes at 6- to 20-m intervals (Hart and Sprague, 1968).

The objective in optimizing a sampling pattern is to provide the exact number of samples needed to represent the grade and dimensions of an orebody; no more and no less. Optimization would be relatively easy if the zone of influence intended for each sample in ore-reserve calculations could be demonstrated to be the actual zone of influence in nature. Zones of influence are often related to adjacent samples on a geometric basis; yet if two adjacent samples cannot be correlated at some acceptable confidence level, neither one can really be expected to have an actual and measurable influence in the zone between them. They might not even belong to the same orebody. Where adjacent samples show a strong correlation, sampling is adequate, a zone of influence can be assigned, and further sampling would be a waste of effort and money. Some approaches to finding actual zones of influence around sample locations use correlation coefficients (Koch and Link, 1970–1971, p. 15–18), the mean square successive difference test (Hazen, 1968), and the variogram function (Blais and Carlier, 1968).

The use of correlation coefficients to control the number of sample locations in a large disseminated orebody is shown in Figure 16-4. The statistical techniques involved are explained by Kuhn and Graham (1972). In this example geologic information and experience were used to design a drilling program on a 122-m (400-ft) square grid with a hole in the center of each square. Grade estimates were needed at a precision of 0.080% Cu. An initial series of holes (1–9) was drilled on 244-m centers, twice the designed grid size. The assays were composited to standard intervals in each hole, and a linear correlation between the data was sought along diagonals in statistical runs 1 to 8. No significant relationship was found, so a second series of holes, A–D, was drilled. Before going to the next series of holes, a–l, correlations were sought in runs 1–24. In run 11, a suitable correlation was found between holes 4 and 5; an acceptable average confidence interval of ±0.08 percent was also calculated for the two holes. On this basis, hole f was cancelled; after drilling the rest of the holes in the series new statistical runs were used to test the need for the holes in the center of each 122-m square. Diagonal runs indicated significant correlations between holes B and 2 and between d and b; hole 102 was cancelled. Another run between holes 5 and D showed that hole 110 was not needed. If further evaluation were to have been considered on the basis of information from the completed pattern, the statistical process could have been carried down to a 61-m grid, scheduling only those holes actually needed.

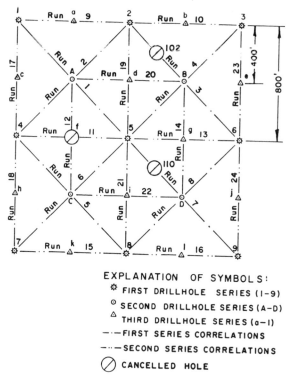

Figure 16-4.   Drill-hole pattern for large disseminated orebodies, based on a method developed at the Christmas mine, Arizona. (From J. G. Kuhn and J. D. Graham, "Application of correlation analysis to drilling programs: a case study—technical notes," *Transactions*, v. 252, no. 2, Fig. 1, p. 142. Copyright © 1972, American Institute of Mining Metallurgical and Petroleum Engineers, Inc. Used by permission of the publisher.)

In classical statistics directional trends are determined and evaluated but the variables have no inherent spatial aspect. The variogram function, part of the *geostatistics* approach to ore estimation developed by Matheron (1963, 1971), is different; it deals with *regionalized variables* that have specific distance and directional characteristics. One premise in geostatistics is easy to see: accurate ore-reserve calculations depend on where samples are taken as well as on how many samples are taken. Another premise, that classical statistics and geologic reasoning cannot provide an equally valid spatial context, is challenged by many statisticians.

Figure 16-5 illustrates one of the basic ideas in geostatistics. Case I and Case II have the same range in values, but in Case I the values are highly erratic and in Case II they are symmetrically distributed around a high-grade center. In terms of classical statistics, the two populations would be identi-

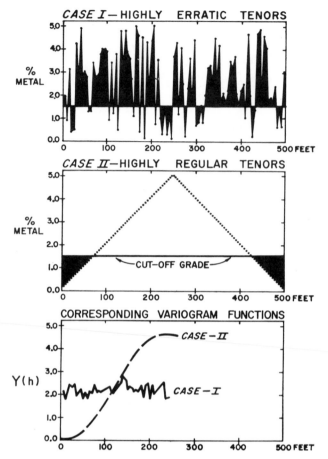

Figure 16-5. Silhouette diagrams of two vastly different populations of identical assay values, with corresponding variogram functions. (From R. A. Blais and P. A. Carlier, "Applications of geostatistics in ore evaluation," CIM special volume No. 9, *Ore Reserve Estimation and Grade Control,* Fig. 2, p. 44, 1968. Used by permission of the Canadian Institute of Mining and Metallurgy.)

cal—they have the same average grade and the same standard error of the mean and would give the same frequency histogram. But the two cases would certainly not be the same from a mine operator's viewpoint. The two variogram functions in the lowest panel of Figure 16-5 show the difference; in Case I there is no specific change with distance, but in Case II there is a regular rate of change. Case I would need a large number of drill holes for adequate sampling, regardless of the statistical treatment; Case II, which uses the geostatistics approach, would need only three. Without considering the equation for a variogram (there are several kinds) it can be seen that the variogram function shown in Figure 16-6 changes in a specific direction as more sample values are added until at a certain distance, called the *range* of the variogram, the influence of the closer samples is lost and the plot approaches an independent (random) behavior. The optimum spacing between drill holes or sample locations in this particular direction would be indicated by the range of the variogram. Samples taken at a greater distance would miss the significant correlation; samples taken at a closer distance would show the correlation but would be unnecessarily close. In actual practice, variograms are much less regular than the one shown, the range of the variogram is not so distinct, and an optimum sampling interval is taken as being slightly closer than the range. Of course some trial sampling must have been done within the range of the variogram so that a variogram can be calculated in the first place. A description of sampling, geostatistical study, and sequential drilling for ore-reserve calculations in the Prony nickel deposit, New Caledonia, is given by Journel (1973, 1974).

16.3. SAMPLE PREPARATION

Some steps in sample preparation may be taken in the field, but most sample preparation is done in the laboratory, within a geologist's area of concern but outside his or her area of direct control. Laboratories are generally very

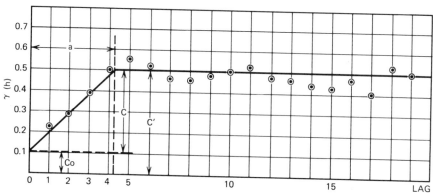

Figure 16-6. Variogram from a large stratabound Canadian uranium deposit. The regionalized variable is the tenor in uranium along a line of drill holes at 15-m intervals. The range (a) of the variogram is at a lag of 4.2 (63 m). The significant geologic dimension is thought to be the width of a buried channel. (After R. A. Blais and P. A. Carlier, "Applications of geostatistics in ore evaluation," CIM Special Volume No. 9, *Ore Reserve Estimation and Grade Control*, Fig. 15, p. 55, 1968. Used by permission of the Canadian Institute of Mining and Metallurgy.)

reliable, and they emphasize the same controls that would be demanded by the geologist who submitted the samples. Still, there is room for error. Duplicate samples, "blank" samples, and check assays by other laboratories are therefore used to determine a level of confidence and to identify any possible bias. It is common practice to ask that sample rejects be kept by the laboratory for an appropriate length of time so that analyses can be repeated if necessary and so that analyses may be made for additional elements.

Whether done in the field or in the laboratory, sample preparation has two objectives: to homogenize each sample and to reduce the amount of sample. Crushing to a smaller particle size comes first. Without crushing, homogenization of all but the very largest volume of sample is impossible; without homogenization, any reduction in volume introduces an immediate bias.

Coarse samples are passed through a small crusher or are broken with a hammer on a hard surface. Whichever method is used the equipment must be cleaned between each crushing. A steel plate is the most commonly used hard surface. It is important to remember that the alloying materials in the steel may contaminate the sample. Crushing is more easily and cleanly done in the laboratory; it may, however, have to be done at a field camp to reduce the bulk for shipment.

A simple rule in sample reduction is that all fragments must be crushed to such a size that the loss of any single particle would not affect the analysis. This rule without numbers depends on the accuracy required, the contrast in value between ore and rock particles, and the size of the sample. Guidelines for the maximum allowable particle size in respect to approximate sample weights are shown in Table 16-1. The table, like most tables, refers to often-cited but seldom-seen "average" conditions. For a homogeneous ore, a somewhat larger particle size would be acceptable. Precious metal ores and

Table 16-1.   Empirical guidelines for reducing weight of ore sample: maximum allowable particle size in reduced sample.

Weight of Sample (kg)	Size (Diameter) of Largest Piece (cm)
1,000	7.6
500	5.0
100	2.3
50	1.5
10	.7
5	.5
1	.2

Data from Aplan, 1973, p. 16.

ores with low or erratic values require finer crushing, and they may even require grinding. In their study of sample preparation methods at the Homestake gold mine, Koch and Link (1972) found that grinding and pulverizing entire core samples rather than the split half core before any sample reduction made a considerable improvement in the accuracy. In this approach, something would be sacrificed as well as gained. The destruction of entire core samples would obviously cancel their use in any further geologic study.

Drill-hole cuttings are recovered in fragments that for most ores are fine enough for immediate mixing and reduction in sample volume. The cuttings are passed through a sample splitter in which equal portions are separated or in which a small portion is accepted and a larger portion is rejected. The most common type of splitter for dry material is the Jones riffle sampler, a container with channels discharging in two directions. By mounting several splitters in a stairlike arrangement, a series of reductions can be made in one pass. The Vezin, or circular, sampler revolves in a falling stream of wet or dry cuttings and takes a selected percentage by passing an opening through the sample stream. Before shipping, wet sludge or cuttings are dried in the sun, under a heat lamp, or over a low fire. An intense fire is avoided because some of the minerals may oxidize or change composition.

Where a mechanical splitter is not available for separating finely crushed material or where the fragments in a bulky sample are too large to be handled, the sample can be reduced by the old but effective method of "coning and quartering." The crushed ore is mixed and shoveled into a conical pile on a clean smooth surface, the pile is flattened into a disk, and then separated into quarters with the edge of the shovel. Alternate quarters are accepted and rejected. The two accepted quarters are remixed, crushed again if necessary, and the coning and quartering process repeated until the sample is reduced to a suitable size.

Guidelines for sample preparation have the same "except for placers" restriction that applies to any generalization applied to mineral deposits and mining. Inasmuch as the recoverable value of placer ore is a direct function of density and particle size, the sample is reduced in size and prepared for analysis by washing and gravity concentration—a selective operation that is essentially the same process used in commercial placer mining and recovery plants. Some finer particles of ore escape, just as they would in mining. The special conditions and procedures for handling placer samples are spelled out by Wells (1973).

16.4. GRADE AND TONNAGE CALCULATIONS

A minable block of ore is described by its average grade and tonnage. Other characteristics, such as range in grade and degree of certainty that the block actually exists, are described in the same terms: grade and tonnage. In order to calculate numbers for the terms, certain basic information is needed.

1. An accurate and adequate sample population: assays.
2. Dimensional measurements: plans and sections.
3. Density measurements for ore, gangue, and wall rock: specific gravity.
4. Geologic limits and projections: lithologic and structural boundaries.
5. Cutoff grade: the lowest grade that can be mined economically.
6. Potential recovery in mining: percentage extraction.
7. Potential dilution in mining: percentage increase in tonnage with little or no increase in total ore mineral content.

The last three kinds of information are related to engineering; nevertheless, they have fundamental geologic elements. The cutoff grade depends very much on the ore mineralogy and mineral distribution within the ore-bearing zone. Mining recovery and dilution are affected by structural conditions. Ultimately, information is needed on potential recovery in metallurgical processing as well as in mining. Geologic and engineering estimates of ore reserves have no significance unless they refer to what can be extracted and marketed. In some mines, a partially oxidized ore requires two types of metallurgical processing with two or more levels of metal recovery; each sample may therefore need two assays for a single metal, one for total metal and one for "oxide" metal or "soluble" metal.

Grade

The first step in treating sample data for an estimate of mining grade is to find composite values for the samples taken within designated widths of channel or lengths of drill core. The designated widths and lengths are obtained from bench heights in open-pit mines, minimum stoping widths in underground mines or—in selective mining—zones that can actually be separated as ore and waste. If the individual samples are all of the same length and of approximately the same density, as in most samples taken from large deposits, an arithmetic average can be used. In complex orebodies and in zoned veins, however, data from samples of various lengths in a channel or drill core are averaged by the formula:

$$\text{Average assay} = \frac{\Sigma\,(\text{widths} \times \text{assays})}{\Sigma\,\text{widths}}$$

The channel sample shown in Figure 16-1 might have the following widths (W) (normal to the vein), assays (A), and calculations:

Sample No.	Width (m)	Pb%	W × A	Ag (g/t)	W × A
1	0.5	1.6	0.8	22	11.0
2	0.3	3.0	0.9	27	8.1
3	0.7	6.0	4.2	55	38.5
4	1.0	9.7	9.7	35	35.0
5	0.5	2.0	1.0	20	10.0
	3.0	22.3	16.6	159	102.6

Average grade   16.6/3 = 5.53% Pb, and 102.6/3 = 34.20 g/t Ag
An unweighted arithmetic average for the five samples would have been 4.46% Pb and 31.80 g/t Ag.

A refinement in averaging assays is to weight the sample data for specific gravity as well as for length, using the formula:

$$\text{Average assay} = \frac{\Sigma(\text{sp gr} \times W \times A)}{\Sigma(\text{sp gr} \times W)}$$

If this had been done in the above calculation, it would have made the

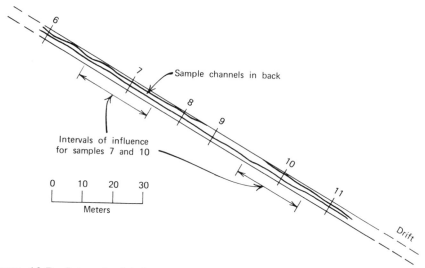

Figure 16-7. Interval of influence concept in weighting assay data from channel samples taken at unequal intervals.

average grade a few hundredths to a few tenths of a percent higher, depending on the density of gangue minerals. In high-grade ore with considerable contrasts in density, the specific gravity factor can change an average assay by several percentage points. Where this additional step is taken, specific gravity determinations are often made on individual samples.

Where a vein of uneven width is sampled at equal intervals in a series of channels, an average grade for the vein can be calculated in the same way an average grade was calculated for a series of unequal-width samples in a single channel. In this instance, each sample value is the average value of one of the channels. In narrow veins, the minimum stoping width enters into the calculation. Suppose, for example, that barren wall rock as well as vein material must be taken where a vein is less than 1.5 m wide. The calculation for five channel samples taken at equal distances along the vein would be:

Channel No.	Stope Width(m)	Vein Width(m)	Channel Average Au (g/t)	Vein W × A
1	1.5	1.0	12.1	12.10
2	1.8	1.8	14.4	25.92
3	1.6	1.6	8.0	12.80
4	1.5	1.1	6.3	6.93
5	1.5	1.2	10.0	12.00
	7.9	6.7		69.75

Average grade of vein 69.75/6.7 = 10.41 g/t Au
Average grade of stope 69.75/7.9 = 8.83 g/t Au

Where channel samples are cut an unequal intervals, some will have to represent greater lengths of vein than others. Each channel must then be weighted by an interval of influence as well as a width in order to obtain an area of influence (W × I). A customary procedure is to assign an interval of influence measuring half the distance to each of the neighboring channels, as shown in Figure 16-7. In this procedure, the vein is assumed to be continuous. The formula is:

$$\text{Average grade} = \frac{\Sigma(I \times W \times A)}{\Sigma(W \times I)}$$

For example:

Distance Between Channels (m)	I	Sampled Vein Width (m)	I × W	Cu%	I × W × A
32 ⎫ 16 + 10 ⎬ 20 ⎭	26	1.6	41.6	3.5	145.6
10 + 6 ⎬ 12 ⎭	16	2.0	32.0	8.5	272.0
6 + 13 ⎬ 26 ⎭	19	2.8	53.2	3.0	159.6
13 + 9 ⎬ 18 ⎭	22	2.2	48.4	5.0	242.0
			175.2		819.2

Average grade for four samples in the vein: 819.2/175.2 = 4.67% Cu.

Where mine workings and drill holes cross a vein or a tabular body in various directions as well as at uneven intervals, the sample locations are projeeted to a plan or section and assigned a polygonal area of influence as well as possible; the customary procedure is to assign half the distance to the next sample, as shown in Figure 16-8.

The averaging procedures described thus far are conventional and, in a sense, old-fashioned. Statistical and computerized procedures, especially trend-surface analysis and the variogram function of geostatistics, are more widely used for large orebodies where there are thousands of samples. Even so, the basic calculations are made by weighting sample data for length, density, and area of influence. The textbook by Koch and Link (1970–1971, p. 234–246) is recommended for information on the statistical methods of grade calculation.

A good case history in statistical methods of ore-reserve calculation is furnished by Sinclair and Deraisme (1974). At the Eagle mine in northern British Columbia, geostatistical methods and variograms were used to estimate reserves for the entire orebody and to select a pattern for additional sampling. "Kriging" (named after D. C. Krige, a South African mining statistician), a widely cited geostatistics approach to assessing local variations within an orebody, is explained and illustrated.

The calculation of grade for industrial mineral and coal reserves presents some special problems because factors other than elemental content are involved. Industrial mineral reserve indices are taken up commodity by commodity in the comprehensive volume on industrial minerals and rocks

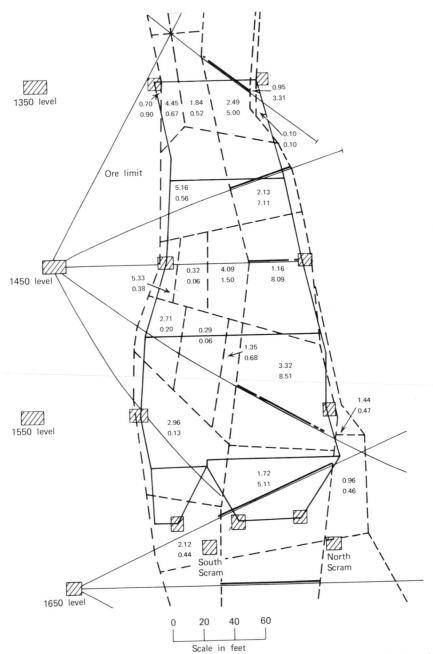

Figure 16-8. Vertical section through workings in the Geco mine, Ontario, showing areas of influence assigned to diamond drill holes for tonnage and grade calculations. (After L. S. Brooks and R. C. E. Bray, "Applications of geostatistics in ore evaluation," CIM Special Volume No. 9, *Ore Reserve Estimation and Grade Control*, Fig. 4, p. 181, 1968. Used by permission of the Canadian Institute of Mining and Metallurgy.)

edited by Lefond (1975). Coal reserves are determined in relation to heat value, moisture, ash, sulfur, hardness, and grindability. Statistical methods of estimation and contouring these conditions between drill holes in a coal seam are described in detail by Gomez and Donaven (1972). Sampling techniques and methods of estimating mining conditions and coal quality are described by Alexander (1975).

Tonnage

The fundamental tonnage calculations are for volume and density. First, boundaries are drawn between ore and waste or between several grades of ore mineralization, as shown in Figure 16-3.

Areas within the ore boundaries are determined as shown in Figure 16-8, or they are calculated on the basis of grade contours or grade-thickness contours (generally by planimeter) or calculated by one of the geometric constructions shown in Figure 16-9. The halfway rule is often used for an area of influence around a sample, but it is modified where warranted by a geological boundary, a marginally mineralized drill hole, or an unreliable sample. Trend-

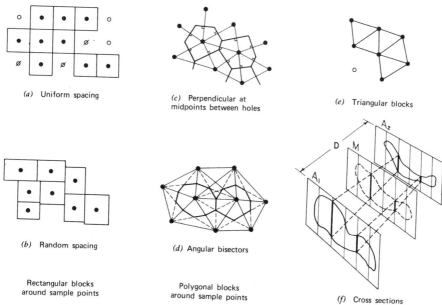

Figure 16-9. Geometric patterns for ore reserve calculations. (a) Uniform spacing. (b) Random spacing. (c) Perpendiculars at midpoints between holes. (d) Angular bisectors. (e) Triangular blocks. (f) Cross sections. (From J. A. Patterson, "Estimating ore reserves follows logical steps," *Eng. Mining Jour.*, v. 160, no. 9, Fig. 4, p. 115, 1959. Copyright © 1959, McGraw-Hill, Inc. Used by permission of the McGraw-Hill Book Company.

surface analysis and similar isoline techniques are good statistical tools for defining blocks of ore with specified average grades. In a statistical plot some of the blocks may not contain a sample point; even so, the blocks may be defined in relation to trends generated from sample points throughout the orebody.

Whatever the technique, areas of influence on plans or sections are given a third dimension to become volumes of ore. The simplest way is to calculate the volume lying between each pair of end sections. Another common technique makes use of the prismoidal formula:

$$V = \frac{A_1 + 4A_2 + A_3}{6} h,$$

where A_1, A_2, and A_3 are areas on three successive and uniformly spaced plans or sections and h is the distance between the end areas. Additional ways of calculating geometric ore volumes are described in detail by Popoff (1966).

A volume of ore—a block to be classified as part of a mine's ore reserves—is given a tonnage value by using a density factor. In metric measurements, specific gravity (density in g/cm^3) is multiplied by volume in cubic meters to obtain metric tons. A volume that has been calculated in cubic feet is divided by a "tonnage factor" (cubic feet per ton) to obtain short tons:

$$\text{Tonnage factor} = \frac{2000}{\text{sp. gr.} \times 62.5} = \text{cubic feet per ton}$$

The density of ore can be calculated from its mineralogy, as explained by Koch and Link (1971, p. 249), by determining the specific gravity of samples, or by weighing the ore from measured excavations. The last method, while more time consuming than the other, is the most accurate if large samples and well-measured sample volumes are used. Still another approach to calculating the weight of an ore block is to use average specific gravity values for ores and rocks, which are found in tabular form in most mining engineering textbooks. A few average specific gravity values for common rocks, gangue, and ore minerals are given in Table 16-2.

16.5. ORE-RESERVE CLASSIFICATIONS

One part of an orebody may have been so thoroughly sampled that we can be certain of its outline, its tonnage, and its average grade; it is *proved* ore. In another part of the same orebody, sampling may not have been so thorough, but we have enough geologic information to be reasonably secure in making an estimate of its tonnage and grade; this is *probable* ore. On the fringes of the orebody, our knowledge may be based on very few samples but we have enough information from other parts of the orebody, enough geologic evidence, and enough knowledge of similar deposits to say that a certain amount of ore and a certain grade are *possible*.

The precise meanings vary from one organization to another, but the general meaning of proved, probable, and possible ore reserves is understood by most mining engineers and geologists. Originally, the terms were defined in relation to vein-type deposits. As shown in Figure 16-10, proved reserves in veins are blocked out (sampled) on four sides by mine workings and surface exposures, probable reserves are blocked out on two or three sides, and possible ore—geological reserves, or "geologist's ore"—is exposed on one side or not exposed at all. The sampled workings are assumed to be closely spaced. Where workings are too far apart to allow comfortable projections, a block might be shown with a core of probable ore and a rim of proved ore. In modern use the terms are also applied to all kinds of ore deposits that have

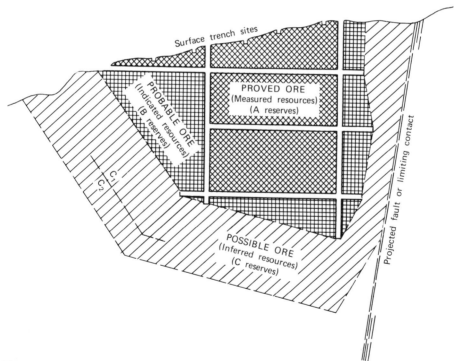

Figure 16-10. Proved, probable, and possible ore designations in the conventional sense of the terms, plotted on a longitudinal section of a vein-type deposit. Approximate equivalent terms are shown from the U.S. Geological Survey designation of identified resources and from a principal European system of ore-reserve classification.

Table 16-2.   Approximate density g/cm³ (specific gravity) of rocks and minerals in place

Rocks			
Andesite	2.4–2.8	Limestone	2.7–2.8
Basalt (dense)	2.7–3.2	Marble	2.6–2.9
Clay	2.2–2.6	Peridotite	3.2–3.4
Coal	1.0–1.8	Rhyolite	2.2–2.7
Diabase	2.8–3.1	Sand (dry)	1.7–2.0
Diorite	2.7–2.9	Sandstone	2.0–3.2
Dolomite	2.7–2.8	Shale	1.6–2.9
Gabbro	2.9–3.1	Schist	2.6–3.0
Gneiss	2.7–2.8	Slate	2.8–2.9
Granite	2.6–2.7	Syenite	2.6–3.0
Gravel (dry)	1.6–2.0	Trachyte	2.5–2.8

Minerals			
Alunite	2.7	Feldspar	2.6–2.8
Anhydrite	2.9	Fluorite	3.1
Anglesite	6.3	Galena	7.6
Ankerite	3.0	Garnierite	2.3–2.8
Argentite	7.3	Goethite	4.2
Arsenopyrite	6.0	Gold	17.5
Azurite	3.8	Graphite	2.2
Barite	4.5	Gypsum	2.3
Bauxite	2.6	Halite	2.2
Beryl	2.7	Hematite	5.2
Bornite	4.9	Ilmenite	4.8
Calaverite	9.0	Magnesite	3.0
Calcite	2.7	Magnetite	5.2
Cassiterite	7.0	Malachite	4.0
Cerargyrite	5.6	Manganite	4.3
Cerussite	6.5	Marcasite	4.9
Chalcedony	2.6	Molybdenite	4.8
Chalcocite	5.7	Monazite	5.1
Chalcopyrite	4.3	Muscovite	2.9
Chromite	4.5	Niccolite	7.5
Chrysocolla	2.1	Orpiment	3.5
Cinnabar	8.1	Pentlandite	4.8
Cobaltite	6.2	Platinum	19.0
Columbite	6.2	Proustite	5.6
Copper	8.8	Psilomelane	4.2
Covellite	4.6	Pyrargyrite	5.8
Cuprite	6.0	Pyrite	5.0
Dolomite	2.9	Pyrochlore	4.3
Enargite	4.5	Pyrolusite	4.8

Table 16-2 (con'd). Approximate density g/cm³ (specific gravity) of rocks and minerals in place

Minerals			
Pyroxene	3.3	Stibnite	4.6
Pyrrhotite	4.7	Sulfur	2.1
Quartz	2.7	Sylvanite	8.1
Rhodochrosite	3.5	Sylvite	2.0
Rhodonite	3.6	Tennantite	4.5
Rutile	4.2	Tenorite	6.0
Scheelite	6.0	Tetrahedrite	4.8
Sericite	2.6	Thorite	4.6
Siderite	3.9	Titanite	3.5
Silver	10.6	Tourmaline	3.1
Smithsonite	4.4	Turquoise	2.7
Sphalerite	4.1	Uraninite	9.4
Spodumene	3.2	Wolframite	7.4
Stephanite	6.2	Zircon	4.5

been delineated and sampled to a certain intended level of accuracy. The actual tonnage and grade are generally expected to be within 90–95 percent of the estimate for proved ore and within 70–80 percent for probable ore. Mining companies usually refer to proved and probable ore as their actual reserves. Possible ore is considered more in the category of an exploration target.

Because the blocking out of reserves is a continuous task during mining, the reserves may change classification with each monthly or yearly report to management. The accuracy of an ore-reserve calculation, an uneasy figure, is also evaluated as the ore is mined, as shown in Figure 16-11. Correction factors are obtained as data accumulate, and the ore-reserve calculations become more reliable unless there is an unexpected change in geologic conditions. The need for constant reevaluation of reserve data can be seen in a study by Wright (1959), who obtained the original drill-hole data from five depleted or nearly depleted uranium mines in the Colorado Plateau region and asked eight geologists to make individual ore-reserve calculations. The estimated tonnages turned out to be from 20 to 600 percent of the tonnage eventually mined; the grade estimated by individual geologists was fairly close to the mined grade in four of the orebodies, but it was 335–642 percent of the mined grade in one orebody.

A system of ore-reserve and mineral-resource classification used by the U.S. Bureau of Mines, the U.S. Energy Research and Development Administration, and by several national geological surveys employs the terms "measured," "indicated," and "inferred." The terms are similar to the proved, probable, and possible terms of most industrial organizations, but they are

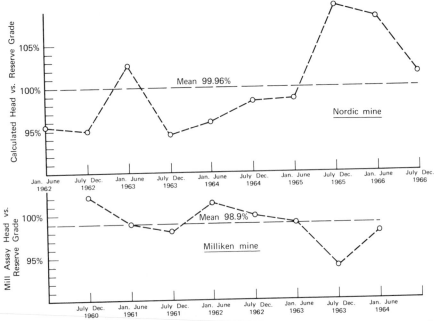

Figure 16-11.   An evaluation of ore reserve grade calculations in the Elliot Lake-Blind River uranium district, Ontario. The proved reserve grades were taken at six-month intervals and compared with the average mill head (mined ore) grade during the following six months. At the Nordic and Milliken mines, the proved reserve grade averaged 99.96 percent and 98.9 percent of the mill head grade. (From R. C. Hart and D. Sprague, "Methods of calculating ore reserves in the Elliot Lake Camp," CIM Special Volume No. 9, *Ore Reserve Estimation and Grade Control*, Fig. 9, p. 259, 1968. Used by permission of the Canadian Institute of Mining and Metallurgy.)

more flexible because they have a greater geological input and apply to the mineral appraisal of regions and districts as well as mines. The terms are defined as follows by the U.S. Geological Survey (1975, p. 4):

Measured. *Identified resources for which tonnage is computed from dimensions revealed in outcrops, trenches, workings, and drill holes and for which grade is computed from the results of detailed sampling. The sites for inspection, sampling, and measurement are spaced so closely and the geologic character is so well defined that size, shape, and mineral content are well established. The computed tonnage and grade are judged to be accurate within limits which are stated, and no such limit is judged to be different from the computed tonnage or grade by more than 20 percent.*

Indicated. *Identified resources for which tonnage and grade are computed partly from specific measurements, samples, or production data and partly from projection for a reasonable distance on the basis of geologic evidence. The sites available for inspection, measurement, and sampling are too widely or otherwise inappropriately spaced to permit the mineral bodies to be outlined completely or the grade to be established throughout.*

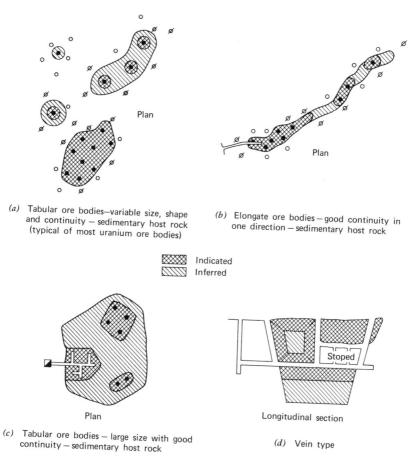

(a) Tabular ore bodies—variable size, shape and continuity — sedimentary host rock (typical of most uranium ore bodies)

(b) Elongate ore bodies — good continuity in one direction — sedimentary host rock

Indicated
Inferred

(c) Tabular ore bodies — large size with good continuity — sedimentary host rock

(d) Vein type

Figure 16-12. The designation of indicated and inferred ore reserves in bodies of uranium ore outlined by drill holes and mine workings. Some geologists would prefer to use the terms "measured" or "proven" for most of the ore shown here as "indicated." Actual practice varies from one organization to another. (a) Tabular orebodies—variable size, shape, and continuity—sedimentary host rock (typical of most uranium orebodies). (b) Elongate orebodies—good continuity in one direction— sedimentary host rock. (c) Tabular orebodies—large size with good continuity— sedimentary host rock. (d) Vein type. (From J. A. Patterson, "Estimating ore reserves follows logical steps," *Eng. Mining Jour.*, v. 160, no. 9, Fig. 2, p. 113, 1959. Copyright © 1959 by McGraw-Hill, Inc. Used by permission of the McGraw-Hill Book Company.)

Inferred. *Identified resources for which quantitative estimates are based largely on broad knowledge of the geologic character of the deposit and for which there are few, if any, samples or measurements. Continuity or repetition is assumed on the basis of geologic evidence, which may include comparison with deposits of similar type. Bodies that are completely concealed may be included if there is specific geologic evidence of their presence. Estimates of inferred reserves or resources should include a statement of the specific limits within which the inferred material may lie.*

Figure 16-12 shows how blocks of uranium ore might appear in terms of indicated and inferred reserves. In tabular "pod" orebodies of variable continuity, a single drill hole in ore is given an amount of indicated ore if it is surrounded by mineralized holes and if it is on a general ore trend, otherwise the ore is considered as inferred. Indicated ore in a larger tabular orebody with an expected continuity in a favorable host rock can be extended a short distance beyond the sampled locations.

European and American approaches to ore-reserve classification are compared by Zeschke (1964, p. 166). Proved or "safe" ore is often classified in continental Europe as type A. Probable and possible ores have a series of designations, B, C_1, and C_2, with C_1 referring to a degree of accuracy of 50–70 percent and C_2 referring to ore that has been assumed or barely outlined. A final designation, D, refers to "prognosticated" ore, the European equivalent of "geologist's ore."

Regardless of the classification given to a block of ore reserve, an *average* ton of ore will never be mined; it only exists on paper. Mining and mineral processing operations are designed for a certain range of conditions, not for an average condition.

CHAPTER 17
COMMUNICATION

Someone with responsibility for locating or developing mineral deposits makes the decision: commit funds and manpower to the next step or abandon the project. And the decision is made with whatever information is available, relevant, and understandable. Relevance, the "here and now" requirement, may screen out so many of the most intriguing geologic observations that some geologists can hardly bring themselves to do the screening. Yet the separation of relevant information and its processing into understandable form is best done by the geologist just as a geophysical message is best separated from signal and explained by a geophysicist.

The relevant geologic information (a result of processing) will eventually be treated as geologic data (an input to further processing) and coupled with economic and engineering data in a computer program; therefore, much of the geologic information will be quantified—put into units of width, composition, and statistical probability—by *someone*. But so much is inherently qualitative, so much is based on inference, and so much is imprecise. Quantify? It is the geologist who knows the limitations of the data; it is the geologist's responsibility to quantify what is actually quantifiable, to differentiate between observation and interpretation, and to state what is measured and what is estimated. This is good field practice, and it is good practice in geologic communications. A geologic contact seen in an outcrop should carry more statistical weight than a projected contact or a contact inferred on the basis of a change in vegetation. Unless the geologist provides expressions that can be evaluated as well as digitized by systems analysts, something important is

489

often lost in the translation and geological absurdities are released to wreak havoc upon the ultimate decisions.

17.1. PREPARING GEOLOGIC DATA FOR COMMUNICATION

Data screening begins in the field with an orientation survey and some decisions on what mapping scale to use, how much data to collect, and what field note format to use. The screen may be composed of information from a preliminary study, an earlier reconnaissance, or a preceding target-area investigation.

Table 17-1 shows the relationship between prior work, an orientation survey, and the remaining steps in an idealized sequence of geologic data screening and processing. If the table were to be shown as a flow chart, it would be laced with feedback lines because each step places some preceding step in a new light.

Some of the steps outlined in Table 17-1 are not ordinarily taken in actual practice, but each step deserves advance consideration because of its potential importance. For example, geologic patterns may be quite clear without any need for statistical treatment, but if a statistical analysis is necessary, it will be most effective if the data have been collected with statistical treatment in mind. It is of no use to find out where samples should have been taken and which areas should have been mapped after the field work is completed.

Some of the need for statistically manageable data is met in the field by making on-the-spot numerical estimates, whether or not the field notes are keyed to a formal quantitative system, such as GEOMAP or GEOLOG. A further way to provide for effective statistical treatment is to obtain the most thorough and unbiased data coverage possible; some of this can be done by a small amount of sampling or mapping in areas between natural data clusters and between lines of data. Collection of some field data from just outside the area of specific interest is also worthwhile; it may help to establish regional background trends and minimizes the statistical "edge effect"—a loss in significance near abrupt data boundaries. Some combination of stratified sampling (sampling with a geologic context) and pattern sampling (sampling for effective coverage) may be needed. Further information on stratified sampling, pattern sampling, and other statistical aspects of geologic data collection is given by Koch and Link (1970, p. 257–335).

"Pattern recognition," a term used in Table 17-1 for a part of the data classification and analysis step, refers to a broad and rather subjective procedure, but it also has a specific statistical meaning. The statistical technique of pattern recognition, useful where hundreds or thousands of data points are available and where little is known of potential geologic associations, is described by Howarth (1973). He gives an example of pattern recognition in the evaluation of stream-sediment geochemistry data taken

Table 17-1.   Steps in Screening and Processing Exploration Data

A. Preliminary. The following information from preceding investigations is used as data for the new sequence.
   1. Working exploration model, based on:
      a. A conceptual model of the orebody and its geologic environment
      b. Patterns of geologic, geophysical, and geochemical response to the conceptual model
      c. Exploration objectives, requirements, and constraints.
   2. The basis for selecting the area to be investigated.
   3. Acceptable exploration methods and techniques.
   4. Priorities among subareas and investigations.
B. Field.
   1. Orientation survey, dealing with:
      a. Response to the exploration model
      b. Choice of scale, level of detail, and degree of accuracy
      c. Identification of noise factors.
   2. Collection of field data, with:
      a. Field quantification insofar as possible
      b. Feedback to the orientation survey for adjustments in plan
      c. Thoroughness of coverage insofar as possible
      d. Coverage in adjacent areas for statistical trend calculations.
C. Data classification and analysis.
   1. Pattern recognition (geological and statistical).
   2. Further statistical treatment, including:
      a. Screening for laws of statistical distribution
      b. Data enhancement if needed
      c. Determination of statistical expressions
      d. Addition of simulated data if needed
      e. Statistical analysis of data.
   3. Comparison and correlation of data with the exploration model.
D. Interpretation.
   1. Identification of relevant information.
   2. Input of relevant information to the report.

around the Dartmoor granite, one of the exposed plutons in the Cornwall-Devon mining region mentioned in Chapter 2. Cluster analysis, mentioned in Chapter 14 as a way of recognizing geochemical patterns, is also discussed by Howarth.

Pattern recognition in the broader geologic sense cannot be so easily dealt with. Nor can it be entirely separated from statistics. Experience, imagination, and perception are personal attributes rather than analytical techniques, but they are often combined with statistical techniques. A pattern in geochemical, geophysical, or geological data may emerge because it is expected or because it fits a perfectly valid but hard-to-describe theory. Statistical support can then be sought, and statistical correlations can be used to determine some of the more elusive correlations and trends. Data enhancement (bringing out selected features) and the addition of simulated data (enlarging the scope of a limited sample population) are useful ways to strengthen apparent patterns provided there is a sound geologic basis for selecting what to enhance and what to simulate.

Data are finally compared with the exploration model and with alternative exploration models that may have become apparent, and then are converted to relevant information for the geologic report, as indicated by the last few steps in Table 17-1. A geologic report is likely to be part qualitative and part quantitative. The distinction must be clearly indicated so that geologic projections will not appear to be too fuzzy and statistical interpretations too arbitrary.

Two important words, *accuracy* and *precision,* would be found in Table 17-1 except that both words would have to be repeated in every step. Quantified field data are not necessarily accurate data. Digitized data carried to several decimal places are not necessarily precise. Furthermore, computer-processed data are neither more accurate nor more precise than the data elements; perhaps they are even less so, because unintentioned "interpretations" may enter the system from several human and machine sources. The geologist who presents the final information may not have been able to control every operation in the screening and processing of the data, but hopefully will have checked the data often enough to be certain that they have not gotten out of hand.

17.2. EXPLORATION COMMUNICATIONS

After information has cleared the relevance hurdle, it still may not be useful. If it is presented late, the decision it was supposed to influence may have already been made. If poorly presented, it may be tolerated only as a courtesy and then ignored. The requirements for timeliness and effective presentation add up to *compatibility*: the information must be compatible with modern methods of communication.

"Whatever happened to the real artistry and philosophy we used to find in geologic reports in the old days?" The frequently used terms "data explosion" and "communications revolution" describe what has happened. The basic means of communication—verbal, written, and graphic—have not changed, but since geologists can obtain accurate data from a remote computer in amounts that would have been impossible before the "revolution," they are expected to provide relevant and accurate information almost as quickly.

Verbal Communications

During an exploration program verbal communication may range from weekly reports by recorder cartridge, telephone, and radio to a formal conference at headquarters or in the field. Major decisions are seldom made without a field conference—a demonstration of what has been seen, inferred, and projected. The field conference is usually well orchestrated, with a planned itinerary, prepared maps and diagrams, organized supporting data, and even a rehearsal—a practice run—through the area.

One particular method of verbal communication must be mentioned because it is so appropriate to an exchange of ideas during an exploration program. A telephone conference, easily arranged for several participants in various locations, has a special advantage because it provides for a group discussion without waiting until the annual exploration manager's meeting and it does not waste travel time. Each participant, a manager, a field geologist, a geophysicist, and a geochemist, for example, can be furnished with copies of appropriate maps and tables. Where a large amount of graphic material is involved it can be mailed in microform and viewed with a microform "reader" during the conference. With relatively noise-free telephone lines, drawings can be sent by telephone for immediate duplication or display. Interactive cathode-ray tube (CRT) terminals are capable of providing the final touch in two-way communication with graphic support.

Almost always a written memorandum should follow a radio or telephone communication of any consequence. It can be a brief hand-written memorandum worded something like: "This is to confirm our telephone conversation on August 15 . . ." The memorandum outlasts memory and personnel changes and it will be available for future reference. It may even be of immediate importance; a return memorandum may state: "I thought you were talking about *vertical* drill holes."

Written Communications

Like verbal communications, written communications seldom stand alone. They need graphic support and they need discussion. Written communications are nevertheless the framework that holds information together. If time is short, they may be handwritten or sent by telegraph; a memorandum of a few pages suffices for many purposes, and it may be more valuable than an extensive report that arrives a week later.

Within the ample literature on technical writing, a book by Tichy (1967) is especially valuable as a reference, and the small *Geowriting* guide of the American Geological Institute (Cochran, Fenner, and Hill, 1973) is helpful. To insure that the writer and the reader are using the same terms in the same way, a copy of the American Geological Institute's *Glossary of Geology*

(Gary, McAfee, and Wolf, 1972) should be available. The U.S. Geological Survey (1958) and Geological Survey of Canada (Blackader, 1968) manuals on report writing are considered by many geologists as their ultimate reference on word usage.

A few types of written communications are especially common in mining and exploration work; these are:

Proposals

Confirmations

Itineraries

Progress reports

Status reports

Comprehensive reports

The first four are often written as memoranda or letters since they are directed to a limited number of people and do not contain a great amount of geologic interpretation. The last two are usually more complete reports.

A *proposal* deals with future eventualities. It is an offer to do something— to make a reconnaissance of an area, to evaluate a dormant mining district, or to test a geologic hypothesis. The writing of proposals and the formulation of supporting data require special skills, so much so that a geologist proposing to do a job for a government will need advice on the format required by the particular agency. A government-oriented research organization will often have a group of scientists who do nothing *but* write proposals. In dealing with mining organizations the format is generally more flexible, especially in the area of geologic investigations. Proposals may become precisely structured and more complex at a later stage when most of the exploration work has been done and the new proposal includes a request for substantial funding. All proposals, however free in their format, are expected to have the following elements:

Identification. (What do you propose to do?)

Motivation. (Why do it?)

Cost. (How much?)

Method. (How do you propose to do it?)

Requirements. (What will be needed?)

A *confirmation memorandum* is especially important in exploration work. The geologist's team, isolated from the head office, may be making decisions in the field and doing whatever is needed to attain the objective. But do they *really* know what the objective is? A confirmation memorandum, essentially a list of assumptions, makes certain that there is no misunderstanding. It may begin with the following wording: "It is my understanding that our objective in examining the Hardscrabble mining district is to search for indications of copper orebodies with a minimum size and grade of . . ." If it is assumed, for example, that orebodies must be amenable to open-pit mining, this should be stated. If iron or fluorspar mineralization, known to occur in the district, is to be excluded from consideration, this assumption should also be stated. Geologists have a tendency to follow all interesting guides to mineralization, and managers sometimes fail to spell out the organization's precise objectives. The time to confirm instructions is before effort and money have been spent— or wasted.

An *itinerary memorandum* stating where the geologist intends to be at any particular time has several purposes. Help may be needed in case of accident. Emergency messages may have to be sent. The head office may be asked by government officials or landowners if they have someone working in area X. It would be embarrassing to answer: "We don't know—he was in area Y last week." Changes in itinerary, after the first memorandum, can be stated in progress reports.

Progress "reports" are often no more than memoranda, and they may be in the form of a handwritten copy or a recorder cartridge. They are, in essence, news briefs: this is what we are doing, this is how our schedule and our funds are being handled, and this is what we plan to do next. If problems seem to be developing, they need to be mentioned. The names of people contacted during exploration work should also be added. Informal property submittals should be attached to the progress report, so that the home office will not be taken off-balance when a prospector drops by to say: "I have already mentioned my diatomite property to your geologist in Lincoln County." Preliminary geological conclusions belong in progress reports, but they are to be clearly identified as preliminary; otherwise they may come back to haunt their maker long after he has modified or abandoned them. One more thing: progress reporting is overdone if it diverts too much time from field work.

Status reports are needed if the project is of long duration, and they are especially needed if a new budgeting period is involved. A status report is likely to be longer and more detailed than any of the earlier memoranda because it will be used for planning the continuation of the project. Also, it is likely to be presented or followed up in a formal conference. A status report summarizes the work that has been done, the results of the work, and the interpretations that have been made. Maps and sections are included, but because they are working sheets with information still to be added, they need not be in polished form. Accounting items—unit costs, distances traversed, and drill holes completed—are important; they will be used as guidelines for the next budget. New ideas—changes in the exploration model—are presented, and specific recommendations are made for further work. If the recommendation is to delay any additional work, the report is more nearly a comprehensive report.

Comprehensive reports, also called "final" or "summary" reports, follow a format such as the one given in Appendix D. Appendix D refers to mineral property examination and evaluation reports, but the topic headings are also applicable to reports on exploration targets and prospects. Further comments on the mineral property evaluation format can be found in Chapter 19. One aspect however deserves mention here. The "key information" page provides material for data storage and retrieval; this page—or the equivalent standard format used within an organization—emphasizes quantitative expressions and key words. The heading "Summary and Digest of Recommendations" also warrants special mention; this may be as far as some readers will need to go into the report, therefore it belongs at the beginning where it can be found without leafing through the body of the report.

Style—the writer's own method of emphasizing the message—cannot be ignored just because the computer is there to emphasize the numbers. A study of *The Elements of Style,* a small book by Strunk and White (1972) is as worthwhile to a geologist as to a literary writer. Style reflects the report writer's attitude, capabilities, and the reliability of the conclusions. Passive or careless sentences tell the reader that the geologist was not really interested in the project. Misused geologic, engineering, and economic terms tell the reader that the geologist may have been working beyond his or her grasp. Expansive writing—"tremendous potential" and "outstanding opportunities"—may get a message across to the reader but at the risk of making him wonder about the reliability and objectivity of the conclusions. Such expressions are better left to the promoter.

Graphic Communications

Maps, sections, and profiles are important elements in most geologic communications. Often, a geologic map constitutes the theme of an entire report. Where geologic conditions are complex, three-dimensional representation becomes necessary. Where concepts are to be explained, the time dimension and some abstract dimensions are added. Statistical information may be given entirely in abstract dimensions, such as frequency and intensity.

Some of the principal ways of showing exploration data are:

Maps

Geological	Surface and subsurface representations, with observed and interpreted data, as shown in Figure 11-9.
Paleo-conditions	Paleogeology, paleogeography, paleostructure, and paleotectonics. Figure 17-1 is from a paleogeographic map showing the relationship of stratiform ore mineralization to a Cretaceous shoreline in North Africa.

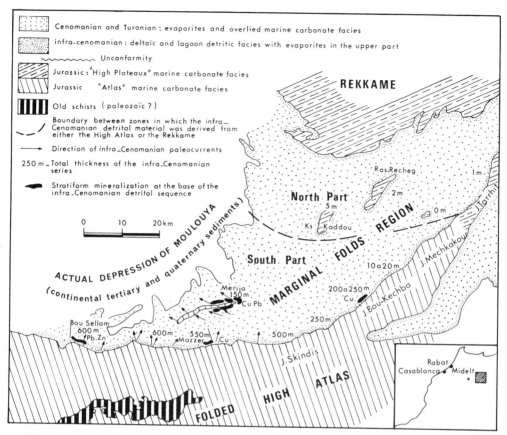

Figure 17-1. Paleogeographic sketch of the marginal fold region of the eastern High Atlas of Morocco during the Late Cretaceous, with locations of areas containing stratiform mineralization. (From J. Caïa, "Paleogeographical and sedimentological controls of copper, lead, and zinc mineralizations in the Lower Cretaceous sandstones of Africa," *Econ. Geology*, v. 71, Fig. 2, p. 411, 1976. Published by Economic Geology Publishing Company. Used by permission of the publisher.)

Isoline   Topography, structure contours, isopachs, isofacies, isolith maps, and isoline representations of geophysical, geochemical, and statistical data. Figure 2-20 and several figures in Chapters 13 and 14 show isoline maps. Figure 17-2 shows how an isoline map can be used to illustrate lithologic and paleogeographic conditions.

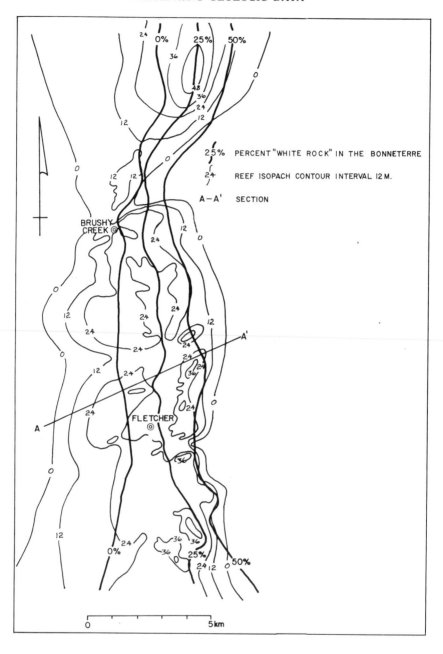

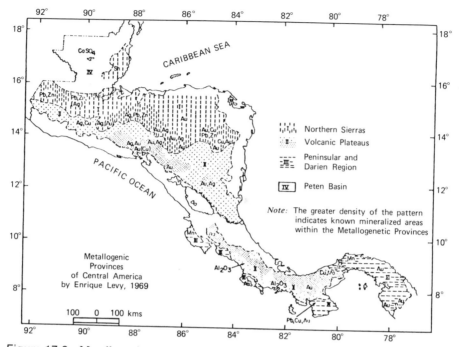

Figure 17-3 Metallogenic map of Central America, showing metallogenic provinces, types of mineralization, and the location of known mineralized areas. (From E. Levy, "Metallogenesis of Central America," Fig. 1, p. 20, 1971. SME Preprint 71-S-1. Used by permission of the American Institute of Mining, Metallurgical Petroleum Engineers, Inc.)

Point data	Mineral occurrences, drill holes, field observation positions, geochemical sample locations, geophysical stations.
Linear data	Photogeologic linears, geophysical vectors, and statistical trends.
Zoning	Alteration and mineralization patterns. Figure 2-27 is an illustration of a mineralization pattern map.
Metallogenic	Genesis and environment. Figure 17-3 is taken from a metallogenic map.

Figure 17-2 Isoline map showing lithologic and paleogeographic conditions in a part of the Southeast Missouri lead district. "White rock" is the back reef facies in the Cambrian Bonneterre formation. The algal reef facies is closely associated with orebodies. (From P. E. Gerdemann and H. E. Myers, "Relationship of carbonate facies patterns to ore distribution and to ore genesis in the Southeast Missouri lead district," *Econ. Geology*, v. 67, Fig. 4, p. 429, 1972. Published by the Economic Geology Publishing Company. Used by permission of the publisher.)

Supporting data   Status of data coverage, land tenure, economic aspects (generally on overlay maps).

Sections

Geological   Vertical and inclined, with factual, projected, and interpreted data, as shown in Figure 17-4, a cross section through the map shown in Figure 17-2.

Generalized   Illustrative of typical and conceptual conditions. An example is shown in Figure 2-7.

Isoline   Direct (based on point data) and indirect (geophysical response related to depth increments), as shown in Figures 13-10 and 13-11.

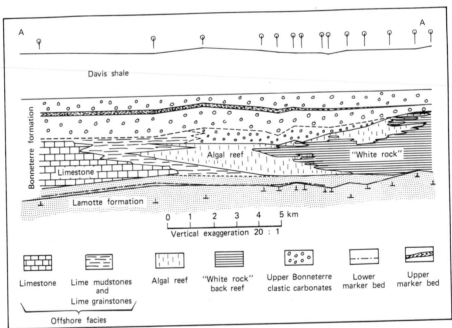

Figure 17-4   Cross section with projected and interpreted data, barrier reef zone in the Bonneterre Formation, Southeast Missouri lead district. Drill-hole data locations are shown. (From P. E. Gerdemann and H. E. Myers, "Relationship of carbonate facies patterns to ore distribution and to ore genesis in the Southeast Missouri lead district," *Econ. Geology*, v. 67, Fig. 5, p. 430, 1972. Published by Economic Geology Publishing Company. Used by permission of the publisher.)

Profiles

Horizontal   Geophysical and geochemical data and digitized geological data. Figure 4-5 is a profile diagram of carbon and oxygen isotope values in a mine working.

Inclined   Symbolized geologic log data, assay logs, geophysical and geochemical logs, as shown in Figures 15-9 through 15-12.

Three-Dimensional Diagrams

Block diagrams   Orthographic, perspective, and cut-away, as shown in Figure 17-5.

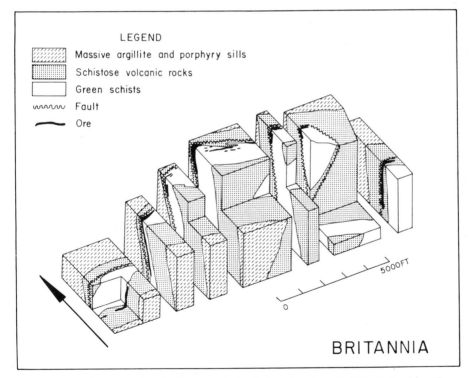

LEGEND

Massive argillite and porphyry sills

Schistose volcanic rocks

Green schists

Fault

Ore

5000FT

0

BRITANNIA

Figure 17-5   Cut-away block diagram of the Britannia mine, British Columbia. (From A. Sutherland Brown, "Mineralization in British Columbia and the copper and molybdenum deposits," *Canadian Mining Metall. Bull.*, v. 62, no. 681, Fig. 8, p. 933, 1969. Used by permission of the Canadian Institute of Mining and Metallurgy.)

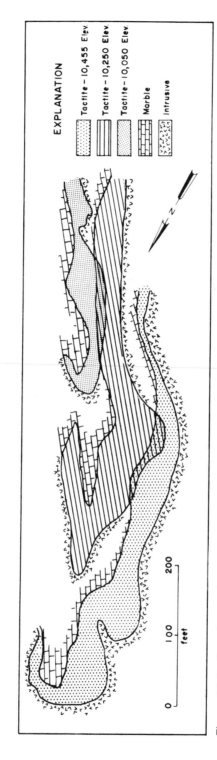

Figure 17-6 Superimposed data from three mine levels, Bishop tungsten district, California, showing orebody control by a structural trough. (After R. F. Gray and others, "Bishop tungsten district, California," in *Ore Deposits of the United States, 1933–1967*, J. D. Ridge, ed., Fig. 6, p. 1544, 1968, by permission American Institute of Mining, Metallurgical and Petroleum Engineers, Inc.)

Composites   Superimposed data with color or pattern identifica-
              tion, as in Figure 17-6.

Stereograms   Stereo pairs and color anaglyphs. The diagrams ap-
              pear in three-dimensional view, as with aerial photo-
              graphs.

Models   Section, plan, and "peg" models.

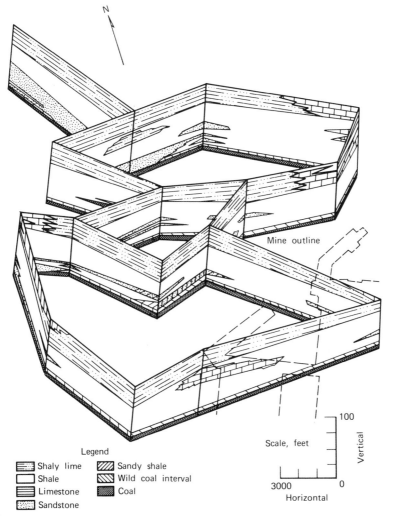

Legend

▦ Shaly lime   ▨ Sandy shale
☐ Shale   ◪ Wild coal interval
▤ Limestone   ■ Coal
▦ Sandstone

Mine outline

Scale, feet

Vertical

100

0

3000

Horizontal

Figure 17-7   Fence diagram of strata above the Pittsburgh coalbed, Marianna No. 58
mine, Pennsylvania. (From Figure 1, p. 4, Report of Investigations 8022, McCulloch,
Jeran, and Sullivan, 1975. Courtesy of the Bureau of Mines, U.S. Department of the
Interior.)

Explanatory Diagrams

Stratigraphy	Composite, correlated, and interpreted sections and "fences" of sections. Figure 17-7 is a "fence" diagram.
Palinspastic maps	Restored or "unrolled" to pretectonic conditions, as in Figure 17-8.
Conceptual	Sections or block diagrams illustrating a sequence or a relationship, as in Figure 17-9.
Petrologic and Mineralogic	Paragenetic sequence, zonal characteristics with distance, and relationships between components. Figures 2-2 and 2-10 are examples.

Statistical Diagrams   Maps, sections, profiles, three-dimensional diagrams, and graphs.

Photographs   Aerial, surface geology, specimen, and microscopic.

The making of geologic maps and supporting diagrams, like the processing of all other forms of geologic information, needs to be carefully monitored by

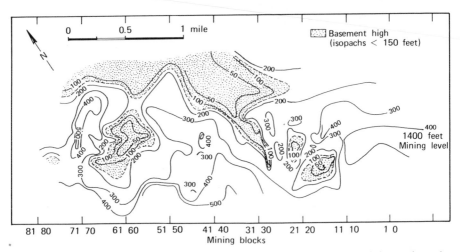

Figure 17-8   Palinspastic or "unrolled" isopach map of the footwall formation plus the C orebody at Mufulira, Zambia. Datum for unrolling was a key bed at the top of the orebody. The mining level shown in the diagram is part of a reference plane. Basement hills are reflected in low isopach values. (From J. G. van Eden, "Depositional and diagenetic environment related to sulfide mineralization, Mufulira, Zambia," *Econ. Geology*, v. 69, Fig. 2, p. 61, 1974. Published by Economic Geology Publishing Company. Used by permission of the publisher.)

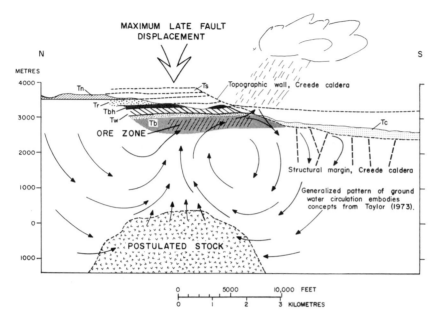

Figure 17-9 Conceptual section, reconstructing the environment of the Creede mining district, Colorado, at the time when mineralization was taking place. (From T. S. Steven and G. P. Eaton, "Environment of ore deposition in the Creede mining district, San Juan Mountains, Colorado," *Econ. Geology*, v. 70, Fig. 7, p. 1034, 1975. Published by Economic Geology Publishing Company. Used by permission of the publisher.)

the geologist responsible for the collection of the data. The real work of making a final drawing from a pencil copy will normally be done by a draftsman who works with the geologist. Many organizations have a graphics department staffed by geological draftsmen who are key professional members of the exploration group and who have a basic knowledge of geology and can help formulate an effective graphic presentation. A draftsman cannot create extra time to do a job that is wanted immediately. Adequate time for drafting, checking, and revision is allotted in all good exploration programs.

Geologic draftsmen are aware (as geologists should be) of their organization's standards for map size, scale, symbols, and line weight. As a broader guide, Appendix B shows generally accepted symbols for geologic maps. A U.S. Geological Survey professional paper on "the logic of geologic maps" (Varnes, 1975) is a recommended reference for all geologic mapmakers. Varnes explains and illustrates the geologic and cartographic characteristics of maps used for engineering purposes, with examples of geotechnical maps from North America and Europe.

As with written communications maps depend on "style" to emphasize a message. A map with a confusing format, unidentified units, and misspelled words will cast doubt upon the entire message. Table 17-2 can be used as a checklist of things that should be shown, clarified, or at least considered on all geologic maps and sections.

Computer graphics, a part of the communications revolution, has become a valuable asset to geologic data reduction and drafting, but its applications are confined to work that is inherently objective and mechanical. Point data and linear data maps, simple isoline plots, some kinds of block diagrams (Fig. 17-10), and most statistical diagrams can be handled economically by computer

Table 17-2. Checklist for Exploration Maps and Sections

---

1. Blunders:
   Numbers should be correct and names spelled correctly.
   Place names in the report should appear on the map.
   Geologic data at the margins of the map should conform with those on adjacent sheets (or the lack of conformity be explained).
   Photographically enlarged or reduced data incorporated on the map should be at the exact scale intended.
   Data from drafting material that is not stable with changes in temperature and humidity should be checked for distortion.
   Control points on overlays should match.
2. Title or subject, location of the area, and an extra identification or indexing notation on the outside or in an upper corner, and a reference to the associated report.
3. Compiler's name, the names of the field mappers, and an index diagram identifying sources of data.
4. Date of field work and date of compilation.
5. Scale: graphic and numerical.
6. Orientation of maps and sections, with magnetic declination shown on maps.
7. Isoline intervals and datum, with sufficient line labels and an explanation of heavier lines, dashed lines, and changes of interval.
8. Legend or explanation, with all units, symbols, and patterns explained.
9. Sheet identification and key, where more than one sheet or a series of overlays is involved.
10. Lines of cross section on the maps and map coordinates or key locations on the sections.
11. Reference grid and reference points.
12. Clarity: All areas enclosed by boundary lines should be labeled, even in several places if necessary, so that they can still be identified if photo-reproduction in black-and-white changes the color pattern into shades of gray. Lines and letters should still be distinguishable after intended reductions in size. A trial reproduction can help in selecting colors and line weights.
13. Size: The size should be compatible with reproduction equipment.

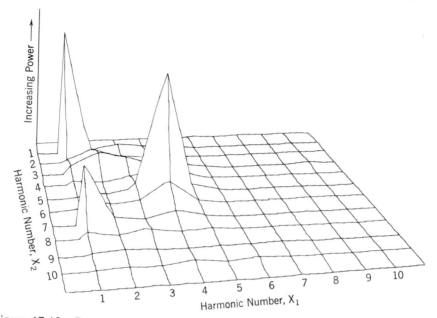

Figure 17-10   Computer-drawn block diagram showing a power spectrum associated with a ripple mark in upper Precambrian sandstone. (From J. C. Davis, *Statistics and Data Analysis in Geology*, Fig. 6.27, p. 369. Copyright © 1973, John Wiley and Sons, Inc. Used by permission of the publisher.)

graphics, but geologic maps and sections are normally too complex to be handled at a reasonable cost. Computer graphic techniques are most effective where large amounts of point data, repetitive calculations, and well-understood (well-explored) relationships are involved. The plotting of laboratory geochemical data, most geophysical data, and most ore-reserve data fits this pattern, but the presentation of most exploration geologic data does not. In actual practice the computer is used where its speed and economy outweigh the need for interpretation and subjective reasoning.

Computer graphic output is obtained in alphanumeric characters on a normal line printer (Fig. 17-11), as lines drawn by an incremental plotter, and as a cathode-ray tube (CRT) display. The cathode-ray tube approach can be used as an interactive system where various parameters are changed and the computer is directed to make new calculations. Picklyk and Ridler (1975) provide a good description of an interactive computer graphics approach to geochemical and stratigraphic studies in the Kirkland Lake mining region, Ontario. For an overview of systems, equipment, and techniques, *Computer Graphics,* edited by Parslow, Prowse, and Green (1969), is recommended.

For an exploration venture or a mining operation, certain aspects of verbal,

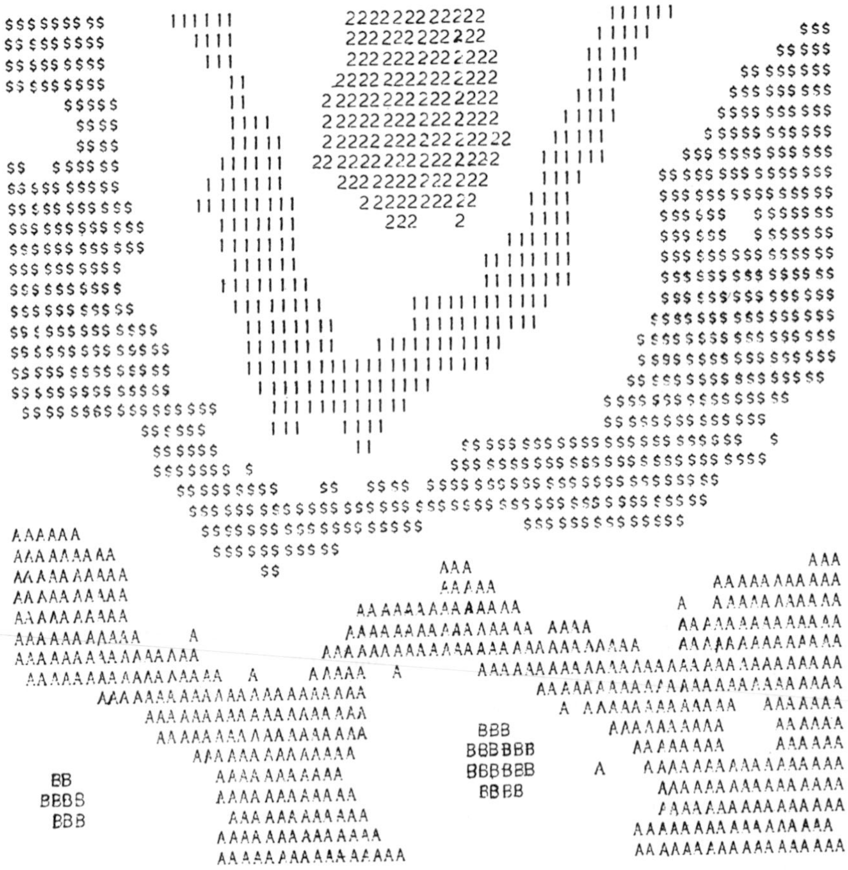

Figure 17-11 Contour map produced on a line printer. Contour interval is approximately 8 m. Contour lines are represented by the edges of the patterns. (From J. C. Davis, *Statistics and Data Analysis in Geology*, Fig. 6.8, p. 321. Copyright © 1973, John Wiley and Sons, Inc. Used by permission of the publisher.)

written, and graphic geologic information require emphasis and certain aspects require elaboration. A geologist who has a feel for the ultimate use of the information knows what to emphasize and what to explain and knows how far a particular format, checklist, or computer program needs to be stretched or compressed. The communications revolution has relieved geologists of some of their routine work, but it has not relieved them of any of their responsibilities.

PART FIVE
THE GEOLOGIST'S ROLE IN EXPLORATION AND MINING

A well-conducted geologic investigation is something to be proud of, and the discovery of a mineral deposit is an even greater source of satisfaction. If the new mineral deposit can be mined within acceptable bounds of economics and technology, so much the better. But there is one more aspect. The deposit must fill someone's particular needs. In essence, the work must be done for a specific purpose and—by inference—according to a specific plan.

In Chapter 18, the plan is called an exploration program; it begins with a need for minerals and, if successful, leads to a newly discovered source of minerals. In Chapter 19, the plan begins with a known mineral occurrence and the need for a mine; the plan, now carried out as a prospect evaluation, is intended to end with a recognized economic mineral deposit. Chapter 20 places the geologist in a longer term situation where a potentially minable deposit is made into a mine and where the mine continues to produce minerals.

Chapter 1 began with a statement about the depletion of mines, Chapter 20 ends with a consideration of the geologist's role in preventing the depletion of orebodies in mines. Actually all of a mining geologist's work is directed toward preventing the depletion of mineral resources—in mines, in nations, and in the entire world.

CHAPTER 18
EXPLORATION PROGRAMS

*Most mineral-resource appraisals integrate two components: an extrapo-
lation of all known results of previous mineral production and mineral
exploration activity, and a theoretical estimation of minerals in existence
in the ground. Both are very hazardous fields. After all, reserves can be
inventoried, but undiscovered resources cannot; they can only be
guessed at. To a large extent, the guesses depend upon subjective
experience and judgment and are subject to constant upgrading. Any
such appraisal is in a constant dynamic state of flux, changing with
geological knowledge and economic conditions. There are no once-and-
for-all resource estimates.*

P. A. Bailly[1]

Given a large number of tries we should in theory find the exact number of
deposits or tons of ore that a resource estimate indicated should be found.
Yet, as Bailly has pointed out, undiscovered resources cannot be inventoried.
To discover new ore deposits we must deal with changing conditions, we
must be exploration geologists, and our strategy must have a geologic basis.

This chapter begins with a look at exploration organizations and their
objectives. Consideration is then given to the patterns and economics of
exploration programs and to the applications of operations research. Some

[1] P. A. Bailly, "The problems of converting resources to reserves," *Mining Enginerring*, v. 28,
no. 1, p. 27–28, 1976. Copyright ©1976, The American Institute of Mining, Metallurgical, and
Petroleum Engineers. Reprinted by permission of the publisher.

examples of exploration strategy and tactics are cited from actual case histories.

If the previous sentence sounds military, so does some of the other exploration vocabulary: reconnaissance, objectives, logistics. The terminology invites comparisons that can give us some insight. Exploration strategies, the grand plans, are developed over maps and models. Exploration tactics call for groups of people to occupy new ground—always within the dictates of weather and terrain. There are massive demands on human endurance and machinery. Morale is important, communications essential, politics often overriding. Grand plans may be left in shambles by small groups operating outside the "rules"—guerrillas moving in and out of the shadows and taking advantage of tactical flaws. At some point a calculated risk must be taken. If no risk is taken, the result may be stagnation, loss of objective, or worse.

An exploration manager knows that strategy, however brilliant, cannot make a mineral discovery; discovery is for field geologists and engineers. As a veteran, the manager understands their capabilities and does not waste the available forces.

To carry the analogy a bit further: wild sorties into the chaos of an exploration rush can bring about a mineral discovery, but often at a too high cost or an inadequate reward. But the military comparisons should not be carried too far; mineral exploration builds rather than destroys.

18.1. EXPLORATION ORGANIZATIONS

Unless they have an uncommon capability for independence, mining geologists work within a private or governmental organization. The organization may be as elementary as that of a small syndicate with a few geologist-partners or as complex as that of a large multinational corporation or a national government mining company with hundreds of geologists plus laboratories, service departments, and research groups.

Private Enterprise Companies

The levels of responsibility that might be found in an exploration group of a corporation that acquires, develops, and operates mines as well as markets the product are shown in Figure 18-1. This type of organization undertakes exploration programs, evaluates prospects, and works continuously to extend reserves at its mines. There are other organizational structures, but most of them have similar characteristics. These characteristics, presented in more detail than can be shown here by Bailly and Still (1973), include the following:

1. A headquarters group of specialists in mineral economics, technology, and geology to assist top management in defining objectives, conceiving

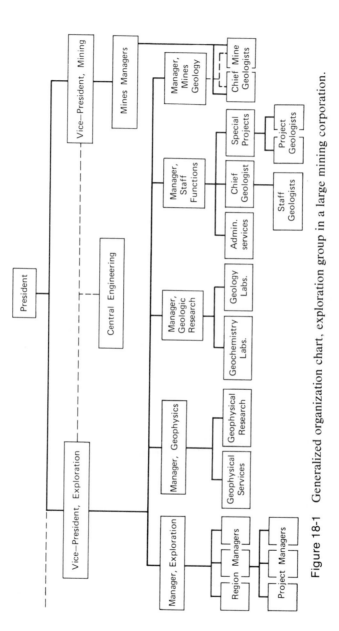

Figure 18-1   Generalized organization chart, exploration group in a large mining corporation.

and approving exploration programs, budgeting funds, and evaluating the results of exploration programs.

2. Regional units assigned to managers who have responsibility for initiating exploration programs and the authority to make relevant decisions. New exploration programs are often conceived at this level rather than at the headquarters level.

3. Project teams, each with a manager who makes a great many geological interpretations and is empowered to make tactical decisions provided they do not require a change in objectives.

4. Geophysical, laboratory, and other services—where not obtained from outside sources—centrally located and sometimes associated with an exploration research group.

5. Resident geologists—geologists at operating mines—reporting to the general manager of the mining unit but also responsible to the exploration group, generally through a manager of operating properties geology. The dual responsibility reflects a common situation. Some of the best exploration targets are associated with known orebodies and are best understood by geologists with local experience.

Some large mining companies have a separate "development" organization to provide the transition from discovery to production. Geologists, and especially geological engineers, are key persons in the innovative environment of this organization where mineralogy relates to metallurgy, structural geology relates to geomechanics, and the geological characteristics of a mineral discovery are the parameters for mine design.

While the organizational structure may appear similar from one company to another, there are considerable differences in the mode of operation. Some companies depend more heavily on their own groups of specialists familiar with particular commodities and exploration models or methods. Other companies emphasize the capabilities of their regional groups. As noted by Miller (1976) the lack of a stereotyped operational system indicates that exploration is in a period of growth—still searching for the best approach to mineral discovery. Miller's comments on corporations, ore discovery, and the geologist, based on a lecture presented to the Society of Economic Geologists, should be read by all who intend to practice mineral exploration as a career.

Obviously, some mining companies are much more successful than others in their exploration efforts, and a certain part of their success can be related to organization and management. But how do we gauge success in exploration? Certainly not by simply adding up the tons of new metal or mineral production brought on stream as a result of exploration, because an orebody may be barely economic and exploration may have been inordinately expensive. In a more acceptable index the present value of exploration expenditures throughout a period of years is compared with the net present value of deposits discovered during the same time. Regan (1971) used this latter index as one criterion in a study of the characteristics of companies "more successful" and "less successful" in their exploration in the United States and Canada. Here are some of his findings in regard to the more successful companies:

1. They are more likely to have a permanent group responsible for the transition from discovery to production.
2. They place more importance on the competence of exploration personnel, and they tend to provide in-company training.
3. They plan further into the future. They are likely to establish goals and objectives for 10–20 years, whereas unsuccessful companies appear to plan in terms of 1–5 years.
4. They maintain significantly higher exploration budgets, long-range factors are important influences in their budgets, and the responsibility for the exploration budget is accepted by top management rather than delegated to a lower level.
5. The quality of communication is better. The "top brass" is aware of the ideas generated by company geologists, and there is an awareness of the need for long-range programs.
6. Morale of exploration personnel—a major "people" factor—is considered to be more important.

Each of the first five characteristics noted above has some indication of continuity and long-range planning. The last characteristic—keeping a "happy ship"—has the same element. Geologists who see their projects and perhaps their jobs turned off and on by short-term economic signals will eventually associate with a more successful company. Companies keyed into one commodity or one major mine are especially prone to have jittery exploration departments. A depressed metal price or a down turn in mine profits requires belt-tightening. Which cost can be most easily cut back? Exploration. Diversified companies, those deriving income from several minerals and mineral-based products, are often in a better position to keep their exploration efforts—and their geologists—adequately supported.

Corporations dominated by manufacturing rather than mining emphasize the market area and the minerals needed for direct plant use. In this atmosphere, the exploration department will most likely report to an operations vice president who, understandably, may be less than fascinated by opportunities for new mining ventures as such. Here geologists are especially oriented toward economics, and they will more often be involved in prospect evaluation and in buying developed orebodies than in regional exploration. There is, of course, the situation in which a manufacturing firm or petroleum company makes a major thrust into the minerals field—call it "diversification" or "external growth"—and crowds its resources behind a small or hastily formed exploration department. The financial support may be more than adequate, but there may be pitfalls. The lines of responsibility may lead directly from a geologist to an executive who, albeit full of enthusiasm, does not yet understand the long-term effort needed in mineral exploration.

There are supercompanies—consortia and joint-venture organizations—that share the high cost and high risk of specific exploration projects. In these organizations geologists commonly wear two hats; they keep an identity with their own firm but have a more direct responsibility to the special company.

An exploration organization may be part of a contracting and consulting firm. In most of these a particular service, such as geophysical surveys or exploration drilling, is emphasized, but some groups offer a complete service from exploration to development and beyond. The firm may be private and independent, may be part of a nonprofit research foundation or a university, or it may be the "outside contracting" division of a mining company. Some, such as Czechoslovakia's Strojexport and Romania's Geomin, are government enterprises that offer exploration services both at home and abroad.

Governmental Exploration

Most of the world's mining geologists work directly for governments, in geological surveys, bureaus of mines, and national mining enterprises. Many others are involved in joint efforts between private and government organizations. There are, in fact, a great many important fields of cooperation—and competition—between government and private interests, and these are handled in almost as many ways as there are countries.

Functions of the U.S. Geological Survey do not normally include exploration for specific ore deposits, but they do extend to exploration research (especially in geochemistry), geologic mapping in areas of potential mining interest, regional geophysics, and a continuing resource appraisal of federal lands. The U.S. Geological Survey also sets guidelines and monitors work in exploration being done by private firms with government loans or for federally leasable minerals. Thus, the Survey's principal objectives are to provide information and to act as the government's mineral manager. There are times, however, when serious mineral shortages call for additional steps; then, the U.S. Geological Survey and the U.S. Bureau of Mines perform enough detailed exploration to identify targets for private development. Where government lands are being considered for classification as wilderness areas or for withdrawal from mining activity, the U.S. Geological Survey and the U.S. Bureau of Mines engage in a type of exploration—an evaluation of the area's potential mineral value.

The Geological Survey of Canada has a pattern similar to that of the U.S.

Geological Survey, and it has the additional responsibility of promoting mineral exploration in Canada's federally administered Northwest Territories and the Yukon. Mexico's Consejo de Recursos Minerales works very closely with private industry, loaning exploration equipment and personnel to small firms where needed and entering into operating agreements with private companies for the mining of investigated areas. The French Bureau de Recherches Géologiques et Minières (BRGM) has a basic scientific research and information function, but it also engages in cooperative exploration programs and joint ventures with mining companies in foreign countries as well as in France.

In socialist countries, exploration may be the responsibility of a central organization, such as the Soviet Union's Ministry of Geology and Conservation of Mineral Resources, or it may be done within a specific government enterprise, such as Yugoslavia's Trepča Lead-Zinc Combine. Trepča's exploration office performs regional exploration, coordinates geological work at more than a dozen mines, and participates in joint ventures with other national mining combines.

In some nations with private enterprise or "mixed" economies, mineral exploration is entirely government managed, as with Argentina's Fabricaciones Militares. In some other nations, exploration for specific commodities is government managed, as with Australia's Atomic Energy Commission and Britain's National Coal Board. Government exploration and mining organizations in these countries generally operate separately from the national geological surveys. Some government-private relationships are more complex, as in Sweden and Spain. In Sweden, LKAB—the government iron corporation—does its own exploration work but it also engages the Geological Survey of Sweden as an exploration contractor; the Geological Survey has—similarly—performed exploration work for the Boliden Mining Company, a private firm. Adaro, the Spanish government exploration company, operates on behalf of private participants, performs its own exploration, and explores on behalf of the Instituto Nacional de Industria, an organization comprising some 70 companies with government, private, and mixed capital.

A formula often found in developing countries is typified in Mauritania's Société Nationale Industrielle et Minière (SNIM), an organization that does exploration work on its own account, serves as a local contractor for foreign companies, and acts as the statutory national partner with foreign companies.

"Supercompanies" have a counterpart in "supergovernments." These are multinational development projects, foreign-aid programs, and the exploration projects of the United Nations Development Program. The project organization is most often integrated with that of the host country but, as with most joint ventures, the participating geologists normally retain an identity with their own national bureaus while on assignment to the international program.

Mineral exploration is predominantly an organization affair, and an exploration geologist will usually practice the profession within an organization; that much is clear. In addition it is likely that a geologist will work for several organizations within the course of a career. Economic cycles, changes in mining's geographic emphasis, and changes in corporate objectives cause a geologist to make entrances and exits in a great many different scenes while wearing the colors of different organizations.

18.2. THE OBJECTIVES OF EXPLORATION

We have an organization. Our colleagues understand geology, technology, and economics. We have information, tools, and methods of analysis. None of these sets us apart from geologists in research and academia. But our objectives do. Find orebodies—the orebodies needed by *our* organization. Find them at a reasonable cost. Resources, even "conditional reserves" will not do, because we cannot mine probabilities. A trifling discovery will not do because it will not be worth the effort.

We might let the mineral deposits—or exploration targets that seem to meet our objectives—come to us as mineral property submittals. A few decades ago, this was what exploration amounted to, and geologists did their field work wherever a particular group of claims or someone's exploration concession had been offered for consideration. Certain areas were considered to be more favorable than others, but the main consideration was that attractive-sounding mineralization had been discovered. The practices of mineral property evaluation (the theme of Chapter 19) are still valid and the examination of prospects is still very much in vogue; but prospect examinations now tend to be part of larger schemes: exploration programs. It is a matter of how discoveries are made; fewer and fewer are being made by conventional surface prospecting alone. And it is a matter of cost; the discovery of a hidden orebody has become too expensive to be attempted without a strategy plan.

The change from discovery by "conventional" surface prospecting methods, to discovery through interdependent geological, geophysical, and geochemical exploration programs has been documented by Derry (1970). Among deposits found in Canada prior to 1950, 85 percent had been found by conventional prospecting. The percentage then decreased substantially to 46 during the period 1951–1955 and 25 during the period 1956–1960; 31 percent of the discoveries made during 1961–1965 were credited to conventional prospecting and only 9 percent were credited to conventional prospecting in the period 1966–1969. Derry has also shown the increase in exploration cost during the same period. Exploration expenditures amounted to 0.8 percent of Canada's metallic mineral production in 1950, 2.4 percent in 1955, 3.2 percent in 1960, and 4.0 percent in 1965. Derry's method and cost figures show a trend, but they are not intended to downgrade the significance of exposed mineralization and value of field geology. The newer and more expensive approaches are still used in combination with field geology, and the most widely accepted field evidence of deeper mineralization is still an occurrence of shallower mineralization.

Whatever the strategy—prospect examination or exploration program—exploration is generally quite active in areas where known mineralization offers strong encouragement. The presence of large mines and large orebodies is so clearly encouraging that a familiar saying in exploration work is: "If you are hunting elephants, go to elephant country." Wise enough and statistically defensible, but the cost may be inordinately high. The most obvious elephant country is sometimes crowded with well-equipped safaris, hunting licenses are likely to be expensive, and the most visible trophies will have already been taken. Many exploration geologists and exploration organizations prefer a more flexible combination of statistical and geological encouragement; they identify potential elephant country and look for the biggest elephants there. An increasing number of important discoveries are being made in just this type of terrain, in areas that had been ignored by less imaginative geologists. The choice between elephant country, potential elephant country, or—sometimes—country that is populated by small game, is controlled by a particular objective. And the objective is a function of many determinants, some of which will be discussed in the next few pages.

Economic Determinants

The economic environment has a lot to do with the size and kind of target sought, and this in turn influences the exploration procedure. If exploration is being done in a remote inland part of Alaska, the minimum acceptable elephant may have to be a veritable mammoth, an orebody with several million tons of high-grade material or several hundred million tons of near-surface low-grade material. Anything smaller or deeper may not be minable under existing or projected conditions of transportation and production cost. With such a high economic threshold and with a high or weakly defined risk, a major part of the exploration budget is often committed immediately to a reconnaissance for the most obvious target. The cost per square kilometer is low, but the total cost is apt to be higher than a small organization can afford. Statistically (see "gambler's ruin" in Section 18.5), this is no place for small companies and individuals, but they are likely to be there anyway, looking for "shows" of mineralization that can be claimed and held at a low cost until a large company enters the scene. Small ambitious exploration groups operate in remote regions with enough success to embarrass statisticians and win admiration or jealousy from corporation geologists.

In an established mining district (older elephant country where the excitement has subsided) such as in Colorado or Montana, the minimum objective has more flexibility. A deeply buried low-grade ore target can be accepted if it is near a large-scale mining operation. A relatively small orebody can also qualify as a target for a large company if mining, processing, and transportation facilities are close at hand, especially if the known orebodies are soon to be exhausted. A high cost in detailed subsurface investigation and land acquisition can sometimes be borne, even for small orebody targets, if a low risk can be identified. In these surroundings, a statistical expression for risk is more appropriate than in unexplored areas because it can be based on quantitative data from a more suitable sample population.

If a target in a mining district has a 50 percent probability of containing a $10 million orebody (net present value), and a target in a virgin area has a 5 percent probability of containing a $100 million orebody, both can be said to have the same prospective value, $5 million. But whereas the first probability expression is well supported, the second expression is simply a way of saying "low." It might just as well be 2 percent or 1 percent, in which case the prospective value would be $2 million or less—approximately the cost of a major exploration program. The statistical example is primitive, but it points out that exploring for small as well as large orebodies may be attractive in well-known surroundings. Prospective value, or the *current exploration limit*, will be discussed again in Section 18.4.

Perhaps a large orebody, an elephant, is not really needed. The objective may be to find an industrial mineral deposit required to supply a chemical plant. As long as enough salt, barite, gypsum, or diatomite can be outlined for plant requirements, that is sufficient. Sometimes an industrial mineral deposit, a coal deposit, or even a specialty metals deposit may be wanted only as "insurance" against future interruptions in normal supplies. The smallness of an acceptable target does not mean that the company is small or that the exploration effort will be short of funds. On the contrary, the fate of an entire industrial complex may depend on the discovery, and a "saturation" search can be supported with expenditures that would seem fantastic if the market value of the ore were the only index.

There is another aspect to defining an acceptable size for an exploration target. The "risk money" needed to find and develop a small but rich orebody will be less and will be committed for a shorter time than that needed to find, *identify*, and develop a large low-grade orebody even though the statistical risk—discovery versus failure—may be the same. In certain instances, typified by exploration for shallow uranium orebodies in the Colorado Plateau region, the main effort is spent in searching for ore. Once enough ore is found to justify an entry, an early return can be obtained on the money invested. A small operation with concurrent mining, development, and orebody delimitation can go ahead until enough ore is outlined to justify an extensive exploration and development program. Exposure to risk is brief. Exploration for bulk low-grade orebodies, at the other end of the spectrum, is done with the knowledge that major costs will be incurred *after* ore mineralization is found. An extensive land position must be maintained, and the ore mineralization must be outlined by drilling until it amounts to millions of tons at the very least. Moreover, the entire mineralized body must be investigated for patterns of grade, mineralogy, and geomechanical conditions before it can be properly called an orebody. Some companies simply do not want to commit their exploration funds for such a long time in a single project.

Suppose a company has ample risk money. It still may regard a local political situation as too unstable to justify an exploration program for large, low-grade orebodies that need a long preproduction period and an inflexible large-scale mining operation. Smaller high-grade orebodies, massive sulfide deposits, for example, may seem to be more attractive targets than a porphyry copper deposit.

Changing Conditions—New Objectives

Mineral exploration objectives are sensitive to the entire interplay of markets, geography, science, and technology identified in Figure 18-2. Market conditions—time, place, and unit value of minerals discussed in Chapter 8—affect the intensity of exploration and the materials sought. Geographic factors, sometimes as obvious as the excitement caused by a bonanza discovery and sometimes as obscure as a trend in politics, have a bearing on where to explore. Access, in the physical sense of a new transportation route or in the

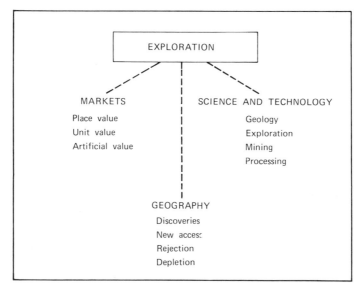

Figure 18-2   Exploration as a function of markets, geography, science, and technology.

abstract sense of a new open-door economic policy, draws exploration from other areas where shallow orebodies are being depleted and mineral development has been discouraged or rejected.

Geologic science and the technology of exploration provide springboards to new field investigations. Certain large orebodies are recognized as deformed stratabound deposits rather than epigenetic deposits. Where are some other occurrences with similar characteristics? Explore. Find them. A new geochemical method is developed. Where should it be most clearly applicable? Go there.

Mining technology may take on a new aspect. Exploration will follow suit. Development work at a mine reaches a record depth; other deep exploration targets in other mines are reassessed. A new hydrometallurgical plant is built in Colorado or Arizona; geologists begin exploring for orebodies in Nevada or New Mexico that could support similar plants.

18.3. PATTERNS OF EXPLORATION PROGRAMS

From an intricate blueprint of major and minor objectives, assume an exploration model, a body of some anticipated size, grade, depth, and mineralogy, in a certain geologic environment and with a certain expected response to the tools and methods of exploration. Now we can search for targets in places where conditions are most likely to fit the model.

The Pima Discovery

In the fall of 1949, two geologists had developed an exploration model (Heinrichs and Thurmond, 1956, Heinrichs 1976). There was a favorable market for copper, and the desert area of the Southwest—Arizona and New Mexico—was "elephant country" with uninvestigated alluvium-covered areas. Geologic information on porphyry copper deposits was accumulating, and the success of large-scale mining operations had been demonstrated. And there was a dominating aspect, the real nucleus of the exploration model. Newly refined geophysical methods were available. The model was attractive enough to encourage the formation of an organization and begin regional study. The organization was small and the area was large—normally an unfavorable ratio—but the opportunity was unique.

The exploration model called for a significant geophysical response that could be recognized through alluvium. Matching this requirement with the available geologic literature, 20 mining districts were selected as favorable areas. One of these, the Pima district, Arizona, appeared to be the most appropriate. Small mines had produced $20 million worth of metals, including copper, and there were apparent structural controls to the mineralization that could be projected into the alluvium-covered pediment. Even better, some of the orebodies were associated with tactite zones which could be expected to cause magnetic and electrical anomalies.

Samples from the surface and from old underground workings in the Pima district supported the expected geophysical contrasts. Simulation experiments and orientation surveys over known mineralization provided tentative geophysical response patterns.

During a magnetic reconnaissance survey an anomalous response was detected in the alluvium 600 m to the east of the Mineral Hill mine in an area of projected geologic favorability. Detailed magnetometer work outlined an important "high." Confirmation was needed. After testing several geophysical techniques, a suitable technique was found and a target was confirmed. Information from resistivity surveys in the target area was combined with geologic projection to obtain an estimate of alluvium thickness: 65 m ± 25%. It was time for the "moment of truth" mentioned in Chapter 15. The first drill hole entered bedrock and oxidized mineralization at 65 m: technical success. Sulfide ore was encountered at 80 m: discovery. Fifteen drill holes followed, then an exploration shaft, underground workings, more drill holes, and finally, development of the Pima Mine.

Much more followed. Other organizations showed increased interest in the area; exploration led to additional discoveries and ultimately to one of the largest concentrations of large-scale copper mines in the world. Six major mines in the Pima district, all brought into production since 1956, now produce 80 million tons of ore per year.

The Pima discovery and the case histories cited in Table 18-1 can be used as illustrations of exploration programs up to a point. Imitation will get us nowhere. Case histories refer to *found* orebodies; we need to seek new ones. Each program has its own objective, and each combination of determinants creates unique situations.

Table 18-1 shows exploration methods that were used in each program, but it does not indicate any single method as being *the* discovery method. Most discoveries cannot even be credited entirely to a program. Many excellent programs have been unsuccessful, and their case histories would make as good reading as case histories of programs that have been successful, but for obvious reasons such histories are harder to find. Some brief but well-stated comments on projects that have gone astray can be found in a series of abstracts in *Economic Geology* (1970, v. 70, p. 244–252); even the titles are intriguing: "Success derived from error . . ." and "A postmortem review . . ." An additional commentary that is well worth reading because it points out the tendency for geologists to expect too much from a tightly formulated exploration program is *The Ore Finders* by Peter Joralemon (1975), the compiler of the purposely over-lineamented metallogenic map shown in Chapter 4.

Table 18-1.   Case Histories in Exploration

		Discovery Year	Climate	Topography	Geological Terrain	Conventional Prospecting	Photogeology	Surface Geology	Subsurface Geology	Airborne Geophysics
Aitik, Sweden	Cu	1956	sA	F	pЄ	X		X	●	●
Arlit, Niger	U	1965	Tr	F	P		●	●	●	
Blind River, Ontario	U	1953	Te	M	pЄ	●		●	●	
Bougainville, Papua New Guinea	Cu	1965	Tr	R	C	X		●	●	
Brunswick, New Brunswick	Pb-Zn	1952	Te	M	P	X	●	●	●	●
Carlin, Nevada	Au	1962	Te	M	P			●	●	
Faro, Yukon Territory	Zn-Pb	1965	sA	R	D	●		●	●	●
Jackpile, New Mexico	U	1951	Te	M	M			X	●	●
Kalamazoo, Arizona	Cu	1965	Te	M	D			X	●	
Kambalda, Australia	Ni	1966	sTr	F	pЄ	●		●	●	
Kidd Creek, Ontario	Zn-Pb-Cu	1963	Te	M	pЄ				X	●
Labrador Trough, Canada	Fe	1952	sA	M	pЄ	●	X	●	●	●
La Caridad, Sonora	Cu	1967	Te	R	D		●	●	●	
Mattagami, Quebec	Zn-Cu	1957	Te	M	pЄ	●	●	X	X	●
Mjövattnet, Sweden	Ni-Cu	1973	sA	M	pЄ	●		●		
Palabora, South Africa	Cu	1962	sTr	F	pЄ	●		●	●	
Pine Point, N.W. Territories	Pb-Zn	1965	sA	F	P			●	●	
Ranger, Australia	U	1970	Tr	F	pЄ		●	X	●	●
Schaft Creek, British Columbia	Cu-Mo	1966	sA	R	M	●		●	●	
Thompson, Manitoba	Ni	1956	sA	M	pЄ		X	X	●	●
Tynagh, Ireland	Pb-Zn	1961	Te	M	P	X		X	●	
Viburnum, Missouri	Pb	1955	Te	M	P			X	●	X

Ground Geophysics					Geochemical				Reference
Electromagnetic	Electric	Magnetic	Gravimetric	Radiometric	Rock	Soil	Stream Sediments	Additional Methods	Reference
●	●	●	X			X		moraine sampling	Malmqvist and Parasnis (1972)
				●					Bigotte and Molinas (1973)
				●					Joubin (1954)
						●	●		MacNamara (1968)
●		X	X			●	●		MacKenzie (1957)
					●				McQuiston and Hernlund (1965)
●		●	●			●	●		Brock (1973)
				●					Fitch and Herndon (1965)
						X			Lowell (1968)
	●	X				X			Woodall and Travis (1970)
●									Donohoo and others (1970)
		●	●						Gross (1968)
	●				X	X	●		Coolbaugh (1971)
●	●		●						Paterson (1966)
●		X	X		X		X	boulders dogs	Nilsson (1973)
				●					Hanekom and others (1965)
	●	X							Seigel and others (1968)
X	X	X	●			●			Ryan (1972)
●	●				X		X		Linder (1975)
●		●	X						Zurbrigg (1963)
●	●					●		boulders	Derry and others (1965)
		X	X						Weigel (1965)

Key

sA = subarctic
Te = temperate
sTr = subtropic
Tr = tropic

R = rugged
M = moderate
F = flat

C = Cenozoic
M = Mesozoic
P = Paleozoic
pЄ = Precambrian
D = Diverse

● = major method
X = minor method

Discovery year refers to most significant economic discovery

The Regional Exploration Sequence

Exploration programs have no universal pattern, but they have some common features that are generalized in Figures 18-3 and 18-4. Three stages can be identified: program design, reconnaissance, and detail. A fourth stage, prospect evaluation, is a part of the overall programs, but it has so many special characteristics that it is treated by itself in Chapter 19. The geologist's role in preproduction work and mine development, an epilogue to the exploration sequence, is discussed in Chapter 20.

The accent here is on ore deposits and industrial mineral deposits. Coal exploration programs follow a similar pattern and make use of many of the same techniques, but coal geology and coal economics require special consideration. The special aspects are emphasized in the proceedings of an international coal exploration symposium (Muir, 1976).

As work flows from one stage to the next, a great many difficult field decisions must be made. These result in rejections and revisions, shown in the "feedback" paths in Figure 18-3. Data from rejected targets, reconnaissance

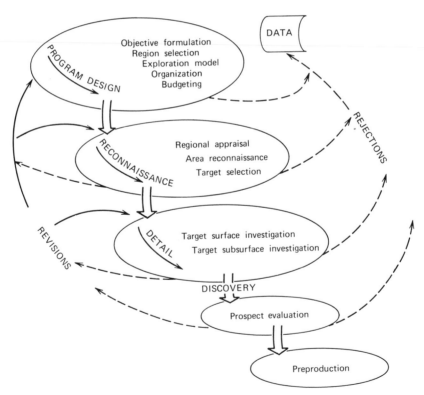

Figure 18-3 Generalized regional exploration sequence.

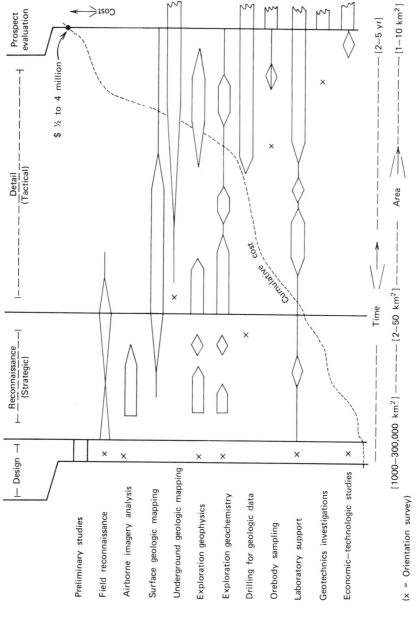

Figure 18-4 Generalized activities in a regional exploration program, assuming an underexplored region with a few existing mines and prospects.

areas, and entire programs are collected in a data bank for future use. The data bank is important because programs laid to rest this year may be resurrected next year when budgets, personnel, concepts, or objectives have changed. In addition, information from rejected targets is part of the regional information and a part of experience. It must be preserved because it will be needed.

Concepts that can change the assumptions used in a prior stage or even in the exploration model itself come to light; we *have* to revise even though it may mean that some of our most comfortable illusions are destroyed and the manager's peace of mind is affected. Long before the word "model" became popular, T. C. Chamberlain referred to a comparable need for flexibility when he argued that field geologists should use "multiple working hypotheses" to avoid following the wrong path too far. He recognized that a hypothesis, such as our exploration model, is needed, otherwise we would wander aimlessly— but "while a single working hypothesis may lead investigation very effectively along a given line, it may in that very fact invite the neglect of other lines equally important" (Chamberlain, 1897, p. 845). Chamberlain's classic article is timeless and worth reading in its entirety.

A premise in Figure 18-3 is that exploration programs begin in large uninvestigated regions because of a need, an opportunity, or an idea and that they lead step by step to orebodies; many do. Others, such as the Pima mine discovery program, have the advantage of a few well-investigated areas within the region. Some begin with an attractive prospect, someone's prior discovery, and the entire program amounts to a target investigation. An exploration program may grow inside out and introduce new ideas from a mine or a prospect into a larger area; an example was mentioned in Chapter 10, the Bou Azzer–Anti-Atlas area reconnaissance. Sometimes there is long-term "busy work" with no apparent pattern or program; this may produce a kind of geological serendipity which, for all of its lack of organization, has been successful on occasion. Finally, exploration programs may be ignored in the excitement of a mineral rush, and this, too, has sometimes been successful.

"Success" would be a hollow word if it did not imply the right to mine a discovered orebody. Whether the exploration organization is governmental or private the means of acquiring mineral rights are critical factors in deciding how to explore.

Acquiring the Right to Explore and Mine

A perspective view of mineral ownership and governmental formulas for mineral management was provided in Section 7.3. Now, a closer view can be taken in terms of an exploration commitment. The reconnaissance stage of a regional exploration program requires an understanding that mineral rights *may* be obtained. Target investigation demands a "land position"—an assurance that mineral rights *can* be obtained. Finally, in the stage of prospect evaluation, the terms of mineral ownership or leasing *must* be spelled out.

Most governments recognize the stages in exploration and provide the three general types of agreements identified by Ely (1970)—reconnaissance permits, exploration licenses, and production concessions.

Reconnaissance permits, other than those dealing with environmental protection, are not required on government lands in the United States. Where they are required, on certain Indian reservations and in many foreign countries, these "hunting permits" convey the right to explore an area but not necessarily on an exclusive basis. They identify the permittee as a legitimate prospector, allow him to make minor excavations, and they may enable him to apply at some later time for exclusive mineral rights in smaller areas. They often include special dispensations for employing foreigners and importing equipment. The reconnaissance area is defined by statute or in a negotiated agreement specifying its size and location. The permit is valid for a certain number of years, generally with a provision for renewal if the permittee continues to be active.

Exploration licenses in some form are part of every nation's mineral law. In the United States they amount to prospecting permits for federally leasable minerals and federally controlled offshore areas. Many individual states also issue exploration permits for their lands. Exploration licenses are exclusive rights to minerals within defined areas, and they usually carry an assurance that if exploration is successful, the licensee will have preference in obtaining a production lease or concession.

Governments generally make sure that licensees will keep their exploration work active by limiting the period of tenure, requiring a certain amount of work, and imposing rentals on the surface area. In addition the permittee may be required to release a certain percentage of the area at the end of specific periods of time.

A group of mining claims staked for "locatable" minerals on U.S. federal land may serve the purpose of an exploration license, but the ultimate right to occupy the land is based upon a mineral discovery rather than upon the recognition of favorable ground.

Production concessions are exclusive rights to mine and sell minerals. These rights may be in the form of mining claims or they may be leases or concessions—written contracts with the government. The rights are usually transferable to new owners and can be treated as private property in transactions that will be described in some detail in Section 19.2.

Clearly, one of the very first steps in exploration is to become familiar with the mineral law of the area. This may be difficult for an engineer or geologist, but the key to an economically sound discovery may rest here as well as with geology. In a country where mining is well established, mineral laws will ordinarily be quite specific, but they also may be plagued with overwhelming detail steeped in history and precedent. In long-settled regions with private mineral ownership, it may take years to piece together enough mineral rights for an exploration program. There are some countries with little mining history but with high hopes of industrial wealth in which the mineral laws amount to a poorly defined statement of aspirations. Then, there are concise and realistic mineral laws, as in Canada's Northwest Territories where three stages of exploration are clearly recognized.

A *prospector's license* gives the right to stake a certain number of claims in each year.

A *prospecting permit* gives an exclusive right to stake claims in remote areas for a period of 3 years. A minimum work requirement calls for spending at least 10 cents per acre (0.4 hectare) in the first year, increasing to 40 cents per acre in the third year. One-quarter of the permit area must be released after the first year, another one-quarter after the second year.

A *lease* can follow 5 years of assessment work ($100 per year per claim) or can be acquired as soon as production reaches 5 tons per day. The term, 21 years, is renewable. Royalties, beginning 3 years after first production, increase stepwise to a maximum of 12 percent on the net (after production costs) value of the ore.

18.4. ECONOMICS OF EXPLORATION PROGRAMS

All mineral exploration is based on the premise that an eventual discovery will result in a mine that makes the effort worthwhile. In Chapter 8 one side of this economic balance was considered: the value of the ultimate orebody. Now consideration can be given to the other side: the cost of searching for the orebody. The chance of missing the orebody—the risk—in 1 or 10 or 1,000 tries is weighed and placed at some point in the balance, as explained in Section 18.2. A very poor shot, even at an elephant, cannot be allowed to cost very much.

The Odds for Success—Measured and Estimated

How many exploration "tries" are successful? A generalized answer can be given that can be supported by a few statistics and some considered opinions. All generalizations require that something be ignored or at least omitted. In this generalization, definitions of deposits, anomalies, and prospects are ignored and a blend of good and bad exploration programs, prospect examinations, serendipity, and wild sorties are poured into a concoction called "sins of omission," which states that out of several hundred favorable sites for ore mineralization examined by geologists, one is likely to become an orebody. If the indications are strong enough for detailed surface investigation, one in 100; if attractive enough to drill, one in 50; after ore-grade mineralization has been verified, one in 10.

With better statistical support, a tabulation published by the International Atomic Energy Agency (1973, p. 295) shows that the investigation of 100,000 anomalies in the United States has resulted in 4,000 prospects among which 700 were classed as deposits. The ratio of anomalies to deposits, 143:1, amounts to a success rating of 0.7 percent.

More "scattergun" statistics. According to Koulomzine and Dagenais (1959), out of 4,865 mining properties held by active companies in Canada, 148 (about 3 percent) were profitable producers. More statistics. Roscoe (1971) calculated an overall trend for mineral discovery versus exploration "tries" in Canada: 1.0 percent in 1951, decreasing to 0.1 percent in 1969.

Statistics with better detail, although from a smaller sample population, can be obtained from the following statement. Five exploration firms working in the southwestern United States examined 352 prospects, conducted geophysical surveys on at least 47 of them, drilled 23, and judged 2 as warranting "possible development" (Perry, 1968). The ratio 176:1 gives a success rating of 0.6 percent.

During 1963–1966 Bear Creek Mining Company, a major American exploration organization, gave serious consideration to 1,649 possible targets. Of these, 60 were drilled and 15 new areas of mineralization were discovered, 8 of which had some tonnage potential. Five of these were significant "mineral deposits," and one was believed to be an orebody (Bailly, 1967). Giving credit to the five "mineral deposits" rather than to the single deposit meeting specific company objectives the success ratio was 330:1, amounting to a rating of 0.3 percent.

The exploration experience of Cominco Ltd., a Canadian mining company, reflects a similar pattern. The company explored more than 1,000 properties in 40 years; 78 of these warranted a major exploration effort. Eighteen mines were finally brought into production, but of these only seven were profitable (Griffis, 1971). The success rating can be called 0.7 percent. The Cominco experience emphasizes a point that can be made from the preceding paragraphs. The ultimate measure of success must be stated in terms of orebodies mined at a profit rather than simply in terms of new mines.

These few statistics on exploration success must be taken as intended by their compilers—as generalizations. The fact that each organization has its own objective at a particular time and place does not enter into the statistics, nor does the fact that rejected prospects are often reexamined—and made into mines—by other companies. According to Bailly (1972), 90 percent of the porphyry copper deposits discovered in North America since 1950 were found in copper mineralized areas recognized before 1950. These areas had been recognized, many had undoubtedly been considered as exploration targets, but the orebodies (the elephants) had been missed.

The opinions of two men with considerable first-hand experience in exploration can be considered: Arthur Brant of Newmont Mining Corporation and D. A. O. Morgan of the Consolidated Goldfields organization. Brant (1968) and Morgan (1969) have explained the factors that enter into successful exploration, and they both state that a working figure for the presence of economically successful prospects among all of those tested should be about one percent.

Most of these statistics and opinions have taken mineral indications, anomalies, and prospects as starting points; something already found. What about a regional exploration program that begins with an area rather than an occurrence? Now we are dealing with two kinds of units. How many starting points could have been assumed in the 80,000 km^2 of potentially ore-bearing terrain covered by Operation Hardrock, a regional exploration project begun in India during 1967? After 10 months of airborne electromagnetic, magnetic, and radiometric surveys, 1,100 anomalies were selected for ground geological reconnaissance. Perhaps these should be called starting points even though the actual number could have been modified by the amount of preliminary screening done. By the summer of 1972 over 700 of the anomalies had been investigated by ground geological, geophysical, and geochemical techniques; 25 of them emerged as targets for further work. By the fall of 1972, 12 of the targets had been investigated by drilling and 6 of these were designated as promising prospects (Subramaniam, 1972). If one of the prospects will eventually make a profitable mine, the original risk might be considered as 1100:1. Even though we are using dimensionless units to describe an incompletely investigated dimensional area, we are recognizing that each aeroanomaly required individual consideration.

Statistical models of mineral distribution will be mentioned in Section 18.5; these provide a useful index for quantifying specific situations, but when they are used in making an inductive leap from preliminary geologic assumptions to solid numbers, most exploration geologists treat them with well-founded suspicion. An exploration program is more likely to be based on numbers that are admittedly uncertain as it moves from regional appraisal (risk 1000:1?) to field reconnaissance (risk 500:1?) to detail (risk 100:1?). The geologic risk numbers are largely subjective, but they are useful because they show a trend in which the odds for success should increase. If the odds do not increase, then one of the feedback paths (revision or rejection) in Figure 18-3 must be taken.

Note that "risk" has collected a modifying term in the last paragraph to become "geologic risk." There are other kinds of risk, and they all have an influence on decisions to begin, to continue, or to abandon an exploration effort. The probability of running into excessive taxation or disastrous restrictions on mining can be called a political risk. Even in the most stable of governmental successions the signal of a new unfavorable attitude toward mineral producers can cause an exploration target to be rejected or at least shelved for the time being. Technologic risk, the probability of encountering overwhelming problems in mine development, mining operations, or mineral processing, adds to uncertainty. Marketing risk adds another dimension; if the metal or mineral price falls far enough to make a particular mineral body uneconomic, then finding the body can hardly be called success.

The Time Aspect

The odds for success have been stated without emphasizing the importance of time; yet, the investigations or examinations that have been equated as "tries" may represent days, years, or even decades of work. A thorough exploration program is shown in Figure 18-4 as taking 2–5 years to reach the prospect evaluation (post-discovery) stage. Once a discovery has been made, a comparable time is needed for prospect evaluation and an even longer time may be needed for evaluation and preproduction work. The investment in an exploration project is exposed to risk during this time, and risk investments are made with high-cost money.

The preproduction stage requires 1, 2, or 3 years where conditions are "easy," but it may take 10 years or more when conditions are difficult. Add preproduction time to the exploration and evaluation time spent in locating and identifying an economic mineral deposit, and a total *lead time* of 15 years or more may be needed. A great many examples could be cited, but two will suffice to illustrate the point.

The existence of a large copper orebody in the White Pine area, Michigan, was recognized in 1929, but production did not begin at the White Pine mine until 1955. Significant mineralization was verified at the head of the Leduc River, British Columbia, in 1948 (claims had been staked there in 1931); a joint-venture company, Granduc Mines Ltd., was formed in 1953, in 1970, after $115 million were spent in exploration and development work, the mine began production.

Mineral exploration has sometimes been characterized as gambling, but the characterization is only a handy figure of speech. Risk is recognized, odds are taken into account, money is "put on the table," and the need for persistence is accepted, but exploration is not a game of chance. Even the "chance discoveries" so widely reported in the history of mining required someone with self-taught or formally acquired skill in geology and engineering to recognize the economic significance of the discovery.

The Current Exploration Limit

Estimated risks in exploration and expected profits from discovery help to establish reasonable limits to what can be spent in an exploration program. Experience and imagination tempered with practicality provide the actual figures and fortunately leave some room for healthy differences of opinion.

Suppose that a risk can be expressed as 1 chance in 100 for the discovery of an orebody meeting the objectives of an organization within a selected target area. The most likely orebody might, for example, contain 20 million tons of ore, the most likely grade might give a revenue of $15.42 per ton, and the most likely operating cost might be $7.34 per ton. Say that nearby mines—the same ones used as sources for the tonnage, revenue, and cost figures—operate at levels of about 1 million tons of ore per year and would need a capital investment of $17 million if they were to be developed a few years from now. We now have a basis for calculating an expected cash flow, a net present value for the orebody, and a reasonable cost of exploration, as illustrated in Figure 18-5.

The cash flow calculation is the same one used for an explanation of such calculations in Section 8.5. The annual cash flow of $5,375,650 would have a

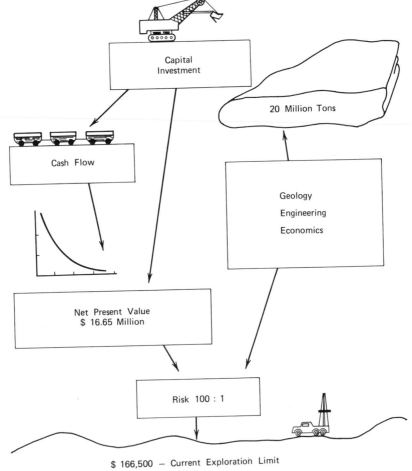

Figure 18-5 The current exploration limit as a function of the expected net present value of an orebody and a stated risk (odds against discovering the orebody).

present value of $33.65 million if taken for 20 years at a company-established discount rate such as 15 percent; the uniform series present value factor of 6.259 used as a multiplier for the annual cash flow was taken from Table 8-11. Subtracting the capital investment of $17 million, a net present value (acquisition value) of $16.65 million is obtained for the expected orebody. The maximum exploration expenditure that can be justified until there is an indication of decreased risk or higher profit can be called the *current exploration limit*; it is 1/100 of the net present value of the orebody, or $166,500. "Enhancement," the term sometimes applied to geologic or economic conditions proving to be better than expected, is an input to the current exploration limit. As soon as enhancement is in evidence the most likely orebody and its net present value is recalculated.

Obviously, a lot of assumptions, a few arbitrary factors, and even an occasional guess have gone into this calculation of a current exploration limit; this is why it includes the word "current." Tomorrow, a measurement may replace an assumption. One good ore intercept in the next drill hole might improve the risk factor to 1 chance in 50 and thereby increase the current exploration limit to $333,000. Any new evaluation of a factor in the geologic analysis or a reevaluation of mining cost should also result in a different exploration limit. The new information may of course be discouraging, in which case it will decrease the current exploration limit and, if the limit has already been reached, end the program.

Alternative assumptions and interpretations can be repeated to the point of making a sensitivity analysis (finding the most important indices to sharpen up) or a risk analysis (finding a probability function for the overall picture). Both of these approaches require a geologist's constant close attention if they are to be kept within reason, because large amounts of numbers have a tendency to become "laws" rather than to remain rational guidelines.

A current exploration limit is just one way of placing a working limit on exploration expenditures. A prospecting profit ratio and several other operations research approaches will be mentioned in Section 18.5. There are rules of thumb, and there is the "gut feeling" that a particular target *must* have been underrated or overrated. Objectives vary, geologic expertise varies, and competition for mineral land blows hot and cold. Thus, in actual practice, the amount of money to be budgeted for an exploration program and the amount to be spent in each step are determined in part by objective analysis, in part by subjective reasoning, and in part by personal qualities such as insight and experience.

The Costs of Exploration

Exploration costs can be considered in the context of an expenditure within an organization or in the more definitive context of a project. An analysis of corporate exploration expenditures (Morgan, 1969) has stated that large mining companies spend 3–20 percent of their annual cash flow (prior to taxation) in exploration and smaller companies may spend a substantially higher percentage. The smaller companies—with smaller mines—must spend more to survive and improve their position. Morgan has given an interesting cost and time estimate for a firm intending to explore for a large mineral deposit: $1.5 million to $3 million per year for 6–10 years. Wargo (1973) studied the expenditures of 21 major mining companies and found that the largest companies each spent on an average $11.5 million a year on exploration, middle-sized companies each about $3.5 million a year on exploration, and the smaller companies each about $850,000 a year on exploration.

The figures seem large, but consider the cost of maintaining a major regional exploration office of 15–20 people (geologists, geophysicists, field and technical assistants, office staff, and laboratory staff). This was estimated by Leaming, Lacy, and Skelding (1969) to be about $300,000–$800,000 per year—and the estimate does not include the cost of target area examinations and prospect evaluations. Regardless the odds for success, whether they are 1 in 500 or 1 in 5, each of the unsuccessful "tries" has its cost.

The cost of a "try"? Obviously, it varies with the entire range of objectives and capabilities. Leaming, Lacy, and Skelding (1969) found that the cost of exploration at 30 American mining properties that were finally developed into mines during the period 1947–1968 ranged from $10,000 at a small lead mine to $4.5 million at a large uranium mine. They found that the cost of exploring 42 rejected properties ranged from $4,000 for a small uranium prospect to $900,000 for a large industrial minerals prospect. In a detailed account of exploration expenditures in a 12-month exploration program in Northwest Queensland, Australia, Griswold and Wallace (1967) provide a more specific figure; an operation involving three geologists; three assistants, and four vehicles cost $64,000. Park and Mahrholz (1967) give a specific cost figure for a 6-month program of reconnaissance and detailed surface follow-up work in Brazil; the cost of exploration in a 1000-km^2 area with heavy plant cover and few access roads was $130,000.

The above expenditures were made at various times in past years. What relationship do these figures have to present-day exploration costs? Heinrichs (1976) provides an indication in his discussion of the 1950 Pima discovery mentioned earlier in this chapter. By the time ore was discovered in the first drill hole, the expenditures had been less than $25,000. "To repeat a similar effort today would cost from $50,000 to $100,000." (Heinrichs, 1976, pp. 3–4)

Suppose that each "try" were to cost $100,000 or $50,000 or even as little as $10,000; the need to maintain an exploration organization and to examine hundreds of interesting situations before discovering a large mineral deposit would support Morgan's generalized cost and time estimates.

Although cost estimates for an exploration budget are made under uncertain circumstances, the budget needs numbers. The approaches are similar to those mentioned in Chapter 8. In order-of-magnitude and preliminary estimates, unit costs are used and analogies are drawn between conditions in the area from which the figures were obtained and the new area. Where costs from a large number of sources or areas have been combined to give a range, costs in the new area will be estimated as average, high, or low, or where conditions are completely different, outside the range. Table 18-2 shows some ranges in mineral exploration costs in the western United States during a specific period of time. For other regions and newer times, adjustments need to be made either by using the cost indices mentioned in Section 8.4 or, better yet, by using a thorough knowledge of the differences involved.

Differences between exploration costs in a source area and those to be estimated for a new area will arise from differences in some of the following factors:

Wages and salaries, additional employment costs, labor policies, and restrictions on employing foreign personnel.

Local customs affecting efficiency, scheduling, and scope of operations.

Taxes and restrictions on importing, leasing, and shipping equipment.

Competition for available exploration services.

Scale, accuracy, and degree of detail.

Existing levels of survey control, maps, and information.

Size of area.

Degree of isolation from towns, roads, labor force, materials, maintenance facilities, and laboratories.

Terrain from the standpoint of access, traversability, and geologic complexity.

Health conditions affecting field schedules and individual efficiency.

Climate and weather affecting seasonal and day-to-day operating schedules.

Allowance for cost of jobs rushed to completion and meeting deadlines.

Local trends in currency inflation.

The analogy principle still holds for definitive cost estimates but it is strongly supported by personal knowledge and experience, cost quotations from contractors, and simulation of possible exploration sequences with various exploration models.

Budget estimates for exploration programs are often made in terms of two categories: contract services and company-controlled services. Contract items may include aerial photography, photogeology, geophysics, laboratory support, consultants, and local labor. Company items include salaries and wages, overhead costs, field expenses, transportation, field office rent, field supplies, and equipment. An additional category, both in budgeting and accounting, covers mineral land acquisition and payments.

Table 18-2.   Mineral Exploration Unit Costs, Western United States, 1977

1. *Manpower* (including company overhead)

Senior geologist	$100–$150/day
Geologist	$ 80–$100/day
Field assistant	$ 40–$ 60/day

2. *Office Studies*

Aerial photography (stereo-coverage)

Government black-and-white	$0.05–$0.25/km²
Contract black-and-white, per km²	

Area (km²)	*1:20,000 scale*	*1:10,000 scale*
10	$30	$35
100	20	30
500	12	20
1000	8	15
3000	5	10

For color, add 20 percent

For near infrared, add 20 percent

Photogeologic interpretation, add to cost of photography

Generalized	$ 1–$ 5/km²
Detail	$ 5–$ 20/km²

Photogrammetric maps, add to cost of photography

25 to 50 km², 1:10,000 scale	$ 60–$120/km²
1: 5,000 scale	$120–$200/km²

3. *Laboratory* (commercial laboratories, including sample preparation)

Geochemical (AA or colorimetric) analysis

—4 to 5 elements	$ 5–$ 10/sample
X-ray diffraction mineralogy	$ 10–$ 50/sample
Semiquantitative spectrography	$ 20–$ 45/sample
Assays	$ 5–$ 12/element
Thin-section preparation	$ 3–$ 5/sample
Polished-section preparation	$ 5–$ 10/sample
Thin-section study	$ 20–$ 30/sample
K-Ar age determination	
—routine service	$300–$400/sample
—priority service (48 hours)	$700–$800/sample
Isotope ratio, rock samples	
—oxygen, carbon, or sulfur	$ 40–$100/sample

4. *Field Geology*

Transportation:

2-wheel drive	$175–$250/month + gasoline + $0.06–$0.08/km
4-wheel drive	$250–$350/month + gasoline + $0.08–$0.10/km
Small fixed-wing aircraft and pilot	$ 50–$ 80/hour

Table 18-2 (con'd). Mineral Exploration Unit Costs, Western United States, 1977

Small helicopter and pilot (4 hour/day minimum)	$125–$300/hour
Cargo helicopter and pilot	$500–$1,500/hour
Field (camp) cost per person	$ 14–$ 25/day
Town (motel-based) cost per person	$ 30–$ 35/day
Unit geologic mapping costs	
Reconnaissance (1:10,000 or smaller)	
—Simple terrain	$ 20–$ 60/km²
—Complex terrain	$ 60–$200/km²
Detail (1:5000 or larger)	
—with few existing prospects	$300–$600/km²

5. *Geochemical Sampling* (without geologic mapping or drilling)

Field sample collection, general	$ 2–$ 20/sample
Field sample collection, reconnaissance	
Stream sediment, accessible terrain	$ 2–$ 20/km²
Stream sediment, difficult terrain	$ 25–$ 50/km²
Field sample collection, detail	
Soil, accessible terrain	$ 70–$200/km²
Soil, difficult terrain	$300–$700/km²

6. Geophysics (including mobilization)

Airborne, fixed-wing	
Multiband camera and infrared scanner	$ 3–$ 7/line km
Radiometric	$ 10–$ 40/line km
Magnetic	$ 6–$ 13/line km
Electromagnetic	$ 11–$ 25/line km
Electromagnetic + magnetic + radiometric	$ 16–$ 25/line km
Helicopter Geophysics: add 50–150 percent to fixed-wing cost	
Ground Geophysics	
Radiometric	$ 10–$ 20/line km
Magnetic	
—Detail	$ 90–$ 200/line km
—Road traverse	$ 4–$ 16/line km
—Reconnaissance	$ 12–$ 60/line km
Gravity	$120–$250/line km
Induced polarization	
—Reconnaissance	$100–$300/line km
—Detail	$400–$950/line km
Electromagnetic	
—Reconnaissance	$ 50–$120/line km
—Detail	$250–$500/line km
Shallow seismic	$200–$750/line km
Borehole geophysics	
Base charge	$0.30–$0.50/m
Logging charge, per parameter	$0.30–$1.00/m

Table 18-2 (con'd).   Mineral Exploration Unit Costs, Western United States, 1977

7. *Drilling*
    Noncoring
        to 300 m $ 10–$ 30/m
        to 1,000 m $ 30–$ 50/m
    Coring
        to 300 m $ 40–$ 50/m
        to 1,000 m $ 50–$ 80/m
    Mobilization, sampling, site preparation,
        miscellaneous cost (access roads and helicopter
        service not included)
        at 100 m   add 60 percent
        at 1,000 m   add 30 percent
8. *Access Roads and Trenches*
    Road building
        Flat to gentle terrain $ 300–$2000/km
        Moderate terrain $2,000–$4,500/km
        Rugged terrain $4,500–$6,000/km
9. *Test Pitting*
    Trenching with bulldozer or backhoe $ 10–$ 30/linear meter
    Test pits to 3 m $ 10–$ 30/meter depth

18.5. THE USE OF MATHEMATICAL MODELING IN EXPLORATION

Operations research is the name for a family of design methods in which mathematical models are applied to the problems of optimization. Statistical analysis provides the major techniques, and computer usage is assumed, but the data can sometimes be handled by using small electronic calculators. Approximations can also be made with nomographs. Cooper, Davidson, and Reim (1973) present a few nomographs that can be used in financial analysis, risk analysis, and some other aspects of operations research during an exploration program.

The principal aims in operations research are to quantify the variables, including the odds for success, in each stage of a complex sequential program and ultimately, to point out the most direct route to success. Operations research has therefore been applied in many situations where administrative decisions must be made, but its use is better established in mining operations than in mineral exploration.

Project Control Techniques

Mining operations often use project control techniques that give a network (arrow diagram) format to all of the activities within a particular program

(Kaas, 1973). Program evaluation and review technique (PERT), a scheduling technique intended for projects with a large degree of uncertainty, involves the probability that work will be done within a certain time; it is more applicable to exploration programs than the critical path method (CPM), which is more deterministic. One of the main benefits in project control techniques comes from laying out the plan. The plan forces one to consider who will be doing what at a particular time, what must follow, what must precede, and where the activities must meet. The critical path in both PERT and CPM is the *necessary* path that must be taken through the most time-consuming activities in the network. For optimization, this path must be made as direct as possible because it controls all the other activities. Once the interrelationships have been plotted on a diagram, computer processing for final optimization is almost an anticlimax.

Credibility

Aside from such project control techniques as PERT used in major programs, operations research methods and mathematical modeling are not as widely accepted in exploration as they are in the more closely defined stages of mine development and production. The potentialities are impressive and so is the literature, but performance has been equivocal. When a brilliantly designed exploration project has failed to find an orebody, the project should be reviewed. There is always something to learn. Where did it go wrong? Too often the answer is found in some ridiculous (on hindsight) inductive leap made early in the project. Why did it go so *far* in the least promising direction? This is where operations research, if it was involved, takes full discredit. The wrong direction may have been based on geologic information that was forced into someone's idealized model and then massaged in a computer until it seemed unassailable. It is not surprising that problems in credibility have risen from premature "magic black box" uses of mathematical modeling.

The same problems existed in the early days of geophysics (the science fiction crew), geochemistry (the alchemists), and even mine geology itself (the guess department)—growing pains.

Imperfect as it is—and caught between true believers and skeptics—mathematical modeling is part of exploration. As long as the simplifying assumptions are understood and the region of fit is defined for each particular project, mathematical modeling contains excellent guidelines. Like any other tool or technique it can also create extensive damage if used carelessly.

Thus exploration geologists must become familiar with the elements of yet another profession so that they can communicate with its practitioners. The chapters on exploration and operations research in volume 2 of the Koch and Link (1971) treatise, *Statistical Analysis of Geological Data*, are good places to start. For more depth, there are review articles and "state-of-the-art" papers in the transactions volumes of APCOM, a series of international symposia on applications of computer methods in the mineral industry. The first of these symposia was held at The University of Arizona in 1961; since then, Stanford University, Colorado School of Mines, Pennsylvania State University, and the American, Canadian, Australian, and South African institutes of mining and metallurgy have acted as sponsors. Each of the transactions volumes is available from the sponsoring organization.

There is something even more promising (or frightening) that lies beyond mathematical modeling; artificial intelligence is an entire field of science involving the potential reasoning and perceptual capabilities of computers. Artificial intelligence is not yet an exploration tool, but it probably will be.

Profit Expectancy Models: Discovery Potential

The history of the mathematical approach to exploration problems goes back to the mid-1950s and to the pioneering work of Allais and Slichter. Because the concepts generated by these two pioneers are basic to all that has followed, they deserve comment before giving consideration to more recent developments.

Allais (1957), professor of economics at the Paris School of Mines, was asked to report on the feasibility of exploring for mineral deposits in the Algerian Sahara. His first step was to estimate the number and value of deposits that might be found in a 1 million-km² study area. Since there was little information in the specific area, Allais decided to extrapolate data from elsewhere. He found a statistical model in the distribution of mineral deposits in the better explored parts of the world, such as France and North America, and he concluded that these deposits were distributed in a way that approximated the Poisson discrete probability function: They were associated with random processes. He then concluded that the values among individual deposits were distributed lognormally. They approximated a frequency distribution curve with positive skew, indicating that a relatively few deposits accounted for a large percentage of the wealth.

On the basis of the Poisson–lognormal model, Allais calculated the probability (P) of success for a three-stage exploration program under the least favorable, median, and the most favorable conditions, as follows:

	Number of deposits of at least 2 billion francs gross value	
P	Estimate	95 % Confidence Interval
1/1000	10 deposits	4–16 deposits
1/500	20	11–29
1/250	40	28–52

After considering an expected exploration cost, Allais established a probability of 0.35 for making a net gain of 50 billion francs and a probability of 0.65 for losing 20 billion francs. This was for the Sahara or for *any* unexplored area of similar size, since geology was not a factor. The prediction was never tested in the field because politics (one of exploration's determinants) intervened and the Algerian venture was abandoned.

The value of Allais's work derives not so much from the numerical results as from three of his observations.

1. In a very large region most of the value from ore production is from a few large deposits. For example, the deposit with the greatest value will often account for 35–40 percent of the total value for all deposits. It is the prize elephant not the total number of deposits that contributes most heavily to the success of the venture.

2. "Skimming," Allais's term for regional appraisal, must be highly selective, since its objective is to eliminate the weaker indications, those having a high probability of using exploration funds without revealing a deposit. If the early skimming stage is not very selective (the screen is too coarse), then the next stage will amount to looking at nearly all of the indications under no better risk conditions than before.

3. Given the strategy of skimming and a Poisson–lognormal model, the probability of loss decreases as greater amounts of capital and a greater number of "tries" are involved.

Allais made reference to "gambler's ruin," the chances of going broke in a run of bad luck during a game where the odds for success are small. Even though exploration is not a game of chance, the concept is worth considering. The biggest reward should accrue (in theory) to the organizations with sufficient staying power. The law of gambler's ruin, in its most common form, is written:

$$P_r = e^{-NP_s}$$

where P_r is the probability of ruin, P_s is the probability of success for a single venture, and N is the number of ventures. Table 18-3 puts the law of gambler's ruin in a handier form; from this table, if each prospect were to have an equal 1 percent chance of being an ore deposit, we would have to spend money on 229 prospects in order to reduce the risk of going broke to 10 percent. Note the word "equal" in assigning "chance" to large numbers of prospects; this is a part of the bothersome gap between exploration work and statistical probability. Politics notwithstanding, neither geologists nor prospects are created equal.

Louis B. Slichter (1955, 1960) of the Institute of Geophysics, University of California at Los Angeles, applied statistical guides to the optimization of grid drilling plans and geophysical flight lines in areas where there are no surface indications of orebodies. In these studies, he calculated patterns that would

Table 18-3.   Number of Prospects to Be Examined to
Reduce Risk of "Gambler's Ruin"[a]

Probability of Success	Probability of Ruin		
	10%	5%	1%
1%	229	298	458
5%	45	58	90
10%	22	28	44
20%	10	13	21

[a] After G. S. Koch and R. F. Link, *Statistical Analysis of Geological Data,* Vol. 2, Table 12.2, p. 191. Copyright ©1970–1971, John Wiley and Sons, Inc. Used by permission of the publisher.

maximize the ratio of discovered ore to the total exploration cost (the "prospecting profit ratio") in a number of hypothetical situations based on information from known mining districts in the western United States and in Ontario.

Slichter confirmed Allais's finding of a lognormal distribution for the values of mineral deposits, but he rejected Allais's conclusion regarding the Poisson occurrence distribution. Slichter concluded that the spatial distribution of mines in his control areas follows an exponential function, thus implying that one known orebody in a unit space does *not* decrease the chance that other orebodies will occur in the same space. Geologists—who generally do not believe in randomly distributed orebodies—would be happy to refer to the space as elephant country.

In the work of Allais and the work of Slichter, a pattern of mineral deposits was taken from better known areas and extrapolated into an unexplored area within which all "cells" or subareas were to have an equal probability of permitting a discovery. This was considered to be an initial or preexploration condition, and it was not intended to be pushed too far. Skimming of an Allais-type area or the reevaluation of Slichter's grid were expected to produce on-site geologic information as a basis for further work.

Profit Expectancy: Search Models

Search models, derived from military operations research work in which targets are moving submarines or stationary life rafts, provide a different approach to the problems of exploration in "blind" areas. One model, the Engel simulator (Griffiths and Singer, 1973), is based on a two-stage search procedure in an area within which a specified number of hidden targets is expected to occur in a negative binomial distribution and noise is expected to have a random or Poisson distribution.

During the first stage of search in an area with Engel simulator characteristics, a sensor system receives signals which may be true (from targets) or false (from noise). True signals will tend to cluster around a valid target; false signals will have no more of a tendency to cluster than would be expected from chance. The second stage of search (assumed to be more expensive) is conducted in areas indicated by apparent clustering. Each stage consists of a number of "coverings," or passes—as many as would maximize the profit from the expected discoveries after deducting the search cost. The calculation is permitted only by "knowing" (estimating) the number of targets to be found.

Search models share the assumptions of Allais's and Slichter's extrapolative spatial models that orebodies are distributed with an initially equal probability of occurrence in each subarea unit. Geological conditions do not enter into any of the profit expectancy models until an initial blind statistical search is done. This disregard for specific geological guidelines is defended in this way: we want to know if we can justify a geological search in area X and if the search is justified, we want to know how thoroughly we can afford to search.

Nevertheless, many geologists are reluctant to use an unqualified probabilistic approach at *any* stage in exploration. Some of this reluctance derives from a suspicion that geology (after all it has done for us) is being ignored just to please a machine. There is a logical philosophical basis for hesitation as well; this has been stated by Mann (1970), who described several *degrees* of randomness in natural processes. He cited combinations of true deterministic relationships, "statistically deterministic" phenomena, and essentially nondirected chance happenings, and concluded that a probabilistic model may be valid at one scale and completely invalid at another. An appreciation of random processes in geology can be obtained from a book edited by Merriam (1976).

Pure probabilistic models appear to work on an assumption that there is no geological information. Geologists find this especially hard to accept. Are there "real-world" areas without any geologic guidelines whatever? Even the Greenland icecap is bordered by areas of investigated geology, mines, and mineralization; there are structural trends to correlate with other trends in the Canadian Arctic, and there is geophysical information.

Geologic Favorability Models

The need for a geologic context prompted geomathematicians and operations researchers to develop models incorporating multivariate statistical techniques—ways of handling several (perhaps an uncertain number) of random variables. The techniques and their relation to exploration are discussed in the chapter on spatial variability of multivariate systems in *Geomathematics* by Agterberg (1974). The variables: the numbers assigned to geological parameters. The models can be called geologic favorability models; some identify the favorable environment for orebodies whereas others identify the favorable areas themselves.

How do we recognize elephant country before we see any elephants? We could follow an approach described by J. M. Botbol of the U.S. Geological Survey. Botbol (1970) took geologic characteristics from 30 well-explored mining districts and obtained a product matrix model in which "typicality" could be determined in relation to copper districts, lead districts, and zinc districts. The model was then applied in less-known districts to determine their favorability for additional exploration and to simulate the effect of acquiring additional data. The procedure is like that followed by field geologists in recalling analogous conditions when looking at a new area. It has an advantage, however, in that the search for supporting data can be concentrated on what appear to be the most significant—highest typicality—geologic features.

Where *is* elephant country? Agterberg (1974, p. 529) used a "similarity index" based on the geologic and geophysical characteristics of the Kidd Creek mine area to identify the best exploration areas or cells (10 km on a side) in the Abitibi Volcanic Belt of the Canadian Shield. Agterberg's input matrix consisted of 52 columns for geological-geophysical parameters and 644 rows for area cells.

In order to identify favorable areas in a more specific way, D. P. Harris and colleagues at the Pennsylvania State University and The University of Arizona developed a multivariate geostatistical (MG) model and a conceptual geostatistical (CG) model.

The basic postulate of the MG model is that the probability of occurrence of V, a measure of mineral wealth, can be determined by a function G operating on numerical values for such variables as age and type of rock, structural forms, and contact relationships. Thus, if the actual form of the function G can be determined for every cell in a well-explored control area, probabilities for V in that area can be generated and applied in a less-known study area.

In the initial development of the MG model, Harris (1966) used the mining region of Arizona and New Mexico as a control area and a part of Utah as the first study area with the result that 19 of 144 cells in Utah were selected for further exploration. Naturally, some of the selected cells had already produced considerable mineral wealth; most, however, had recorded only minor past production, and their indicated high probability for mineral discovery could be considered a function of the geologic variables.

The conceptual geostatistical (CG) model (Harris and Brock, 1973) is extrapolative, like the MG model, but the geologic input is based on the opinions—subjective probability statements—of expert geologists familiar with the control area. In one example, 20 geologists (10 from industry, 5 from government, and 5 from universities) were interviewed individually and asked to assign probabilities to the occurrence or absence of twelve key geologic features in randomly chosen cells within the control area. The model was then applied to similar but less-explored study areas.

Some approaches to selecting favorable exploration area involve combined geochemical and geological data, as in a forecasting model used in predicting the base-metal potential of northwestern England (Cruzat and Meyer, 1974), a specific mineralizability model used in estimating copper-zinc-lead resources in the Oslo region, Norway (Brinck, 1972), and target rating models applied to the Wisconsin zinc mining area, areas in Quebec, and part of the Canadian Cordillera (De Geoffroy and Wignall, 1973).

In extrapolating measurable conditions from known areas into less-known areas and interpolating between the better-known portions of an area, geologic favorability models do not normally give any weighting to economic conditions, technology, and the tools of exploration. Most exploration geologists accept this limitation with the understanding that it is the occurrence of mineral bodies, not their likelihood of being found and developed, that is being predicted. The emphasis on geologic input is welcomed by most exploration geologists, but with some reservations about using the distribution of known orebodies to find less obvious orebodies. New orebodies, sought on the basis of new geologic ideas, cannot be expected to follow old patterns so precisely.

Simplification, a necessary step in treating geologic data, is often questioned because it may destroy the original context of the information. It is like the familiar admonition against equating apples and oranges (simplified, both are fruit). Rhyolite and basalt are both volcanic rocks; White Pine and Bisbee have little in common, but they are both copper mining districts. Then, there is the problem of simplifying a relative intensity or abundance to suit a statistical format or, for that matter, a metallogenic map. Sulfide mineralization may consist almost entirely of pyrite with a few grains of chalcopyrite in one place; in another place the amounts may be nearly equal; yet both places will warrant a symbol that says: "pyrite is more abundant than chalcopyrite." Geophysical and geochemical data are easier to accept as quantified input to geologic favorability models because they are more often obtained in digital form, but even for these data the context may be forgotten after the data disappear into the brackets of a formula.

Program Design Models

What happens when new maps, samples, and drill core come in from the field? These provide feedback, and they call for revision of the model; they may even call for a new model and a radical change in the program. Most models cited so far have the words "feedback" and "decision making" in their descriptions, but they refer to planning rather than to field operations. A few exploration program design models go further. One of these, based on the Bayesian decision analysis method (Davis, Kisiel, and Duckstein, 1973), allows for the incorporation of new information as it arrives in sequence.

Bayes' theorem, including a step-by-step definition of conditional probabilities, is used by Harris and by De Geoffroy and Wignall in their geologic favorability approaches and in some respect by most model builders who deal with existing data. Bayesian decision analysis simply takes the theorem from the broad needs of the strategic stage into the immediate and specific needs of the tactical stage. The plan of a Bayesian process is commonly shown in a decision tree with branching nodes representing probabilities.

Bayesian modeling begins with *initial probability* estimates—probabilities that certain events will follow others. One of these might be the probability that a geophysical anomaly or an alteration envelope of a certain width will occur if an orebody exists. The impact of new information, all of the alternative data that might be expected from a drill hole, for example, is simulated to obtain *conditional probabilities*—probabilities that certain of the initial estimates can or cannot follow others in specific situations. Alternative situations might call for all holes to be in ore, half in ore, some in sericitic alteration, or all to be "blank." With the initial and conditional probabilities as input to Bayes' formula, *posterior probabilities*—revised estimates—are calculated. Now the posterior probabilities become *prior probabilities* to be treated as initial probabilities in another round of simulated information, conditional probabilities, and posterior probabilities. The procedure is repeated through the decision tree.

Bayesian decision analysis helps to anticipate conditions at decision points in an exploration program. These conditions, as prior probabilities, form a basis for simulating the effect of taking the next step or one of several alternative steps. Even though the information obtained in the real step will almost never be the perfect information assumed in the simulation, a statistically advantageous course of action should become apparent.

Something else will become apparent, even before the analysis begins. Since a large number of parameters and decisions can produce enough combinations (branches of the decision tree) to swamp a computer, some simplification is required; and simplification is just as dangerous here as in any of the methods and models previously mentioned. It can add up to being forced into Chamberlain's "single working hypothesis" or onto one of two branches in the decision tree. Real-world geologic information is seldom a binary "presence or absence"; we sometimes wish it were, but then geology would be a rather dull subject.

Target Delineation Models

With some quantitative reason for exploring a region, an idea of how closely to search, some geologic favorability numbers, and the pleasure of simulating an entire program in the office, we *should* stand a good chance (statistically speaking, of course) of finding some concrete geologic evidence of an orebody. Suppose we do; does it actually amount to an orebody? The remainder of the investigation still involves mathematical modeling, as mentioned in Chapter 15 in relation to sampling patterns. It is here that geostatistics is used in its most specific sense, and here that probability functions can be applied to ore deposits in a closely controlled environment.

Probability and Inferred Probability

Statistical probability, the key to this section on mathematical modeling has freed geologists from the long-recognized trap of inappropriate deterministic models. But freedom, scientific as well as political, carries the hazard of falling into unexpected traps. Geologists have tried, unsuccessfully, to learn enough about mineral deposits to say (deductively and deterministically): "The presence of these conditions will cause an orebody to exist. Dig here." Now geologists are tempted to say, with statistical inference: "The presence of five orebodies in area A means that we have a 35 percent probability of finding five orebodies in similar area B. Dig in area B, not in area X." However, we may really not know that much about sample area A, let alone areas B and X. The trap: our ignorance may be hidden, even from us, by a shield of impressive numbers.

Since most exploration geologists know that the shield of numbers may hide all sorts of imperfections, statistical probability is seldom used by itself to adjust for uncertainty. There are, however, intelligent compromises. One such compromise, called inferred probability in management literature, makes use of a small but carefully selected amount of reliable information and the user's experience—"savvy." As dangerous as this sounds (one of the basic rules in statistics is "don't burn conflicting data"), it is permissible as long as the departure from the rules is identified. In this approach, apparent grains of truth are picked out and saved before they become thoroughly mixed with bias and noise in the computer. Some of the data, poor core recovery, for example, are labeled "suspect" in order to reduce their impact if they *must* be used. Above all, the persons supplying the data are appraised—since each person has his or her own objectives and abilities at a particular time.

Incomplete data, such as "convenience" samples taken from the most accessible outcrop, can often be recognized and isolated for special consideration or improvement before they do much damage. There is still a problem. Will the risk involved in using the available but incomplete data outweigh the cost and delay involved in completing the data? Simulation of tentative courses of action in the Bayesian decision model can help to provide the answer. If the weak data must be strengthened, this should become evident.

In their current form probabilistic approaches to evaluating mineral wealth are more acceptable to governmental planning agencies than to private exploration firms. Government mineral policy needs numbers, but the numbers do not have to be tested next week or next month in a drill hole at an exact place or to an exact depth. Yet, the trend toward large numbers in the overall economics of exploration is bringing operations research into wider use. The acquisition of mineral property requires ever larger advance commitments in money (bonus bidding) and manpower (performance guarantees). This involves things we cannot know with certainty or even project deterministically. We have to use probability.

CHAPTER 19
THE EXAMINATION AND EVALUATION OF PROSPECTS AND MINES

As geologists gain field experience, they learn to respect their less formally educated counterpart, the prospector. It is a sobering experience to find old mine workings in a specific target location chosen by sophisticated reasoning and in rugged terrain that seems accessible only by helicopter. A geochemical anomaly in stream sediments may be found to have an insultingly simple explanation—it leads to an unsuspected prospect pit. Now it is necessary to investigate mineralization that someone else has already found. The job is called prospect examination.

If the exposed mineralization is favorable, an evaluation of the prospect may follow, just as an evaluation would follow the discovery of mineralization during new work in an exploration program—except for one thing. The mineral rights may be held by someone outside the geologist's own organization. The mine or prospect is then a *mineral property,* and another dimension—property acquisition—is added.

Mineral property evaluation is part of the exploration process. In the next stage, preproduction (Chapter 20), many of the techniques from the exploration stage continue to be used, but the mine is being readied for production, expenditures enter a new taxation category, and the bookkeeping label changes to "development."

19.1. OLD MINES—NEW EXPLORATION

At some point most exploration programs involve the examination of prospects or mines. The mine workings may be obvious or may even be currently
active. They may be no more than small areas of disturbed ground seen on
aerial photographs, or they may be evidenced only as "Smith's Prospect"
mentioned in old records or shown on a map. In any event, mine workings
focus attention on the most obvious geologic guide to ore: mineralization.

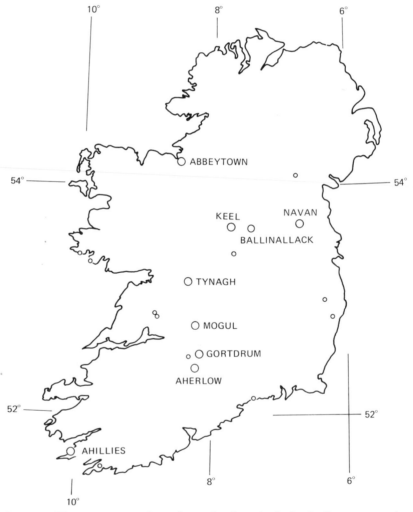

Figure 19-1 The new generation of metal mines in Ireland. Prospects and older
mines are shown in small circles.

Although the new generation of metal mines in Ireland (Fig. 19-1) may be properly credited to modern exploration programs, attention was first drawn to the area by old mine workings. The Tynagh orebody, the first notable modern discovery, was hidden beneath 2–12 m of glacial drift, and it was found with the aid of geochemical work, induced-polarization geophysics, and a geologically directed drilling program. But there was another factor in selecting the area: older mine dumps and shafts were in evidence along the nearby Tynagh fault.

Discovery of the orebody at Mogul in Ireland, Europe's largest lead-zinc mine (Fig. 19-2) resulted from a sequence of modern geologic mapping, geochemical soil sampling, and geophysical work; yet Mogul's village, Silvermines, has been a mining site for a thousand years, with at least 50 companies having operated in the area during the last 400 years.

Gortdrum, an Irish copper-silver orebody discovered in a well-executed program of geology, geochemistry, and geophysics, is on a recognizable trend that projects from the nineteenth-century Oola Hills mines 3 miles to the west. At one point, geochemical work at Gortdrum led to an old limestone quarry with exposed copper staining and to a feature with the historic name of Miner's Hill. Eventually, old mine timbers were discovered in the modern pit. If nature had not hidden the scars of mining, the Gortdrum exploration program might have begun with a mine examination.

Mine examination was involved in the discovery of the Sar Cheshmeh

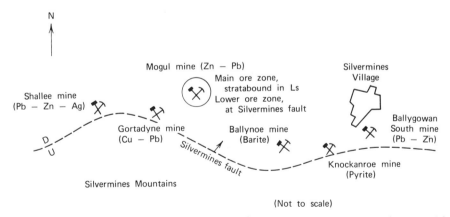

Figure 19-2   The Silvermines area, Ireland. The Mogul orebodies were discovered in drill holes in 1963. Workings along the Silvermines fault were active in prehistoric times, during the seventeenth century, and during the mid-nineteenth century. The Ballygowan and Shallee mines were reopened and worked for a short time during the 1950s. The Ballynoe barite deposit, discovered in the late 1950s, is in the same stratigraphic unit as the main ore zone at Mogul.

orebody in Iran during the late 1960s. Geology and geochemistry furnished guides to a porphyry copper body, but previous mining had taken place at the actual site, and there was evidence in the region of copper mining dating from the fifth millenium B.C.—fossil elephant country.

Discovery of the Panguna copper orebody on Bougainville Island, a factor in the surge of exploration into the Southwest Pacific in the early 1970s, was based on old information and new ideas. Geological reasoning and the examination of analogous deposits in the Philippines provided a model for exploration on Bougainville, but the specific choice of the Panguna area was based on reports describing gold mine workings with minor copper mineralization.

"Teaser" Districts

Areas with dozens or hundreds of small mines and prospects are understandably attractive, so attractive that they may consume an inordinate amount of time and money before they are rejected. These are the "teaser districts"— sites of repeated investigation and disappointment. In some districts where mineralization is inherently weak or where some basic condition for a major orebody is lacking, all of the luck and hard work in the world cannot create a suitable orebody. In others, the elusive major orebody is there—somewhere— but the clues must be found from newer concepts or a more imaginative prospect examination. It would be unwise to cite any "teaser" district by name because the label is too fragile. A district so named may turn out to be the setting for next year's outstanding example of success based on creative geologic analysis, economic reassessment, or the use of new exploration tools.

Safety

In the investigation of mines and prospects, there is something which must *always* be weighed against the temptation to gain information quickly: personal safety. If the information thought to exist in an abandoned underground working is critical to an exploration program, the cost and time needed for safe access—for stabilization, ventilation, and pumping—belong in the exploration budget. It takes some experience to recognize danger in old mines as opposed to benign "spookiness." The safety guidelines given in Table 12-1 bear consideration.

19.2. MINERAL PROPERTY EVALUATION

If immediate acquisition of mineral rights is not needed, a descriptive "note and consider" *examination report* can be written for eventual use with other data on the area. It is the area, not the prospect or mine, that will be evaluated. If, however, the site is a mineral property to be acquired from an individual, an organization, or a government, it becomes a tentative purchase that may or may not be a bargain. The information will be put to immediate and specific use, and the examination report is only a part of—or a preliminary to—a comprehensive "decision and action" *evaluation report.*

Inherited Data: Signal and Noise

In the evaluation of a raw prospect or a mining concession with only an indication of economic mineralization the work of defining an orebody is left to the acquirer, who begins work with little data—and little misinformation. This is not true with the acquisition of an explored prospect or a partly developed mine. Data as well as property are bought. Here is something of potential value, but abundant data do not necessarily amount to abundant information. Since the drill-hole logs, shipment records, and geologic data were not collected by the evaluating geologist, they must be questioned for accuracy, tested where possible, and differentiated into "signal," "message," and "noise." This is a vexing job. Factual and inferential geologic mapping may be so intermingled that maps of some potentially critical areas now caved or otherwise inaccessible must be discounted. "Good news for stockholders" in the last few reports from a now-idle mine has sometimes been based on desperate geologic projections which, unless considered in the proper context, can lead to disastrous conclusions. Even worse, bulletins of respected government agencies may in good faith quote from secondhand data worthy of Mark Twain's comment that a mine is a hole in the ground owned by a liar. It is no wonder that skepticism is common among experienced mining property examiners.

Where information comes from past evaluations by other mining groups, it must be reviewed in the light of each group's objectives, competence, and limitations. One cannot assume that "if the Mohawk Corporation did not want to drill deeper, neither should we." Mining is more dynamic than that; economic conditions change, older concepts are replaced, and new tools of exploration give access to better information. Perhaps the Mohawk Corporation geologists did not suspect that the barren volcanic series is underlain by a favorable limestone.

An earlier "negative" report may, in fact, stress the very conditions warranting investigation. The postmineralization fault that put an end to previous work may be the best place to begin work on the basis of a new idea or under a new assumption.

New interpretations should always be sought. Prior interpretations have a tendency to be adopted by successive evaluators who want to get on with the job of collecting further information until, like entrenched rumors, the interpretations and assumptions come to be accepted without question.

Prior Data: Prior Context

There is an important question to ask when reviewing the results of earlier evaluations: "Was the previous acquisition agreement very different from the one we are now contemplating?" An earlier evaluation may have been cut short because the payments demanded by the property owners were too high. The earlier investigation may have been made with the hope of verifying an envisioned bonanza. No bonanza—no deal.

Normally, the "asking price" for a mineral property will decrease after the first few rejections; when the "mountain of ore" reduces to a less impressive hill, so do the property owner's demands. Normally—but not *always*. Astute prospectors are aware that a mildly interesting zone of mineralization can become a critically located orebody under changed geologic and economic conditions; on this basis, a "waiting game" may work out so well that an outrageous price will eventually be met if the buyer becomes desperate.

A geologist may become involved in a situation like that described by Arnold Hoffman (1947) in the book *Free Gold*, his tribute to Canadian mining and miners. A strategically located prospect in the Porcupine area, Ontario, was bought for $8,000 by F. W. Schumaker. In 1911, he optioned his property, known as the Vet, to another investor who, in turn, offered it to the owners of the adjacent Dome mine for $75,000. When Dome refused the offer, Schumaker announced that his own basis for any subsequent negotiation, should they become interested, would be $150,000. In later years, the importance of the Vet claims became more apparent to the Dome organization; the $75,000 price now seemed appropriate. This amount was offered, Schumaker repeated that his asking price was $150,000, not $75,000, and added that if this was refused, the *next* price would be $300,000. Negotiations collapsed. During the next few years, the Dome workings continued toward the Vet claims and the company decided to offer the $150,000. Schumaker again explained that the price was $300,000, as promised, and that the next price would be $600,000. A nerve-wracking series of offers and refusals followed, with the geologic importance of the claims and the asking price for the claims improving at each encounter until the property was sold in 1936 for nearly $2 million. The negotiating formula could be called geometric progression; Canadian mining men call it "Schumakering."

Verifying the Inherited Data

If we question the inherited data, we should find useful information to verify, "noise" to discard, and "loose ends" to pursue. The questions:

1. *What is the "catch"*? Production was stopped, kept at a low rate, or was never begun. If there is a salient reason, it may be entirely geologic or it may at least have some geologic roots.

2. *Why our organization?* If the property was rejected by other organizations, their objectives and their limitations may have been different from ours.

3. *What data did our predecessors miss?* Their drill core recovery may have been poor, mapping inadequate, sampling insufficient.

4. *What data cannot be verified?* Drill core—even if adequate—may not have been kept, stopes may have been mined out, workings caved or flooded.

5. *What is new?* Information from the property itself, from the district, or from analogous deposits may change the conceptual model of earlier times. New exploration tools and techniques are available. Perhaps the economic situation or the conditions of purchase have changed. Also, the property status may have changed. Maybe the owners now control *all* of the ground rather than just a part of it.

Some of the data will remain conjectural, no matter how much checking is done. Other data can be verified by a minor amount of resampling and remapping and used in the same way as new data, but with a considerable saving in time and cost.

The Objectives of Evaluation

In evaluating mineral properties, a geologist may act on behalf of a variety of organizations and individuals, including some not normally involved in mineral exploration.

There are, of course, potential purchasers of the property or of an entire company's assets; for them, the objective will be to obtain an adequate return from the capital to be invested. Adequate return will be expressed in terms of profitability, and profitability will be calculated on the basis of deposit's size, grade, and technologic characteristics. There may be maxima as well as minima. If "risk capital" has limits, so does the size of the objective.

Mineral property evaluations may be made for the current owners of a property. Perhaps they must decide whether to retain, sell, or abandon the property; they will want to know how much additional exploration work should precede the decision, they may want to establish a realistic asking price for the property, or they may want to know if they can justify a search for financial support.

An evaluation may be made for potential investors, partners, or suppliers of materials or of credit. The investment may be part of a complex pattern, as in the financing of the Ertsberg copper mine, Indonesia, in which 30 organizations became involved. Each participant, including an American mining company, seven American banks, a German bank, a group of Japanese smelting companies, and five insurance companies, required an evaluation to be made and related to individual objectives.

Long-term purchasers of minerals from a mine may ask for an evaluation. An industrial plant designed to use a mineral with certain specifications must have an assured supply at a reasonable price. For example, a barium chemical plant may require lump barite containing at least 95% $BaSO_4$ and having no more than 1% Fe_2O_3; substitutes, such as finely ground flotation concentrates, cannot be used because they will be lost as flue dust; it will therefore be necessary to know that a particular orebody or mining district *can* furnish the specific material for 5, 10, or 15 years.

Governments often call upon geologists to evaluate specific mines for a variety of reasons, such as taxation, regional land-use planning, and government-private participation agreements. In many areas it is the overall pattern of mines—existing and projected—that is important, but sometimes the critical site may be dominated by a single, large, undeveloped orebody. One prospect in Upper Volta, West Africa, the Tambao manganese deposit containing more than 12 million tons of high-grade manganese ore, was evaluated repeatedly by the Voltan government, the United Nations Development Program, and by French and Canadian foreign aid teams before development was finally decided upon in a joint venture between the government and a group of Japanese, West German, American, and French companies. A particular objective in several of the evaluations was to establish a basis for building 320 km of new railroad through an isolated but potentially important mining region to the existing rail terminus (Fig. 19-3).

The objectives to be met in mineral property evaluation, for all of their outward diversity, have a common imperative: identify a potential orebody and place a value on it. There is often one more demand: identify the factors that warrant further investigation. If further investigation is actually needed, the need must be pointed out and, if necessary, argued. It may turn out to be quite an argument. Somewhere in each mineral property situation there is an acceptable balance between courting disaster by acting upon rash assumptions and losing opportunities by delaying the action. Just where this balance will be found is often as much a matter of personalities and economic pressures as it is a matter of geologic projections.

The Personal Element

Objectives differ, and so do the standards of conduct among those who formulate them. An intended misuse of the information from an evaluation report must be identified and avoided as soon as possible. The quagmire of professional discredit, even discredit by association, is too deep to escape after falling in.

The fact that a variety of persons, firms, and governmental agencies is involved in prospect evaluations has another aspect; there is often a "gallery" of observers—legitimate and otherwise—to contend with. Some of the ob-

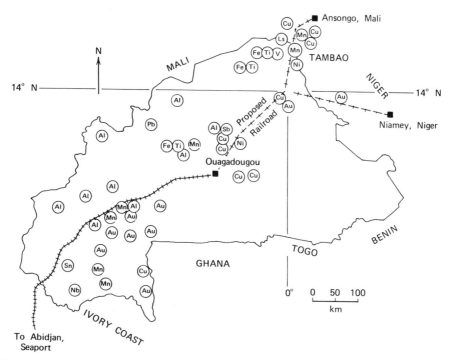

Figure 19-3 Upper Volta, with reported mineral prospects, showing the rail terminus at Ouagadougou, the proposed railroad to the Tambao manganese deposit, and proposed lines to Niamey and Ansongo.

servers are helpful, some create confusion; the latter may have to be dismissed as diplomatically as possible.

Investors in a mineral property generally leave the details of field work to the geologist. Still, the investors have a perfectly valid reason to accompany the geologist during some part of the evaluation. They want to know the facts, in numbers if possible, and they will be impatient (sometimes justifiably) with studies of minutia that seem too academic—"I'm interested in a mine, not indigenous limonite." The geologist may need to provide a certain amount of education as well as evaluation; time spent in this way is generally well spent, because investors will learn *why* they are risking funds on a geologic projection.

The owner of a mineral property, often a professional prospector, may serve as a guide and will almost always be an observer. He is understandably concerned that his hard-won showings of mineralization be given a thorough appraisal. "The geologist for Launcelot Consolidated sniffed around here for a short time, and spent most of that time within a few steps of his jeep; he didn't listen to me, he didn't tell me what he was trying to do, and he didn't even look at the best mineralization." Sometimes it happens that way; the cavalier attitude of some geologists creates immense problems for their colleagues. An experienced prospector has a wealth of valuable information to give, even though it will be from an optimistic viewpoint. The best relationship with a mineral property owner involves careful listening—careful response, too. After searching every cliff and ravine on the property together, an evaluating geologist and the prospector-guide will often become good friends. How much should you tell the prospector about your own conceptual model of the mineralization underlying his mining claims? Caution! With some forethought, ideas can be exchanged without painting unintentioned pictures of bonanza ore and fantastic profits that will complicate further property negotiations.

Another matter: a supreme call for tact comes with the rejection of the property. Whether this is decided on the spot or at a later time, the owner deserves an explanation. An explanation plus some suggestions regarding alternative small-scale mining or other companies that might be interested, *if appropriate,* can pay dividends in future cooperation. If there is little hope of any profit, the owner needs to know this as well. In gaining the respect of prospectors in a region, a geologist or a mining firm can build a good information network. News of fair treatment travels far; news of shabby treatment travels *very* far.

There is the special case of surface landowners. They may not own the mineral rights and may understandably be hostile to the idea that the government controls the minerals. You are an uninvited guest, and a suspiciously acting one at that. Become a diplomat and a careful borrower of the land. Respect for the land is part of the code of exploration ethics given in Appendix G.

Some observers are hard to accept and even harder to dismiss with any degree of diplomacy. Among these are "courthouse miners"— opportunists looking for a way to twist property statutes—and "paper hangers" who will locate mineral claims in any area of apparent activity. Their "discovery" is simply a discovery that you are in the area. Many mining districts have a resident "Fraction Jack" who is quick to occupy any possible gap between mineral claims. Onerous or difficult as it may seem, the best defense against this segment of the "gallery" is secrecy. Recall the instructions to Russian field parties in 1752, cited in Chapter 1, in which geologists were forbidden to have anything to do with Ivan Zubarev, the "Fraction Jack" of his day.

The Conditions of Evaluation: Property Agreements

Mineral property evaluation involves a formal agreement—past, imminent, or contemplated—between buyer and seller or between permittee and government. Earlier permission to "look around" becomes an option to purchase; the large area of a prospecting permit reduces to a smaller area in a mining lease. The time for generalities has passed. The geologist's task: find out if the objectives can be satisfied within the terms of the agreement.

Mining property agreements in their final detail are generally written by lawyers and "land men" acting on behalf of management. The geologist is involved, however, from the time the area of interest is outlined until the agreement is reviewed for its geologic soundness. For this, the geologist must know something of the mineral ownership patterns discussed in Chapters 7 and 18 and must know something of property agreements.

Private Mineral Rights. Most property agreements involving private mineral owners are made at an early stage in evaluation when the chances *against* developing an orebody are so large that it would be folly to make a large initial payment. Exploration dollars are intended to be spent in obtaining geologic information, not in property payments. A property owner who accepts geologic uncertainty as a fact will agree to take the profits from future option payments and royalties. In any event, an agreement goes onto paper and this calls for a geologist's preliminary evaluation based on whatever information is available. If the agreement does not allow for a range in geologic possibilities, it can turn into a nightmare. It is the geologist's responsibility to see that this does not happen; after all, it will be the *geologist's* nightmare.

A typical option-purchase agreement for an unexplored mineral property in the western United States will most often provide for payments to be made at yearly or monthly intervals after an initial free period for preliminary evaluation. The payments are generally small at first and increase as the project moves toward the development stage.

With an initial free period and a graduated schedule of payments, the mining firm can afford immediate geologic work directed toward reducing the level of uncertainty as it exercises each option to continue. The owner is assured, by the increasing payments—or sometimes by a requirement that a certain amount of work be done—that the property will not be left idle. But here is a point for consideration: steeply increasing payments and overdemanding work requirements may force a decision to abandon the property before adequate information can be obtained.

If the evaluation takes a discouraging turn and the firm chooses to drop its interest in the property, the owner retains the prior payments as compensation for having given an exclusive right to examine the property. The arrangement for abandoning interest in a mineral property often includes a provision for the owner to receive the core samples and copies of the geologic reports and assay data.

An "end price" for final transfer of the property to new ownership is desirable, but in some properties it cannot be established and the agreement will take the form of a lease with royalties paid to the owner as the ore is mined. A minimum annual royalty payment, which serves the same purpose as an option payment, is often established to assure the owner that mining will continue.

In areas where competition for undeveloped mineral properties is intense, outright initial payments may be made, but these are still no more than a small fraction of the total price agreed upon. Individual mining claims are sometimes bought outright by the geologist or by a company representative if the price does not exceed a "discretionary-funds" level of a few thousand dollars. The legend of a mining company paying an astronomical amount of money *in one lump sum* for an uninvestigated mining property is kept alive by the news media: "Prospector Receives $4 Million for his Claims." This makes good headlines, and the human interest side of the story can be powerful—years of unrewarded labor, faith overcoming adversity, scorn and pity from the easier living people on Main Street, and creditors snapping at the prospector's heels until a Man from The Big Corporation opens a checkbook and asks "How do you spell your name?" Creative journalism. Still, there would be some truth in many such stories if it were pointed out that $4 million is the total price, receivable during a series of investigations over a period of, perhaps, many years—*if* the expected orebody really exists.

Government Mineral Rights. Where property agreements are made directly between mining organizations and governments, the provisions for a mining lease or mining rights are commonly spelled out in the statutes governing mineral claim staking or in the wording of prospecting permits and exploration concessions that have been obtained at an earlier stage in the exploration program. Where a known mine or a designated mineral reserve is involved, the procedure is more like that followed between private citizens; a series of payments must be made and royalties on the mineral product must be agreed to. On U.S. federal lands and on lands belonging to some states, the right to mine "leasable" minerals in designated reserve areas is awarded on the basis of sealed competitive bonus bids.

Whether the rights to mineral property are to be acquired from private owners or from a government, a geologist's preliminary estimate of profitability (sometimes no more than a guess based on an analogous situation) is demanded at the very earliest stage. It is an uncomfortable demand, but the commitments are so large—especially in competitive bonus bids—that it must be met. Otherwise, an agreement may have to be made under guidelines set by someone with no appreciation for the inherent geologic conditions.

Satisfying the Objectives and Conditions: Expressions of Value

Profitability indexes for mining ventures, introduced as part of the economic framework in Chapter 8, were recalled as broadly stated determinants for exploration models in Chapter 18. Now they are needed again, but in a more definitive way. We now have *ore* mineralization, and this can be used as an expression of the orebody itself. Geotechnical information—however preliminary—can now be applied to potential mining methods. Mineralogy now has bearing on recoverable ore value, and the apparent limits of the orebody can be stated in terms of ore reserves. We have a better basis for estimating a rate of mining and a life span for the orebody.

The steps outlined in Table 19-1 can be guides to placing profitability indexes on undeveloped mineral properties. We begin by establishing working estimates for the geologic, economic, and technological characteristics in the deposits. With these and the fixed parameters applying to deposits in general we estimate cash flow and, finally, some expressions of value.

But do we really have enough information to determine "dilution factor" and "marketing method"? In fact, do we have enough information to provide any estimates at all? The objectives with their precise requirements say no. The conditions, limits to available time and money, say yes, we have no choice. The conditions generally speak louder. We need additional information, but obtaining the information requires additional expenditure. Because expenditures must be justified, they must be based on earlier estimates. Fortunately, we have such labels as "order-of-magnitude," "preliminary," and "detailed," given in Table 8-5, to prevent our earliest estimates from being confused with later ones based on better information.

An order-of-magnitude estimate might be derived from a few days' examination at a new prospect. With this first indication of profitability and with an expression of assumed risk, the cost of additional information can be justified step by step through increasingly refined estimates. Somewhere along the line the investigation attains the status of a detailed feasibility study. As mentioned in section 8.4 the final decision to develop a mine may require months or even years of work, hundreds of drill holes, experimental mining, and testing of the ore in a metallurgical pilot plant. The cost of a complete feasibility study may run into millions of dollars where a large capital investment is at stake. The budget for a feasibility study of the Cerro Colorado copper project, Panama, beginning in 1976, was $20 million. This amount would be equivalent to the capital investment in some entire mines, but compare it with the estimated cost of the entire Cerro Colorado project: $800 million.

Sensitivity and Risk Analyses. So many estimates are piled upon estimates in Table 19-1 that we dare not jump to conclusions with just one group of profitability expressions. We must have an idea of the range to be expected. This is sometimes provided by a sensitivity analysis (Grant, 1970), a technique whereby the most critical factors in determining profitability can be identified and selected for further study. In a sensitivity analysis as many factors as can be quantified are related, one at a time, to the index

Table 19-1.   Steps in Estimating Profitability of an Undeveloped Mineral Property

---

I. *Establish working estimates*
   1. Tonnage and grade in place
      a. Consider ore reserve classifications
      b. Consider distribution pattern in grade and mineralogy
      c. Consider effects of several alternative cutoff grades
   2. Future cost index and commodity price levels for the apparent period of mine life
   3. Future environmental considerations
   4. Estimate mining cost and probable cost range
      a. Select a general mining method
      b. Consider dilution and percentage extraction of ore in mining
   5. Concentration cost and probable cost range
      a. Select a general method of concentration
      b. Consider concentration ratios and percentage recovery for each intended product
   6. Post-concentrator costs and probable cost range
      a. Select transportation and marketing methods
   7. Total capital investment and probable range
      a. Select the most likely mining rate
      b. Select the most likely mine plant installation, development method, access, and concentration and supporting facility
      c. Consider how much of the total is depreciable and amortizable
   8. Mine life and preproduction period in years
   9. Revenue from sale of product
      a. Consider royalty payments on gross revenue
II. *Summarize, from working estimates, the following:*

	Example
Ore tonnage in place (Tp)	100 million tons
Ore grade in place (Gp)	0.8% Cu
Broken ore tonnage	
(Tp) × (% extraction) × (dilution factor)	
Example: $(100 \times 10^6) \times (0.9) \times (1.1)$	99 million tons
Mining grade = Gp/(dilution factor)	
Example: (0.8)/(1.1)	0.73% Cu
Concentration ratio and recovery	45:1 and 80%
Operating cost	
(mining cost) + (concentrating cost)	
+ (post-concentrator)	$3.85 per ton
Mining rate (tons per day)	18,000 tons per day
Mining rate (tons per year)	6 million tons per year
Total capital investment (mine, plant, preproduction	
development, concentrator, facilities)	
(depreciable and amortizable portion)	$80 million
Mine life	16.5 years
Revenue per ton concentrate	$300 per ton
Revenue per ton broken ore	$6.67 per ton

Table 19-1 (con'd).   Steps in Estimating Profitability of an Undeveloped Mineral Property

---

III. *Obtain fixed (or assigned) parameters, such as:*

	Example
Depletion rate	= 15% (limit 50% net)
Federal and state income tax	= 55%
Depreciation and amortization	= straight-line method
Assigned hurdle rate for DCFROI	= 15%

IV. *Obtain annual cash flow estimate,* as demonstrated in Chapter 8

	Example (based on II and III)
Revenue	$40.02 million
Operating costs	− 23.10
Operating income	16.92
Depreciation and amortization	− 4.85
Net before depletion	12.07
Depletion	− 6.00
Taxable income	6.07
Income tax	− 3.34
Net income	2.73
Add depreciation and amortization	+ 4.85
Add depletion	+ 6.00
Cash flow	$13.58 million

V. *Obtain profitability expressions,* as demonstrated in Section 8.42.

	Example (continuation of IV)
Payout period	5.9 years
DCFROI	15.4%
Net present value	$1.48 million

VI. *Identify critical data* warranting further investigation
   1. Determine sensitivity of profitability expressions to changes in the factors considered in Part I.
   2. Consider probability distribution for values established in Part I.

VII. *Relate* profitability expressions, sensitivity analysis, and probability (risk) analysis to objectives and pending decisions

---

expressions. For example: a 5, 10, or 15 percent change in estimated tonnage, average grade, or mining cost will have certain effects on the net present value of an orebody. In further detail, wall-rock dilution, percentage recovery in milling, or the rate of production may prove to be especially critical. With a computer to process the data a wide variety of variables and constants can be

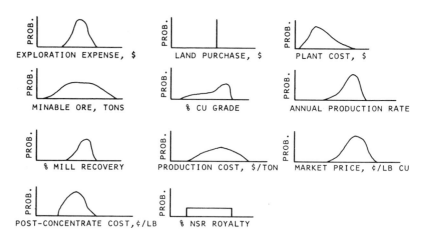

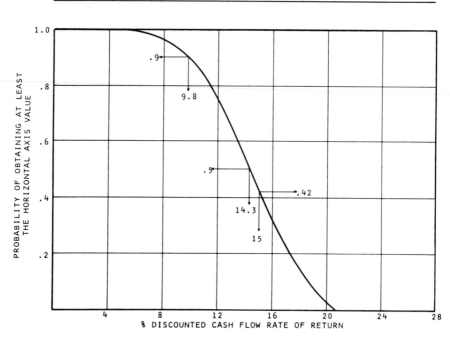

Figure 19-4   Risk analysis for a hypothetical prospect. There is a 50 percent chance for the discounted cash flow being above 14.3 percent and a 50 percent chance for the discounted cash flow being above 15 percent. (From D. T. O'Brian, "Financial analysis applications in mineral exploration and development." Fig. 2, 1969. SME Preprint 69-AR-76. Used by permission of the American Institute of Mining, Metallurgical and Petroleum Engineers, Inc.)

studied. A formal sensitivity analysis is not normally justified in order-of-magnitude estimates, but the concept is often applied.

Another type of analysis may be made as data are obtained (Fig. 19-4). This is risk analysis, in which probability distribution models are set up (subjectively) for each quantifiable factor and successive estimates of profitability are made on the basis of a Monte Carlo simulation, using randomly selected numerical values from each of the probability models (Roman and Becker, 1973). Computer processing is essential because so many repetitive calculations are needed. This approach, like sensitivity analysis, helps to identify the estimates that need the greatest refinement in successive stages of the investigation (Fig. 19-4).

19.3. KEY GEOLOGICAL FEATURES IN MINES AND PROSPECTS

Expressions of value are made of derived numbers—derived in large part from relevant geologic information. Consider some of the technologic and economic guidelines; the geologic element is there, and geologic context is needed. *Marketing*: The mineralogy of the deposit in question and the mineralogy of material being marketed from other deposits are determinants. *Availability of capital*: There will be competing opportunities for investment in other and perhaps more favorable mineral occurrences. *Taxation*: Governments establish taxes on mining to the extent that most orebodies, including the one in question, will be able to pay. There are, of course, guidelines more obviously related to geology: some mineral bodies are too complex to mine at a profit, others are almost ideally uniform and well situated for low-cost mining.

Priorities

Prospect evaluation would be an exciting academic exercise if we could use the preceding paragraph as our sole guideline: All geologic information is potentially relevant. But our objectives and conditions are not academic. This means we have to decide the relative importance of information so that we can assign priorities.

Appendix D, a format and checklist for mineral property evaluation reports, draws attention to relevant and potentially relevant information, but the list is so comprehensive that the relative importance of the individual topics is not apparent. Checklists in other books emphasize specific types of deposits, such as placer deposits (Wells, 1973), coal deposits (Muir, 1976), and marine minerals (Cruikshank, 1973), but these, like Appendix D, do not indicate which features are of most immediate importance. Suppose that the evaluation must be made in a limited time, say in a 3-week or 3-month period. What geologic information *must* be obtained first?

Given enough data, we could use sensitivity analysis and risk analysis to isolate critical economic factors and these techniques would help to identify the most important geologic information. We may, however, need to collect information from sparse data. In this situation, we can use the specific objectives of the evaluation as a coarse screen to catch the critical items.

If the objective is to locate a base-metal orebody capable of supporting a 10-year operation at a minimum mining rate of 1,000 tons per day, one of the key geologic features will certainly be the size of the orebody; if the geologic limits to the potential orebody are such that no more than 1 million tons of ore can be inferred, we need go no further. If the geologic limits are extensive enough to accommodate at least 3 million tons of ore, the average grade of ore is the next critical factor. If the tonnage and grade are acceptable, the shape of the orebody is the next critical factor in line. Then, in a reasonable sequence, geologic information on the patterns of grade, mineralogy, and geomechanics conditions can be investigated. Once the minimum objectives are met, a host of additional factors can be considered. It amounts to this: if the prospect or mine will not meet the minimum objectives, even under the most favorable conditions, further geologic information will be of only indirect interest, perhaps useful in assembling a district information file, but not important enough to warrant any additional expenditure.

Ira Joralemon,[1] the distinguished consulting geologist, once compared mine examination to testing the links in a chain of observations. If any link—geologic, technologic, or economic—is too weak for the purpose, the examination is over. He (1928, p. 536) illustrated the comparison thus:

> If the examination is made for a company so big that only a profit of half a million a year or more will have an appreciable effect on dividends, a 3-ft. vein of $10 ore can be eliminated at a glance. The fact that a careful operation might yield a profit of a few thousand dollars a month has no bearing on the subject.

The highest priority field work in mine evaluation generally involves ore sampling and geologic mapping within the orebody, but there is an important place for supporting geochemical sampling and geophysical mapping as well because we are dealing with chemical and physical *limits* to ore-grade material and its favorable environment. Limits could be "assay boundaries," such as an imaginary surface between 0.3% Cu and 0.29% Cu in a porphyry copper deposit. Limits could be structural—truncation of the San Rafael vein at El

---

[1] Ira B. Joralemon, "The weakest link; or saving time in a mine examination," *Engineering and Mining Journal*, v. 123, no. 13, p. 536, 1928. Copyright ©1928, *Engineering and Mining Journal*. Reprinted by permission of the publisher.

Oro, Mexico, by the Esperanza fault (Fig. 2-26), for example. Limits could be lithologic, as with ore in the limestones at Gilman, Colorado (Fig. 2-7); or they could be stratigraphic, as in the Ore Shale unit within the lower Roan Group at the northern end of the Zambian copper belt.

Special Objectives

Frequently other factors and other types of field work are just as important as the determination of grade, tonnage, and morphology. If the objective is to evaluate a coking coal deposit, the first critical factor is the coking character-istics. Testing of a few large bulk samples would be important. Next, the mineralogical nature of any sulfur present, the blending characteristics with other coal, and the behavior of the coal in preparation plants would be investigated. Once the required characteristics are established, there will be some basis for going to the expense of estimating the reserve tonnage and determining the structural details of the deposit. Neither of the last two factors would be worth investigating if the product—coke—were not accepta-ble.

19.4. EVALUATION PROGRAMS AND PROJECTS

Earlier in this chapter, prospects were considered from several viewpoints: evidence of former activity, mineralization discovered in an exploration target area, and as mineral property. In the first two considerations, ore mineraliza-tion is taken as one of several guides to orebodies and the field work at a prospect is part of a broad and rather flexible program schedule. The program continues as long as there are other prospects to investigate and other ideas to be tested. For a mineral property evaluation, on the other hand, evaluation is the entire program; the prospect or mine is the only target and the schedule is simpler. The schedule at a mineral property may also be tighter because a formal property agreement may be imminent.

At those mineral properties warranting several stages of investigation, each stage amounts to a separate evaluation upon which to base the next. Each stage is normally more expensive than the last, each option period requires a higher level of property payments, and each decision requires a more refined estimate of profitability. The stage-by-stage process becomes a formal pro-ject, with a need for a continuing budget and the project control techniques mentioned in Chapter 18.

But most mineral properties do not get this far. They are rejected in a preliminary evaluation. In Chapter 18, it was noted that five exploration firms in the southwestern United States examined 352 prospects and rejected 329 of them before drilling. This is typical, and it is not new; as mentioned in Chapter 1, the Guggenheim Exploration Company considered 1,605 mine propositions during 1910–1911, made preliminary examinations on 266, continued with examinations on 74, and recommended 3 for purchase.

Preliminary Evaluation

A preliminary evaluation is brief; it amounts to a few days or a few weeks at most. Many are so discouraging that they last for only half a day. If the exposed mineralization cannot possibly lead to an acceptable orebody the evidence is often quite clear. Making these evaluations—"chasing cats and dogs"—occupies a large portion of a district geologist's time. Yet one certain minor prospect just *may* turn out to be the subtle guide to a major orebody, and this will pay for years of half-day examinations.

No matter how brief a preliminary evaluation, it follows certain steps in which *some* time (even as little as one hour) is spent on each of these: regional context, information review, orientation, verification, new investigations, analysis and report, and follow up.

Regional Context. This may have taken years to attain, or it may have to be gleaned from a book on "the Geology of Sierra Antigua" carried in a briefcase and read en route. Thorough familiarity with the region is a great advantage. A folder of appropriate regional maps and air photographs is valuable. A total ignorance of the region is calamitous.

Information Review. Under the best circumstances, all of the existing information will have become available before field work begins and there will have been an opportunity for a thorough review. Some organizations make particularly good use of their geologist's time by demanding copies of all pertinent information from the property owners prior to arranging for a property evaluation.

It often happens, however, that the prior reports and maps will be given to the geologist by the property owner during the first day in the field; this means that some time (hours to days) must be set aside for reviewing them before field work gets too far underway.

Under the worst circumstances, and this sometimes happens, reports and maps are "somewhere in the Plomo Lead Company's office files" and they cannot be obtained until after field work has finished.

Orientation. The property has some relation to terrain, access routes, and to other mines and prospects. Even though this may be described in the available information, a first-hand "overview" is helpful. It may involve no more than a hike along a few ridges, a jeep tour, or an aircraft flight, but it adds meaning to the reports and to the field work. Key traverses can be planned. The most significant exposures may turn out to be just over the hill from the main prospect pit.

Verification. Once the key factors are identified and some supporting geologic information is found in prior reports, verification—field checking—can begin. We can now look for better answers to the questions posed in Section 19.2. If sample data are available, visit the sampled locations and take a few check samples. Perhaps the sampling was not entirely appropriate; it may have been based on a cutoff grade that is no longer applicable. Sampled pillars and faces may show that a zonal change in mineral ratios rather than a lack of mineralization was the reason for abandonment. New samples for metallurgical testing or mineralogical study may be needed.

If drill core is available, check it against available logs. If the core has been lost, visit the drill-hole sites anyway. They may have been planned according to a structural concept, for example, whereas you are working with a stratigraphic concept.

Where workings are inaccessible the mine dumps may help verify statements made in the reports. Mineralogy in the ore zone should be evidenced, if only in a few pieces of low-grade material. Of course, the presence of too many high-grade specimens in conspicuous places may mean that the dump has been "salted." In general, a dump provides a good check on reported lithology; a major quartzite zone may be shown on mine maps, but close inspection of the light-gray material on the dump may show it to be a metavolcanic rock or a silicified limestone.

If the volume of the dump exceeds the apparent volume of the mapped workings, "leasers" (lessees) may have mined after the map was made, or perhaps a working that was reported to stop just short of a projected ore zone may have actually gone through the zone without encountering ore.

New Investigations. Verification or rejection of existing information is seldom an adequate basis for evaluation—even in a preliminary investigation. New concepts deserve supporting field work, but the mapping control does not have to be precise. The objective of a preliminary investigation is to determine what precise work is justified. Sketching on aerial photographs and topographic maps or some compass-and-pace field sketches supported by surface photographs can illustrate new ideas.

Sampling for geochemical analysis is a quick way of gaining information on extended ore zones that were not previously identified. In sampling, as in preliminary mapping, the objective is to identify potentially favorable conditions so that detailed geochemistry and the related expense can be justified.

Analysis and Report. The field work for a completely unacceptable prospect will have been brief, and most of the geological analysis may have been done mentally while walking around the property. A report on the property will also be brief, but it should contain sufficient information to be useful in regional studies and complete enough to make another evaluation unnecessary.

If field work has included sampling, laboratory results need study. If some existing data have been verified and significant new information added, these need geological analysis even if the property is rejected. Rejection of a property that has been worth sampling and mapping requires some explanation. The report should be detailed enough to show how the conclusion was reached.

Follow-up. If further work is to be recommended, some time is needed for collecting data on its applicability and scope before leaving the property.

> *Surface mapping.* Select an appropriate scale and outline an area for airphoto coverage.
>
> *Underground mapping and sampling.* Determine the condition of the workings and estimate the cost of making them accessible.
>
> *Geophysics.* Select some appropriate traverse lines, construct approximate geologic sections, and take samples for laboratory testing.
>
> *Geochemistry.* Sample as many lithologic, soil, and stream-sediment zones as are likely to be involved so that background data can be estimated.
>
> *Drilling.* Take note of possible sites, access, water supply, and any potentially difficult rock types or structural conditions.
>
> *General.* Provide some information on seasonal conditions, housing, and supplies.

Follow-up and Ultimate Evaluations

Follow-up work amounts to repeated examinations, each ending with a "continue or abandon" decision. The "ultimate" evaluation (in quotations because there is always something left undone) leads to preproduction work. The decision to abandon a property becomes more difficult as the follow-up commitment deepens. There is the temptation to go just one step further, to follow the time-honored prospector's advice: "*Never* quit until you've put in one more round." There is an even greater temptation to hold the property for better conditions—new economic factors, new technology, more exploration money. Large companies are capable of doing this, and it is often done. Small company investors and their accountants, however, are not likely to be so patient. If a large sum of money has been spent, the expenditure is "frozen"—to be recovered in profits from an uncertain mine—unless the project is abandoned and the expenditure deducted from taxable income.

Increasing levels of property payments are a factor as well, because they serve their intended purpose of preventing long-term inaction.

Scheduling the Work

The general idea in scheduling property evaluations, whether preliminary or detailed, is to provide as much time as possible at the site of field work and to waste as little time as possible in travel and other nonproductive activities. The most expensive means of access (helicopter) may actually be the least expensive way to provide the most effective work.

Office time also has its place. Sufficient time must be scheduled for review and analysis of the data; a rule of thumb, one day in the office for one day in the field, is sometimes used.

An estimate of average field time versus total time (southwestern United States) has been made by Thomas W. Mitcham, a consulting geologist with considerable experience in prospect examination and evaluation. In his estimate (personal communication), field time means the time physically on the property. Total time includes office preparation, travel, meeting with owners, extraction and assembly of existing data, literature study, and report preparation, in addition to field time. Thus,

Type of Examination	Field Time as Percentage of Total Time
Zero data base; 0.5 field day	10
Zero data base; 2 field days	30
Zero data base; 10 field days	60
Moderate data base	15
Major data base	8

Note that the smallest percentage of field time is found at the two ends of the list: in very brief examinations (travel is the main time consumer) and in examinations with a major data base. The latter situation applies to mining properties with abundant literature and to mineral properties with long histories of mining and development.

One final note on scheduling. Mineral property evaluations are seldom so well organized. Contacts are frequently made with mineral property owners by chance rather than by design, and the evaluation may be totally unscheduled. One examination or evaluation may lead to another because the geologist happens to notice a pit or an old headframe in the area. Sometimes, as in exploration programs, geological serendipity has been responsible for starting investigations that result in a mine.

19.5. MINERAL PROPERTY REPORTS

An examination report may be nothing more than completion of a brief data card or a memorandum of a few pages for inclusion in district exploration studies. The evaluation report is more likely to be an exhaustive presentation with recommendations for specific action. In any event, one major objective is to provide necessary information to the people who make major decisions—and to provide this information in a concise, quantitative way. Detail, required by those who implement the decision, must be included, but must not be allowed to bury the main issues.

No single format can be used without modification. But in order to compare individual properties and to accommodate computer-using systems of data management a specific pattern may be followed within an organization.

The format and checklist in Appendix D are intended as a guide rather than as a requirement for routine mention of every topic. The inclusion of irrelevant topics would make tedious reading. A better approach is to select topics from the format after identifying the "readership" in the same way a journalist does. A journalist provides answers to "who, what, where, and when" at the earliest possible opportunity and then follows with details that may be skimmed by most readers but read thoroughly by others who have more time or a particular interest. The "suspense" approach of fiction has no place in journalism or in mineral property reports; a reader who delights in waiting until the last chapter to see "who done it" will select a novel rather than a report.

The "Readership"

There are those who must translate the entire report into action and those who need only selected portions to guide their particular parts of the action. The first group includes the principal client or the organization's management who must relate the report to organizational objectives, choose between this investment opportunity and opportunities in other projects, and allocate the necessary funds and manpower. For this group, a summary and digest of recommendations is of first importance; but because their decision is made on a technical and professional basis, they will also study *some* of the supporting information and will want to locate the appropriate parts of the report in a table of contents rather than by searching from cover to cover.

Among the second and larger group, those who need to know the gist of the recommendations for context but who must study selected portions of the report in more detail, are the reporting geologist's associates and those of the immediate manager—these people will have studied other properties and they will ask "*why* this one?". The second group also contains specialists who may have already been called upon to furnish information for the evaluation and who will next be asked to design further work in such matters as mining, processing, environmental protection, and marketing. They will be looking for material on which to base their respective contributions. Accountants and cost engineers belong to the "selected portions" group; a record of expended money, time, and manpower is their concern.

An information recipient with a longer term view is the exploration geologist of the next year or the next decade; to this geologist, the report is a record of geological relationships, a source of regional data, and a basis for reevaluation of the property. A mineral property report is never dead, but it can be very sick if it is described in unintelligible terms.

Quantitative Data

In the suggested format and checklist, the first page, "key information," is intended as a summary for transferring data into computer language. It is important to concentrate on quantitative expressions wherever feasible and to employ key words. But quantification without adequate basis and key words without real significance will provide computer "garbage"; in fact, irrelevant and unsubstantiated information does not belong *anywhere* in the report. The former will be regarded as "padding" and the latter (unless identified as such) as downright dangerous.

Illustrations, such as maps, graphs, photographs, and cross sections, are major parts of the report. Whether from previous reports or newly made, the degree of generalization (fact versus interpretation) must be stated. In every instance, allowance must be made for data transmission and micromedia handling. Much of these data and most tabular data provide essential detail, but they belong in appendices. Interpretation belongs in the body of the report.

With a completed evaluation report and an accepted recommendation to go ahead with mine making, a sequence in exploration has been completed. The new phase, preproduction, and the ultimate phase, mining, involve decisions that dwarf all previous decisions. The expenditures are orders of magnitude greater than in an exploration program or a prospect evaluation. The geologist and the geological engineer enter a new scene with a much larger cast of players.

CHAPTER 20
MINING OPERATIONS: MINING GEOLOGY

Assume that a prospect has been recommended for development. A potential orebody has been sufficiently defined to reduce the odds against success to an acceptable range. A decision is made and mine-making begins.

If it made sense to separate "mining geologist" from "exploration geologist," this is where the separation would be made and the mining geologist would be identified as the continuing man-on-the-scene. From the mine-making (preproduction) stage on, the mining geologist's work is done in more detail than in exploration, there is a demand for greater accuracy, and the work has an immediate engineering context.

Actually, the separation is not warranted. The most valuable members of an exploration team are those who understand the complexities that lie between a mineral discovery and a mine because they have spent some time unraveling the detail in a productive mineral deposit from the "inside." Similarly, the most capable mine staff geologists are those who have had enough time in exploration to relate the characteristics of other orebodies and comparable mineralization patterns to their mine. Some mining companies provide the experience for their personnel by encouraging young exploration geologists to accept a temporary assignment with the resident staff at a mine or with the project geologist's staff at a preproduction site; the same companies often provide an opportunity for their mine geologists to participate in exploration programs.

20.1. PREPRODUCTION WORK: THE PROJECT GEOLOGIST

A preproduction project begins with the decision to develop a mine and it ends with the first day of mine operation at the designed capacity. Some of the work can still be called exploration and evaluation because an orebody is being outlined and tested, but the outlining and testing are now related to *how* the deposit is to be developed and mined rather than *if* it should be mined.

The preproduction interval was identified in Chapter 7 in relation to the life cycle of a mine (Fig. 7-6). The conditions influencing the preproduction interval are now listed in Table 20-1. Physical conditions, such as the actual dimensions of the deposit, are essentially static inasmuch as they are no different today than during any previous era in human history; they belong to the geologic time scale and to the project geologist's direct field of responsibility. The remaining conditions are dynamic; they involve the changing aspects of technology and economics in which today's major mines are made from the small mines and undeveloped prospects of earlier years. The dynamic conditions of mine-making are the engineer's responsibility, but the geologist shares part of the responsibility as well.

The project geologist at a developing mine spends a major amount of time investigating the physical conditions directly associated with the deposit itself and interpreting the conditions in terms of technology and economics. Perhaps the deposit is larger than it was originally thought to be; perhaps it is broken by small faults or folds into more segments than expected. The design engineers must take these geologic conditions into account. The orebody may have the same average grade as indicated by earlier sampling, but the distribution of values and the mineralogy may now appear to be more complex on the basis of a tighter drilling pattern. The metallurgical engineers must be advised. Hydrologic and geotechnic conditions already may have been estimated in a general way on the basis of a few observations, but now the specific aquifers must be investigated and the wall-rock structures must be mapped in detail before establishing a development pattern and selecting a mining method.

Some of a project geologist's work extends to the investigation of physical conditions in the surrounding area and even in the entire region. An environmental condition is being changed by mine development; the original condition must be measured and the disturbance must be monitored. Ore from other deposits may eventually be processed with ore from the principal mine; the deposits must be evaluated. Sources of coal and other supporting minerals for the mine, mill, and smelter must be located and evaluated. New transportation routes and new sources of water or hydroelectric power may have to be investigated.

Target characteristics, the dynamic conditions that determine how well a project geologist can interpret the parameters of the deposit, are some of the main day-to-day concerns. Certain geophysical techniques may be ideally

Table 20-1. Conditions Influencing Preproduction Interval in Mining

---

Physical conditions
1. Geometric
   Size, shape, attitude
   Depth
   Continuity (multiplicity of orebodies and ore shoots)
2. Geologic
   Grade (average, range, and uniformity)
   Mineralogy
   Petrology
   Geothermal pattern
3. Hydrologic
4. Geotechnic
5. Geographic
   Topography
   Climate
   Environment
   Proximity to other deposits

Technologic conditions
1. Target characteristics
   Response to geologic analysis, geophysical and geochemical measurement, exploratory drilling, and statistical representation
2. Mining
   Suitability for conventional methods, for new methods, and requirements for pioneering design work
3. Processing
   Amenability to treatment, complexity of treatment, and need for pilot plant work
4. Transportation
   Existing facilities and construction conditions for new facilities
5. Power and Water
   Existing facilities, outlook, and conditions for development
6. Labor and management
7. Technologic information
   Guidelines from district and similar operations
8. Unforeseen difficulties

Economic conditions
1. Marketing
   Value, specifications, predictability
2. Land
   Pattern, tenure
3. Financial
   Investment characteristics
4. Political
   Taxation, attitudes, continuity
5. Conservation
   Goals, limitations

suited for delimiting the deposit but they may not be appropriate because of geologic or cultural noise; other techniques must then be tried. Drill-hole data may be most needed in zones where drilling conditions are particularly difficult; the right drilling techniques must be sought.

Some of the technologic conditions most certain to affect a preproduction program can only be described as unforeseen difficulties. These are best explained in the burlesqued scientific statement known as Murphy's law: "If anything *can* go wrong, it *will*." The geologist can expect a responsible role in dealing with unforeseen difficulties because most difficulties in mining operations are likely to have geologic sources. An unexpected flow of water is a problem for the project geologist, as is the encounter with an unknown fault zone or an unmapped dike. The position of project geologist is normally assigned to an experienced person with a broad range of capabilities— someone who can deal with unforeseen difficulties.

Very few of the conditions listed in Table 20-1 lie outside the project geologist's consideration. The geologist may be asked to provide information on conditions as remote from geology as market predictability and political attitudes because these conditions are influenced by the geologic distribution of comparable mineral deposits in the district, in the nation, and in the world.

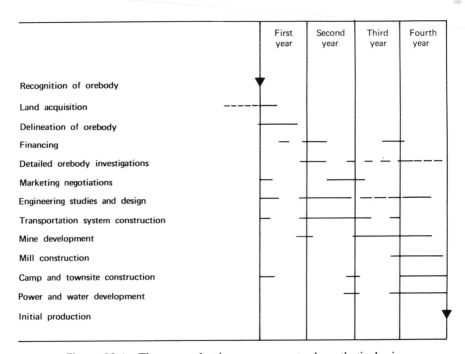

Figure 20-1   The preproduction sequence at a hypothetical mine.

The project geologist's assignment generally lasts for several years. Its duration is determined by the preproduction sequence, shown in Figure 20-1 in terms of a hypothetical four-year project, and in Figure 20-2 in relation to the Bougainville copper mine in the South Pacific. Mines with near-optimum physical conditions, few technologic difficulties, and favorable economic circumstances may go into production with only a 1- or 2-year preproduction program. Most mines and the "average" mine (as hard to find as an "average" man) have preproduction periods lasting from 2 to 5 years. Large-scale mines with low unit value ore and complex processing problems need 5–7 years of preproduction work. Mines being developed under difficult physical conditions, high technologic hurdles, and large economic problems may need 10–15 years of preproduction work, and they require a staff of first-rate project geologists.

A good description of the work of a project geologist and geologic staff is provided by Blais and Stebbins (1962) in their description of the development of the huge iron ore deposits at Knob Lake, Canada. The project at Knob Lake involved much more than the development of an open-pit mine; it involved the construction of processing plants, the erection of two hydroelectric plants, the building of a townsite, the construction of a 580-km railroad through difficult terrain, and the installation of port facilities. All of these accomplishments were supported by intense geologic work and extensive mapping, geophysical surveys, trenching, and drilling.

20.2. THE GEOLOGICAL DEPARTMENT

When most mines were small and most orebodies were high in grade (by modern standards), the formula was: one mine—one geologist, and the resident geologist's principal job was to assure the mine's continuity by finding a ton of new ore for every ton mined. Ore shoots might die out or be offset in all kinds of structural situations, but the job of keeping the stopes in ore could still be handled by one person, a geologist or mining engineer who maintained an image (on maps and mentally) of the situation in every working face and at every drill site. Some mines, even some relatively large operations, did not have a resident geologist as such. With enough ore in sight to maintain production for several years and with some yet-undefined boundaries to the orebody, it did not take a geologist's expertise to probe into new ground with headings and drill holes. Structural and stratigraphic problems could often be solved by mining engineers, and mining engineers were better people to have on the staff because they could do a great many other jobs as well. If an especially difficult geologic problem arose, it could be turned over to a consultant or to an experienced geologist from company headquarters.

Some mines still operate with a single resident geologist, with a geologist who works within the engineering department, or with none at all, but the

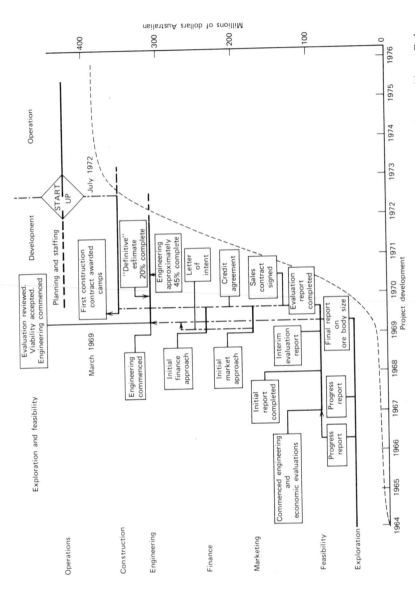

Figure 20-2 The preproduction sequence at the Bougainville copper project, Papua New Guinea. (From R. B. Hope, "Engineering management of the Bougainville project," Civil Eng. Trans., v. 13, no. 1, Fig. 4, p. 46, 1971, by permission, The Institute of Civil Engineers, Australia.)

scale and complexity of most modern-day mining operations have made the geological department a necessity. Although the geological department may be important, it usually is a small department, comprising a chief geologist, three of four mine geologists, and a few technicians. A larger geological department may, in addition, have the administrative responsibility for work, such as geotechnical investigations and ore-grade control, that is shared with other staff departments.

Responsibilities

The most direct responsibilities of a mine geological department are to interpret the geology of the known orebodies and to find more ore. The search for new ore is *exploration* in the strict sense of the word, but the bookkeeping and taxation label is *development* if the new reserves can be identified as outlying parts of the main orebodies.

Other responsibilities relate to the design, planning, and operation of the mine and the metallurgical plant, to exploration in the immediate district, and to conservation practice in land use. The resident geologists at individual mines that belong to large organizations have some broader responsibilities connected with mining plans and exploration or research projects that lie outside the immediate district but still need expert geologic guidance. In each field of responsibility, there are long-term (years), intermediate-term (months), and short-term (day-to-day) elements.

Orebody Reevaluation

Orebodies that were outlined in earlier exploration and preproduction work are verified, orebody limits are defined, and new ore reserves are sought. The geological department is usually responsible for providing an inventory of ore reserves, generally at yearly intervals. Changes and discontinuities in the ore mineralization are investigated, and if a major fault zone or the bottoming of an orebody is involved, it may become a dominating concern of the geological department for months or years. In the short-term aspect, ore-grade control and the determination of ore boundaries in stopes and benches will often be the responsibility of the geological department.

Mine Design and Planning

Geologic factors involved in design and development plans are identified and interpreted. Where development is planned for new ground with untested geological and geotechnical characteristics or when changes in the mining method are anticipated, the geologic input is critical. Monitoring of ground water and periodic reading of geotechnical instruments are often a responsibility of the geological department. Unexpected problems with ground control and slope stability demand immediate geologic investigation and immediate advice to the operations and engineering departments.

Metallurgical Design and Planning

Where metallurgical treatment methods are being improved or methods are being projected for future operations, samples are tested from all accessible parts of the orebodies and the samples must be provided with a geologic context. Mineralogical zoning, both in ore and gangue, is important. Production goals—forecasts of the monthly or yearly recovery of metals or minerals—require geological advice. In the short-term aspect, the processing plant may need daily advice on changes in the characteristics of ore from the active stopes or benches. Where ore blending or selective mining is practiced in order to give a uniform mill feed, the resident geologist's advice is needed at the mine, at the concentrator, and sometimes even at the smelter.

District Exploration

One of the principal responsibilities of a mine geological department is to understand and interpret the pattern of mineralization throughout the district. Nearby prospects are examined and evaluated, active mining properties belonging to other companies are evaluated insofar as possible without trespassing, and the exploration activity of other companies is monitored insofar as legitimate (spying is not a legitimate activity). The need for districtwide information may suddenly become important when a joint venture, a mining lease, or a purchase of mineral land is considered. After a joint venture in exploration or mining has been agreed upon, the resident geologist may then become the company's representative and will be asked to spend a certain amount of time on the project.

Supporting minerals, such steam coal or metallurgical coke or limestone for cement and metallurgy may be needed; the resident geologist evaluates the sources. In some instances, byproducts from the principal mining and smelting operation may have to be combined with minerals from other deposits or used in special mining operations; sulfuric acid from a smelter may, for example, be used to make phosphate fertilizer or to leach oxidized copper orebodies; the phosphate rock deposit or the amenable copper deposit is sought.

Land Use, Conservation, and Environment

Land will be classified by the company or by the government for mining and nonmining use. The resident geologist's responsibility is to make certain that the classification is valid, that enough land is controlled for future expansion, and that the mineral rights are maintained by annual labor or by meeting other requirements. Mine dumps, tailings disposal areas, plant sites, housing areas, and roads are not intentionally placed in areas that may be mined. The investigation of such areas may require drilling to prove that ore is *not* there—an expenditure that is sometimes difficult to explain to management.

Land that is designated for multiple surface use, for multiple mineral extraction, and for postmining restoration must have certain geologic characteristics, and these characteristics must be identified before plans are made. New construction sites and new transportation routes are required during the life of most mining operations; geological and geotechnical investigations for site preparation are made by the resident geologist alone or in conjunction with a consultant.

The existing environment—land, water, air, and biota—has geologic aspects, and most of the possible changes in the environment that result from mining have geologic aspects. Environmental investigations and environmental monitoring are definite responsibilities of the resident geologist insofar as ground subsidence and ground water are concerned, and the responsibility may cover the monitoring of surface water as well.

The Broader Responsibilities

In addition to the function (planned or unplanned) of providing young geologists and engineers with postgraduate training for positions in mine management, mine engineering, and exploration, a mine geological department may provide a "think tank" for generating new ideas in science and technology. A great many of the most revolutionary concepts in exploration geology are born from the detailed work of mine geologists and in the great natural laboratories composed of mine workings and drill holes. Samples for laboratory research work usually come from large mines, where the sampling can be done with care by resident geologists who understand the intended context. The results of laboratory research are also evaluated at the very earliest opportunity by the resident geology staff—for the same reason: context. Exploration techniques developed in the research departments of mining companies are commonly tested in the company mines, where a particular geophysical or geochemical response can be related to measured conditions and verified by further investigations.

An experienced resident geologist can better understand the ore association at a mine than could wider ranging exploration geologists. The resident geologist thus becomes an in-company consultant on target investigations and prospect evaluations where conditions resemble home ground. A mining company with mines in a wide variety of geologic terrain has wide access to well-qualified experts.

The Work Pattern

A part of the satisfaction in a resident geologist's work comes from its variety. In the morning, a principal concern may be the hydrologic aspects of an in situ leaching proposal. In the afternoon, the geologist may be in an underground exploration drift, studying an exposure of ore that defies all previous ideas of zoned mineralization. By the end of the day, the emphasis may have changed to a critical stratigraphic sequence expected in the district's deepest drill hole. At night, the geologist may be awakened by a telephone call from the drilling foreman with the news that Murphy's law is damnably valid—the steel in the deepest drill hole has twisted off just above the critical zone.

The basis for a mine geologist's interpretations and decisions is a thorough picture, in maps, sections, and models, of the orebodies and the district. The picture requires geologic mapping, the assembling of drill-hole data, sampling, and long-term laboratory study; the work is continuous, but it is not routine. Continuity and the opportunity for scientific depth are, in fact, the characteristics of resident mine geologic work that appeal to many geologists.

An illustration of the work pattern in the geological department at a modern mine is explained in a paper by Matulich and others (1974) on the Kidd Creek mine, Ontario. The mine is sufficiently well established to typify actual production-stage work and yet is young enough to escape the inflexible practices built by tradition.

The Kidd Creek operation comprises an open-pit mine, an underground mine, a 10,000-ton-per-day concentrator, an electrolytic zinc plant, and a 28-km railroad. There are two major orebodies, stratabound massive sulfide zinc-copper-silver deposits in felsic volcanics. Figure 2-23 shows a cross section of one orebody, the pit, and some of the underground workings. Elements in the discovery case history, one of the outstanding exploration stories of the 1960s, are cited in Table 18-1.

The chief geologist at Kidd Creek, as head of a staff department, is responsible to the general manager. Departmental work, carried out by four geologists, four technicians, and one core sampler, consists of:

1. Geologic examination and interpretation of cuttings from rotary drill holes (blastholes) in the pit. Assay data from the same blastholes are compiled by the grade-control section of the engineering department.

2. Logging, petrographic examination, and trace-element study of chip samples (overburden and bedrock material) from the rotary overburden drill. Outcrops are scarce in the vicinity of the mine, and the overburden drilling provides a method of probing to bedrock.

3. Diamond drilling in the pit, the underground mine, and the immediate area. Exploration drilling and primary ore definition (preliminary) drill-

ing are the responsibility of the geology department. Secondary ore definition (production) drilling is the joint responsibility of the geological department and the grade-control section of the engineering department.

4. Geologic mapping in the pit and the underground mine and of the few available rock exposures. Data from mapping surface exposures are combined with data from overburden drilling; data from pit-face mapping is integrated with lithologic data from blasthole sampling. Special mapping for geomechanics studies is done at the request of the engineering department.

5. Ore-reserve calculations. For thoroughness and context, two methods are used, one by the geology department and one by the grade-control section of the engineering department. Ore-reserve computerization is provided by the operations research section of the engineering department.

6. Geological research. One geologist is assigned to research work. Projects are initiated within the department or in the exploration division of the parent company or other mine departments. The research laboratory is equipped for incident-light and transmitted-light petrographic studies. Chemical analyses are performed by the mine chemistry laboratory and by independent laboratories.

7. Special projects are referred to the mine geological department by the exploration division of the parent company. Some of the projects— consultation on a mining problem or on a property evaluation, for example—are handled in conjunction with the engineering department.

8. Immediate area exploration. In addition to overburden drilling and outcrop mapping, the Kidd Creek geology staff conducts an entire surface and underground exploration program. Much of this work includes electromagnetic and magnetic surveys made in cooperation with the company exploration department.

The work of a geology department at an entirely different kind of mine, a uranium mine in Colorado Plateau sandstone-type ore, is described by Hohne (1963). At the Kermac mines, near Grants, New Mexico, the orebodies are erratic, discontinuous, and nonhomogeneous; the wall rock is soft and friable and acts as an aquifer. Mining under these conditions is inherently difficult, and the mine planners are closely guided by geologic information and geologic advice. Grade-control engineers—graduate geologists, geological engineers, or mining engineers—work closely with the mine geologists. As at Kidd Creek, the geological department maps all underground workings, supervises the drilling for additional ore, and provides ore reserve calculations; but there is an additional and very interesting job for the geologists at the Kermac mines. At the beginning of each month, they draw up a weekly "prediction sheet" for the tonnage and grade expected from each working place; then at the end of each week, they record the actual production and make a comparison as a measure of their predictions. Geologists who accept such well-documented records of their performance must necessarily be quite competent; they can hardly be accused of getting their ideas from a crystal ball.

20.3. CONSULTING GEOLOGISTS AND CONSULTING FIRMS

Individual consultants and well-staffed consulting organizations may be called upon at any stage in exploration and mining, but they are most often called upon when special problems arise at an operating mine or when an independent evaluation of a mining situation is needed. In either event, the consultant has a much more direct "line" to top management than the resident geologist has. Things get done.

The consultant's assignment might be to work on a problem that could just as well be handled by the mine geological department, except that the problem is of such high priority that a small, busy resident staff cannot afford the time. Geological guidance for an expansion program or for the development of an adjacent mine would fit in this category. The consultant's organization, in this situation, acts as an extension of the resident department.

On the other hand, the problem may be so extraordinary that a specialist is needed. Hydrologic problems, geotechnical problems, and the geologic aspects of a change in mining method—from conventional mining to solution mining, for example—are of this kind. The need for outside advice is often pointed out by the resident geologist, who is likely to recognize the complexity of the situation, who knows how to isolate the problem so that it can be assigned to a specialist, and who knows how to interpret the specialist's findings. Sometimes there is a delicate professional aspect in the decision to call for help or to "go it alone"; a resident geologist may recognize personal limitations but may not want others to recognize them. If a request for a consultant's help is delayed because it may reflect unfavorably upon someone's skill, it probably will do just that.

Consultants are sometimes employed to evaluate information supplied to one company by another company in the course of negotiations for mergers, loans, and joint ventures. The consultant's independence and objectivity as well as expertise are stock-in-trade; therefore the consultant may want to check almost all mapped and recorded information in the field and in the mine, not necessarily because of a suspicion of the local geologists but because the evaluation is expected to represent an original interpretation of the data.

Whatever the purpose of a consultant's participation in geological work at a mine, there is a close association with the resident staff. It is a valuable association. New ideas are born and new points of view are exchanged, and the exchange is of considerable advantage to the local staff, to the consultant, and to the company. Geological theories that have been developed by the local staff may eventually appear in the consultant's report, and they may appear in terms of exploration targets and recommendations for action. It is the consultant's responsibility to give proper credit to the geologist who provided the ideas and information, and most consultants are careful to do this. Only rarely will resident geologists find that they are being "upstaged" with the results of their own work.

Geological work at an operating mine is such a demanding assignment and the work covers such a broad spectrum that the views of experienced consulting geologists and experts must be sought whenever the occasion arises. Mining geology has developed into a powerful profession, but mining geologists—better than anyone else—know that mineral deposits are still too complex to be understood by any particular group of scientists or explained by any prevailing theory. A mining geologist needs all the help he can get.

APPENDIX A
COMMON ABBREVIATIONS IN FIELD AND LABORATORY NOTES

These abbreviations are usually seen in charts, diagrams, and field notes. Exploration and mining organizations generally adopt more comprehensive sets of abbreviations that emphasize their particular commodity interests and their "home" terrain. This system is synthesized from lists provided by several North American mining companies and from a list compiled by Chace (1956). Fundamental rules for abbreviating terms in field and mine geologic mapping are discussed by Chace.

Computer-based systems for recording field data make use of precise abbreviations. The systems are *internally* consistent, but there is little consistency in coding *between* the computer-based systems; therefore, the older and more generally recognized abbreviations given in this appendix continue to be used.

In GEOLOG (Blanchet and Goodwin, 1972), a computer-based system designed for mining geology, dolomite-calcite mineralization would be given in a two-letter code thus:

DX   dolomite and calcite, proportion unspecified
Dϕ dolomite alone
D>   dolomite more abundant than calcite
D#   dolomite and calcite in equal proportions
D<   dolomite less abundant than calcite
CA   calcite alone

The coding is precise, but it would not apply to an "outcrop input" document used in the Canadian Grenville project (Laurin and others, 1972) where calcite is CC.

In the following lists, the mineral, rock, and miscellaneous terms are alphabetized according to their abbreviations so that existing field and laboratory notes can be read. Some common alternative abbreviations are shown in parentheses. Where several abbreviations can be expected to follow the same general pattern, only a few examples are given.

Color abbreviations and general scientific abbreviations are not included. Standard dictionary listings are normally used, and in order to avoid conflicts between standard abbreviations and similar geologic abbreviations, the modifying terms precede the mineral or rock term.

Letter symbols for age and name of rock units generally follow the guidelines established on maps of government geological surveys. In U.S. Geological Survey practice, letter symbols consist of an initial capital letter for the name of the system and one or more lower case letters for the name of the rock unit, as Cpv (Carboniferous—Pottsville Formation).

1. MINERALS AND MINERAL ASSEMBLAGES

Mineral names are not capitalized unless they refer to naturally occurring elements (Au), groups of minerals (FeOx), and alteration (Prop). Only the most common ore minerals are included in this list. Abbreviations that have been used for other ore minerals are shown in Table 2-1, "Economic Minerals in the Subsurface Environment."

ab (al)	albite	carb	carbonate minerals
Ag	silver	cc	chalcocite
alu (at, al)	alunite	chl (ch)	chlorite
amph	amphibole	Chl	chloritization
an	anorthite	chr	chrysocolla
anh (ah, an)	anhydrite	cht	chert
ank	ankerite	cp (cpy)	chalcopyrite
apa (ap)	apatite	Cu	copper
Arg	argillization	CuOx	copper oxides
asp (aspy)	arsenopyrite	cup	cuprite
Au	gold	cv	covellite
az	azurite		
		dck (dt)	dickite
bar (ba)	barite	dg	digenite
bent	bentonite	dol	dolomite
bio (bi)	biotite	Dol	dolomitization
bit	bitumen		
bn (bo)	bornite	en	enargite
cal (ca)	calcite	ep	epidote

FeOx	iron oxides	Phyl	phyllic alteration
feld (fs)	feldspar	pla (pg)	plagioclase
Feld	feldspathization	pn	pentlandite
fl	fluorite	po	pyrrhotite
fm	famatinite	Pot	potassic alteratio
		Prop	propylitization
glauc	glauconite	prp	pyrophyllite
goe (go)	goethite	pu (pru)	proustite
gn	galena	py	pyrite
gr (gt, gar)	garnet	pyx (px)	pyroxene
gyp	gypsum		
Grei	greisenization	qtz (qz, q)	quartz
hbl (hb)	hornblende	rc	rhodochrosite
hem (he)	hematite	rd	rhodonite
		rt	rutile
ill (it)	illite		
ilm (il)	ilmenite	ser (sr, sc)	sericite
		Ser	sericitization
ja	jarosite	serp (sert)	serpentine
jasp (jas)	jasper	sid (si, sd)	siderite
		sil	silica
k-spar (kf)	K-feldspar	Sil	silicification
kao (ka)	kaolinite	sl	sphalerite
		sp (spec)	specularite
li (lim, lm)	limonite		
		tm (tl)	tourmaline
mag (mgO)	magnetite	tn	tennantite
mal (mc)	malachite	tt (td, tet)	tetrahedrite
mb (mo)	molybdenite	tu	turquoise
mc (mar)	marcasite	tz	topaz
mont (mon)	montmorillonite		
MnOx	manganese oxide	wf	wulfenite
musc (mv)	muscovite	wz	wurtzite
or	orthoclase	Zeo	zeolites

2. ROCKS

The abbreviation is capitalized where used as a noun (Rhy) and not capitalized where used as an adjective (rhy Volcs). Mineral terms sometimes prefixed (qtz-ser Sch).

And	andesite	Bas	basalt
Apl	aplite	Bx	breccia
Ark	arkose		

Clst	claystone	Mdst	mudstone
Cgl (Congl)	conglomerate	Monz (Mz)	monzonite
		Mp	monzonite porphyry
Dac	dacite		
Dap (Dac Por)	dacite porphyry	Peg	pegmatite
Dio (Di)	diorite	por	porphyritic
Dol	dolomite	Por (Ppy)	porphyry
Dp	diorite porphyry		
		Qd (Qdi)	quartz diorite
Fangl	fanglomerate	Qm	quartz monzonite
		Qmp	quartz monzonite
Gd (Grd)	granodiorite		porphyry
Gn (Gns)	gneiss	Qp	quartz porphyry
Gr	granite	Qzt (Qtzt, Qte)	quartzite
Grnst	greenstone		
Gwke (Gw)	graywacke	Rhy	rhyolite
		Rk	rock
Hfls	hornfels	Rx	rocks
ign (ig)	igneous	Sch	schist
intr	intrusive	Sh	shale
		sed	sedimentary
Lat	latite	Slts	siltstone
Ls (Lst)	limestone	Ss (Sdst)	sandstone
Lut	lutite		
		volc	volcanic
Mb (Mrb)	marble	Volcs	volcanic rocks
met	metamorphic		

3. MISCELLANEOUS

abu	abundant	cav	cavernous
Alt	alteration	Cbl	cobble
alt	altered	cmtd	cemented
ang	angular	co gr (csg)	coarse-grained
apx	approximate	crb	carbonaceous
arg	argillaceous	Ct	contact
av	average		
		diss	disseminated
bdd	bedded		
bdg	bedding	Exp	exposure
Bldr	boulder		
bkn	broken	Fa	fault
bndd	banded	fi gr (fng)	fine-grained
		Fm	formation
calc	calcareous	fol	foliated

fos	fossiliferous	mot	mottled
Fr	fracture		
frag	fragmented	Otc	outcrop
fri	friable		
FW	footwall	Pbl (peb)	pebble
		Phen	phenocryst
Gos	gossan		
Gms	groundmass	rx	recrystallized
HW	hanging wall	slicks	slickensides
		Spec	specimen
Incl	inclusion		
Int	intrusion	Veg	vegetation
		Vlt	veinlet
Jt (Jnt)	joint	Vn	vein
lam	laminated	xbdd (xb)	cross-bedded
		Xl	crystal
mas	massive	xln	crystalline
Mbr	member		
med gr	medium-grained	wx (wth)	weathered

APPENDIX B
SYMBOLS FOR FIELD AND MINE GEOLOGIC MAPPING

The following symbols, based on mapping practice in North American mining companies, are in general agreement with mapping practice in the U.S. Geological Survey and the Geological Survey of Canada. The symbols are primarily for use at a scale of 1:10,000 or larger. Symbols in European usage are similar: they are shown in detail by Amstutz (1971, p. 132–142). Special-purpose symbols, sometimes needed for mining work can be found in the following references:

Engineering geological mapping—United Nations (1976)

Geomorphologic mapping—Demek (1972)

Geotechnical mapping—Geological Society (London) (1972); Varnes (1975)

Hydrologic mapping—United Nations (1970b)

Photogeologic mapping—Allum (1966).

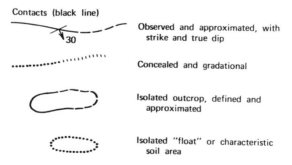

Contacts (black line)

Observed and approximated, with strike and true dip

Concealed and gradational

Isolated outcrop, defined and approximated

Isolated "float" or characteristic soil area

Lithology (black, coded color, or shades of solid color)

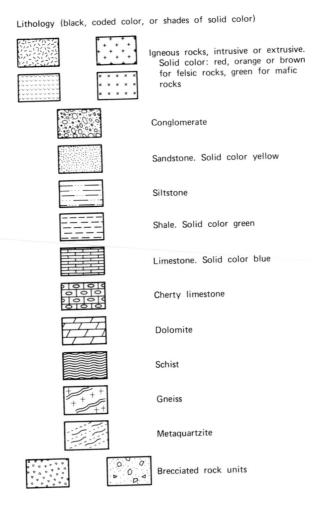

Igneous rocks, intrusive or extrusive. Solid color: red, orange or brown for felsic rocks, green for mafic rocks

Conglomerate

Sandstone. Solid color yellow

Siltstone

Shale. Solid color green

Limestone. Solid color blue

Cherty limestone

Dolomite

Schist

Gneiss

Metaquartzite

Brecciated rock units

Bedding (black)

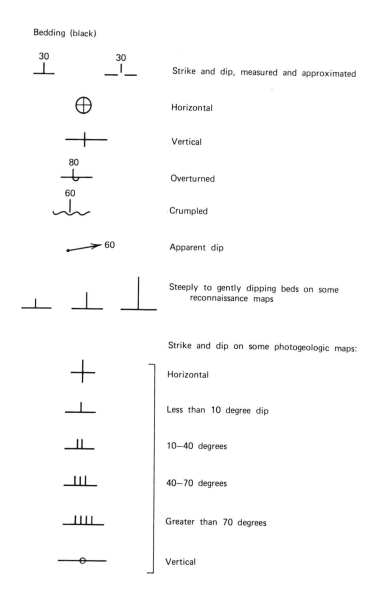

Strike and dip, measured and approximated

Horizontal

Vertical

Overturned

Crumpled

Apparent dip

Steeply to gently dipping beds on some reconnaissance maps

Strike and dip on some photogeologic maps:

Horizontal

Less than 10 degree dip

10–40 degrees

40–70 degrees

Greater than 70 degrees

Vertical

Planar Features (black)

Foliation

Cleavage

Joints

Multiple jointing

Linear Features (black)

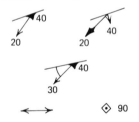

Plunge of lineation in plane of foliation and plunge of striations on fault (blue line)

Rake or pitch of lineation in plane of foliation

Horizontal and vertical lineation

Faults (blue line)

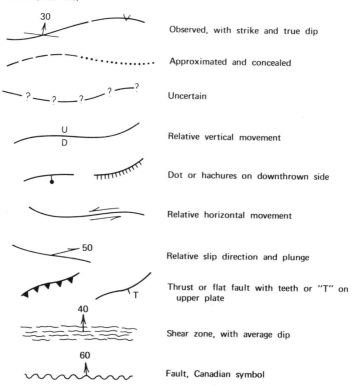

Observed, with strike and true dip

Approximated and concealed

Uncertain

Relative vertical movement

Dot or hachures on downthrown side

Relative horizontal movement

Relative slip direction and plunge

Thrust or flat fault with teeth or "T" on upper plate

Shear zone, with average dip

Fault, Canadian symbol

Broken Ground (black, blue, or red, depending on the association)

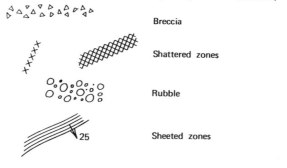

Breccia

Shattered zones

Rubble

Sheeted zones

Folds (black line)

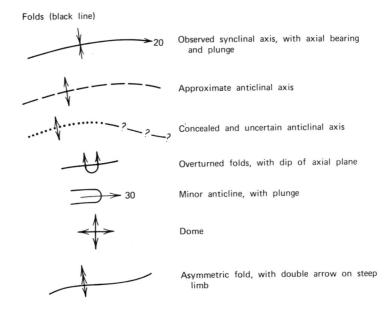

Observed synclinal axis, with axial bearing and plunge

Approximate anticlinal axis

Concealed and uncertain anticlinal axis

Overturned folds, with dip of axial plane

Minor anticline, with plunge

Dome

Asymmetric fold, with double arrow on steep limb

Mineralization (red)

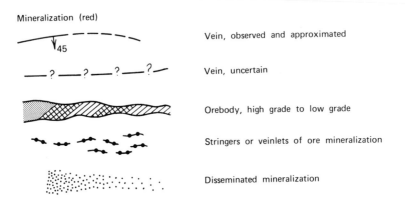

Vein, observed and approximated

Vein, uncertain

Orebody, high grade to low grade

Stringers or veinlets of ore mineralization

Disseminated mineralization

Alteration (black, coded color, or shades of solid color)

Altered areas, keyed to type

Mine Workings (black)

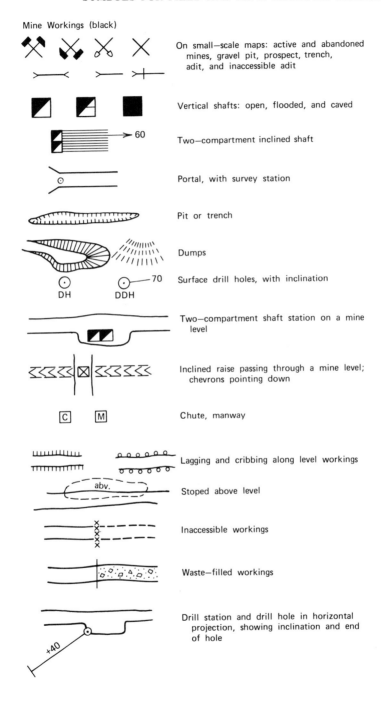

On small—scale maps: active and abandoned mines, gravel pit, prospect, trench, adit, and inaccessible adit

Vertical shafts: open, flooded, and caved

Two—compartment inclined shaft

Portal, with survey station

Pit or trench

Dumps

Surface drill holes, with inclination

Two—compartment shaft station on a mine level

Inclined raise passing through a mine level; chevrons pointing down

Chute, manway

Lagging and cribbing along level workings

Stoped above level

Inaccessible workings

Waste—filled workings

Drill station and drill hole in horizontal projection, showing inclination and end of hole

APPENDIX C
SOURCES OF PRELIMINARY DATA AND INFORMATION FOR EXPLORATION PROJECTS

I. GUIDES TO LIBRARY INFORMATION SOURCES

Brewer, J. G., The literature of geography: a guide to its organization and use: Hamden, Conn., The Shoe String Press, 208 p., 1973.

Includes sources of regional information and maps.

Corbin, J. B., An index to state geological survey publications issued in series: Metuchen, N.J., The Scarecrow Press, 1965.

Falk, A. L., and Miller, R. L., Worldwide directory of national earth-science agencies: U.S. Geological Survey Circ. 716, 32 p., 1975.

Lists national agencies dealing with geology, cartography, and minerals, with addresses.

Geoscience Documentation, 1969—, List of geoscience serials: *Geoscience Documentation,* v. 1, no. 1, July 1969.

The list has been updated in each subsequent monthly issue.

Given, I. A., Sources of information, *in* Cummins, A. B., and Given, I. A., eds., SME mining engineering handbook: New York, Am. Inst. Mining Metall. Petroleum Engineers, v. 2, sec. 35, p. 35-1-35-34, 1973.

Lists departments of mines, geologic surveys, societies, institutes, and their publications, by country and by U.S. state. Also lists major periodicals, directories, and yearbooks.

Hoy, R., Sources of information, *in* Lefond, S. J., ed., Industrial minerals and rocks, ed. 4: New York, Am. Inst. Mining, Metall. Petroleum Engineers, p. 1290-1305, 1975.

Lists industrial minerals publications and publishers.

Kaplan, S. R., Guide to information sources in mining, minerals, and geosciences: New York, Interscience Publishers, 599 p., 1965.

Part I lists names, addresses, function, and publications of national, state, and private associations dealing with mining; U.S. and foreign bureaus of mines are included. Part II describes available literature in books and journals by country and subject.

National Referral Center, A directory of information resources in the United States— physical sciences and engineering: Washington, D.C., National Center for Science and Technology, 1971.

Ward, D. C., and Wheeler, M. W., eds., Geologic reference sources: Metuchen, N.J., The Scarecrow Press, 453 p., 1972.

Covers general information by subject (e.g., economic geology), regional information by country and state. The map bibliography is concerned with small-scale maps and atlases for each country rather than large-scale "quadrangles."

Wood, D. N., ed., Use of earth science literature: London, Butterworth and Co., 459 p., 1973.

This could be called "everything you might possibly want to know about geologic information sources." Detailed information is included on methods of literature search, with lists of regional geologic handbooks and maps, by region and country.

II. EXAMPLES OF DOCUMENTED AND GENERALLY AVAILABLE INFORMATION

A. Bibliographies and Indexes in Current Use

Applied Science and Technology Index, monthly-annually: New York, H. W. Wilson Co.

Bibliographic Index, semiannually: New York, H. W. Wilson Co.

Bibliography and Index of Geology, monthly: Boulder, Colorado, The Geological Society of America.

Bibliography of North American Geology, 1919–1970, monthly-annually: Washington, U.S. Geological Survey.

Business Periodicals Index, monthly: New York, H. W. Wilson Co.

Covers business policies, marketing, finance, and associated economics.

Canadian Index to Geoscience Data, with periodic update: Ottawa, Geological Survey of Canada.

Published and open-file data, by province.

Catalog of the United States Geological Survey Library, with supplements: Boston, G. K. Hall Co., 1964.

Classed Subject Catalog of the Engineering Societies Library (New York): Boston, G. K. Hall Co., 1963.

Commonwealth Index of Unpublished Scientific and Technical Translations: London, British Library, Lending Division.

---

* Available on magnetic tape and/or in remote terminal access.

Current Contents—Engineering and Technology, monthly, and *Index to Scientific Reviews,* semiannually: Philadelphia, Institute for Scientific Information.

Dictionary Catalog of the U.S. Department of Interior Library, with supplements: Boston, G. K. Hall Co., 1967.

Engineering Index, monthly-annually: New York, The Engineering Index Co.

F. and S. International Index—Industries, companies, countries, monthly-annually: Cleveland, Ohio, Predicasts Inc.

Includes "mining" as a topic and as entries under countries and companies.

Geotitles Weekly: London, Geosystems (Lea Associates).

Cumulative in *Geotitles Repertorium* (annual) and on *Geoarchives* tapes.

Government Reports Index, monthly: Springfield, Virginia, National Technical Information Service.

Includes government-sponsored research projects and translations.

Publications of the U.S. Geological Survey, monthly;

Publications of the U.S. Bureau of Mines, monthly;

Catalog of United States Government Publications, monthly: Washington, Government Printing Office.

Many countries publish similar monthly or annual announcements.

Science Citation Index, quarterly: Philadelphia, Institute for Scientific Information.

TRANSDEX—Bibliography and index to the United States Joint Publications Research Service (JPRS), monthly: New York, CCM Information Corporation.

Translations Register-Index, annually: Chicago, National Translations Center, John Crerar Library.

VINITI Program Index, semiannually to biannually: Boston, G. K. Hall Co.

Translations of Russian research summaries and bibliographies.

World Index of Scientific Translations, annually: Delft, The Netherlands, European Translations Center.

Chronic, J. B., Bibliography of theses written for advanced degrees in geology and related sciences at universities and colleges in the United States and Canada through 1957: Boulder, Colorado, Pruett Press, 1958.

Chronic, J. B., Bibliography of theses in geology, 1958–1963: Washington, American Geological Institute, 1964.

Long, H. K., A bibliography of earth science bibliographies of the United States: Washington, American Geological Institute, 1971.

Ward, D. C., Bibliography of theses in geology: Geoscience Abstracts, v. 7, no. 12, pt. 1, p. 103–129, 1965.

Ward, D. C., Bibliography of theses in geology, 1967–1970: Boulder, Colorado, Geol. Soc. America Spec. Paper 143, 1973.

Ward, D. C., and O'Callaghan, T. C., Bibliography of theses in geology, 1965–66: Washington, American Geological Institute, 1969.

B. Sources of Abstracts

Abstracts of North American Geology, monthly, 1966–1971: Washington, U.S. Geological Survey.

Earlier abstract publications were *Geoscience Abstracts,* 1959–1966, and *Geological Abstracts,* 1953–1958, of the American Geological Institute.

Bulletin signalétique, monthly: Paris, Centre de Documentation du Centre National de la Recherche Scientifique.

Published in numerous sections. Those of particular interest to exploration geologists are:

   Bull. 221, Gitologie, économie minière
   Bull. 222, Roches cristallines
   Bull. 223, Roches sedimentaires, geologie marine
   Bull. 224, Stratigraphie, géologie regionale, géologie générale
   Bull. 225, Tectonique
   Bull. 226, Hydrogéologie, géologie de l'ingenieur, formations superficielles.

Chemical Abstracts, weekly: Columbus, Ohio, American Chemical Society.

Topics include minerals, mining, geology, and specific metals.

Dissertation Abstracts International, monthly: Ann Arbor, Michigan, University Microfilms.

Economic Geology, Geology of ore deposits (abstracts of Russian Academy of Science articles) in several issues each year.

Geoabstracts, bimonthly: Norwich, England, University of East Anglia.

With a worldwide geographical and subject index in seven parts:

   A. Landforms and the Quaternary
   B. Climatology and hydrology
   C. Economic geography (including minerals)
   D. Social and historical geography
   E. Sedimentology
   F. Regional and community planning
   G. Remote sensing and cartography.

Geocom Bulletin/Programs, monthly: London, Geosystems (Lea Associates).

Abstracts and information on mathematical geology, exploration techniques, and computer methods in geoscience.

Geomechanics Abstracts, quarterly: Oxford, England, The Pergamon Press.

Successor to *Rock Mechanics Abstracts.* Combined in 1974 with issues of the *International Journal of Rock Mechanics and Mining Sciences.*

Geotechnical Abstracts, monthly: Essen, German National Society of Soil Mechanics and Foundation Engineering.

Includes engineering geology, site investigation, soil-and-rock mechanics. With "Geodex" punched keyword cards.

I.M.M. Abstracts, bimonthly: London, Institution of Mining and Metallurgy.

Includes economic geology, mining, mineral processing.

Mineralogical Abstracts, quarterly: London, Oxford University Press.

USSR Scientific Abstracts, periodically, and *East European Scientific Abstracts,* periodically: Washington, Joint Publications Research Service; Government Printing Office.

C. Guides to Regional Geologic Information

In addition to the sources cited in A and B, there are the following guides to regional information:

Beatty, W. B., Mineral resources data in the western states: Palo Alto, California, Stanford Research Institute, 1962.

Earth Sciences Research Catalog: Tulsa, Oklahoma, University of Tulsa.
For the entire United States; indexed by area.

Geological Field Trip Guidebooks for North America: Washington, American Geological Institute, 1968.

Ridge, J. D., Selected bibliographies of hydrothermal and magmatic mineral deposits: Boulder, Colorado, Geological Society of America Mem. 75, 198 p., 1958.

Ridge, J. D., Annotated bibliographies of mineral deposits in the Western Hemisphere: Boulder, Colorado, Geological Society of America Mem. 131, 681 p., 1972.

Ridge, J. D., Annotated bibliographies of mineral deposits in Africa, Asia (exclusive of the USSR), and Australia: Oxford, England, The Pergamon Press, 1976, 545 p.

Many countries have abstract journals devoted to their specific regions, for example,

British Geological Literature, new series, quarterly: London, Brown's Geological Information Service Ltd.

Most countries and states publish bibliographies, indexes and summaries of information in their areas from time to time. Beginning in 1963, summary volumes on mineral resources were prepared for each western state in the United States by the U.S. Geological Survey in cooperation with state geological and mining agencies; these reports contain extensive bibliographies. Example:

Mineral and Water Resources of New Mexico: New Mexico Bureau of Mines and Mineral Resources Bull. 87, 437 p., 1965.

Gazetteers are published by most national and state governments. The U.S. Board of Geographic Names has compiled gazetteers for each foreign country.

D. Sources of Unedited Geologic Data

Many state and national bureaus of mines and geological surveys have preliminary reports, project files, and raw numerical data on open file for public inspection.

Open-file reports of the U.S. Geological Survey can be examined in Reston, Va., Denver, Colo., Menlo Park, Calif., and in the nearest district office to the area involved. They can also be examined in the offices of the appropriate state bureaus of mines, and copies can sometimes be made for the user at local blueprinting service companies.

Open-file reports of the Geological Survey of Canada can be examined in Ottawa and at the appropriate provincial departments of mines. In some cases, U.S. and Canadian geological survey open-file data in printed form, micromedia, or on magnetic tapes can be ordered by mail from the government or from a service company.

E. Status Reports on Mines and Mining Districts

The Minerals Yearbook, annually: Washington, U.S. Bureau of Mines.
Contains state and country summaries, with news of developments at major mines as well as commodity reviews.

Mineral Trade Notes, monthly: Washington, U.S. Bureau of Mines.
Includes news of developments in foreign mining areas.

Annual summaries and yearbooks are published by most national and state bureaus of mines and also by some regional mining associations. These generally contain district-by-district reviews of mining progress. Examples:

Canadian Minerals Yearbook, annually: Ottawa.
Geology, Exploration and Mining in British Columbia, annually: Vancouver.
Mining Yearbook, Colorado Mining Association, annually: Denver.

Annual review issues of mining magazines contain summary comments on major mines and districts. The coverage may be worldwide or may be confined to a particular "home" area. Good sources of international status reports are the annual review numbers of the London *Mining Journal* (by commodity and country), *Engineering and Mining Journal* (by commodity), and *Mining Engineering* (by commodity). *Annales des mines* carries several annual summaries, such as the fairly detailed "Panorama de l'industrie minière du continent africain."

F. Mining Company Information

Coal Mine Directory of the United States and Canada, annually: New York, McGraw-Hill Book Co.

E/MJ International Directory of Mining and Mineral Processing Operations, annual: New York, Engineering and Mining Journal.

Jane's World Mining: Who Owns Whom: New York, McGraw-Hill Book Co.
Information on joint ventures, subsidiaries, investments, and management control.

Mines Register, annually: New York, American Metal Market Co.
Western Hemisphere mines.

Mining Companies of the World, annually: London, Mining Journal Books Ltd.

Mining International Year Book, annually: London, Walter B. Skinner Co.
Includes information on capitalization, operating properties, and some reserve figures.

Pit and Quarry Directory of Nonmetallic Minerals Industries, annually: Chicago, Pit and Quarry Publications.

For more detail on the operations of corporate firms, there are annual reports available from companies, "10-K" reports on file in major libraries, and the *F. and S. International and Business Periodicals Index* mentioned in IIA.

For current developments, three additional sources are:

Moody's Industrial Manual—American and Foreign, weekly, annually: New York, Moody's Investor's Service

Standard and Poor's Corporation Records, daily, monthly, annually: New York, Standard and Poor's Corporation.

The Wall Street Journal Index, monthly: Princeton, N.J., Dow Jones Books.

G. Mineral Commodity Orientation and Production Data

In addition to the commodity reviews in magazines and national minerals yearbooks there are:

Brobst, D., and Pratt, W., United States mineral resources: U.S. Geological Survey Prof. Paper 820, 722 p., 1973.

Canada Department of Energy, Mines and Resources, Mineral Resources Branch, Mineral reports, mineral surveys, and mineral information bulletins, time to time: Ottawa

Canada Geological Survey, Economic geology reports, time to time: Ottawa.

Chemical Market Abstracts, monthly: New York, Foster D. Snell Inc.

For "chemical" industrial minerals.

Department of Trade and Industry, Mineral dossiers, time to time: London.

Examples: No. 1, Fluorspar, 1971; No. 9, Tin, 1974.

Industrial Minerals, monthly: London, Metal Bulletin Ltd.

Keystone Coal Industry Manual, annually: New York, McGraw-Hill Book Co.

LeFond, S. J., ed., Industrial minerals and rocks, ed. 4: New York, Am. Inst. Mining Metall. Petroleum Engineers, 1360 p., 1975.

Metals Analysis and Outlook, twice annually: London, Charter Consolidated Ltd.

For major metals, a market analysis and forecast.

Metals Statistics, annually: New York, American Metal Market Co.

For each major metal, a profile, news highlights, and international trade.

U.S. Bureau of the Census, Census of the mineral industries, each 4 years: Washington.

Production statistics by county rather than mining district.

U.S. Bureau of Mines, Commodity data summaries, annually: Washington.

U.S. Bureau of Mines, Materials surveys, published from time to time: Washington.

For individual commodities, such as copper, nickel, lead.

U.S. Bureau of Mines, Mineral facts and problems, each 5 years: Washington.

Yearbook, American Bureau of Metal Statistics: New York.

H. National and State Mineral Policy

Up-to-date information on United States mineral policy is published in

Annual Report of the Secretary of Interior under the Mining and Minerals Policy Act of 1970: Washington, Government Printing Office.

Includes an appendix with status reports on supplies of each major mineral.

For foreign mining districts, *Mineral Trade Notes* and the annual review numbers of mining magazines mentioned under "Status Reports" carry news of developments in mineral policy.

AGID News, the quarterly newsletter of the Association of Geoscientists for International Development (Memorial University, St. John's, Newfoundland), also reports on changes in mineral policy involving "third-world" nations.

Journals of regional trade interest, such as *Africa Research Bulletin* (London), contain special sections on minerals and mineral policy.

A source of current bibliographic citations to government pamphlets and reports dealing with economic aspects of world regions and areas is:

Public Affairs Information Service Bulletin, monthly: New York, Public Affairs Information Service Inc.

For a very practical summary of history, economy, and political conditions in foreign countries, there are:

American University, Area handbooks series, time to time: Washington, Government Printing Office.

Compiled for individual countries by the American University.

U.S. Department of State, Background notes, time to time: Washington, Government Printing Office.

Brief summaries prepared by the U.S. Department of State for individual countries: up-dated frequently.

For mineral land laws in the United States, most western states issue "legal guides to prospectors" and digests of mining laws. For American mining law in general, there is a treatise compiled by the Rocky Mountain Mineral Law Foundation, American law of mining: Albany, N.Y., Mathew Bender Co. (5 volumes), 1960 with up-date.

For foreign mining laws, a series of U.S. Bureau of Mines Information Circulars, 1970–1974, includes:

I.C. 8482, Western Hemisphere
I.C. 8514, East Asia and Pacific
I.C. 8544, Near East and South Asia
I.C. 8610, Africa
I.C. 8631, Europe.

III. INFORMATION RETRIEVAL SYSTEMS AND LITERATURE

The specific use of information retrieval systems in geology and mining is discussed in the following:

Bergeron, R., Burk, C. F., Jr., and Robinson, S. C., eds., Computer-based storage, retrieval, and processing of geological information, Section 16, International Geological Congress, 24th Session, Canada, 1972: Ottawa, 222 p., 1972.

Calkins, J. A., Kays, O., and Keefer, E. K., CRIB—The mineral resources data bank of the U.S. Geological Survey: Washington, U.S. Geological Survey Circ. 581, 39 p., 1973.

Eimon, P. I., and Morris, D. B., An introductory review—data storage and retrieval, in Weiss, A., ed., A decade of digital computing in the mineral industry: New York, Am. Inst. Mining, Metall. Petroleum Engineers, pp. 277–283, 1969.

Fabbri, A. G., Design and structure of geological data banks for regional mineral potential evaluation: Canadian Mining Metall. Bull., v. 68, no. 70, pp. 91–98, 1975.

Hubaux, A., A new geological tool—the data: Earth Science Reviews, v. 9, pp. 159–196, 1973.

Hutchison, W. W. (chmn.), Computer-based systems for geological field data: Canada Geol. Survey Paper 74-63, 99 p., 1975.

Laffitte, P., Traité d'information géologique: Paris, Masson et Cie., 624 p.

Three periodicals concerned with information retrieval are of special interest to geologists:

Bulletin signalétique 101—Science de l'information, documentation, monthly: Paris, Centre de Documentation de Centre National de la Recherche Scientifique. Includes a section on geology.

Geoscience Documentation, bimonthly: London, Geosystems (Lea Associates). Journal for geology librarians and information specialists. Includes, in v. 1, no. 1, July 1969, a complete list of geoscience periodicals; up-dated in each subsequent issue.

Proceedings of the Geoscience Information Society, annual: Washington, American Geological Institute.

Some information retrieval systems in current use are:

Canadian Index to Geoscience Data, citations to Canadian localities and data: Ottawa, Canadian Centre for Geoscience Data.

CAN/SDI, Canadian Selective Dissemination of Information, a profile service: Ottawa, Canada Geological Survey.

COMPENDEX, and ORBIT, based on the *Engineering Index:* Santa Monica, California, System Development Corporation.

CRIB, Computerized Resources Information Bank. Used within the U.S. Geological Survey, Washington, D.C.

DATRIX, Direct Access to Reference Information, theses and dissertations: Ann Arbor, Michigan, University of Michigan.

Geo-Archives: London, Geosystems (Lea Associates Ltd.)

GEODAT, numerical results produced by laboratories in the Geological Survey of Canada, chemical, spectrographic, and age data. Available to users in the private sector. Geological Survey of Canada, Ottawa.

Geo Ref, a geoscience-oriented service provided by the American Geological Institute and the Geological Society of America; files date from 1966.

GRASP, Geologic Retrieval and Synopsis Program. Used within the U.S. Geological Survey, Washington, D.C.

MANIFILE, data on the world's nonferrous metallic deposits: Winnipeg, University of Manitoba.

RASS, Rock Analysis Storage System. Used within the U.S. Geological Survey. Files not available to the public but some data are released on magnetic tape. Washington, D.C.

SSIE, Smithsonian Science Information Exchange; information on research in progress: Washington, Smithsonian Institute.

IV. SOURCES OF MAPS AND AERIAL PHOTOGRAPHS

A. Topographic Maps

About 90 percent of the United States is covered by 1:62,500 (15-minute quadrangle) to 1:24,000 (7½-minute quadrangle) topographic mapping. Indexes to topographic mapping in each state are published quarterly by the U.S. Geological Survey. These and the topographic maps are obtainable by mail from the U.S. Geological Survey offices in Arlington, Virginia, for areas east of the Mississippi and in Denver, Colorado, for the western states. Copies of U.S. Geological Survey topographic maps and advance prints of preliminary quadrangle maps are also available (although not by mail) from district U.S. Geological Survey offices and from state geological surveys and bureaus of mines at the addresses shown by Wood (1973), Given (1973), and Ward and Wheeler (1972).

Smaller scale maps, such as the AMS 1:250,000 series, state-by-state 1:500,000 contour maps, and U.S. area sheets of the International Map of the World at 1:1 million, are available from the U.S. Geological Survey.

Small-scale topographic maps of foreign areas are also available in the United States. Total world coverage at 1:2 million and 1:1 million and partial coverage at 1:500,000, 1:250,000, and 1:200,000 are available from the Defense Mapping Agency Topographic Center in Washington. Aeronautical charts with contours and shaded relief topography are available for the entire world at 1:1 million (operational navigation charts) and for a large part of the world at 1:500,000 (pilotage charts) from the National Ocean Survey, Aeronautical Chart and Information Center, Washington.

Larger scale topographic maps for foreign areas, 1:100,000 to 1:10,000, are available from the national topographic surveys and cartographic offices listed in the Worldwide Directory of National Earth-Science Agencies, U.S. Geological Survey Circ. 716, 1975. A United Nations publication, *World Directory of Map Collections*, 1976, gives the location and extent of map and aerial photograph library holdings in 45 countries. New map coverage, topographic and some geologic, is listed annually in the publication *Bibliographie cartographique internationale*. For quick purchasing, there are private international map companies that publish catalogs with index maps showing available coverage at various scales. Two of these companies are Geocenter in Stuttgart, Germany, and Edward Stanford Ltd. in London, England.

B. Geologic and Geophysical Maps

Government geologic mapping in the United States covers most of the country at a scale of 1:500,000 (state maps), about 40 percent of the country at 1:250,000, and about 25 percent of the country at 1:62,500 to 1:24,000. Unlike topographic mapping, some of this has been done by the state geological surveys; in addition, some areas have been mapped for universities by candidates for advanced degrees. Even though the maps are scattered through federal, state, and scientific association publications, most states have an updated index to geologic mapping compiled by the U.S. Geological Survey or by the state bureau of mines.

The principal U.S. Geological Survey large-scale map series are:

Coal Investigation Maps

Geologic Quadrangle Maps. This series is a continuation of the *Geologic Folios* published between 1894 and 1946.

Geophysical Investigations Maps. This series includes aeromagnetic and radiometric maps at 1:62,500 and 1:24,000 scale.

Hydrologic Investigations Maps

Mineral Investigations Field Studies Maps. This series includes preliminary tectonic, metallogenic, mineral deposits, and geological maps.

Mineral Investigations Resource Maps. These are mineral deposit maps.

Miscellaneous Geological Investigation Maps. This series includes photo-geologic maps, foreign country maps, and paleotectonic maps.

Oil and Gas Investigations Maps.

Large-scale foreign geologic maps are listed in many of the publications listed under "bibliographies and indexes in current use." The primary subject listing is generally by region or area, the second or tertiary heading is "maps." *Bulletin signalétique* (Section 224) and *Geoabstracts* (Section G) also list large-scale geologic maps.

Foreign geologic map coverage, 1:250,000–1:10,000, is summarized for individual countries from time to time in the quarterly journal *Geological Newsletter,* published by the International Union of Geological Sciences, Haarlem, The Netherlands. The Commission for the Geological Map of the World, 74 Rue de la Federation, Paris, publishes up-to-date bulletins on the activities of national geological surveys, including lists of new maps.

Geologic maps and indexes to geologic mapping can be obtained from the national bureaus listed in *Geological Newsletter* and in U.S. Geological

Survey Circ. 716 or, as with topographic maps, quick service can be obtained (providing the maps are available) from the private map companies listed under "Topographic Maps" or from Telberg Geological Map Service, Sag Harbor, New York 11963.

Magnetic, radioactivity, and gravity maps are available from several government geological surveys. Common scales for national gravity maps are 1:500,000 and 1:250,000; for aeromagnetic and aeroradioactivity maps, 1:50,000 and 1:25,000.

C. Aerial Photography and Spacecraft Imagery

Aerial photography coverage in the United States is shown on the U.S. Geological Survey quarterly indexes to topographic mapping for each state. Smaller scale indexes to aerial photography coverage of the entire country are also published from time to time. Indexes and advice on coverage by government agencies for specific areas can be obtained from the National Cartographic Information Center, U.S. Geological Survey, National Center (STOP 507), Reston, Virginia 22092.

Where photographs are held by another government agency, such as the Forest Service or the Soil Conservation Service, flight line maps, photo index sheets, and copies of the photographs can be obtained directly from the agency. Forest Service and Soil Conservation Service district offices, some universities, and some state geological surveys have aerial photographs on open file for their specific areas; these may be inspected and the exposure numbers can be selected for immediate purchase by mail without waiting for index sheets.

The U.S. Geological Survey EROS Data Center, Sioux Falls, South Dakota 57198, is the source for copies of geological survey aerial photographs, NASA photography and imagery, Landsat imagery, and Skylab photography and imagery. The abbreviations here are: EROS = Earth Resources Observation Systems; NASA = National Aeronautics and Space Administration; and Landsat = the former ERTS, Earth Resources Technology Satellite. Satellite imagery is available on magnetic tape and in photographic form. Standard catalogs and film strips as well as transparencies, paper prints, enlargements, and state image maps are available. A geographic search and inquiry system provides free information on specific photographic coverage. EROS applications assistance facilities and data reference files are located at more than a dozen offices throughout the United States.

Foreign aerial photography is generally obtainable from the national mapping, cartographic, and geographic offices listed in U.S. Geological Survey Circ. 716. Most countries publish indexes to aerial photographic coverage. In some countries, photographs are on open file at a central library; in others, access to aerial photography is closely restricted.

V. INFORMATION BY DIRECT INQUIRY

The headquarters and field offices of government geological surveys, bureaus of mines, and departments of natural resources are the prime locations for obtaining unpublished information and advice regarding new exploration areas. There are, in addition, some less obvious sources of geologic and mining information, such as:

Newspapers (archives and current interest "clipping" services)

Railroads (land and development offices)

Regional development associations and commissions (example: Tennessee Valley Authority)

Research foundations and institutes (example: Stanford Research Institute)

State historical societies (archives)

State mining associations

Universities.

Preliminary information for foreign exploration programs can be obtained from:

Country and regional trade associations

Foreign embassy commercial officers in the United States

National mining corporations (example: SONAREM, Algeria)

United Nations Development Program, Office of External Relations, United Nations, New York

United States embassies in the foreign area

United States State Department foreign area officers.

APPENDIX D
FORMAT AND CHECKLIST FOR MINERAL PROPERTY EVALUATION REPORTS

1. **KEY INFORMATION** (a single page abstract for filing and for data retrieval)

   a. *Property name* and cross reference to alternative names.
   b. *Commodities*: current and potential.
   c. *Index* and *project numbers.*
   d. *Location*: country, province, district. Coordinates or land grid location plus relation to a town or natural feature.
   e. *Type of working,* with stage of development, production, and reserves.
   f. *Reserves,* as summary tonnage, with classification (proved plus probable, etc.).
   g. *Geology*: Morphologic type of deposit, host rock, type and trend of mineralization.
   h. *Dates* of examination and report plus date when action is needed.
   i. *Names* of personnel involved and persons bringing property to the attention of the organization.
   j. *Objective*: reason for interest in the property.
   k. *Recommendation,* in three or four words.
   l. *Time and manpower* spent in the field examination, laboratory, and office.
   m. *Distribution of the report.*

2. TABLE OF CONTENTS

With subheadings, position of tables and illustrations, and list of appendexes

3. SUMMARY AND DIGEST OF RECOMMENDATIONS

a. *Summary*: as quantitative as possible, with mention of the major attraction and major handicap in further consideration of the property.
b. *Exploration target definition,* with apparent size and grade, economic potential, level of risk, and degree of interest to the organization (major or minor, immediate or eventual).
c. *Action recommended,* and first step to be taken.

4. GENERAL

a. *Location*: in more detail than in key information, and with index map.
b. *Access*: road types and condition. Road log from nearest town, highway junction, port, or railhead.
c. *Geography and environment*: geographic region. Quantitative data on topographic relief, drainage, and climate. Comments on seasonal aspects of exploration and mining. Health conditions. Existing environmental baseline data in regard to:
   1. Soil, water atmosphere, and erosion
   2. Biological conditions
   3. Land use
   4. Aesthetics and features of archaelogical and cultural interest.
   5. Special ecological conditions and problems.
d. *History and past production*: source and reliability of information. Mining and exploration activity in the district and at the property, with cost and profit data, financing experience, and past property agreements.
e. *Current development,* with maps and as much quantitative data as possible. Condition of workings, pits, and drill holes. Costs of exploration, development, and mining. District development with special emphasis on new technologic and economic guidelines.

5. OWNERSHIP AND TERMS OF ACQUISITION OR PARTICIPATION

a. *Size of property,* with number of claims or leasing units. Index map.
b. *Ownership and legal status*. Names and file numbers of leases, claims, and concessions. Location of government offices maintaining

records. Methods of tenure, with special conditions, such as rentals and royalties. Data on the organization having present ownership. Title validity in regard to liens, taxes, and annual labor requirements. Owner's terms for transfer of interest in the property, with royalties, payments, and exploration guarantees. Level of detail discussed to date and with whom. Owner's plans for future development.

c. *District land status*, with map. Mineral rights controlled by other organizations. Status of rights to additional minerals and to surface area for supporting facilities. Water rights and impending water-use problems. Recent and anticipated government land withdrawals.

6. GEOLOGY (with small maps and explanatory maps. Large and detailed maps belong in the appendexes.)

a. *Geologic nature of the exploration target*: approximate size and grade, with an estimate of the risk and enhancement factors to be expected in searching for an acceptable orebody.

b. *Quantity and quality of information*: mapping and reconnaissance procedures, with traverse lines shown on a map. Evaluation of older information incorporated in the analysis. Location of uninvestigated or under-investigated areas, with comments on the level of additional information needed.

c. *Regional geology,* with maps and sketches. Principal rock units and stratigraphic sequence. Major structural features. Metallogenic setting and ore controls. Mineral deposits in the region with characteristics analogous to the property in question.

d. *Geology of the deposit,* with maps, sections, and sketches. Classification and age of mineralization, with parameters used and relation of factual data to interpretation. General geology, with stratigraphic sequence, lithologic units, structural features, and brief geologic history. Ore and gangue mineralization, texture, paragenesis, zoning, supergene ore zones, and near-surface leaching. Types and extent of wall-rock alteration. Attitudes, dimensions, and grades of individual orebodies. Postmineral deformation. Guides to ore: structural, lithologic, stratigraphic, and mineralogic.

e. *Geochemical and geophysical characteristics.* Description of work done and consideration of further exploration methods and patterns. Comments on response to be sought, with threshold levels, rock-type effects, and background "noise."

f. *Outcrops and covered areas*: geomorphic and mineralogic indications of the mineralization. Significant covered zones, with extent, thickness, and type of cover.

g. *Location of geologic records*: field notes, drill-hole logs, geochemical samples, specimens, and prepared mineralogic and petrologic sections.

7. SAMPLING OF THE ORE

a. *Methods,* with names of samplers, core and cuttings recovery, handling and preparation. Laboratory analytical methods and names of laboratories.
b. *Patterns,* with basis for selection.
c. *Guidelines* for future sampling.
d. *Storage location* of logs, core, cuttings, sample rejects, and pulps. Location of sample record notebooks.

8. RESERVES

a. *Overall tonnage and grade,* with further categorization into classes (measured, indicated, inferred, or similar). Probability expressions and limits for each class.
b. *Range in grade and thickness,* with distribution of ore shoots and with variations in mineralogy.
c. *Alternative tonnage and grade* relative to alternative cutoff grades, waste-to-ore ratios, and mining methods.
d. *Byproduct grades* in reserve and potentially marketable "waste" products.
e. *Method of calculation* and derivation of tonnage factor. Assumptions and methods of projection.

9. GEOTECHNICAL DATA

a. *Geomechanics* of orebody, wall rock, and overburden (consolidated and unconsolidated), as needed for selection of mining system.
b. *Ground-water* and hydrologic data
c. *Geothermal* conditions
d. *Exploration characteristics,* including zones of difficult drilling and low core recovery, location of workings and solution openings.
e. *Guidelines* for obtaining geotechnical data needed in mine design.

10. MINING AND PROCESSING

a. *Mining methods,* present and proposed. Analogous mines. Mineral conservation regulations. Zones for the earliest development.
b. *Milling methods,* with alternatives for type and grade of concentrate and for levels of recovery.

c. *Metallurgical methods* and alternatives.

d. *Land requirements and environmental impact* in regard mining, waste disposal, plant, utilities, and support facilities. Reclamation guidelines. Multiple mineral and land-use guidelines.

11. ECONOMIC CONSIDERATIONS

a. *Marketing*. Ore buyers and mineral traders, with comments on quoted market price versus expected price. Projected commodity prices. Market specifications, especially in regard to industrial minerals, byproducts, and saleable "waste" products. Smelter and custom mill location and purchase schedules. Cost and time in market development for unique products.

b. *Mining and processing costs,* with upper and lower limits, apparent trends, source of cost information, and reference to similar operations.

c. *Capital requirements,* fixed and revolving, at several alternative levels of production. Comparison with similar installations. Equipment on hand, condition, and adaptability to proposed operation. Requirements for pioneering design, pilot plant, and experimental operations.

d. *Facilities*: transportation, power, fuel, and water, with unit costs and with amounts available, both current and projected. Cost of possible new installations. Supply and repair facilities. Townsite requirements and cost.

e. *Labor and supervisory personnel*: availability, efficiency, and general cost. Labor laws and recent labor-management history. Needs for training. Availability of temporary construction labor and contractors.

f. *Rate of mining and mine life,* with alternatives for consideration in profitability analysis (*h*).

g. *Sociopolitical factors*. Opportunities and hazards. Stability of government, efficiency of government bureaus, and degree of government participation in mining. Laws and attitudes regarding foreign and domestic mining investments. Cost of establishing subsidiary companies. Taxation policies. Repatriation of profits. Export duties on products and import restrictions on equipment. Mining subsidies and bonus programs. Restrictions on employment of personnel. Sociological impact of new industry. Tariffs, import restrictions, and government attitudes in the capital source country.

h. *Profitability analysis,* with minimum acceptable size and grade. Alternative levels of production, and alternative methods of mining and processing with net present value, DCF return on investment, and payout period for each major alternative. Risk and enhancement probability estimates.

i. *Cost, time, and manpower* involved in exploration program to be recommended in next section.

12. CONCLUSIONS AND RECOMMENDATIONS

a. *Specific opportunities,* with attractions and handicaps.

b. *Recommendations for action*: rejection, acquisition for a holding property, or acquisition for immediate steps toward production, with reasons for the proposed action.

c. *Acquisition value*: payment limits and alternatives, with provisions for participation agreements.

d. *Exploration steps recommended,* with objectives, decision guidelines, and preliminary budget for each step and with an outline of action for at least the first step.

13. REFERENCES

a. *Selectivity and thoroughness of search,* to indicate any further library search needed.

b. *Topics to include*: the property, the district, and sources of broader information quoted in the report.

c. *Maps and aerial photographs*: index numbers and sources.

14. APPENDIXES—for all lengthy or detailed information and special information, such as:

Names, addresses, and phone numbers of principal informants, government officials, officers in the organization holding mineral rights, legal counsel, consultants, contractors, and laboratories. Copies, abstracts, and translations of appropriate laws and regulations.

APPENDIX E
GEOLOGIC TIME TERMS

Stratigraphic terms for system and series (rock terms) and for period and epoch (time terms) have nearly the same meaning from continent to continent. Terms for stage (rock) and age (time) are, however, influenced to a large extent by local geology. A complete listing of local terms would be too extensive for this book. It should be helpful, however, to know the commonly encountered terms shown in the following list.

Stage terms within each series are given in sequential lists in the Europe and North America columns; terms appearing on the same line are not necessarily equivalent. Frequently cited names for orogenies are shown in bold type, with an approximate age date (millions of years) given for the orogenic climax.

A bibliography on stratigraphic terminology, a listing of publications on stratigraphic nomenclature (system-by-system), and a bibliography of regional geology (country-by-country) are provided by Sarjeant and Harvey (1973).

ERA	SYSTEM or PERIOD	SERIES or EPOCH	STAGE or AGE and **OROGENIC CLIMAX** Europe	North America
Cenozoic	Quaternary	Holocene		-2- **Cascadian**
		Pleistocene		
	Tertiary	(Neogene)		
		Pliocene	Astian	
			Plaisancian	
		Miocene	Pontian	
			(Sahelian)	
			Sarmatian	
			Tortonian	
			Helvetian	
			Burdigalian	
			Aquitanian	
		(Paleogene)	**Alpine**	-25-
		Oligocene	Chattian	
			Rupelian	
			Tongrian	
				-40-
		Eocene	**Pyrenean**	Jackson
			Priabonian	Claiborne
			Lutetian	Wilcox
			Ypresian	Midway
		Paleocene	Thanetian	
			Montean	
			Laramide	-65- **Laramide**

Mesozoic

Cretaceous

Upper (Senonian)
- Maestrichtian
- Campanian
- Santonian
- Coniacian
- Turonian
- Cenomanian

Austrian

Lower
- Albian
- Aptian
- Barremian
- Hauterivian
- Valanginian
- Berriasian

– 100 –

– 135 – **Nevadan**

Jurassic

Upper (Malm)
- Purbeckian
- Portlandian
- Kimmeridgian
- Oxfordian

Middle (Dogger)
- Callovian
- Bathonian
- Bajocian

Lower (Lias)
- Toarcian
- Pliensbachian
- Sinemurian
- Hettangian

Triassic

Upper (Keuper)
- Rhaetian
- Norian
- Carnian

Middle (Muschelkalk)
- Ladinian
- Anisian

Lower (Buntsandstein)
- Scythian

ERA	SYSTEM or PERIOD	SERIES or EPOCH	STAGE or AGE and **OROGENIC CLIMAX**	
			Europe	North America
Paleozoic	Permian	Upper (Zechstein)	**Pfalzian** Tartarian Kazanian	–230– **Appalachian** Ochoan
		Middle	Kungurian Artinskian Sakmarian	Guadalupian Leonardian Wolfcampian
		Lower (Rothliegendes)	**Hercynian,** **(Variscan,** **Armorican)**	–285– **Marathon**
	Carboniferous	(Pennsylvanian) (Silesian)	Stephanian **Asturian** Westphalian Namurian	Virgilian Missourian Desmoinesian Atokan Morrowan
		(Mississippian) (Dinantian)	**Sudetic** Visean Tournaisian Strunian	–320– **Ouachitan** Chesterian Meramecian Osagean Kinderhookian
	Devonian	Upper	**Bretonian** Famennian Frasnian	–350– **Acadian**

Period	Epoch	Stage	Ma	Orogeny
Silurian	Middle	Givetian		
		Eifelian		
		Coblenzian		
		Gedinnian		
		Caledonian		
	Lower	Downtonian		
		Ludlovian		
		Wenlockian		
		Llandoverian		
			–405–	
		Cayugan		
		Niagaran		
		Albian		
			–440–	**Taconic**
		Richmondian		
		Maysvillian		
		Edenian		
		Mohawkian		
		Chazyan		
Ordovician	Cincinnatian	Ashgillian		
	Champlainian	Caradocian		
		Llandeilian		
		Skiddavian		
		Sardic		
			–500–	**Vermont**
		Croixian		
		Albertan		
		Waucoban		
Cambrian		Potsdamian		
		Acadian		
		Georgian		
		Assyntian		
		Baikalian		
			–570–	
Precambrian				
Proterozoic				
(Algonkian)				
Archeozoic				
(Archean)				
			–1000–	**Grenville**
			–2400–	**Algoman**
				(Kenoran)
				Laurentian

APPENDIX F
INTERNATIONAL SYSTEM (SI) OF UNITS AND OTHER METRIC UNIT EQUIVALENTS

Prefixes

10^9 = giga (G)	10^{-1} = deci (d)
10^6 = mega (M)	10^{-2} = centri (c)
10^3 = kilo (k)	10^{-3} = milli (m)
10^2 = hecto (h)	10^{-6} = micro (μ)
10^1 = deka (da)	10^{-9} = nano (n)
1 = (unit)	10^{-12} = pico (p)

Metric Units	English Units		Reciprocals
LENGTH			
millimeter (mm)	0.03937	inch	25.40
centimeter (cm)	0.3937	inch	2.540
meter (m)	3.281	feet	0.3048
kilometer (km)	0.6214	miles	1.609
AREA			
square centimeter (cm²)	0.1550	in.²	6.452
square meter (m²)	10.76	ft²	0.0929
hectare (ha) (10,000 m²)	2.471	acres	0.4047
hectare (ha)	0.003861	mi²	259.0
square kilometer (km²)	0.3861	mi²	2.590
VOLUME			
cubic centimeter (cm³)	0.06102	in.³	16.39
cubic meter (m³)	1.308	yd³	0.7646

631

Metric Units		English Units	Reciprocals
liter (l)	0.2642	gallon (U.S.)	3.785
liter (l)	0.2200	gallon (U.K.)	4.546
	MASS		
gram (g)	0.03215	ounce troy (20 dwt)	31.10
gram (g)	0.6430	pennyweight (dwt)	1.555
gram (g)	0.03527	oz avoirdupois	28.35
kilogram (kg)	2.205	lb avoirdupois	0.4535
ton (t)	1.102	short ton (2000 lb)	0.9072
ton (t)	0.9842	long ton (2240 lb)	1.016
	MISCELLANEOUS METRIC EQUIVALENTS		
cm/second	0.01968	ft/min	50.81
m³/second	22.82	million gal/day	0.04382
m³/minute	264.2	gal/min	0.003785
t-km	1.774	short tons-mi	0.5637
g/cm³	62.43	lb/ft³	0.01602
kg/cm²	14.22	lb/in² (psi)	0.07031
N (Newton − 1 kg m/s²)	0.2248	lbf (pound force)	4.448
g/t	0.02917	oz troy/short ton	34.28
g/t	0.583	dwt/short ton	1.714
g/t	0.653	dwt/long ton	1.531
deg C (degree Celsius)	9/5	degree Fahrenheit	5/9
rad (radian)	57.29	degrees	0.0175
2π rad	400	grades (360 degrees)	

APPENDIX G
ENVIRONMENTAL GUIDELINES FOR MINERAL EXPLORATION*

The ultimate responsibility for protecting the quality of our environment should be a cooperative endeavor to be shared by government, by private enterprise, and by the individual citizen. As an aid to cooperation among these groups, the Department of Mineral Resources offers these Mineral Exploration Environmental Guidelines.

In all exploration activities we will comply with the following guidelines:

1. Know and observe federal, state, and local laws and regulations pertaining to mineral exploration. Special attention should be given to controls outlined in special use permits when such permits are required. In cases where the intent of the laws is to prevent unnecessary damage to the environment, the intent of the law should be observed above and beyond the letter of the law whenever possible.

2. Respect private property. Always obtain permission from property owners before starting operations or using any water or timber on their property.

3. Establish and maintain cordial relations with property owners and government officials having jurisdiction over land on which you are operating; respect their rights and obligations. At all times, whenever possible, discuss planned actions with owners and officials prior to starting operations and regularly throughout exploration activities.

*Courtesy: Arizona Department of Mineral Resources

4. Respect gates and fences so as to minimize disturbance of livestock. Leave all gates as found; open or closed. Observe signs posted by land owners. Repair any damage caused to fences and gates.

5. Avoid disturbance of vegetation and wildlife, to the extent feasible.
   a. Keep vehicles on established roads whenever feasible.
   b. Do not blaze trees or cut brush when establishing survey lines, except where necessary.
   c. Construct roads and drill sites so as to minimize cuts and scars in the landscape.
   d. Use drill holes instead of open pits for prospect and location work whenever possible.
   e. Use minimum size equipment so as to reduce landscape damage caused by movement of equipment and construction of access roads.
   f. Whenever possible, choose the location of drill sites so that such sites and their access will be generally inconspicuous.

6. Restore to the maximum extent reasonable the natural setting of areas where you have worked.
   a. Plug or cap drill holes.
   b. Fill test pits and excavations when no longer needed.
   c. Level or contour abandoned drill sites and excavations in a reasonable manner to limit erosion and aid natural restoration, and reseed.
   d. Contour and construct water bars on access roads no longer needed to encourage natural restoration and limit damaging erosion.
   e. Reseed former access routes.
   f. All reseeding should be done with the type seed recommended by the appropriate Land Management Agency.

7. Avoid water, air, and litter pollution.
   a. Practice careful use of chemicals so as to avoid spillage that may cause water pollution.
   b. Contain and handle petroleum produces so as to avoid possible water pollution. Do not dump used petroleum products on the ground, but **remove from the area** and dispose of properly.
   c. Avoid stream pollution from camp activities. Be especially careful with soap, detergents, etc.
   d. Use proper mufflers on motorized equipment so as to avoid noise pollution.
   e. Remove all wire, flagging, stakes, and other extraneous material upon completion of surveys.
   f. Remove all litter from the area. Do not burn or bury litter, but remove and dispose of properly.
   g. Leave the area cleaner than when you arrived.

h. Remove physical evidence of claim location when claims are abandoned.

8. Prevent forest, brush, or range fires. Carry fire fighting equipment such as shovels, picks, and extinguishers in all field vehicles. Observe all fire prevention controls. Be aware of fire hazard levels. Be certain campfires are out before leaving an area. Be especially careful with cigarettes. Motorized equipment must have an acceptable spark arrester and/or muffler.

9. Observe all hunting and fishing regulations when on field assignments.

10. Promptly initiate negotiations with the owner or land management agency for settlement of any claims for damages resulting from exploration activities.

APPENDIX H
INDEX MAPS FOR MINING SITES
MENTIONED IN TEXT

Figure H-1.   North America

Figure H-2.   Western United States

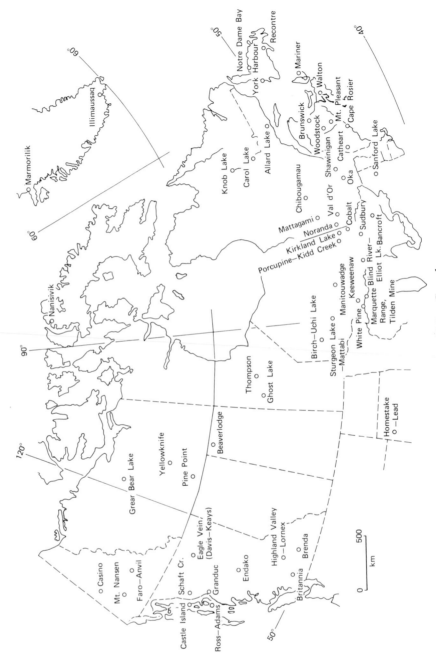

Figure H-3.   Canada

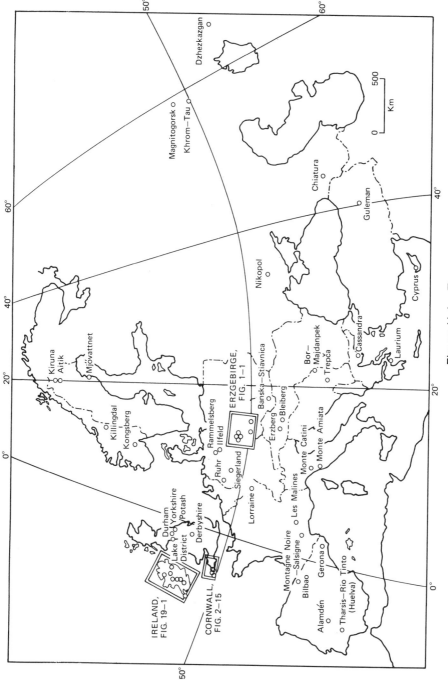

Figure H-4. Europe

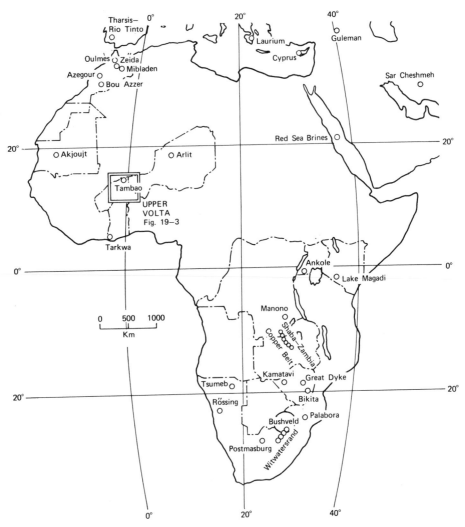

Figure H-5. Africa

Figure H-6.   South America

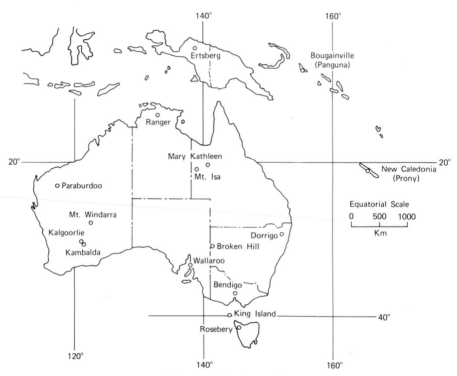

Figure H-7.   Australia

REFERENCES

Abe, H., and Aoki, M., 1975, Experiments on the hydrothermal alteration of mordenite rocks in sodium carbonate solution, with reference to analcimization around Kuroko deposits: Econ. Geology, v. 70, p. 770–780.

Abel, J. F., 1970, Tunnel support—where does the geology come in?: Civil Eng., February, p. 69–71.

Agapito, J. F., 1974, Rock mechanics applications to the design of oil shale pillars: Mining Eng., v. 26, no. 5, p. 20–25.

Agarwal, J. C., Schapiro, N., and Mallio, W. J., 1976, Process petrography and ore deposits: Mining Cong. Jour., v. 62, no. 3, p. 28–35.

Agricola, Georgius, 1556, De re metallica (English translation by H. C. Hoover and L. H. Hoover, 1912): New York, Dover Publications, 1950.

Agterberg, F. P., 1974, Geomathematics: Amsterdam, Elsevier Scientific Publishing Company, 595 p.

Alexander, G. T., 1975, Direct-cast stripping of thin coal seams: Canadian Mining Metall. Bull., v. 68, no. 761, p. 97–107.

Alexander, L., Eichen, L., Haselden, F. O., Pasucci, R. F., and Ross-Brown, D. M., 1974, Remote sensing—environmental and technical applications: Dames and Moore Eng. Bull. 45, p. 5–47.

Allais, M., 1957, Methods of appraising economic prospects of mining exploration over large territories: Algerian Sahara case study: Management Sci., v. 3, p. 285–345.

Allum, J. A. E., 1966, Photogeology and regional mapping: Oxford, England, Pergamon Press Ltd., 107 p.

Almela-Samper, A., Alvarado, M., Coma, J., Felgueroso, C., and Quintero, I., 1959, Estudio geologico de la region de Almadén: Inst. Geol. y Min. de España Bol., v. 73, p. 197–327.

645

American Institute Mining and Metallurgical Engineers, 1933, Ore deposits of the western states (Lindgren volume): New York, 797 p.

Amstutz, G. C., 1971, Glossary of mining geology: Stuttgart, Ferdinand Enke Verlag, 196 p.

Amstutz, G. C., and Bernard, A. J., 1973, Ores in sediments: Berlin, Springer Verlag.

Anderson, C. A., 1955, Oxidation of copper sulfides and secondary sulfide enrichment: Econ. Geology, 50th Anniversary Volume, p. 324–340.

Andrews-Jones, D. A., 1968, The application of geochemical techniques to mineral exploration: Colorado School of Mines, Mineral Industries Bull., v. 11, no. 6, p. 1–31.

Aplan, F. F., 1973, Evaluation to indicate processing approach, in Cummins, A. B., and Given, I. A., eds., 1973. SME Mining engineering handbook: New York, Am. Inst. Mining Metall. Petroleum Engineers, Sec. 27.3, p. 27–15–27–28.

Aplan, F. F., McKinney, W. A., and Pernichele, A. D., eds., 1974, Solution mining—Proceedings of the Dallas symposium: New York, Am. Inst. Mining, Metall. Petroleum Engineers.

Archer, A. R., and Main, C. A., 1971, Casino, Yukon—a geochemical discovery of an unglaciated Arizona-type porphyry, in Geochemical exploration: Montreal, Canadian Inst. Mining Metall. Spec. Vol. 11, p. 67–77.

Argall, G. O., and Wyllie, R. J. M., 1975, World mining glossary of mining, processing, and geological terms: San Francisco, Miller Freeman Publications, Inc., 432 p.

Atkinson, B. K., 1974, Experimental deformation of polycrystalline galena, chalcopyrite, and pyrrhotite: Inst. Mining Metall. (London) Trans., v. 83, p. B19–B28.

Badgley, P. C., 1959, Structural methods for the exploration geologist: New York, Harper Brothers, 280 p.

Bailey, E. H., and Everhart, D. L., 1964, Geology and quicksilver deposits of the New Almaden district, Santa Clara County, California: U.S. Geol. Survey Prof. Paper 360, 206 p.

Bailly, P. A., 1967, Mineral exploration and mine developing problems related to use and management of other resources and to U.S. public land laws, especially the Mining Law of 1872, in Public land law review conference, University of Idaho: Moscow, University of Idaho, p. 51–99.

Bailly, P. A., 1968, Exploration methods and requirements, in Pfleider, E. P., ed., Surface mining: New York, Am. Inst. Mining, Metall. Petroleum Engineers, p. 19–42.

Bailly, P. A., 1972, Mineral exploration philosophy: Mining Cong. Jour., v. 58, no. 4, p. 31–35.

Bailly, P. A., 1976, The problems of converting resources to reserves: Mining Eng., v. 28, no. 1, p. 27–37.

Bailly, P. A., and Still, A. R., 1973, Exploration for mineral deposits—purpose, procedure, methods, management, in Cummins, A. B., and Given, I. A., eds., SME Mining engineering handbook: New York, Am. Inst. Mining Metall. Petroleum Engineers, Sec. 5.1, p. 5–2—5–12.

Baltosser, R. W., and Lawrence, H. W., 1970, Application of well logging techniques in metallic mineral mining: Geophysics, v. 35, no. 1, p. 143–152.

Barbier, M. J., 1974, Continental weathering as a possible origin of vein-type uranium deposits: Mineralium Deposita, v. 9, p. 271–288.

Barnes, H. L., ed., 1967, Geochemistry of hydrothermal ore deposits: New York, Holt, Rinehart and Winston.

Bartholomé, P., ed., 1974, Gisements stratiformes et provinces cuprifères: Liege, Société géologique de Belgique, 427 p.

Bateman, A. M., 1950, Economic mineral deposits: New York, John Wiley & Sons.

Bates, R. L., 1960, Geology of the industrial rocks and minerals: New York, Harper & Row.

Baud, L., 1956, Les gisements et indices de manganese de l'Afrique Equatoriale Française, in Gonzalez Reyna, J., ed., Symposium sobre yacimientos de manganeso: Internat. Geol. Cong., 20th, Mexico City, Pubs., v. 2, p. 9–38.

Bentz, A., and Martini, H. J., 1968, Lehrbuch der Angewandten Geologie: Stuttgart, Enke Verlag.

Berge, J. W., 1971, Iron formation and supergene iron ores of the Goe Range area, Liberia: Econ. Geology, v. 66, p. 947–960.

Berkman, D. A., ed., 1976, Field geologist's manual; Carlton South, Victoria, Australia, Australasian Institute of Mining and Metallurgy Mon. 10, 291 p.

Berner, H., Ekström, T., Lilljeqvist, R., Stephansson, O., and Wikström, A., 1975, GEOMAP, in Hutchison, W. W., ed., Computer-based systems for geological field data: Canada Geol. Survey Paper 74–63, p. 8–18.

Berning, J., Cooke, R., Hiemstra, S. A., and Hoffman, U., 1976, The Rössing uranium deposit, South West Africa: Econ. Geology, v. 71, p. 351–368.

Berry, L. G., ed., 1971, The silver arsenide deposits of the Cobalt-Gowganda region, Ontario: Canadian Mineralogist, v. 11, part I, special issue, 429 p.

Beus, A. A., 1966, Geochemistry of beryllium: San Francisco, W. H. Freeman and Co., 401 p.

Bierman, H., and Smidt, S., 1966, The capital budgeting decision, ed. 2: New York, The Macmillan Company.

Bigotte, G., and Molinas, E., 1973, How French geologists discovered Niger uranium deposits: World Mining, v. 9, no. 4, p. 34–39.

Blackader, R. G., 1968, Guide for the preparation of geological maps and reports: Canada Geol. Survey Misc. Rept. 16, 147 p.

Blainey, G., 1969, The rush that never ended: Melbourne, Victoria, Melbourne University Press, p. 116–117.

Blais, R. A., and Carlier, P. A., 1968, Applications of geostatistics in ore evaluation, in Ore reserve estimation and grade control: Canadian Inst. Mining Metall., Spec. Vol. 9, p. 41–68.

Blais, R. A., and Stebbins, J. B., 1962, Role of mine geology in the exploitation of iron deposits of the Knob Lake Range, Canada: Am. Inst. Mining, Metall. Petroleum Engineers Trans., v. 223, p. 15–23.

Blanchard, R., 1939, Interpretation of leached outcrops: Chem. Metall. Mining Soc. South Africa Jour., v. 39, no. 11, p. 344–372.

Blanchard, R., 1968, Interpretation of leached outcrops: Nevada Bur. Mines Bull. 66, 196 p.

Blanchard, R., and Boswell, P. F., 1928, Status of leached outcrops investigation: Eng. Mining Jour., v. 125, no. 9, p. 373–377.

Blanchard, R., and Boswell, P. F., 1930, Limonite products derived from bornite and tetrahedrite: Econ. Geology, v. 25, p. 557–580.

Blanchard, R., and Boswell, P. F., 1934, Additional limonite types of galena and sphalerite derivation: Econ. Geology, v. 29, p. 671–690.

Blanchet, P. H., and Goodwin, C. I., 1972, "Geolog system" for computer and manual analysis of geologic data from porphyry and other deposits: Econ. Geology, v. 67, p. 796–813.

Bluekamp, P. R., and Beinlich, E. G., 1973, Drifting and raising by rotary drilling, in Cummins, A. B., and Given, I. A., eds., SME Mining engineering handbook: New York, Am. Inst. Mining Metall. Petroleum Engineers, Sec. 10.6, p. 10–89—10-99.

Bolstad, D. D., and Mahtab, M. A., 1974, A Bureau of Mines direct reading azimuth protractor: U.S. Bureau of Mines Inf. Circ. 8617.

Borlase, W. C., 1758, The natural history of Cornwall: Oxford.

Botbol, J. M., 1970, Characteristic analysis as applied to mineral exploration, in Decision-making in the mineral industry: Montreal, Canadian Inst. Mining Metall., Spec. Vol. 12, p. 92–99.

Bowie, S. H. U., Davis, M., and Ostle, D., eds., 1972, Uranium prospecting handbook: London, Inst. Mining Metall., 346 p.

Boyd, J., 1975, Speech given at the Mining and Metallurgical Society of America annual meeting in New York City: Typescript.

Boyle, R. W., 1968, The geochemistry of silver and its deposits: Canada Geol. Survey Bull. 160.

Boyle, R. W., 1970, Regularities in wall-rock alteration phenomena associated with epigenetic deposits, in Pouba, Z., and Stemprok, M., eds., Problems of hydrothermal ore deposition: Stuttgart, E. Schweizerbart'sche Verlagsbuchhandlung.

Boyle, R. W., ed., 1971, Geochemical exploration: Montreal, Canadian Inst. Mining Metall. Spec. Vol. 11, 593 p.

Boyle, R. W., and Garrett, R. G., 1970, Geochemical prospecting—a review of its status and future: Earth Sci. Rev., v. 6, p. 51–75.

Boyle, R. W., Shaw, D. M., and Webber, G. R., eds., 1972, Exploration geochemistry in glaciated terrains, in Sec. 10, Geochemistry, Internat. Geol. Cong., 24th, Canada, 1972: Ottawa, p. 361–404.

Bradshaw, P. M. D., ed., 1975, Conceptual models in exploration geochemistry: Jour. Geochem. Explor., v. 4, no. 1, p. 1–207.

Bradshaw, P. M. D., Clews, D. R., and Walker, J. L., 1971, Exploration geochemistry: Rexdale, Ontario, Barringer Research Ltd., 49 p.

Braithwaite, R. L., 1974, The geology and origin of the Rosebery ore deposit, Tasmania: Econ. Geology, v. 69, p. 1086–1101.

Brant, A. A., 1965, A review and discussion of present geophysical methods applied in mining exploration: Econ. Geology, v. 60, p. 819–821.

Brant, A. A., 1968, The pre-evaluation of the possible profitability of exploration prospects: Mineralium Deposita, v. 3, p. 1–17.

Breiner, S., 1973, Applications manual for portable magnetometers: Palo Alto, California, Geometrics Co.

Brinck, J. W., 1972, The prediction of mineral resources and long-term price trends in the non-ferrous metal mining industry, in Sec. 4, Mineral deposits, Internat. Geol. Cong., 24th, Canada, 1972: Ottawa, p. 3–15.

British Sulphur Corporation, Ltd., 1964, A world survey of phosphate deposits, ed. 2: London, 206 p.

Brobst, D., and Pratt, W., eds., 1973, United States mineral resources: U.S. Geol. Survey Prof. Paper 820, 722 p.

Brobst, D. A., and Wagner, R. J., 1967, Barite, in Mineral and water resources of Missouri: Missouri Div. Geol. Survey and Water Resources Repts., series 2, v. 43, p. 99–106.

Brock, B. B., 1972, A global approach to geology: Capetown, South Africa, A. A. Balkema, Publishers, 365 p.

Brock, J. S., 1973, Geophysical exploration leading to the discovery of the Faro deposit: Canadian Mining Metall. Bull., v. 66, no. 738, p. 97–117.

Brooks, D. B., and Andrews, P. W., 1974, Mineral resources, economic growth, and world population: Science, v. 185, no. 4145, p. 13–19.

Brooks, L. S., and Bray, R. C. E., 1968, A study of the tonnage and grade calculations at the Geco Division of Noranda Mines, in Ore reserve estimation and grade control: Montreal, Canadian Inst. Mining Metall. Spec. Vol. 9, p. 181.

Brooks, P. R., 1972, Geobotany and biogeochemistry in mineral exploration: New York, Harper & Row, 290 p.

Brown, J. S., ed., 1967, Genesis of stratiform lead-zinc-barite-fluorite deposits: Lancaster, Pa., Economic Geology Publishing Co., Monograph 3.

Brown, J. S., 1970, The Mississippi Valley type lead-zinc ores: Mineralium Deposita, v. 5, p. 103–119.

Brubaker, S., 1975, In command of tomorrow, resource and environmental strategies for Americans: Baltimore, Resources for the Future, Johns Hopkins University Press, 177 p.

Brundsen, D., Doornkamp, J. C., Fookes, P. G., Jones, D. K. C., and Kelly, J. M. H., 1975, Large scale geomorphological mapping and highway engineering design: Quart. Jour. Eng. Geology, v. 8, p. 227–253.

Buddington, A. F., 1935, High-temperature mineral associations at shallow to moderate depth: Econ. Geology, v. 30, p. 205–222.

Burgin, L., 1966, Guide to marketing of commodities: Colorado School of Mines, Mineral Industries Bull., v. 9, nos. 1, 4, 6.

Burk, C. A., and Drake, C. L., eds., 1974, The geology of continental margins: New York, Springer Verlag, 1009 p.

Burns, K. L., 1969, Computer-assisted geological mapping: Australasian Inst. Mining Metall. Proc., no. 232, p. 41–47.

Burt, D. M., 1975, Beryllium mineral stabilities in the model system $CaO-BeO-SiO_2-P_2O_5-F_2O_{-1}$ and the breakdown of beryl: Econ. Geology, v. 70, p. 1279–1292.

Burwell, H. B., 1950, Brown phosphate rock in Tennessee, in Southeastern mineral symposium, 1949: Knoxville, University of Tennessee Press, p. 388–397.

Bush, J. B., and Cook, D. R., 1960, The Chief Oxide-Burgin area discoveries, East Tintic district, Utah: a case history. Part II, Bear Creek Mining Company studies and exploration: Econ. Geology, v. 55, p. 1507–1540.

Cabri, L. J., and LaFlamme, J. H. G., 1976, The mineralogy of the platinum-group elements from some copper-nickel deposits of the Sudbury area, Ontario: Econ. Geology, v. 71, p. 1159–1195.

Caïa, J., 1976, Paleogeographical and sedimentological controls of copper, lead, and zinc mineralizations in the Lower Cretaceous sandstones of Africa: Econ. Geology, v. 71, p. 409–422.

Call, R. D., 1975, Personal communication: Tucson, Arizona, Pincock, Allen & Holt, Inc.

Cameron, E. M., ed., 1967, Proceedings, Symposium on Geochemical Prospecting: Canada Geol. Survey Paper 66-54, 282 p.

Cameron, E. N., ed., 1973, The mineral position of the United States 1975–2000: Madison, University of Wisconsin Press, 159 p.

Campbell, H., 1948, West Bay fault, in Wilson, M. E., ed., Structural geology of Canadian ore deposits: Montreal, Canadian Inst. Mining Metall., p. 244–259.

Canadian Institute of Mining and Metallurgy, 1968, Ore reserve estimation and grade control: Montreal, Spec. Vol. 9, 321 p.

Canney, F. C., ed., 1969, International symposium on geochemical prospecting: Colorado School of Mines Quart., v. 64, p. 1–520.

Cannon, H. L., 1960, The development of botanical methods of prospecting for uranium on the Colorado Plateau: U.S. Geol. Survey Bull. 1085-A, 50 p.

Cassidy, S. M., ed., 1973, Elements of practical coal mining: New York, Am. Inst. Mining Metall. Petroleum Engineers, 614 p.

Chace, F. M., 1956, Abbreviations in field and mine geologic mapping: Econ. Geology, v. 51, p. 712–723.

Chamberlain, T. C., 1897, The method of multiple working hypotheses: Jour. Geology, v. 5, p. 837–848.

Chapman, E. P., 1970, Why feasibility studies for very large low grade deposits?: World Mining, v. 26, no. 6, p. 16–20.

Chapman, R. W., 1974, Calcareous duricrust in Al-hasa, Saudi Arabia: Geol. Soc. America Bull., v. 85, p. 119–130.

Chase, C. G., Herron, E. M., and Normark, W. R., 1975, Plate tectonics: commotion in the ocean and continental consequences, in Donath, F. A., ed., Annual review of earth and planetary sciences, Vol. 3: Palo Alto, California, Annual Reviews, Inc., p. 271–292.

Checkland, S. G., 1967, The mines of Tharsis: London, George Allen and Unwin Ltd.

Chow, V. T., 1972, Hydrologic maps, in Fairbridge, R. W., ed., Encyclopedia of geochemistry and environmental sciences: New York, Van Nostrand Reinhold, p. 519–530.

Clavel, M., and Leblanc, M., 1971, Les minéralisations de la region de Bou Azzer–El Graara: Maroc, Service géolog. Notes et Mem. no. 229, p. 219–243.

Cloke, P. L., 1966, The geochemical application of Eh-pH diagrams: Jour. Geol. Education, v. 14, no. 4, p. 140–148.

Cloud, P., 1973, Materials, their nature and nurture: Earth Sci. Newsletter, Dec., p. 2–7.

Coates, D. F., 1970, Rock mechanics principles: Dept. Energy, Mines, and Resources, Mines Branch Mon. 874, 450 p.

Cochran, W., Fenner, P., and Hill, M., 1973, Geowriting, a guide to writing, editing, and printing in earth science: Washington, D.C., Am. Geol. Inst., 80 p.

Cole, M. M., and Owen Jones, E. S., 1974, Remote sensing in mineral exploration, in Barrett, E. C., and Curtis, L. C., 1974, Environmental remote sensing, applications and achievements: London, Edward Arnold Publishers Ltd., p. 51–65.

Compton, R. R., 1962, Manual of field geology: New York, John Wiley & Sons, 378 p.

Cook, D. R., ed., 1957, Geology of the East Tintic Mountains and ore deposits of the Tintic mining districts: Utah Geol. Soc. Guidebook to the geology of Utah, no. 12.

Cooker, W. B., and Nichol, I., 1975, The relation of lake sediment geochemistry to mineralization in the northwest Ontario region of the Canadian Shield: Econ. Geology, v. 70, p. 202–218.

Coolbaugh, D. F., 1971, La Caridad, Mexico's newest and largest porphyry copper deposit, an exploration case history: Am. Inst. Mining, Metall. Petroleum Engineers Trans., v. 250, no. 2, p. 133–138.

Cooper, D. O., Davidson, L. B., and Reim, K. M., 1973, Simplified financial and risk analysis for minerals exploration, in Sturgul, J. R., ed., Annual symposium on computer applications in the minerals industry, 11th.: Tucson, University of Arizona, v. 1, p. B1–B14.

Cooper, J. R., and Huff, L. C., 1951, Geological investigations and geochemical prospecting experiment at Johnson, Arizona: Econ. Geology, v. 46, p. 731–756.

Cornwall, H. R., 1966, Nickel deposits of North America: U.S. Geol. Survey Bull. 1223, 63 p.

Cousins, C. A., 1969, The Merensky Reef of the Bushveld igneous complex, in Wilson, H. D. B., ed., Magmatic ore deposits: Lancaster, Pa., Economic Geology Publishing Co., Econ. Geology Mon. 4.

Cratchley, C. R., and Denness, B., 1972, Engineering geology in urban planning with an example from the city of Milton Keynes: in Sec. 13, Engineering geology, Internat. Geol. Cong., 24th, Canada, 1972: Ottawa, p. 13–22.

Creasey, S. C., 1950, Geology of the St. Anthony (Mammoth) area, Pinal County, Arizona, in Arizona zinc and lead deposits, pt. 1: Tucson, Arizona Bureau of Mines Bull. 156, p. 63–84.

Crook, T. R., 1933, History of the theory of ore deposits: London, Thomas Murby and Co.

Cruickshank, M. J., 1973, Marine mining, in Cummings, A. B., and Given, I. A., eds., SME Mining engineering handbook: New York, Am. Inst. Mining Metall. Petroleum Engineers, Sec. 20, p. 1–200.

Cruzat, A. C. E., and Meyer, W. T., 1974, Predicted base-metal resources of northwest England: Inst. Mining Metall. (London) Trans., v. 83, p. B131–134.

Culbert, R. R., 1976, A multivariate approach to mineral exploration: Canadian Mining Metall. Bull., v. 69, no. 766, p. 39–52.

Cumming, J. D., and Wicklund, A. P., 1975, Diamond drill handbook: Toronto, J. K. Smit and Sons Diamond Products Ltd., 541 p.

Cummins, A. B., and Given, I. A., eds., 1973, SME Mining engineering handbook: New York, Am. Inst. Mining, Metall. Petroleum Engineers, 2 vols., 3500 p.

Curtin, G. C., King, H. D., and Mosier, E. L., 1974, Movement of elements into the atmosphere from coniferous trees in subalpine forests of Colorado and Idaho: Jour. Geochem. Explor., v. 3, p. 345–363.

Dadson, A. S., and Emery, D. J., 1968, Ore estimation and grade control at the Giant Yellowknife mine, in Ore reserve estimation and grade control: Canadian Inst. Mining Metall. Spec. Vol. 9, p. 221.

Daily, A. F., 1968, Placer mining, in Pfleider, E. P., ed., Surface mining: New York, Am. Inst. Mining Metall. Petroleum Engineers, p. 928–954.

Damberger, H. H., 1971, Coalification pattern of the Illinois Basin: Econ. Geology, v. 66, p. 488–494.

Darnley, A. G., 1973, Airborne gamma-ray techniques—present and future, *in* Uranium exploration methods: Vienna, Internat. Atomic Energy Agency, p. 67–108.

Dass, A. S., Boyle, R. W., and Tupper, W. M., 1973, Endogenic haloes of the native silver deposits, Cobalt, Ontario, Canada, *in* Jones, M. J., ed., Geochemical exploration in 1972: London, Inst. Mining Metall., p. 25–36.

Davenport, P. H., and Nichol, I., 1973, Bedrock geochemistry as a guide to areas of base-metal potential in volcano-sedimentary belts of the Canadian Shield, *in* Jones, M. J., ed., Geochemical exploration in 1972: London, Inst. Mining Metall., p. 45–57.

Davis, D. R., Kisiel, C. C., and Duckstein, L., 1973, Bayesian methods for decision-making in mineral exploration and exploitation, *in* Proceedings, Internat. symposium on computer applications in the minerals industry, 11th, 1973: Tucson, University of Arizona, p. B55–67.

Davis, J. C., 1973, Statistics and data analysis in geology: New York, John Wiley & Sons, 550 p.

Davis, J. F., 1973, A practical approach to uranium exploration drilling from reconnaissance to reserves, *in* Uranium exploration methods: Vienna, Internat. Atomic Energy Commission, p. 109–128.

Davis, S. N., and DeWiest, R. J. M., 1966, Hydrogeology: New York, John Wiley & Sons, 463 p.

Dawson, K. R., 1974, Niobium and tantalum in Canada: Canada Geol. Survey Econ. Geol. Rept. 29, 154 p.

Deere, D. U., Hendron, A. J., Patton, F. D., and Cording, E. J., 1967, Design of surface and near-surface construction in rock, *in* Symposium on rock mechanics, 8th, Minnesota, Proc: New York, Am. Inst. Mining Metall. Petroleum Engineers, p. 237–302.

De Geoffroy, G., and Wignall, T. K., 1973, Design of a statistical data processing system to assist regional exploration planning: Canadian Mining Jour., v. 94, no. 11, p. 30–35, and no. 12, p. 35–36.

De Launay, L., 1913, Traité de métallogénie. Gîtes minéraux et métallifères: Paris, Beranger, Editeurs, 3 Vols.

Demek, J., ed., 1972, Manual of detailed geomorphological mapping: Prague, Czechoslovak Academy of Sciences, Academia Publishing House, 376 p.

Demnati, A., and Naudy, H., 1975, Gamma-ray spectrometry in central Morocco: Geophysics, v. 40, no. 2, p. 331–343.

Derry, D. R., 1970, Exploration expenditure, discovery rate, and methods: Canadian Mining Metall. Bull., v. 63, no. 695, p. 362–366.

Derry, D. R., 1971, Supergene remobilization at the Tynagh mine, Ireland, of Northgate Exploration, *in* Valera, D. R., ed., Proceedings of meeting on remobilization of ores and minerals: Cagliari, Ente Minerario Sardo, p. 211–217.

Derry, D. R., 1973, Ore deposition and contemporaneous surfaces: Econ. Geology, v. 68, p. 1374–1380.

Derry, D. R., Clark, G. R., and Gillatt, N., 1965, The Northgate base metal deposit at Tynagh, County Galway, Ireland: Econ. Geology, v. 60, p. 1218–1237.

Desbrough, G. A., 1970, Silver depletion indicated by microanalysis of gold from placer occurrences, western United States: Econ. Geology, v. 65, p. 304–311.

Dick, R. C., 1975, In situ measurement of rock permeability: influence of calibration error on test results: Assoc. Eng. Geologists Bull., v. 12, no. 3, p. 193–211.

Dines, H. G., 1956, The metalliferous mining region of southwest England: London, Her Majesty's Stationery Office, 2 vols.

Dixon, C. J., and Pereira, J., 1974, Plate tectonics and mineralization in the Tethyan region: Mineralium Deposita, v. 9, no. 3, p. 186–198.

Donohoo, H. V., Podolsky, G., and Clayton, R. H., 1970, Early geophysical prospecting at Kidd Creek mine: Mining Cong. Jour., v. 56, no. 5, p. 44–53.

Dooley, J. R., Jr., 1974, Uranium-lead ages of the uranium deposits of the Gas Hills and Shirley Basin, Wyoming: Econ. Geology, v. 69, p. 527–531.

Dravo Corporation, 1974, Analysis of large scale non-coal underground mining methods: Springfield, Virginia, National Technical Information Service, U.S. Department of Commerce.

Duncan, D. C., and Swanson, V. E., 1965, Organic-rich shale of the United States and world land areas: U.S. Geol. Survey Circ. 523, 30 p.

Dunham, K. C., 1964, Neptunist concepts in ore genesis: Econ. Geology, v. 59, p. 1–21.

Dunlop, A. C., and Meyer, W. T., 1973, Influence of late Miocene-Pliocene submergence on regional distribution of tin in stream sediments, southwest England: Inst. Mining Metall. (London) Trans., v. 82, p. B62–B64.

Dyck, W., 1972, Radon methods of prospecting in Canada, in Bowie, S. H. U., Davis, M., and Ostle, D., eds., Uranium prospecting handbook: London, Inst. Mining Metall., p. 212–243.

Dyck, A. V., 1975, Borehole geophysics applied to metallic mineral prospecting: a review: Canada Geol. Survey Paper 75-31, 65 p.

Edwards, R. P., 1976, Aspects of trace metal and ore distribution in Cornwall: Inst. Mining Metall. (London) Trans., v. 85, p. B83–B90.

Eidel, J. J., Frost, J. E., and Clippinger, D. M., 1968, Copper-molybdenum mineralization at Mineral Park, Mohave County, Arizona, in Ridge, J. D., ed., Ore deposits of the United States, 1933–1967: New York, Am. Inst. Mining Metall. Petroleum Engineers, p. 1258–1281.

Ekström, T. K., Wirstam, A., and Larsson, L., 1975, COREMAP—a data system for drill cores and boreholes: Econ. Geology, v. 70, p. 359–368.

Élie de Beaumont, L., 1847, Note sur les émanations volcaniques et métallifères: Soc. géol. France Bull., v. 4, Séance du 5 juillet, p. 1249.

Elliott, I. L., and Fletcher, W. K., eds., 1975, Geochemical exploration 1974: New York, Elsevier North-Holland, 720 p.

Ely, N., 1970, Mining laws of the world. Part 1. Western hemisphere: Washington, U.S. Bur. Mines, Inf. Circ. 8482.

Ely, N., 1971, Mining laws of the world. Part 2. East Asia and Pacific: U.S. Bur. Mines Inf. Circ. 8514.

Ely, N., 1972, Mining laws of the world. Part 3. Near East and South Asia: Washington, U.S. Bur. Mines, Inf. Circ. 8544.

Ely, N., 1973, Mining laws of the world. Part 4. Africa: Washington, U.S. Bur. Mines Inf. Circ. 8610.

Ely, N., 1974, Mining laws of the world. Part 5, Europe: Washington, U.S. Bur. Mines Inf. Circ. 8631.

Emberger, A., 1969, Problème des remobilisations dans les gîtes de plomb et de zinc, *in* Valera, R., ed., Proceedings of the meeting on remobilization of ores and minerals: Cagliari, Ente Minerario Sardo, p. 37–57.

Emmons, W. H., 1936, Hypogene zoning in metalliferous lodes, *in* Report, Internat. Geol. Cong., 16th, Washington, 1933: v. 1, p. 417–432.

Emmons, W. H., 1940, The principles of economic geology, ed. 2 New York, McGraw-Hill Book Co.

Erickson, R. L., 1973, Crustal abundance of elements, and mineral reserves and resources, *in* Brobst, D., and Pratt, W., eds., United States mineral resources: U.S. Geol. Survey Prof. Paper 820, p. 21–25.

Erickson, R. L., Marranzino, A. P., Oda, U., and Janes, W. W., 1964, Geochemical exploration near the Getchell mine, Humboldt County, Nevada: U.S. Geol. Survey Bull. 1198-A, 26 p.

Espie, F. F., 1971, The Bougainville copper project: Australasian Inst. Mining Metall. Proc., no. 238, p. 1–10.

Eugster, H. P., 1970, Chemistry and origin of the brines of Lake Magadi, Kenya: Mineralog. Soc. America Spec. Paper 3, p. 213–235.

Evans, A. M., 1975, Mineralization in geosynclines—the Alpine enigma: Mineralium Deposita, v. 10, no. 3, p. 254–260.

Eyde, T. H., 1974, Obtaining geologic information from deep mineral exploration targets utilizing oilfield rotary rigs: Am. Inst. Mining, Metall. Petroleum Engineers Trans., v. 255, no. 1, p. 53–58.

Fairbridge, R. W., ed., 1966, Encyclopedia of oceanography: New York, Van Nostrand Reinhold Co., 1021 p.

Fairbridge, R. W., ed., 1968, The encyclopedia of geomorphology: New York, Van Nostrand Reinhold Co., 1295 p.

Fairbridge, R. W., ed., 1972, The encyclopedia of geochemistry and environmental sciences: New York, Van Nostrand Reinhold Co., 1321 p.

Farmer, I. W., 1968, Engineering properties of rocks: London, E. and F. N. Spon Ltd., 180 p.

Feather, C. E., and Koen, G. M., 1975, The mineralogy of the Witwatersrand reefs: Minerals Sci. Eng., v. 7, no. 3, p. 189–224.

Fedorchuk, V. P., 1975, Commercial types of mercury deposits: Internat. Reology Review, v. 17, p. 989–1006.

Felder, F., 1974, Shawinigan nickel-copper property—a case history of reconnaissance geochemical discovery in the Grenville province of Quebec, Canada: Jour. Geochem. Explor., v. 3, p. 1–23.

Ferguson, N. F., ed., 1974, Geologic mapping for environmental purposes: Boulder, Co., Geological Society of America, 40 p.

Ferguson, S. A., 1966, The relationship of mineralization to stratigraphy in the Porcupine and Red Lake areas, *in* Precambrian Symposium: Toronto, Ontario, Geol. Assoc. Canada Spec. Paper 3, p. 99–119.

Fields, H. H., Perry, J. H., and Deul, M., 1975, Commercial-quality gas from a multipurpose borehole located in the Pittsburg coal bed: U.S. Bur. Mines Rept. Inv. 8025, 14 p.

Fischer, R. P., 1973, Vanadium, in Brobst, D. A., and Pratt, W. P., eds., United States Mineral Resources: U.S. Geol. Survey Prof. Paper 820, p. 679–688.

Fischer, R. P., 1974, Exploration guides to new uranium districts and belts: Econ. Geology, v. 69, p. 362–376.

Fischer, R. P., and Stewart, J. H., 1961, Copper, vanadium, and uranium deposits in sandstones: Econ. Geology, v. 56, p. 509–529.

Fitch, A. J., and Herndon, J. P., 1957, Anaconda's Jackpile mine: Mining Cong. Jour., v. 43, no. 6, p. 57–60.

Flawn, P. T., 1970, Environmental geology: New York, Harper & Row, 313 p.

Foose, R. M., 1968, Surface subsidence and collapse caused by ground water withdrawal in carbonate rock areas, in Proc. Section 12. Engineering Geology in Country Planning, Internat. Geol. Cong., 23rd, Czechoslovakia, 1968: Prague, Academia, p. 155–166.

Forrester, J. D., 1946, Principles of field and mining geology: New York, John Wiley & Sons.

Fortescue, J. A. C., 1972, The relationships between practical and research aspects of geochemical prospecting in glacial terrain, in Sec. 10, Geochemistry, Internat. Geol., Cong., 24th, Canada, 1972: Ottawa, p. 361–369.

Fortescue, J. A. C., 1974, Exploration geochemistry and landscape: Canadian Mining Metall. Bull., v. 67, no. 751, p. 80–87.

Fox, W., ed., 1967, A technical conference on tin: London, Internat. Tin Council.

Fox, W., ed., 1969, A second technical conference on tin: Bangkok. Internat. Tin Council.

Franklin, J. M., Kasarda, J., and Poulsen, K. H., 1975, Petrology and chemistry of the alteration zone of the Mattabi massive sulfide deposit: Econ. Geology, v. 70, p. 63–79.

Frohling, E. S., and McGeorge, R. M., 1975, How stepwise financing can turn your prospect into an operating mine: Mining Eng., v. 27, no. 9, p. 30–32.

Fryklund, V. C., Jr., 1964, Ore deposits of the Coeur d'Alene district, Shoshone County, Idaho: U.S. Geol. Survey Prof. Paper 445, 103 p.

Gallagher, M. J., 1970, Portable X-ray spectrometers for rapid ore analysis, in Jones, M. J., ed., Mining and petroleum geology, Commonwealth Mining Metall. Cong., 9th, Proc.: London, Inst. Mining Metall., p. 691–729.

Galopin, R., and Henry, N. F. M., 1972, Microscopic study of opaque minerals: Cambridge, England, W. Hefner and Sons, 322 p.

Garnett, R. H. T., 1967, The underground pursuit and development of tin lodes, in Fox, W., ed., A technical conference on tin: London, Internat. Tin Council, Vol. 1, p. 141–202.

Garrels, R. M., Mineral species as a function of pH and oxidation-reduction potentials, with special reference to the zone of oxidation and secondary enrichment of sulfide ore deposits: Geochim. et Cosmochim. Acta, v. 5, no. 4, p. 153–168.

Garrels, R. M., and Christ, C. L., 1965, Solutions, minerals, and equilibria: New York, Harper & Row, 450 p.

Gautier, R., 1970, Perspectives nouvelles dans le traitment des minerais d'uranium et l'obtention des concentrés directement utilisables, in Proceedings, Symposium on recovery of uranium, Sao Paulo: Vienna, Internat. Atomic Energy Agency, p. 17–32.

Gentry, D. W., and Hrebar, M. J., 1976, Procedures for determining economics of small underground mines: Colorado School of Mines, Mineral Industries Bull., v. 19, no. 1, 18 p.

Geological Society (London), 1970, Working party report, the logging of rock cores for engineering purposes: Quart. Jour. Eng. Geology, v. 3, p. 1–24.

Gerdemann, P. E., and Myers, H. E., 1972, Relationship of carbonate facies patterns to ore distribution and to ore genesis in the southeast Missouri lead district: Econ. Geol., v. 67, p. 429–430.

Gilbert, G., ed., 1957, Structural geology of Canadian ore deposits (Congress volume): Montreal, Canadian Inst. Mining Metall.

Gill, J. E., ed., 1972, Sec. 4, Mineral deposits, Internat. Geol. Cong., 24th, Canada, 1972: Ottawa, p. 107–160.

Gilmour, P., 1965, The origin of massive sulfide mineralization in the Noranda district, northwestern Quebec: Geol. Assoc. Canada, Proc., v. 16, p. 63–81.

Gilmour, P., 1971, Strata-bound massive pyritic sulfide deposits—a review: Econ. Geology, v. 66, p. 1239–1243.

Gingrich, J. E., 1975, Results from a new uranium exploration method: Am. Inst. Mining Metall. Petroleum Engineers Trans., v. 258, no. 1, p. 61–64.

Ginzburg, I. I., 1960, Principles of geochemical prospecting (translated from the Russian by V. P. Sokoloff): London, Pergamon Press, 311 p.

Goddard, E. N., chm., 1963, Rock-color chart: New York, Geol. Soc. America.

Goldenberg, L. A., 1968, Organization of state geological expeditions in Russia in the 18th century: Internat. Geol. Cong., 23rd, Prague, Pubs., sec. 13, p. 345–354.

Gomez, M., and Donaven, D. J., 1972, Forecasting the properties of coal seams in place: U.S. Bur. Mines Rept. Inv. 7680, 53 p.

Gonzalez Reyna, J., ed., 1956, Symposium sobre yacimientos de manganeso: Internat. Geol. Cong., 20th, Mexico City, Spec. Pub., 5 vols.

Goodman, R. E., 1976, Methods in geological engineering in discontinuous rocks: St. Paul, Minn., West Publishing Company, 472 p.

Gott, G. B., and Botbol, J. M., 1973, Zoning of major and minor metals in the Coeur d'Alene mining district, Idaho, U.S.A., in Jones, M. J., ed., Geochemical Exploration 1972: London, Inst. Mining Metall., p. 1–12.

Gott, G. B., and Botbol, J. M., 1975, Possible extension of mineral belts, northern part of the Coeur d'Alene district, Idaho: U.S. Geol. Survey, Jour. Res., v. 3, no. 1, p. 1–7.

Govett, G. J. S., and Govett, M. H., eds., 1976, World mineral supplies—assessment and perspective: Amsterdam, Elsevier Scientific Publishing Co., 472.

Govett, G. J. S., and Pantazis, Th. M., 1971, Distribution of Cu, Zn, Ni, and Co in the Troodos Pillow Lava Series, Cyprus: Inst. Mining Metall. (London) Trans., v. 80, p. B27–B46.

Graham, A. R., Buchan, R., and Kozak, C., 1975, A new rapid method for element determination in drill core: Canadian Mining Metall. Bull., v. 68, no. 764, p. 94–100.

Granier, C. L., 1973, Introduction à la prospection géochimique des gîtes métallifères: Paris, Masson et Cie, 143 p.

Grant, E. L., and Ireson, W. G., 1970, Principles of engineering economy, ed. 5: New York, The Ronald Press Company, 640 p.

Grant, F. S., 1972, Review of data processing and interpretation methods in gravity and magnetics: Geophysics, v. 37, p. 647–661.

Grant, F. S., and West, G. F., 1965, Interpretation theory in applied geophysics: New York, McGraw-Hill Publishing Co.

Grant, R. E., 1970, Computers as tools during mineral property investigations, *in* Decision-making in the mineral industry: Canadian Inst. Mining Metall. Spec. Vol. 12, p. 270–279.

Graton, L. C., 1933, The depth zones in ore deposition: Econ. Geology, v. 28, p. 513–555.

Graton, L. C., 1940, Nature of the ore-forming fluid: Econ. Geology, v. 35, Suppl. to no. 2, p. 197–358.

Gray, R. F., Hoffman, V. J., Bagan, R. J., and McKinley, H. L., 1968, Bishop tungsten district, California, *in* Ridge, J. D., ed., Ore deposits of the United States, 1933–1967: New York, Am. Inst. Mining Metall. Petroleum Engineers, p. 1532–1554.

Greenslade, W. M., Brittain, R., and Baski, H., 1975, Dewatering for underground mining—"the anatomy of anomalous conditions": Mining Cong. Jour., v. 61, no. 11, p. 34–38.

Griffis, A. T., 1971, Exploration: changing techniques and new theories will find new mines: World Mining, v. 7, no. 7, p. 54–60.

Griffith, S. V., 1960, Alluvial prospecting and mining: New York, Pergamon Press, 245 p.

Griffiths, J. C., and Singer, D. A., 1973, The Engel simulator and the search for uranium, *in* Proceedings, Internat. symposium on applications of computer methods in the mineral industry, 10th, Johannesburg: Johannesburg, South African Inst. Mining Metall., p. 9–25.

Griggs, A. B., 1945, Chromite-bearing sands of the southern part of the coast of Oregon: U.S. Geol. Survey Bull. 945-E, p. 113–150.

Griswold, W. T., and Wallace, W. K., 1967, Organization and costs for mineral exploration in Australia: Paper presented at Pacific Southwest Mineral Industry Conference (AIME), Pacific Grove, California, May 8, 1967 (privately published).

Gross, G. A., 1968, Iron ranges of the Labrador geosyncline, *in* Geology of iron ore deposits in Canada: Canada Geol. Survey Econ. Geol. Rept. 22, v. 3, 179 p.

Gross, W. H., 1975, New ore discovery and source of silver-gold veins, Guanajuato, Mexico: Econ. Geology, v. 70, p. 1175–1189.

Guilbert, J. M., 1971, Known interactions of tectonics and ore deposits in the context of new global tectonics: Am. Inst. Mining Metall. Petroleum Engineers, Preprint 71-S-91, 19 p.

Guilbert, J. M., and Lowell, J. D., 1974, Variations in zoning patterns in porphyry ore deposits: Canadian Inst. Mining Metall. Bull., v. 67, no. 742, p. 99–109.

Guild, P. W., 1968, Metallotects of North America: Paper presented at the Society of Economic Geologists meeting, New York, Feb. 27, 1968 (typescript).

Gustafson, L. B., and Hunt, J. P., 1975, The porphyry copper deposit at El Salvador, Chile: Econ. Geology, v. 70, p. 857–912.

Guy-Bray, J. V., 1972, New developments in Sudbury geology: Geol. Assoc. Canada Spec. Paper 10, 124 p.

Gy, P., 1968, Theory and practice of sampling broken ores, *in* Ore reserve estimation and grade control: Montreal, Canadian Inst. Mining Metall. Spec. Vol. 9, p. 5–10.

Haapala, P. S., 1953, Morococha anhydrite: Soc. Geol. Peru Bull., v. 26, p. 21–32.

Hadson, T., Askevold, G., and Plafker, G., 1975, A computer-assisted procedure for information processing of geologic field data: U.S. Geol. Survey, Jour. Res., v. 3, no. 3, p. 369–375.

Hails, J., and Carr, A., eds., 1975, Nearshore sediment dynamics and sedimentation: New York, John Wiley & Sons, 296 p.

Hajek, H., 1966. Über das Auftreten roteisensteinführender Porphyroid-horizonte im Steirischen Erzberg: Archiv. f. Lagerstättenforschung in den Ostalpen, v. 4, p. 4–39.

Hale, L. A., ed., 1967, Anatomy of the western phosphate field. Intermountain Assoc. Geologists, 15th Ann. Field Conf.: Salt Lake City, Utah, Intermountain Assoc. Geologists, 287 p.

Hall, V. S., 1975, Environmental geology: a selected bibliography: Boulder, Colorado, Geol. Soc. America (microfiche, 98 frames).

Hall, W. E., and Friedman, I., 1969, Oxygen and carbon isotopic composition of ore and host rock of selected Mississippi Valley deposits: U.S. Geol. Survey Prof. Paper 650-C, p. C140–C148.

Hallof, P. G., 1966, The use of resistivity to outline sedimentary rock types in Ireland, *in* Case histories, Vol. 1 of Mining geophysics: Tulsa, Oklahoma, Soc. Explor. Geophysicists, p. 18–27.

Hallof, P. G., 1972, The induced polarization method, *in* Sec. 9, Exploration geophysics, Internat. Geol. Cong., 24th, Canada, 1972: Ottawa, p. 64–81.

Hallof, P. G., and Winniski, E., 1971, A geophysical case history of the Lakeshore orebody: Geophysics, v. 36, p. 1232–1249.

Hammer, D. F., and Peterson, D. W., 1968, Geology of the Magma mine area, Arizona, *in* Ridge, J. D., ed., Ore deposits of the United States, 1933–1967: New York, Am. Inst. Mining Metall. Petroleum Engineers, p. 1282–1310.

Handin, J., 1966, Strength and ductility, *in* Clark, S. P., Jr., ed., Handbook of physical constants: Geol. Soc. America Mem. 97, p. 223–289.

Hanekom, H. J., 1965, The geology of the Palabora igneous complex: South Africa Geol. Survey Mem. 54.

Harbaugh, J. W., 1964, A computer method for four-variable trend analysis illustrated by a study of oil-gravity variations in southeastern Kansas: Kansas Geol. Survey Bull. 171, 58 p.

Harris, D. P., 1966, A probability model of mineral wealth: Am. Inst. Mining Metall. Petroleum Engineers Trans., v. 235, p. 199–216.

Harris, D. P., and Brock, T. N., 1973, A conceptual Bayesian geostatistical model for metal endowment, *in* Proceedings, International symposium on computer applications in the minerals industry, 11th: Tucson, University of Arizona, p. B113–B184.

Hart, R. C., and Sprague, D., 1968, Methods of calculating ore reserves in the Elliot Lake Camp, *in* Ore reserve estimation and grade control: Montreal, Canadian Inst. Mining Metall., Spec. Vol. 9, p. 251–260.

Hawks, H. E., and Webb, J. S., 1962, Geochemistry in mineral exploration: New York, Harper & Row, 415 p.

Hazen, S. W., Jr., 1968, Ore reserve calculations, *in* Ore reserve estimation and grade control: Montreal, Canadian Inst. Mining Metall., Spec. Vol. 9, p. 11–32.

Heath, D. C. G., Kalcov, G. D., and Inns, G. S., 1974, Treatment of inflation in mine evaluation: Inst. Mining Metall. (London) Trans., v. 83, p. A20–A33; discussion p. A151–A154.

Heenan, P. R., 1972, The discovery of the Ming zone, Consolidated Rambler Mines Limited, Baie Verte, Newfoundland: Canadian Mining Metall. Bull., v. 66, no. 729, p. 83.

Heinrich, E. W., 1966, The geology of carbonatites: Chicago, Rand McNally and Co., 555 p.

Heinrichs, W. E., Jr., 1976, Pima district, Arizona—a historical and economic perspective: Am. Inst. Mining Metall. Petroleum Engineers, Annual Meeting 1976, Preprint 76-I-85, 20 p.

Heinrichs, W. E., Jr., and Thurmond, R. E., A case history of the geophysical discovery of the Pima mine, Pima County, Arizona, *in* Geophysical case histories: Tulsa, Oklahoma, Soc. Explor. Geophysicists, v. 2, p. 600–612.

Henderson, J. F., and Borwn, I. C., 1966, Geology and structure of the Yellowknife greenstone belt, District of MacKenzie: Canada Geol. Survey Bull. 141, 87 p.

Hernon, R. M., and Jones, W. R., 1968, Ore deposits of the Central mining district, Grant County, New Mexico, *in* Ridge, J. D., ed., Ore deposits of the United States, 1933–1967: New York, Am. Inst. Mining Metall. Petroleum Engineers, p. 1211–1238.

Hewett, D. F., 1972, Manganite, hausmannite, braunite: features, modes of origin: Econ. Geology, v. 67, p. 83–102.

Heyl, A. V., 1964, Oxidized zinc deposits of the United States, pt. 3, Colorado: U.S. Geol. Survey Bull. 1135-C, 98 p.

Heyl, A. V., 1968, The Upper Mississippi Valley base metal district, *in* Ridge, J. D., ed., Ore deposits of the United States, 1933–1967: New York, Am. Inst. Mining Metall. Petroleum Engineers, p. 431–459.

Heyl, A. V., Landis, G. P., and Zartman, R. E., 1974, Isotopic evidence for the origin of Mississippi Valley type mineral deposits—a review: Econ. Geology, v. 69, p. 992–1006.

Hobbs, S. W., Griggs, A. B., Wallace, R. E., Campbell, A. B., 1965, Geology of the Coeur d'Alene district, Shoshone County, Idaho: U.S. Geol. Survey Prof. Paper 478, 139 p.

Hodgson, R. A., Gay, S. P., and Benjamins, J. Y., eds., 1976, Proceedings, First Internat. Conference on the new basement tectonics, Salt Lake City, June 3–7, 1974: Salt Lake City, Utah Geol. Assoc. Publ. No. 5, 636 p.

Hoek, E., and Bray, J., 1974, Rock slope engineering: London, Inst. Mining Metall., 309 p.

Hoffman, Arnold, 1947, Free gold: New York, Associated Book Service.

Hohne, F. C., 1963, Production geology methods at the Kermac mines: New Mexico Bur. Mines Mem. 15, p. 248–255.

Hollister, V. F., 1975a, The porphyry molybdenum deposit of Compaccha, Peru, and its geologic setting: Mineralium Deposita, v. 10, no. 2, p. 141–152.

Hollister, V. F., 1975b, An appraisal of the nature and source of porphyry copper deposits: Minerals Sci. Eng., v. 7, no. 3, p. 225–233.

Hollister, V. F., Potter, R. P., and Barker, A. L., 1974, Porphyry type deposits in the Appalachian orogen: Econ. Geology, v. 69, p. 618–630.

Hope, R. B., 1971, Engineering management of the Bougainville project: Inst. Civil Engineers Australia, Civil Eng. Trans., v. 13, no. 1, p. 45–52.

Horn, D. R., ed., 1972, Papers from a conference on ferromanganese deposits on the ocean floor, Palisades, New York, Jan. 20–22, 1972: Washington, National Science Foundation.

Hornbrook, E. H. W., 1969, Biogeochemical prospecting for molybdenum in west-central British Columbia: Canada Geol. Survey Paper 68-56, 41 p.

Hornsnail, R. F., and Fox, P. E., 1974, Geochemical exploration in permafrost terrain with particular reference to the Yukon Territory: Canadian Mining Metall. Bull., v. 67, p. 741, p. 56–60.

Hosking, K. F. G., 1963, Geology, mineralogy, and paragenesis of the Mount Pleasant tin deposits: Canadian Mining Jour., v. 84, no. 4, p. 95–102.

Hosking, K. F. G., 1965, The search for tin: Mining Mag., v. 113, p. 261–273, 308–383, and 448–461.

Hotz, P. E., and Willden, R., 1964, Geology and mineral resources of the Osgood Mountains quadrangle, Humboldt County, Nevada: U.S. Geol. Survey Prof. Paper 431, 128 p.

Howarth, R. J., 1973, The pattern recognition problem in applied geochemistry, in Jones, M. J., ed., Geochemical exploration 1972: London, Inst. Mining Metall., p. 259–274.

Hudson, T., Askevold, G., and Plafker, G., 1975, A computer-assisted procedure for information processing of geologic data: U.S. Geol. Survey Jour. Res., v. 3, no. 3, p. 369–375.

Huff, L. C., 1970, A geochemical study of alluvium-covered copper deposits in Pima County, Arizona: U.S. Geol. Survey Bull. 1312-C.

Huff, L. C., and Marranzino, A. P., 1961, Geochemical prospecting for copper deposits hidden beneath alluvium in the Pima district, Arizona: U.S. Geol. Survey Prof. Paper 424-B, p. B308–B310.

Hughes, F. E., and Munro, D. L., 1965, Uranium deposits at Mary Kathleen, in McAndrew J., ed., Geology of Australian ore deposits, ed. 2: Melbourne, Australasian Inst. Mining Metall., p. 256–263.

Hunkin, G. G., 1975, The environmental impact of solution mining for uranium: Mining Cong. Jour., v. 61, no. 10, p. 27.

Hunt, T. S., 1873, The geognostical history of the metals: Am. Inst. Mining Metall. Engineers Trans., v. 1, p. 331–342.

Hutchinson, C. S., 1974, Laboratory handbook of petrographic techniques: New York, John Wiley & Sons, 527 p.

Hutchinson, R. W., 1973, Volcanogenic sulfide deposits and their metallogenic significance: Econ. Geology, v. 68, p. 1223–1247.

Hutchinson, R. W., and Hodder, R. W., 1972, Possible tectonic and metallogenic relationships between porphyry copper and massive sulfide deposits: Canadian Mining Metall. Bull., v. 65, no. 2, p. 34–39.

Hutchison, W. W., ed., 1975, Computer-based systems for geological field data: Canada Geol. Survey Paper 74-63, 100 p.

Hyatt, D. E., 1973, The environmental impact statement: a new requirement in planning the mining operation: Colorado School Mines Mineral Inf. Bull., v. 16, no. 3, 10 p.

Ineson, P. R., 1969, Trace-element aureoles in limestone wallrocks adjacent to lead-zinc-barite-fluorite mineralization in the northern Pennine and Derbyshire ore fields: Inst. Mining Metall. (London), v. 78, p. B29–B40.

Information Canada, 1975, Toward a mineral policy for Canada—opportunities for choice: Ottawa, 55 p.

Ingham, A. E., ed., 1975, Sea surveying: New York, John Wiley & Sons, 548 p.

Institution of Mining and Metallurgy (London), 1965, Symposium on opencast mining, quarrying, and alluvial mining: London, 772 p.

Institution of Mining and Metallurgy (London), 1974, Geological, mining, and metallurgical sampling: London.

International Atomic Energy Agency, 1973, Uranium exploration methods: Vienna, 320 p.

Jacobsen, J. B. E., 1975, Copper deposits in space and time: Minerals Sci. Eng., v. 7, no. 4, p. 337–371.

Jaeger, J. C., and Cook, N. G. W., 1969, Fundamentals of rock mechanics: London, Methuen, 513 p.

James, A. H., 1946, Profile technique useful in mapping stope geology: Eng. Mining Jour., v. 147, no. 11, p. 74–75.

James, H. L., and Sims, P. K., 1973, Precambrian iron formations of the world: Econ. Geology, v. 68, p. 913–1173.

James, H. L., 1966, Chemistry of the iron-rich sedimentary rocks: U.S. Geol. Survey Prof. Paper 440-W, 61 p.

Jenkins, J. C., 1969, Practical applications of well logging to mine design: Am. Inst. Mining Metall. Petroleum Engineers, Preprint 69-F-73, p. 11.

Jenks, W. F., 1975, Origins of some massive pyritic ore deposits in western Europe: Econ. Geology, v. 70, p. 488–498.

Jonasson, I. R., and Allen, R. J., 1973, Snow: a sampling medium in hydrogeochemical prospecting in temperate and permafrost regions, in Jones, M. J., ed., Geochemical exploration 1972: London, Inst. Mining Metall., p. 161–176.

Jones, M. J., ed., 1973a, Geochemical exploration 1972: London, Inst. Mining Metall., 458 p.

Jones, M. J., ed., 1973b, Prospecting in areas of glacial terrain: London, Inst. Mining Metall., 138 p.

Jones, M. J., ed., 1975a, Prospecting in areas of glacial terrain 1975: London, Inst. Mining Metall., 154 p.

Jones, M. J., ed., 1975b, Minerals and the environment: London, Inst. Mining Metall., 804 p.

Joralemon, Ira B., 1928, The weakest link; or saving time in a mine examination: Eng. Mining Jour., v. 125, p. 536–540.

Joralemon, P., 1967, The fifth dimension in ore research: Am. Inst. Mining Metall. Petroleum Engineers, SME Preprint 67-I-302, 18 p.

Joralemon, P., 1975, The ore finders: Mining Eng., v. 27, no. 12, p. 32–35.

Joubin, F. R., 1954, Uranium deposits of the Algoma district: Canadian Inst. Mining Metall. Trans., v. 57, p. 431–437.

Journel, A. G., 1973, Geostatistics and sequential exploration: Mining Eng., v. 25, no. 10, p. 44–48.

Journel, A. G., 1974, Geostatistics for conditional simulation of ore bodies: Econ. Geology, v. 69, p. 673–687.

Kaas, L. M., 1973, Project control techniques, *in* Cummings, A. B., and Given, I. A., eds., SME Mining engineering handbook: New York, Am. Inst. Mining Metall. Petroleum Engineers, sec. 30.6, p. 30–42—30–48.

Kartashov, I. P., 1971, Geological features of alluvial placers: Econ. Geology, v. 66, p. 879–885.

Kelley, W. C., 1958, Topical study of lead-zinc gossans: New Mexico Bur. Mines and Mineral Resources Bull. 46, p. 35.

Kerr, J. W., 1974, Tips on organizing arctic geological field work: Canada Geol. Survey Paper 74-12, 12 p.

Kihlstedt, P. G., 1975, Energy and mineral exploitation techniques, lecture notes, Swedish Mining Mission to the U.S., May 3–22, 1975: Stockholm, Swedish Export Council, mimeographed, 10 p.

King, C. A. M., 1959, Beaches and coasts: New York, St. Martin's Press, 403 p.

Kirkemo, H., Anderson, C. A., and Creasey, S. C., 1956, Investigation of molybdenum deposits in the conterminous United States: U.S. Geol. Survey Bull. 1182-E, 90 p.

Klingmueller, L. M. L., 1971, Evaluating different method of geological mapping: Mines Mag. (Colorado), Sept., p. 6–7.

Klovan, J. E., 1975, R- and Q-mode factor analysis, *in* McCammon, R. S., ed., Concepts in geostatistics: New York, Springer-Verlag, p. 21–69.

Knight, C. L., 1957, Ore genesis—the source bed concept: Econ. Geology, v. 52, p. 808–817; discussion, 1958, v. 53, p. 219, 339–340, 493, 494, 622–625, 890–893; discussion, 1959, v. 54, p. 745–748.

Koch, G. S., and Link, R. F., 1970–1971, Statistical analysis of geological data: New York, John Wiley & Sons, 2 vols.

Koch, G., and Link, R. F., 1972, Sample preparation variability in diamond drill core from the Homestake mine, South Dakota: U.S. Bur. Mines Rept. Inv. 7677, 15 p.

Koch, G. S., Schuenemeyer, J. H., and Link, R. F., 1974, A mathematical model to guide the discovery of ore bodies in a Coeur d'Alene lead-silver mine: U.S. Bur. Mines Inf. Circ. 7989, 43 p.

Koschman, A. H., and Bergendahl, M. H., 1968, Principal gold producing districts of the United States: U.S. Geol. Survey Prof. Paper 610, 283 p.

Koulomzine, T., and Dagenais, R. W., 1959, Statistical determination of the chances of success in mineral exploration in Canada: Canadian Mining Jour., v. 80, April, p. 107–110.

Krauskopf, K. B., 1967, Introduction to geochemistry: New York, McGraw-Hill Book Co.

Kreiter, V. M., 1968, Geological prospecting and exploration: Moscow, Mir publishers, 383 p.

Krynine, D., and Judd, W. R., 1957, Principles of engineering geology and geotechnics: New York, McGraw-Hill Book Co., 730 p.

Kuhn, J. G., and Graham, J. D., 1972, Application of correlation analysis to drilling programs: a case study—technical notes: Am. Inst. Mining Metall. Petroleum Engineers Trans., v. 252, no. 2, p. 140–145.

Kunasz, I. A., 1970, Geology and geochemistry of the lithium deposit in Clayton Valley, Esmeralda County, Nevada: unpublished Ph.D. thesis, The Pennsylvania State University, University Park, 114 p.

Kunzendorf, H., 1973, Non-destructive determination of metals in rocks by radioisotope X-ray fluorescence instrumentation, in Jones, M. J., ed., 1973, Geochemical exploration 1972: London, Inst. Mining Metall., p. 401–414.

Kuo, S., and Follinsbee, R. E., 1974, Lead isotope geology of mineral deposits spatially related to the Tintina trench, Yukon Territory: Econ. Geology, v. 69, p. 806–813.

Kutina, J., 1963, Symposium, problems of postmagmatic ore deposition: Prague, Czechoslavakia Geol. Survey, Vol. 1, 588 p.

Kvalheim, A., ed., 1967, Geochemical prospecting in Fennoscandia: New York, John Wiley & Sons, 350 p.

Lahee, F. H., 1961, Field geology, ed. 6 New York, McGraw-Hill Book Co., 926 p.

Lasky, S. G., 1950, Mineral resource appraisal by the U.S. Geological Survey: Colorado School School Mines Quart., v. 45, no. 1-A, p. 1–27.

Lathram, E. H., Tailleur, I. L., Patton, W. W., Jr., and Fischer, W. A., 1973, Preliminary geologic application of ERTS imagery in Alaska, in Symposium on significant results obtained from Earth Resources Technology Satellite-1: NASA Spec. Pub. SP-327, v. 1, sec. A, p. 257–264.

Laurin, A. F., Sharma, K. N. M., Wynne-Edwards, H. R., and Franconi, A., 1972, Application of data processing techniques in the Grenville province, Quebec, Canada, in Sec. 16, Computer-based storage, retrieval and processing of geological information, Internat. Geol. Cong., 24th, Canada, 1972: Ottawa, p. 22–35.

Laznicka, P., and Wilson, H. D. B., 1972, The significance of a copper-lead line in metallogeny, in Sec. 4, Mineral deposits, Internat. Geol. Cong., 24th, Canada, 1972: Ottawa, p. 24–37.

Leake, R. C., and Aucott, J. W., 1973, Geochemical mapping and prospecting by use of rapid automatic X-ray fluorescence analysis of panned concentrates, in Jones, M. J., ed., Geochemical exploration 1972: London, Inst. Mining Metall., p. 389–400.

Leaming, G. F., Lacy, W. C., and Skelding, F. H., 1969, Nonfuel mineral resources and the public lands—report prepared for the United States Public Land Law Review Commission: Tucson, University of Arizona, v. 2, p. 308–310.

Learned, R. E., and Boissen, R., 1973, Gold—a useful pathfinder element in the search for porphyry copper deposits in Puerto Rico, in Jones, M. J., ed., Geochemical exploration 1972: London, Inst. Mining Metall., p. 93–103.

Lefond, S. J., ed., 1975, Industrial minerals and rocks, ed. 4 New York, Am. Inst. Mining Metall. Petroleum Engineers, 1360 p.

Leggett, R. F., 1967, Soil: its geology and use: Geol. Soc. America Bull., v. 78, p. 1433–1460.

Leggett, R. F., 1973, Cities and geology: New York, McGraw-Hill Book Co., 624 p.

Leney, G. W., 1966, Field studies in iron ore geophysics, in Mining geophysics, Vol. 1, Case histories: Tulsa, Oklahoma, Society of Exploration Geophysicists, p. 391–417.

Leonard, J. W., 1973, Mill design—coal, *in* Cummins, A. B., and Given, I. A., eds., SME Mining engineering handbook: New York, Am. Inst. Mining Metall. Petroleum Engineers, Sec. 28.8, p. 28-44—28-49.

Leopold, L. B., Clarke, F. E., Hanshaw, B. B., and Balsley, J. R., 1971, A procedure for evaluating environmental impact: U.S. Geol. Survey Circ. 645, 13 p.

Leopold, L. B., Wolman, M. G., and Miller, J. P., 1964, Fluvial processes in geomorphology: San Francisco, W. H. Freeman Co., 522 p.

Lepeltier, C., 1969, A simplified statistical treatment of geochemical data by graphical representation: Econ. Geology, v. 64, p. 538-550.

Lepp, H., ed., 1975, Geochemistry of iron: New York, Halstead Press Division, John Wiley & Sons, 464 p.

Le Roy, L. W., Le Roy, D. O., and Raese, J. W., eds., 1977, Subsurface geology—petroleum, mining, construction: Golden, Colorado, School of Mines, 800 p.

Lessing, P., and Smosna, R. A., 1975, Environmental impact statements—worthwhile or worthless?: Geology, May, p. 241-242.

Levinson, A. A., 1974, Introduction to exploration geochemistry: Calgary, Applied Publishing Company, 608 p.

Levy, E., 1971, Metallogenesis of Central America: Am. Inst. Mining Metall. Petroleum Engineers, SME Preprint 71-S-1, 20 p.

Lewis, F. M., and Bhappu, R. S., 1975, Evaluating mining ventures via feasibility studies: Mining Eng., v. 27, no. 10, p. 50-55.

Li, K. C., and Wang, C. Y., 1955, Tungsten, ed. 3: New York, Reinhold Publishing Co., 506 p.

Lindgren, W., 1911, The Tertiary gravels of the Sierra Nevada of California: U.S. Geol. Survey Prof. Paper 73.

Lindgren, W., 1913, Mineral deposits: New York, McGraw-Hill Book Co.

Lindgren, W., 1933, Mineral deposits, ed. 4: New York, McGraw-Hill Book Co.

Lobeck, A. K., 1958, Block diagrams: Amherst, Mass., Emerson-Trussell Book Co., 212 p.

Locke, A., 1926, Leached outcrops as guides to copper ore: Baltimore, Williams and Wilkins, 175 p.

Loofbourow, R. L., 1973, Ground water and ground-water control, *in* Cummins, A. B., and Given, I. A., eds., SME Mining engineering handbook: New York, Am. Inst. Mining Metall. Petroleum Engineers, sec. 26, p. 26-2—26-55.

Lovering, T. S., 1949, Rock alteration as a guide to ore, East Tintic district, Utah: Econ. Geology Mon. 1, 65 p.

Lovering, T. S., and Morris, H. T., 1960, The Chief Oxide-Burgin area discoveries, East Tintic district, Utah; a case history; pt. I, U.S. Geological Survey studies and exploration: Econ. Geology, v. 55, p. 1116-1147.

Lowell, J. D., 1968, Geology of the Kalamazoo orebody, San Manuel district, Arizona: Econ. Geology, v. 63, no. 6, p. 645-654.

Lowell, J. D., 1974, Regional characteristics of porphyry copper deposits of the Southwest: Econ. Geology, v. 69, p. 601-617.

Lowell, J. D., and Guilbert, J. M., 1970, Lateral and vertical alteration-mineralization zoning in porphyry ore deposits: Econ. Geology, v. 65, p. 373-408.

Ludeke, K. L., 1973, Vegetative stabilization of tailings disposal berms: Mining Cong. Jour., v. 59, no. 1, p. 32-39.

Lukanuski, J. N., 1975, Locomotive-type post-ore fanglomerate as exploration guides for porphyry copper deposits: New York, Am. Inst. Mining Metall. Petroleum Engineers, SME Preprint 75-S-35.

Lukashev, K. I., 1974, The problem of technical progress and mineral resources: Impact of Science on Society, v. 24, no. 3, p. 225–235.

Lynd, L. E., and Lefond, S. J., 1975, Titanium minerals, in Lefond, S. J., ed., Industrial minerals and rocks: New York, Am. Inst. Mining Metall. Petroleum Engineers, p. 1149–1208.

Mackenzie, G. S., 1957, History of mining exploration, Bathurst-Newcastle district, New Brunswick: Canadian Mining Metall. Bull., v. 51, no. 551, p. 156–161.

Mackin, J. H., and Schmidt, D. L., 1956, Uranium and thorium-bearing minerals in placer deposits in Idaho, in Page, L. R., Stocking, H. E., and Smith, H. B., compilers, Contributions to the geology of uranium and thorium—United Nations Internat. Conference on Peaceful Uses of Atomic Energy: U.S. Geol. Survey Prof. Paper 300, p. 375–380.

MacNamara, P. M., 1968, Rock types and mineralization at Panguna porphyry copper prospect, upper Kaverong Valley, Bougainville island: Australasian Inst. Mining Metall. Proc., no. 228, p. 71–79.

Maddex, P. J., 1972, How "changing" transportation can affect mining profits: New York, Am. Inst. Mining Metall. Petroleum Engineers, SME Preprint 72-H-78, 10 p.

Mahtab, M. A., Bolstad, D. D., and Kendorski, F. S., 1973, Analysis of the geometry of fractures in San Manuel copper mine, Arizona: U.S. Bur. Mines Rept. Inv. 7715, 24 p.

Malan, R. C., 1968, The uranium mining industry and geology of the Monument Valley and White Canyon districts, Arizona and Utah, in Ridge, J. D., ed., Ore deposits of the United States 1933–1967: New York, Am. Inst. Mining Metall. Petroleum Engineers, p. 798.

Malmqvist, D., and Parasnis, D. S., 1972, Aitik: geophysical documentation of a third-generation copper deposit in North Sweden: Geoexploration, v. 10, p. 149–200.

Mann, C. J., 1970, Randomness in nature: Geol. Soc. America Bull., v. 81, no. 1, p. 95–104.

Marsden, R. W., 1975, Politics, minerals, and survival: Madison, University of Wisconsin Press, 136 p.

Martin-Kaye, P., 1974, Application of side-looking radar in earth resource surveys, in Barrett, E. C., and Curtis, L. F., Environmental remote sensing: Applications and achievements: London, Edward Arnold Publishers Ltd., p. 31–48.

Matheron, G., 1963, Principles of geostatistics: Econ. Geology, v. 58, p. 1246–1266.

Matheron, G., 1971, The theory of regionalized variables and its applications: Fontainebleau, France, Les cahiers du centre de morphologie mathématique, no. 5, 211 p.

Matulich, A., Amos, A. C., Walker, R. R., and Watkins, J. J., 1974, The Ecstall story—the geology department: Canadian Mining Metall. Bull., v. 67, no. 747, p. 56–63.

Mayo, E. B., 1958, Lineament tectonics and some ore districts of the Southwest: Mining Eng., v. 9, no. 11, p. 1169–1175.

Mayo, E. B., 1976, Intrusive fragmental rocks directly or indirectly of igneous origin: Arizona Geol. Soc. Digest, v. 10, p. 347–430.

McCammon, R. B., 1968, The dendrograph, a new tool for correlation: Geol. Soc. America Bull., v. 79, p. 1663–1670.

McCarthy, J. H., 1972, Mercury vapor and other components in the air as guides to ore deposits: Jour. Geochem. Explor., v. 1, p. 143–162.

McCulloch, C. M., Jeran, P. W., and Sullivan, C. D., 1975, Geologic investigations of underground coal mining problems: U.S. Bur. Mines Rep. Inv. 8022, 30 p.

McGregor, K., 1967, The drilling of rock: London, C. R. Books Ltd. (MacLaren), 306 p.

McKelvey, V. E., 1973, Mineral potential of the United States, in Cameron, E. N., ed., The mineral position of the United States 1975–2000: Madison, University of Wisconsin Press, p. 67–82.

McKinstry, H. E., 1948, Mining geology: Englewood Cliffs, N.J., Prentice Hall, 680 p.

McLeod, A., 1973, Blackwater mine, Australia: Mining Cong. Jour., v. 59, no. 1, p. 56–60.

McQuiston, F. W., Jr., and Hernlund, R. W., 1965, Newmont's Carlin gold project: Mining Cong. Jour., v. 52, no. 11, p. 26–30, 32, 38–39.

Meadows, D. H., Meadows, D. L., Randers, J., and Behrens, W. W. III, 1972, The limits to growth: New York, Universe Books, 205 p.

Mehrtons, M. B., Tooms, J. S., and Troup, A. G., 1973, Some aspects of geochemical dispersion from base metal mineralization within glaciated terrain in Norway, North Wales, and British Columbia, Canada, in Jones, M. J., ed., Geochemical exploration 1972: London, Inst. Mining Metall., p. 105–115.

Meinzer, O. E., ed., 1942, Hydrology: New York, McGraw-Hill Book Co., 712 p.

Mendelsohn, F., ed., 1961, The geology of the northern Rhodesian copperbelt: London, MacDonald and Co. Ltd.

Merriam, D. F., ed., 1976, Random processes in geology: Berlin, Springer-Verlag, 168 p.

Metz, R. A., 1972, Rapid geologic mapping in large tonnage open pit mines: Mines Mag. (Colorado), Aug., p. 10–12.

Meyer, C. S., 1968, Ore deposits at Butte, Montana, in Ridge, J. D., ed., Ore deposits of the United States 1933–1967: New York, Am. Inst. Mining Metall. Petroleum Engineers, p. 1373–1416.

Meyer, W. T., and Peters, R. G., 1973, Evaluation of sulfur as a guide to buried sulfide deposits in the Notre Dame Bay area, Newfoundland, in Jones, M. J., ed., Prospecting in areas of glacial terrain: London, Inst. Mining Metall., p. 55–66.

Miller, L. J., 1976, Corporations, ore discovery, and the geologist: Econ. Geology, v. 71, p. 836–847.

Miller, R. J. M., 1961, Wall-rock alteration at the Cedar Bay mine, Chibougamau district, Quebec: Econ. Geology, v. 56, p. 321–330.

Mills, J. W., and Eyrich, H. T., 1966, The role of unconformities in the localization of epigenetic mineral deposits in the United States and Canada: Econ. Geology, v. 61, p. 1232–1257.

Mitcham, T. W., 1974, Origin of breccia pipes: Econ. Geology, v. 69, p. 412–413.

Mitchell, A. H. G., and Garson, M. S., 1976, Mineralization at plate boundaries: Minerals Sci. Eng., v. 8, no. 2, p. 129–169.

Mookherjee, A., 1970, Dykes, sulfide deposits, and regional metamorphism: criteria for determining their time relationship: Mineralium Deposita, v. 5, no. 2, p. 120–144.

Moore, J. M., 1975, A mechanical interpretation of vein and dyke systems of the S.W. England orefield: Mineralium Deposita, v. 10, p. 374–388.

Morgan, D. A. O., 1969, A look at the economics of mineral exploration, in Jones, M. J., ed., Proceedings, Commonwealth Mining Metall. Cong., 9th: London, Inst. Mining Metall., v. 2, p. 305–365.

Morris, H. T., and Lovering, T. S., 1952, Supergene and hydrothermal dispersions of heavy metals in wall rocks near ore bodies, Tintic district, Utah: Econ. Geology, v. 47, no. 7, p. 685–716.

Morrissey, C. J., and Whitehead, D., 1969, Origin of the Tynagh residual orebody, Ireland, in Jones, M. J., ed., Proceedings, Commonwealth Mining Metall. Cong., 9th: London, Inst. Mining Metall., v. 2, p. 131–145.

Motica, J. E., 1968, Geology and mineral deposits in the Uravan mineral belt, southwestern Colorado, in Ridge, J. D., ed., Ore deposits of the United States, 1933–1967: New York, Am. Inst. Mining Metall. Petroleum Engineers, v. 1, p. 805–814.

Muir, W. L. G., ed., 1976, Coal exploration: Proceedings of the First International Coal Exploration Symposium, London, England: San Francisco, World Coal Publishers, 664 p.

Mulligan, R., 1965, Geology of Canadian lithium deposits: Canada Geol. Survey Econ. Geol. Rept. 21, 131 p.

Murad, E., 1974, Hydrothermal alteration of granitic rocks and its possible bearing on the genesis of mineral deposits in the southern Black Forest, Germany: Econ. Geology, v. 69, p. 532–544.

Murchison, D. G., and Westoll, T. S., eds., 1968, Coal and coal bearing strata: London, Oliver and Boyd.

Naldrett, A. J., 1973, Nickel sulfide deposits—their classification and genesis, with special emphasis on deposits of volcanic association: Canadian Mining Metall. Bull., v. 66, no. 739, p. 45–63.

Newman, D. G., 1976, Engineering economic analysis: San Jose, California, The Engineering Press, 469 p.

Newhouse, W. H., 1942, Ore deposits as related to structural features: Princeton, N.J., Princeton University Press, 280 p.

Niccolini, P., 1970, Gîtologie des concentrations minérales stratiformes: Paris, Gauthier-Villars Editeurs, 792 p.

Nichol, A. W., 1975, Physicochemical methods of mineral analysis: New York, Plenum Press.

Nichol, I., 1973, The role of computerized data systems in geochemical exploration: Canadian Mining Metall. Bull., v. 66, no. 729, p. 59–68.

Nichol, I., Garrett, R. G., and Webb, J. S., 1969, The role of some statistical and mathematical methods in the interpretation of regional geochemical data: Econ. Geology, v. 64, p. 204–220.

Niggli, P., 1954, Rocks and mineral deposits: San Francisco, W. H. Freeman and Co., 559 p.

Nilsson, Gunnar, 1973, Nickel prospecting and discovery of the Mjövattnet mineralization, northern Sweden: a case history of the use of combined techniques in drift-covered glaciated terrain, *in* Jones, M. J., ed., Prospecting in areas of glacial London, Inst. Mining Metall., p. 97–109.

Noble, J. A., 1950, Ore mineralization in the Homestake gold mine, Lead, South Dakota: Geol. Soc. America Bull., v. 61, p. 221–251.

Noble, J. A., 1976, Metallogenic provinces of the cordillera of western North America and South America: Mineralium Deposita, v. 11, no. 2, p. 219–233.

Nolan, T. B., 1935, The underground geology of the Tonopah mining district, Nevada: University of Nevada Bull., v. 29, no. 5, 49 p.

Norman, J. W., 1976, Photogeological fracture trace analysis as a subsurface exploration technique: Inst. Mining Metall. (London) Trans., v. 85, p. B52–B62.

Norton, J. S., 1973, Lithium, cesium, and rubidium—the rare alkali metals, *in* Brobst, D. A., and Pratt, W., eds., United States mineral resources: U.S. Geol. Survey Prof. Paper 820, p. 365–378.

Notley, K. R., and Wilson, E. B., 1975, Three-dimensional mine drawings by computer graphics: Canadian Mining Metall. Bull., v. 68, no. 754, p. 60–64.

Obert, L. A., 1973, Rock mechanics, *in* Cummins, A. B., and Given, I. A., eds., SME Mining engineering handbook: New York, Am. Inst. Mining Metall. Petroleum Engineers, Sec. 6.2, p. 6-13—6-52.

Obert, L. A., and Duvall, W. I., 1967, Rock mechanics and the design of structures in rock: New York, John Wiley & Sons.

Obial, R. C., and James, C. H., 1973, Use of cluster analysis in geochemical prospecting, with particular reference to southern Derbyshire, England, *in* Jones, M. J., ed., Geochemical exploration 1972: London, Inst. Mining Metall., p. 237–257.

O'Brian, D. T., 1969, Financial analysis applications in mineral exploration and development: New York, Am. Inst. Mining Metall. Petroleum Engineers, SME Preprint 69-AR-76.

Olade, M. A., and Fletcher, W. K., 1975, Primary dispersion of rubidium and strontium around porphyry copper deposits, Highland Valley, British Columbia: Econ. Geology, v. 70, p. 15–21.

O'Neil, T. J., 1974a, The minerals depletion allowance: its importance to nonferrous metal mining: Mining Eng., v. 26, no. 10, p. 61–64.

O'Neil, T. J., 1974b, The minerals depletion allowance: its effect on future supply and financing: Mining Eng., v. 26, no. 11, p. 39–41.

Ovchinnikov, L. N., Sokolov, V. A., Fridman, A. I., and Yanitskii, I. N., 1973, Gaseous geochemical methods in structural mapping and prospecting for ore deposits, *in* Jones, M. J., ed., Geochemical exploration 1972: London, Inst. Mining Metall., p. 177–182.

Page, N. J., and Creasey, S. C., 1975, Ore grade, metal production, and energy: U.S. Geol. Survey, Jour. Res., v. 3, no. 1, p. 9–13.

Parasnis, D. S., 1973, Mining geophysics: Amsterdam, Elsevier Scientific Publishing Company, 395 p.

Park, C. F., Jr., 1972, The iron ore deposits of the Pacific Basin: Econ. Geology, v. 67, p. 339–349.

Park, C. F., Jr., 1975, Earthbound: San Francisco, Freeman, Cooper and Company, 279 p.

Park, C. F., Jr., and MacDiarmid, R. A., 1975, Ore deposits, ed. 3: New York, McGraw-Hill Book Co., Inc., 522 p.

Park, C. F., Jr., and Mahrholz, W. W., 1967, Organization and costs of exploration in Latin America: Paper presented at the Southwest Pacific Minerals Conference, Am. Inst. Mining Metall. Petroleum Engineers, May 1967, typescript.

Parks, R. D., 1957, Examination and evaluation of mineral property, ed. 4: Cambridge, Mass., Addison-Wesley Press, Inc., 504 p.

Parr, C. J., and Ely, N., 1973, Mining law, in Cummins, A. B., and Given, I. A., eds., SME Mining engineering handbook: New York, Am. Inst. Mining Metall. Petroleum Engineers, sec. 2, p. 2-2—2-54.

Parslow, R. D., Prowse, R. W., and Green, R. E., 1969, Computer graphics: New York, Plenum Publishing Corporation, 247 p.

Paterson, N. R., 1966, Mattagami Lake mines—a discovery by geophysics, in Mining geophysics, Vol. 1, Case histories: Tulsa, Soc. Explor. Geophysicists, p. 185-196.

Patterson, J. A., 1959, Estimating ore reserves follows logical steps: Eng. Mining Jour., v. 160, no. 9, p. 111-115.

Pelissonnier, H., 1972, Les dimensions des gisements de cuivre du monde: Bur. Recherches Géol. et Minières Mem. 57.

Pelton, W. H., and Hallof, P. G., 1972, Applied potential method in the search for massive sulfides at York Harbour, Newfoundland: Am. Inst. Mining Metall. Petroleum Engineers Trans., v. 252, p. 121-124.

Pelton, W. H., and Smith, P. K., 1976, Mapping porphyry copper deposits in the Philippines with IP: Geophysics, v. 41, no. 1, p. 106-122.

Pemberton, R. H., 1962, Airborne EM in review: Geophysics, v. 27, no. 5, p. 691-713.

Perel'man, A. I., 1967, The geochemistry of epigenesis (translated from the Russian by N. N. Kohanowski): New York, Plenum Press, 266 p.

Perry, A. J., 1968, Organization and costs for mineral exploration in Southwest U.S.A.: Paper presented to the American Mining Congress, Annual Meeting, San Francisco, Oct. 20, typescript, 12 p.

Peterson, A., 1967, Ranstad—a new uranium processing plant: Panel on Uranium Recovery, Proc., Vienna, Internat. Atomic Energy Agency.

Petrascheck, W. E., 1968, Kontinental Verschiebung und Erzprovinzen: Mineralium Deposita, v. 3, p. 56-65.

Petrascheck, W. E., ed., 1973a, Metallogenetic and geochemical provinces: New York, Springer-Verlag, 183 p.

Petrascheck, W. E., 1973b, Orogene und kratogene Metallogenese: Geol. Rundschau, v. 62, no. 3, p. 617-626.

Petrascheck, W. E., 1976, Mineral zoning and plate tectonics in the Alpine-Mediterranean area, in Strong, D. F., ed., Metallogeny and plate tectonics: Toronto, Geol. Assoc. of Canada, Special Paper 14, p. 351-358.

Pettijohn, F. J., 1957, Sedimentary rocks, ed. 2: New York, Harper & Row, 718 p.

Pfleider, E. P., ed., 1968, Surface mining: New York, Am. Inst. Mining Metall. Petroleum Engineers, 1061 p.

Pfleider, E. P., and Weaton, G. F., 1973, Production methods and economics—ferrous (iron) ore mining, in Cummins, A. B., and Given, I. A., eds., SME Mining engineering handbook: New York, Inst. Mining Metall. Petroleum Engineers, Sec. 17.4.1, p. 17-20—17-89.

Phillips, C. H., Cornwall, H. R., and Rubin, M., 1971, A Holocene orebody of copper oxides and carbonates at Ray, Arizona: Econ. Geology, v. 66, p. 495–498.

Phillips, W. J., 1973, 1974, Mechanical effects of retrograde boiling and its probable importance in the formation of some porphyry ore deposits: Inst. Mining Metall. (London) Trans., v. 82, p. B90–B98 and v. 83, p. B42–B43.

Phizackerley, P. H., and Scott, L. O., 1967, Major tar sand deposits of the world, in World Petroleum Conference, 7th, Mexico: Proc., v. 3, p. 551–571.

Picklyk, D. D., and Ridler, R. H., 1975, Computer graphics—an interactive approach to volcanic geochemistry and stratigraphy; illustrated by data from the Kirkland Lake area: Canadian Mining Metall. Bull., v. 68, no. 754; p. 44–51.

Pierce, W. G., 1944, Cobalt-bearing manganese deposits of Alabama, Georgia, and Tennessee: U.S. Geol. Survey Bull. 940-J, p. 265–285.

Pirkle, E. C., and Yoho, W. H., 1970, The heavy mineral orebody of Trail Ridge, Florida: Econ. Geology, v. 65, p. 17–30.

Popoff, C. C., 1966, Computing reserves of mineral deposits; principles and conventional methods: U.S. Bur. Mines Inf. Circ. 8283, 113 p.

Pošepný, F., 1894, The genesis of ore deposits: Inst. Mining Metall. Engineers Trans., v. 23, p. 197–369.

Pretorius, D. A., 1975, The depositional environment of the Witwatersrand goldfields; a chronological review of speculations and observation: Minerals Sci. Eng., v. 7, no. 1, p. 18–47.

Prinz, W. C., 1967, Geology and ore deposits, Phillipsburg, Montana: U.S. Geol. Survey Bull 1237, 66 p.

Pryor, E. J., 1965, Mineral processing, ed. 3: New York, Elsevier North-Holland.

Quin, B. F., Brooks, R. R., Boswell, C. R., and Painter, J. A. C., 1974, Biogeochemical exploration for tungsten at Barrytown, N.Z.: Jour. Geochem. Explor., v. 3, p. 43–51.

Radabaugh, R. E., Merchant, J. S., and Brown, J. M., 1968, Geology and ore deposits of the Gilman (Red Cliff, Battle Mountain) district, Eagle County, Colorado, in Ridge, J. D., ed., Ore deposits of the United States, 1933–1967: New York, Am. Inst. Mining Metall. Petroleum Engineers.

Ragan, D. M., 1973, Structural geology, an introduction to geometrical techniques, ed. 2: New York, John Wiley & Sons, 308 p.

Ramdohr, P., 1969, The ore minerals and their intergrowths, ed. 3: Oxford, Pergamon Press Ltd., 1174 p.

Ramović, M., 1968, Principles of metallogeny: Sarajevo, Yugoslavia, Geographical Institute, University of Sarajevo, 271 p.

Rayment, B. D., Davis, G. R., and Willson, J. D., 1971, Controls to mineralization at Wheal Jane, Cornwall: Inst. Mining Metall. (London) Trans., v. 80, p. B224–B237.

Regan, M. D., 1971, Management of exploration in the metals mining industry: unpublished M.S. thesis, Massachusetts Institute of Technology, Cambridge, 137 p.

Reedman, J. H., 1974, Residual soil geochemistry in the discovery and evaluation of the Butiriku carbonatite, southeast Uganda: Inst. Mining Metall. (London) Trans., v. 83, p. B1–B12.

Reeves, R. G., ed., 1975, Manual of remote sensing: Falls Church, Va., American Society of Photogrammetry, 2 vols., 2144 p.

Renfro, A. R., 1974, Genesis of evaporite-associated stratiform metalliferous deposits—a sabkha process: Econ. Geology, v. 69, p. 33–45.

Richard, D. J., and Walraven, F., 1975, Airborne geophysics and ERTS imagery: Minerals Sci. Eng., v. 7, no. 3, p. 234–278.

Ridge, J. D., ed., 1968, Ore deposits of the United States, 1933–1967: New York, Am. Inst. Mining Metall. Petroleum Engineers, 2 vols., 1880 p.

Robbins, J. C., 1973, Zeeman spectrometer for measurement of atmosphere mercury vapor, in Jones, M. J., ed., Geochemical exploration 1972: London, Inst. Mining Metall., p. 315–324.

Roberts, R. J., Radtke, A. S., and Coats, R. R., 1971, Gold-bearing deposits in north-central Nevada and southwestern Idaho: Econ. Geology, v. 66, p. 14–33.

Rocky Mountain Mineral Law Foundation, 1960, American law of mining: Albany, N.Y., Matthew Bender Company, 5 vols. with supplements.

Roddick, J. A., and Hutchison, W. W., 1972, A computer-based system for geological field data on the Coast Mountains Project, British Columbia, Canada, in Section 16. Computer-based storage, retrieval and processing of geological information, Internat. Geol. Cong., 24th, Canada, 1972: Ottawa, p. 36–46.

Roman, R. J., and Becker, G. W., 1973, Computer program for Monte Carlo economic evaluation of a mineral deposit: New Mexico Bur. Mines Mineral Resources Circ. 137, 23 p.

Romanowitz, C. M., Bennett, H. J., and Dare, W. L., 1970, Gold placer mining—placer evaluation and dredge selection: U.S. Bur. Mines Inf. Circ. 8462, 56 p.

Roorda, H. J., and Queneau, P. E., 1973, Recovery of nickel and cobalt from limonites by aqueous chlorination in sea water: Inst. Mining Metall. (London) Trans., v. 82, p. 679.

Roscoe, S. M., 1968, Huronian rocks and uraniferous conglomerates: Canada Geol. Survey Paper 68-40, 205 p.

Roscoe, W. E., 1971, Probability of an exploration discovery in Canada: Canadian Mining Metall. Bull., v. 64, no. 707, p. 134–137.

Rose, A. W., 1972, Statistical interpretation techniques in geochemical exploration: Am. Inst. Mining Metall. Petroleum Engineers Trans., v. 252, no. 3, p. 233–238.

Rose, E. R., 1965, Geology of titanium and titaniferous deposits of Canada: Canada Geol. Survey Econ. Geol. Rept. 25, 177 p.

Ross-Brown, D. M., and Atkinson, K. B., 1972, Terrestrial photogrammetry in open pits; 1. Description and use of the phototheodolite in mine surveying: Inst. Mining Metall. (London) Trans., v. 81, p. A205–213.

Ross-Brown, D. M., Wickens, E. H., and Markland, J. T., 1973, Terrestrial photogrammetry in open pits; 2. An aid to geologic mapping: Inst. Mining Metall. (London) Trans., v. 82, p. A115–130.

Routhier, P., 1963, Les gisements métallifères: Paris, Masson et Cie, 2 vols., 1282 p.

Routhier, P., 1976, A new approach to metallogenic provinces—the example of Europe: Econ. Geology, v. 71, p. 803–811.

Routhier, P., and associates, Laboratoire de Géologie Apliquée, Université de Paris, 1973, Some major concepts of metallogeny: Mineralium Deposita, v. 8, no. 3, p. 237–258.

Rudio, R. M., Reclamation for the exploration geologist in Montana: Am. Inst. Mining Metall. Petroleum Engineers Trans., v. 256, no. 4, p. 288–293.

Rui, I. J., 1973, Structural control and wall rock alteration at Killingdal mine, central Norwegian Caledonides: Econ. Geology, v. 68, p. 859–883.

Russell, M. J., 1975, Lithogeochemical environment of the Tynagh base-metal deposit, Ireland, and its bearing on ore deposition: Inst. Mining Metall. (London) Trans., v. 84, p. B128–B133.

Ryan, A. P., 1975, Mercury—U.S. demand good in '74; new producer in wings: Eng. Mining Jour., v. 176, no. 3, p. 139–141.

Ryan, A. P., 1976, Mercury—prices nosedived to a 25-year low in '75: Eng. Mining Jour., v. 177, no. 3, p. 153–154.

Ryan, G. B., 1972, Ranger 1: a case history, in Bowie, S. H. U., Davis, M., and Ostle, D., eds., Uranium prospecting handbook: London, Inst. Mining Metall., p. 296–300.

Rye, D. M., Doe, B. R., and Delevaux, M. H., 1974, Homestake gold mine, South Dakota—lead isotopes, mineralization ages, and sources of lead in ores of the northern Black Hills: Econ. Geology, v. 69, p. 814–822.

Saager, R., and Sinclair, A. J., 1974, Factor analysis of stream sediment geochemical data from the Mount Nansen area, Yukon Territory, Canada: Mineralium Deposita, v. 9, p. 243–252.

Saegart, W. E., Sell, J. D., and Kilpatrick, B. E., 1974, Geology and mineralization of La Caridad porphyry copper deposit, Sonora, Mexico: Econ. Geology, v. 69, p. 1060–1077.

Sainsbury, C. L., 1968, Tin and beryllium deposits in the central York Mountains, western Seward Peninsula, Alaska, in Ridge, J. D., ed., Ore deposits of the United States, 1933–1967: New York, Am. Inst. Mining Metall. Petroleum Engineers, p. 1555–1572.

Salsbury, M. H., Kerns, W. H., Fulkerson, F. B., and Branner, G. C., 1964, Marketing ores and concentrates of gold, silver, copper, lead, and zinc in the United States: U.S. Bur. Mines Inf. Circ. 8206, 148 p.

Sapozhnikov, D. G., ed., 1970, Manganese deposits of the Soviet Union: Jerusalem, Israel Program for Scientific Translations, 522 p.

Sarjeant, W. A. S., and Harvey, A. P., 1973, Stratigraphy (historical geology) including regional geology, in Wood, D. N., ed., Use of earth science literature: London, Butterworth and Company, p. 179–253.

Sawkins, F. J., 1972, Sulfide ore deposits in relation to plate tectonics: Jour. Geology, v. 80, p. 377–397.

Sawkins, F. J., and Rye, D. M., 1974, Relationship of Homestake-type gold deposits to iron-rich Precambrian sedimentary rocks: Inst. Mining Metall. (London) Trans., v. 83, p. B56–B59.

Sayers, R. W., Tippett, M. C., and Fields, E. D., 1968, Duval's new copper mines show complex geologic history: Mining Eng., v. 20, p. 55–62.

Schieferdecker, A. A. G., ed., 1959, Geological nomenclature: Delft, The Netherlands, Royal Geological and Mining Society of The Netherlands, 521 p.

Schmitt, H. A., 1936, On mapping underground geology: Eng. Mining Jour., v. 137, p. 558.

Schmitt, H. A., 1939, Outcrops of ore shoots: Econ. Geology, v. 34, p. 654–673.

Schmitt, H. A., 1966, The porphyry copper deposits in their regional setting, *in* Titley, S. R., and Hicks, C. L., eds., Geology of the porphyry copper deposits, southwestern North America: Tucson, The University of Arizona Press, p. 17–33.

Schmitt, R. G., 1976, Exploration for porphyry copper deposits in Pakistan using digital processing of Landsat-1 data: Jour. Research, U.S. Geol. Survey, v. 1, no. 4, p. 27–34.

Schneiderhöhn, H., 1941, Lehrbuch der Erzlagerstättenkunde: Jena, Gustav Fischer Verlag.

Schneiderhöhn, H., 1962, Erzlagerstätten Kurzvorlesung zur Einführung und Wiederholung. 4 Auflage: Stuttgart, Fischer-Verlag.

Schnellman, G. A., and Scott, B., 1970, Lead-zinc mining areas of Great Britain, *in* Jones, M. J., ed., Mining and petroleum geology: London, Mining and Metall. Cong., 9th, p. 325–365.

Schot, E. H., 1971, The Rammelsberg mine, a synopsis, *in* Müller, G., ed., Sedimentology of parts of central Europe, Internat. Sedimentological Cong., 8th, Guidebook, Heidelberg: Frankfurt am Main, Verlag Waldemar Kramer, p. 264–272.

Schwartz, G. M., 1959, Hydrothermal alteration: Econ. Geology, v. 54, p. 161–183.

Schwartz, M. O., and Friedrich, G. H., 1973, Secondary dispersion patterns of fluoride in the Osor area, Province of Gerona, Spain: Jour. Geochem. Explor., v. 2, p. 103–114.

Scott, J. H., Daniels, J. J., Gasbrouck, W. P., and Guu, J. Y., 1975, Hole-to-hole geophysical measurement for mineral exploration, *in* Society of Professional Well Log Analysts, Annual Logging Symposium, 16th, Houston, 1975: Sec. KK, p. 1–16.

Scott, J. H., and Tibbets, B. L., 1974, Well logging techniques for mineral deposit evaluation; a review: U.S. Bur. Mines Inf. Circ. 8627, 45 p.

Seigel, H. O., Hill, H. L., and Baird, J. G., 1968, Discovery case history of the Pyramid ore bodies, Pine Point, Northwest Territories, Canada: Geophysics, v. 33, no. 4, p. 645–656.

Senftle, F. E., Wiggins, P. F., Duffey, D., and Philbin, P., 1971, Nickel exploration by neutron capture gamma rays: Econ. Geology, v. 66, p. 583–590.

Shaffer, J. W., 1975, Bauxitic raw materials, *in* Lefond, S. J., ed., Industrial minerals and rocks: New York, Am. Inst. Mining Metall. Petroleum Engineers, p. 443–462.

Shawe, D. R., 1968, Geology of the Spor Mountain beryllium district, Utah, *in* Ridge, J. D., ed., Ore deposits of the United States, 1933–1967: New York, Am. Inst. Mining, Metall. Petroleum Engineers, p. 1148–1162.

Shibaoka, M., and Smyth, M., 1975, Coal petrology and the formation of coal seams in some Australian sedimentary basins: Econ. Geology, v. 70, p. 1463–1473.

Shillabeer, J. H., Sparling, J. H., Masson, A. G., and Roach, R. J., 1976, Environmental planning at the design stage of a new mine: Canadian Mining Metal. Bull., v. 69, no. 768, p. 63–67.

Shvartsev, S. L., 1972, Geochemical prospecting methods in the regions of permanently frozen ground, in Section 10. Geochemistry, Internat. Geol. Cong., 24th, Canada, 1972: Ottawa, Internat. Geol. Cong., p. 380–384.

Sibson, R. H., Moore, J. M., and Rankin, A. H., 1975, Seismic pumping—a hydrothermal fluid transport mechanism: Geol. Soc. (London) Jour., v. 1313, p. 653–659.

Siegel, F. R., 1974, Applied geochemistry: New York, John Wiley & Sons, 353 p.

Sillitoe, R. H., 1972, A plate tectonic model for the origin of porphyry copper deposits: Econ. Geology, v. 67, p. 184–197.

Sillitoe, R. H., 1976, A reconnaissance of the Mexican porphyry copper belt: Inst. Mining Metall. (London) Trans., v. 85, p. B170–B189.

Sillitoe, R. H., Halls, C., and Grant, J. N., 1975, Porphyry tin deposits in Bolivia: Econ. Geology, v. 70, p. 913–927.

Sinclair, A. J., 1976, Applications of probability graphs in mineral exploration: Toronto, Association of Exploration Geochemists, Special Volume 4, 95 p.

Sinclair, A. J., and Deraisme, J., 1974, A geostatistical study of the Eagle copper vein, northern British Columbia: Canadian Mining Metall. Bull., v. 67, no. 746, p. 131–142.

Singer, D. A., and Drew, L. J., 1976, The area of influence of an exploratory hole: Econ. Geology, v. 71, p. 642–647.

Singer, S. F., 1972, The predicament of the Club of Rome: EOS, Am. Geophys. Union Trans., v. 53, no. 7, p. 697–700.

Skipsey, E., 1970, Role of the geologist in modern coal mining, with special reference to the East Midlands Coal Field of Britain, in Jones, M. J., ed., Proceedings, Commonwealth Mining and Metall. Congr., 9th: London, Inst. Mining Metall., v. 2, p. 369–384.

Slichter, L. B., 1955, Geophysics applied to prospecting for ore: Econ. Geology, 50th Anniversary Volume, p. 885–969.

Slichter, L. B., 1960, The need for a new philosophy of prospecting: Mining Eng., v. 12, no. 6, p. 570–576.

Smirnov, V. I., 1971, Essays on metallogeny (translated by E. A. Alexandrov): Flushing, New York, Queens College Press, 96 p.

Smith, E. E. N., 1974, Review of current concepts regarding vein deposits of uranium, in Formation of uranium ore deposits: Vienna, Internat. Atomic Energy Agency, p. 515–527.

Smith, G. W., 1973, Engineering economy; analysis of capital expenditures, ed 2: Ames, Iowa, The Iowa State University Press.

Snyder, F. G., 1969, Precambrian iron deposits in Missouri, in Wilson, H. D. B., ed., Magmatic ore deposits: Lancaster, Pa., Econ. Geology Publishing Co., Econ. Geology Mon. 4, p. 231–239.

Sönge, P. G., 1974, Sedimentary ore deposits; a review of recent trends: Geol. Soc. South Africa Trans., v. 77, pt. 2, p. 159–168.

Sørensen, H., 1973, Ore deposits of agpaitic nepheline syneites, in Morin, P., ed., Les roches plutoniques dans leur rapports avec les gîtes mineraux: Paris, Masson et Cie., Editeurs, p. 332–341.

Sowers, G. B., and Sowers, G. F., 1970, Introductory soil mechanics and foundations: New York, The Macmillan Co., 556 p.

Spence, C. C., 1970, Mining engineers and the American West: New Haven, Yale University Press, 407 p.

Spurr, J. E., 1923, The ore magmas: New York, McGraw-Hill Book Co., 2 vols.

Spurr, J. E., 1927, Discussion: Lindgren, W., 1927, Magmas, dikes, and veins: Am Inst. Mining Metall. Engineers Trans., v. 74, p. 109.

Stach, E., 1975, Stach's textbook of coal petrology: Berlin, Gebrüder Borntraeger, 428 p.

Staley, W. C., 1964, Introduction to mine surveying: Stanford, California, Stanford University Press, 275 p.

Stancioff, A., Pasucci, R., and Rabchevsky, G., 1973, ERTS application to mineral exploration in Sonora, Mexico, *in* Proceedings, Symposium on management and utilization of remote sensing data: Falls Church, VA., Am. Soc. of Photogrammetry, p. 313–332.

Stanley, R., 1961, Chromium in Southern Rhodesia: Salisbury, Southern Rhodesia Mines Dept., 21 p.

Stanton, R. L., 1972, Ore petrology: New York, McGraw-Hill Book Co., Inc., 713 p.

Stather, N. J., and Prindle, F. L., 1970, Milling practices at Bunker Hill, *in* Rausch, D. O., and Mariacher, B. C., eds., World symposium mining and metallurgy of lead and zinc: New York, Am. Inst. Mining Metall. Petroleum Engineers, p. 348–372.

Stemprok, M., ed., 1965, Symposium—problems of postmagmatic ore deposition: Prague, Czechoslovakia Geol. Survey, Vol. 2, 595 p.

Steven, T. S., and Eaton, G. P., 1975, Environment of ore deposition in the Creede mining district, San Juan Mountains, Colorado. I. Geologic, hydrologic, and geophysical setting: Econ. Geology, v. 70, p. 1034.

Stout, K. S., 1975, Soil and rock mechanics illustrated: Montana Bureau of Mines and Geology Bull. 96, 190 p.

Strong, D. F., ed., 1976, Metallogeny and plate tectonics: Toronto, Geol. Assoc. Canada Spec. Paper 14, 660 p.

Strunk, W., and White, E. B., 1972, The elements of style, ed. 2: New York, The Macmillan Company, 78 p.

Stumpel, E. F., 1974, The genesis of platinum deposits; further thoughts: Minerals Sci. Eng., v. 6, no. 3, p. 120–141.

Subramaniam, A. P., 1972, Operation Hardrock—a mineral exploration project based on airborne geophysics, *in* Section 9, Geophysics, Internat. Geol. Cong., 24th, Canada, 1972: Ottawa, Internat. Geol. Cong., p. 121–134.

Sullivan, C. J., 1948, Ore and granitization: Econ. Geology, v. 43, p. 471–498.

Sumner, J. S., 1976, Principles of induced polarization for geophysical exploration: Amsterdam, Elsevier Scientific Publishing Company, 274, p.

Sutherland Brown, A., 1969, Mineralization in British Columbia and the copper and molybdenum deposits: Canadian Mining Metall. Bull., v. 62, no. 681, p. 933.

Sutherland Brown, A. (ed.), 1976, Porphyry deposits of the Canadian cordillera: Montreal, Canadian Inst. Mining Metall. Special Volume 15, 510 p.

Szechy, K., 1973, The art of tunnelling, ed 2: Budapest, Akademiai Kiado, 1097 p.

Taggart, A. F., 1951, Elements of ore dressing: New York, John Wiley & Sons.

Takeuchi, Y., ed., 1971, Proceedings, Internat. Mineralogical Association—Internat. Association for the Genesis of Ore Deposits, Meeting, 1970: Soc. Mining Geologists Japan Trans., Spec. Issue No. 3.

Tarling, D. H., and Tarling, M. P., 1973, Continental drift: Baltimore, Penguin Books.

Telford, W. M., Geldart, L. P., Sheriff, R. E., and Keys, D. A., 1976, Applied geophysics: Cambridge, Cambridge University Press, 700 p.

Terzaghi, K., 1943, Theoretical soil mechanics, New York, John Wiley & Sons, 510 p.

Terzaghi, K., and Peck, R. B., 1967, Soil mechanics in engineering practices, ed. 2: New York, John Wiley & Sons, 566 p.

Thomas, L. J., 1973, An introduction to mining: New York, Halsted Press Division, John Wiley & Sons, 421 p.

Thomas, M. F., 1974, Tropical geomorphology: New York, John Wiley & Sons, 332 p.

Thrush, P. W., ed., 1968, A dictionary of mining, mineral, and related terms: Washington, U.S. Bur. Mines, 1269 p.

Tichy, H. J., 1967, Effective writing for engineers, managers, and scientists: New York, John Wiley & Sons, 337 p.

Tinsley, C. R., 1975, Economics of deep ocean resources—a question of manganese or no manganese: Mining Eng., v. 27, no. 4, p. 31–34.

Tischendorf, G., 1973, The metallogenetic basis of tin exploration in the Erzgebirge: Inst. Mining Metall. (London) Trans., v. 82, p. B9–B24.

Titley, S. R., and Hicks, C. L., 1966, Geology of the porphyry copper deposits, southwestern North America: Tucson, University of Arizona Press, 287 p.

Tixier, M. P., and Alger, R. P., 1970, Log evaluation of nonmetallic mineral deposits: Geophysics, v. 35, no. 1, p. 124–142.

Tjia, H. C., 1968, Fault plane markings, in Proceedings, Section 13, Other subjects, Internat. Geol. Cong., 23rd, Czechoslovakia, 1968: Prague, Academia, p. 279–284.

Tollon, F., 1972, Zonalité métallogénique dans le region de Salsigne, in Section 4, Mineral deposits, Internat. Geol. Cong., 24th, Canada, 1972: Ottawa, p. 180–187.

Tooms, J. S., 1970, Some aspects of exploration for marine mineral deposits, in Jones, J. J., ed., Proceedings, Commonwealth Mining Metall. Cong., 9th: London, Inst. Mining Metall., v. 2, p. 285–301.

Tooms, J. S., 1973, Dispersion from an exhalative submarine body, in Jones, M. J., ed., Geochemical exploration 1972, London, Inst. Mining Metall., p. 193–202.

Towse, D., 1970, Evaluation of cement raw materials: New York, Inst. Mining Metall. Petroleum Engineers, SME Preprint 70-S-30.

Tsusue, A., and Ishihara, S., 1975, Residual iron-sand deposits of southwest Japan: Econ. Geology, v. 70, p. 706–716.

Turneaure, F. S., 1955, Metallogenic provinces and epochs: Econ. Geology, 50th Anniversary Volume, p. 38–91.

Turner, A. K., 1976, Computer aided environmental impact analysis: Colorado School of Mines, Mineral Industries Bull., v. 19, nos. 2 and 3.

United Nations, 1970a, Survey of world iron ore resources: New York, United Nations, UNESCO, 479 p.

United Nations, 1970b, International legend for hydrogeological maps: New York, United Nations, UNESCO, 101 p.

United Nations, 1976, Engineering geological maps: a guide to their preparation: New York, United Nations, UNESCO, 79 p.

U.S. Geological Survey, 1958, Suggestions to authors of reports of the United States Geological Survey: Washington, U.S. Government Printing Office, 255 p.

U.S. Geological Survey, 1975, Mineral resource perspectives 1975: U.S. Geol. Survey Prof. Paper 940, 24 p.

Vachtl, J., ed., 1968, Symposium on kaolin deposits: Prague, Internat. Geol. Cong., 23rd, Czechoslovakia, 3 vols.

van Eden, J. G., 1974, Depositional and diagenetic environment related to sulfide mineralization, Mafulira, Zambia: Econ. Geology, v. 69, p. 59–79.

Van Horn, J. C., 1968, Financial management and policy: Englewood Cliffs, N.J., Prentice-Hall, Inc.

Varnes, D. J., 1975, The logic of geologic maps, with reference to their interpretation and use for engineering purposes: U.S. Geol. Survey Prof. Paper 837, 48 p.

Vermaak, C. F., 1976, The Merensky Reef—thoughts on its environment and genesis: Econ. Geology, v. 71, p. 1270–1298.

Viljoen, R. P., Viljoen, M. L., Grootenboer, J., and Longshaw, T. G., 1975, ERTS-1 imagery; its applications in geology and mineral exploration: Minerals Sci. Eng., v. 7, no. 2, p. 132–168.

Vine, J. D., 1975, Lithium in sediments and brines: U.S. Geol. Survey Jour. Research, v. 3, no. 4, p. 479–485.

Visser, D. J. L., and von Gruenewaldt, G., eds., 1970, Symposium on the Bushveld igneous complex and other layered intrusions: Johannesburg, Geol. Soc. South Africa Spec. Pub. 1.

Vogely, W. A., 1976, Economics of the mineral industries: New York, Am. Inst. Mining Metall. Petroleum Engineers, 863 p.

Wainerdi, R. E., and Uken, E. A., eds., 1971, Modern methods in geochemical analysis: New York, Plenum Press, 397 p.

Walker, R. R., Matulich, A., Amos, A. C., Watkins, J. J., and Mannard, G. W., 1975, The geology of the Kidd Creek mine: Econ. Geology, v. 70, p. 80–89.

Walthier, T. N., 1976, The shrinking world of exploration: Mining Eng., v. 25, no. 4, p. 27–31 and no. 5, p. 46–50.

Walton Smith, F. G., ed., 1974, Handbook of marine science. Section 1. Oceanography. Vol. 1, Physical—chemistry, physics, geology, engineering: Cleveland, Ohio, CRC Press, Inc., 640 p.

Wang, C-S., Capp, J. P., Wayne, M. T., and Boshkov, S. H., 1973, In-situ gasification and liquifaction mining systems, in Cummins, A. B., and Givens, I. A., eds., SME Mining engineering handbook: New York, Am. Inst. Mining metall. Petroleum Engineers, Sec. 21.9, p. 21–78—21–96.

Ward, M. H., 1973, Engineering for in-situ leaching: Mining Cong. Jour., v. 59, no. 1, p. 21–27.

Ward, S. H., 1967, The electromagnetic method, in Mining geophysics, Vol. 2, Theory: Tulsa, Society Explor. Geophysicists, p. 224–372.

Ward, S. H., and Rogers, G. R., 1967, Introduction, in Mining geophysics, Vol. 2. Theory: Tulsa, Oklahoma, Soc. Explor. Geophysicists, p. 6.

Wargo, J. G., 1973, Trends in corporate mineral exploration expenditures: Mining Eng., v. 25, no. 5, p. 43–45.

Warner, L. A., Holser, W. T., Wilmarth, V. R., and Cameron, E. N., 1959, The occurrence of nonpegmatitic beryllium in the United States: U.S. Geol. Survey Prof. Paper 318, 198 p.

Waterman, G. C., and Hazen, S., 1968, Development drilling and bulk sampling, in Pfleider, E. P., ed., Surface mining: New York, Am. Inst. Mining Metall. Petroleum Engineers, p. 89.

Watmuff, I. G., 1974, Supergene alteration of the Mt. Windarra nickel sulfide ore deposit, Western Australia: Mineralium Deposita, v. 9, p. 199–222.

Weaver, R. C., and Call, R. D., 1965, Computer estimation of oriented fracture set intensity, *in* Transactions, Symposium on computer applications in mining and exploration: Tucson, University of Arizona, College of Mines, p. BB1–BB17.

Webb, J. S., 1973, Applied geochemistry and the community: Inst. Mining Metall. (London) Bull., no. 800, July, p. 23–24.

Webber, B. J., 1972, Supergene nickel deposits: Am. Inst. Mining Metall. Petroleum Engineers Trans., v. 252, no. 3, p. 333–347.

Wedepohl, K. H., ed., 1958, Handbook of geochemistry: Heidelberg, Springer-Verlag, loose-leaf, with supplements.

Weigel, W. W., 1965, The inside story of Missouri's exploration boom: Eng. Mining Jour., v. 166, no. 11, p. 77–86, 170, 172.

Weiss, N., ed., 1977, Mineral processing handbook: New York, Am. Inst. Mining Metall, Petroleum Engineers.

Welling, C. G., 1976, Ocean mining systems: Mining Cong. Jour., v. 62, no. 9, p. 43–48.

Wells, J. H., 1973, Placer examination—principles and practices: U.S. Bureau of Land Management Tech. Bull 4, 209 p.

Wenk, H.-R., 1976, Electron microscopy in mineralogy: New York, Springer-Verlag, 564 p.

West, E. G., 1973, Factors in the future demand for metals with special reference to usage in the United Kingdom: Inst. Mining Metall. (London) Trans., v. 82, p. A45–A51.

White, L., 1975, In-situ leaching opens new uranium reserves in Texas: Eng. Mining Jour., v. 176, no. 7, p. 75.

Whittaker, B. N., 1974, An appraisal of strata control practice: Inst. Mining Metall. (London) Trans., v. 83, p. A87.

Whitten, E. H. T., 1966, Structural geology of folded rocks: New York, John Wiley & Sons, 678 p.

Wilderman, G. H., 1973, Exploration drilling techniques used as a "field determinator" and data gatherer in mill and plant design: Canadian Mining Metall. Bull., v. 66, no. 740, p. 110–116.

Willemse, J., 1969, The vanadiferous magnetic iron ore of the Bushveld igneous complex, *in* Wilson, H. D. B., ed., Magmatic ore deposits, a symposium: Lancaster, Pa., Economic Geology Publishing Co., Econ. Geology Mon. 4, p. 187–208.

Williams, R. S., and Carter, W. D., eds., 1976, ERTS-1, a new window on our planet: U.S. Geol. Survey Prof. Paper 929, 362 p.

Williamson, I. A., 1967, Coal mining geology: London, Oxford University Press, 266 p.

Wilson, H. D. B., ed., 1969, Magnatic ore deposits, a symposium: Lancaster, Pa., Economic Geology Publishing Co., Econ. Geology Mon. 4, 366 p.

Wilson, M. E., ed., 1948, Structural geology of Canadian ore deposits: Montreal, Canadian Inst. Mining Metall.

Wisser, E., 1927, Oxidation subsidence at Bisbee, Arizona: Econ. Geology, v. 22, p. 761–790.

Wisser, E., 1966, The epithermal precious metal province of northwest Mexico: Nevada Bur. Mines Rept. 13, p. 63–92.

Wolf, K. H., ed., 1976, Handbook of strata-bound and stratiform ore deposits: Amsterdam, Elsevier Scientific Publishing Co., 7 vols., 2945 p.

Wolfe, J. A., 1973, Tectonic fingerprint in Philippine porphyry deposits: New York, Am. Inst. Mining Metall. Petroleum Engineers, SME Preprint 73-S-37.

Wolff, E., 1969, Handbook for the Alaskan prospector, ed. 2: Fairbanks, Mineral Industry Research Laboratory, University of Alaska.

Wolle, M. S., 1953, The bonanza trail: Bloomington, Indiana, The Indiana University Press, 256 p.

Wood, D. N., ed., 1973, Use of earth science literature: London, Butterworth and Co., 459 p.

Woodall, R., and Travis, G. A., 1970, The Kambalda nickel deposits, Western Australia, in Jones, M. J., ed., Proceedings, Commonwealth Mining Metall. Cong., 9th: London, Inst. Mining Metall., v. 2, p. 517–533.

Woodruff, S. D., 1966, Methods of working coal and metal mines: New York, Pergamon Press, 3 vols.

Worst, B. G., 1960, The Great Dyke of Southern Rhodesia: Rhodesia Geol. Survey Bull. 47, 239 p.

Wray, W. B., Jr., 1970, FORTRAN IV CDC 6400 computer program for constructing isometric diagrams: Kansas Geol. Survey Computer Contr. 44.

Wyllie, R. J. M., and Argall, G. O., eds., 1975, World mining glossary of mining, processing, and geological terms: San Francisco, Miller Freeman Publications, Inc., 432 p.

Yarroll, W. H., and Davis, F. T., 1975, The economics of small milling operations: Colorado School of Mines, Mineral Industries Bull., v. 18, no. 6, 16 p.

Zablocki, C. J., 1966, Some applications of geophysical logging methods in mineral exploration drill holes: Soc. Professional Well Log Analysts Seventh Annual Well Logging Trans., Section U, p. 1.

Zeschke, G., 1964, Prospektion und feldmässige Beurteilung von Lagerstätten: Vienna, Springer Verlag, 307 p.

Zietz, I., and Andreason, G. E., 1966, Remanent magnetization and aeromagnetic interpretation, in Mining geophysics, Vol. 2. Theory: Tulsa, Soc. Exploration Geophysicists, p. 569–590.

Zimmerman, D. O., 1967, Gossans at Northern Leases, Mount Isa, as visible geochemical anomalies, in Woodcock, J. T., Madigan, R. T., and Thomas, R. G., eds., Proceedings, Commonwealth Mining and Metall. Cong., 8th, Melbourne: v. 6, p. 359–368.

Zimmerman, O. T., 1968, Elements of capital cost estimation: Cost Eng., v. 13, no. 4, p. 4–18.

Zurbrigg, H. F., 1963, Thompson mine geology: Canadian Mining Metall. Bull., v. 56, no. 614, p. 451–460.

Zurflueh, E. G., 1967, Applications of two-dimensional linear wavelength filtering: Geophysics, v. 32, p. 1015–1035.

Zurkowski, I., 1975, Strip mining of coal, in Van Tassel, A. J., ed., The environmental price of energy: Lexington, Ma., D. C. Heath, p. 223–263.

INDEX